MAGNETIC RECONNECTION

Magnetic reconnection is at the core of many dynamic phenomena in the universe, such as solar flares, geomagnetic substorms, and tokamak disruptions. This major work, written by two world leaders on the subject, gives the first comprehensive overview of a fundamental process. It includes an illuminating account of the basic theory and a wide-ranging review of the physical phenomena created by reconnection.

Most of the universe is in the form of a plasma threaded by a magnetic field. When twisted or sheared, the field lines may break and reconnect rapidly, converting magnetic energy into heat, kinetic energy, and fast-particle energy. The book begins with an accessible introduction to all aspects of the theory behind this phenomenon and goes on to review a wide range of applications – from laboratory machines, the Earth's magnetosphere, and the Sun's atmosphere to flare stars and astrophysical accretion disks. Finally, the authors provide a succinct account of particle acceleration by electric fields, stochastic fields, and shock waves, and how reconnection can be important in these mechanisms.

The clear and logical style makes this book an essential introduction for graduate students and an authoritative reference for researchers in solar physics, astrophysics, plasma physics, and space science.

Eric Priest holds the Gregory Chair of Mathematics at St. Andrews University, Scotland. His research interests are primarily in solar plasma physics problems and extend to other applications in astrophysics. He is a Fellow of the Royal Society of Edinburgh and a Member of the Norwegian Academy of Sciences and Letters. He has served on many national and international committees, including the International Astronomical Union's Commission on Solar Activity (of which he was president from 1988 to 1991), and has chaired the Royal Society of Edinburgh's Mathematics Committee. He is currently chair of the UK's Particle Physics and Astronomy Research Council's Astronomy Committee. In 1982 he published *Solar Magnetohydrodynamics*, and he has edited 11 other books.

Terry Forbes is Research Professor in the Department of Physics and the Institute for the Study of Earth, Oceans, and Space at the University of New Hampshire in Durham. He has also held positions at the Los Alamos Laboratory and St. Andrews University. His research interests focus on magnetospheric substorms in the Earth's geomagnetic tail and solar flares. He is currently an associate editor of the journal *Solar Physics*.

MAGNETIC RECONNECTION

MHD Theory and Applications

ERIC PRIEST

Department of Mathematical and Computational Sciences
University of St. Andrews

TERRY FORBES

Institute for the Study of Earth, Oceans, and Space
University of New Hampshire

CAMBRIDGE UNIVERSITY PRESS
Cambridge, New York, Melbourne, Madrid, Cape Town, Singapore, São Paulo

Cambridge University Press
The Edinburgh Building, Cambridge CB2 2RU, UK

Published in the United States of America by Cambridge University Press, New York

www.cambridge.org
Information on this title: www.cambridge.org/9780521481793

First published 2000
This digitally printed first paperback version 2006

A catalogue record for this publication is available from the British Library

Library of Congress Cataloguing in Publication data
Priest, E. R. (Eric Ronald), 1943–
Magnetic reconnection : MHD theory and applications / Eric Priest,
Terry Forbes.
p.   cm.
ISBN 0-521-48179-1
1. Magnetic reconnection.   2. Space plasmas.   3. Plasma (Ionized gases)
4. Magnetohydrodynamics
I. Priest, Eric R. 1943–   , II. Forbes, Terry, G. 1946–   .   III. Title.
QC809.P5P75   1999
523.01´886 – dc21   99-14939
CIP

ISBN-13 978-0-521-48179-3 hardback
ISBN-10 0-521-48179-1 hardback

ISBN-13 978-0-521-03394-7 paperback
ISBN-10 0-521-03394-2 paperback

To Andrew, David, Matthew, and Naomi

Contents

Preface

Magnetic reconnection is a fundamental process with a rich variety of aspects and applications in astrophysical, space, and laboratory plasmas. It is one that has fascinated us both for much of our research careers, so that we have from time to time felt drawn to return to it after working on other topics and to ponder it anew or view it from a different angle.

Indeed, it was reconnection that brought us together in the first place, since one of us (TGF) went to work as a postdoc with the other in 1980 on the subject of reconnection in solar flares. Our initial meeting in Edinburgh at the start of this collaboration was rather amusing, since a friend had misleadingly described Eric Priest as a tall old man with ginger hair, and so the inaccuracy of this description did not exactly help us to find each other in the crowded airport!

At present the whole field of reconnection is a huge, vibrant one that is developing along many different lines, as can be seen by the fact that a recent science citation search produced a listing of 1,069 published articles written on this subject in only the past three years. We are therefore well aware of the impossibility of comprehensively covering the whole field and apologise in advance to those who may be disappointed that we have not found space to discuss their work on reconnection. We have attempted to cover the basics of the various aspects of magnetic reconnection and to give brief accounts of the applications at the present point in time. Because of the vastness of the field and our own limited knowledge, we decided to focus only on the magnetohydrodynamic (MHD) aspects of reconnection, but these aspects do provide a foundation for treatments using kinetic theory.

One of the difficulties in writing a book of this complexity is the choice of notation. It is impossible to use all the terms uniquely unless we resort to using Chinese characters! Thus, we have found it necessary to use a few characters in more than one way and occasionally to modify the standard notation, but we have attempted to minimise this and hope there is no loss of clarity. For example, α is a universal character for many purposes, but we have used it to represent the force-free parameter in most of the book and an Euler potential in Section 8.1. Again, p is used for pressure and so we have adopted $\bar{p}$ instead of p for momentum in Chapter 13. A comprehensive list of the notation is given in Appendix 1.

The introductory chapter describes the early history, basic ideas, and equations, while Chapter 2 continues to set the scene by describing the many ways in which singularities, called current sheets, can form in an ideal MHD plasma. Then Chapter 3 describes the way such a sheet is resolved by magnetic diffusivity in a one-dimensional manner, in other words, the diffusion of magnetic fields by so-called magnetic annihilation. Chapter 4 presents the steady two-dimensional behaviour of a current sheet, namely, magnetic reconnection by the classical mechanisms of Sweet, Parker, and Petschek. Then in Chapter 5 the newer generation of fast reconnection models is summarised. Chapters 6 and 7 proceed to investigate the unsteady aspects of reconnection, both by resistive instabilities such as the tearing mode, by the collapse of an X-point, and by unsteady fast reconnection. Various facets of the relatively new area of three-dimensional reconnection are outlined (Chapter 8), including regimes of so-called spine, fan, separator, and quasi-separatrix layer reconnection. Four chapters (9–12) follow on some of the particular manifestations of reconnection in the laboratory, the Sun, the magnetosphere, and in astrophysics. Finally, we have added a chapter (13) on particle acceleration which demonstrates how MHD concepts, such as reconnection, shocks, and turbulence, are linked with kinetic concepts to understand the origin of energetic particles.

There are many people to whom we are deeply thankful: to our families and friends for patience during our preoccupied time of being "with book"; to Gill for endless patience and good humour in typing the many versions as we have wrestled to express our ideas in print; to Duncan for his reprographics skill; to Klaus for invaluable help with the figures and for providing the cover picture; to our colleagues in St. Andrews and Durham for help and advice in numerous ways (namely, Aaron, Alan C., Alan H., Andrew C., Andrew W., Bernie, Clare, Dan, David, Duncan, Gordon, Ineke, Jack, Joe, Jun, Kathy, Katie, Keith, Kelly, Marty, Moira, Phil, Robert, Thomas, Tony, Valera, and Yuri); and to friends elsewhere who have aided greatly our understanding of the subject (namely, Spiro Antiochos, Ian Axford, Arnold Benz, Mitch Berger, Joachim Birn, Alan Boozer, Philippa Browning, Jorg Büchner, David Burgess, Dick Canfield, Peter Cargill, Ian Craig, Len Culhane, Russ Dahlburg, Pascal Démoulin, Brian Dennis, Luke

Drury, John Finn, Richard Harrison, Jim Hastie, Michael Hesse, Martin Heyn, Jean Heyvaerts, Raymond Hide, Gordon Holman, Gunnar Hornig, Randy Jokipii, John Kirk, Jim Klimchuk, Mike Lockwood, Dana Longcope, Boon-Chye Low, Piet Martens, Leon Mestel, Jim Miller, Keith Moffatt, Gene Parker, Harry Petschek, Reuven Ramaty, Karl Schindler, Reinhard Schlickeiser, Steve Schwartz, Bengt Sonnerup, Andrew Soward, Slava Titov, Saku Tsuneta, Aad van Ballegooijen, Grisha Vekstein, Nigel Weiss, and John Wesson) – although any mistakes or misconceptions are entirely our fault.

We look forward at this exciting time to great advances in understanding many aspects of magnetic reconnection – especially the three-dimensional aspects, which are so important for interpreting observations, and the particle acceleration aspects, which remain so mysterious. We hope the reader will share our fascination for the myriads of intriguing ways that the effects of reconnection are revealed in our beautiful universe.

Eric Priest and Terry Forbes
November, 1998

1

Introduction

Like most fundamental concepts in physics, magnetic reconnection owes its appeal to its ability to unify a wide range of phenomena within a single universal principle. Virtually all plasmas, whether in the laboratory, the solar system, or the most distant reaches of the universe, generate magnetic fields. The existence of these fields in the presence of plasma flows inevitably leads to the process of magnetic reconnection. As we shall discuss in more detail later on, reconnection is essentially a topological restructuring of a magnetic field caused by a change in the connectivity of its field lines. This change allows the release of stored magnetic energy, which in many situations is the dominant source of free energy in a plasma. Of course, many other processes besides reconnection occur in plasmas, but reconnection is probably the most important one for explaining large-scale, dynamic releases of magnetic energy.

Figures 1.1–1.4 illustrate the rich variety of plasma environments where reconnection occurs or is thought to occur. The evidence of reconnection in laboratory fusion machines such as the tokamak [Fig. 1.1(a)] and the reversed-field pinch [Fig. 1.1(c)] is so strong that there is no longer any controversy about whether reconnection occurs, but only controversy about the way in which it occurs (§9.1). However, as one considers environments which are further away from the Earth, the evidence for reconnection becomes more circumstantial. Most researchers who study the terrestrial aurorae [Fig. 1.2(a)] believe that they are directly or indirectly the result of reconnection in the Earth's magnetosphere, but the evidence for similar phenomena in other planetary magnetospheres [Fig. 1.2(b)] is much smaller (§10.6). It has also been argued that reconnection lies at the root of phenomena called disconnection events which occur in comet tails [Fig. 1.2(c)].

1

Fig. 1.1. Laboratory Plasmas. (a) Drawing of the Joint European Torus (JET) located in England. The scale is given by the figure at lower right (courtesy of JET Joint Undertaking). (b) Plasma injection gun of the Swarthmore Spheromak (courtesy of M.R. Brown). (c) Reversed-field-pinch device at the University of Wisconsin (courtesy of R. Dexter, S. Prager, and C. Sprott). (d) Interior photograph of the MRX reconnection experiment at the Princeton Plasma Physics Laboratory taken during operation. In the region at the top, between the two dark toroidal coils, a magnetic X-line is outlined by a bright emission, which comes primarily from the Hα Balmer line of hydrogen (courtesy of M. Yamada).

Even though the Sun is a rather distant object compared with the aurora, some of the best evidence for reconnection is to be found there. Reconnection provides an elegant, and so far the only, explanation for the motion of chromospheric ribbons and flare loops during solar flares [Fig. 1.3(b)]. At the same time, it also accounts for the enormous energy release in solar flares. The ejection of magnetic flux from the Sun during coronal mass ejections and prominence eruptions [Fig. 1.3(a)] necessarily requires reconnection; otherwise, the magnetic flux in interplanetary space would build up indefinitely (§11.4.1). Reconnection has also been proposed as a

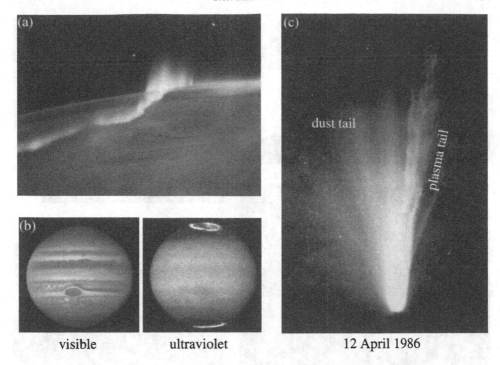

visible ultraviolet 12 April 1986

Fig. 1.2. Planets and Comets. (a) Aurora Australis photographed over Antarctica by the space shuttle Discovery (NASA). (b) The Jovian aurora as observed by the Wide-Field Planetary Camera 2 on the Hubble Space Telescope. Auroral ovals are visible at both poles in the ultraviolet image, which was taken 15 min after the visible image (courtesy of John T. Clarke). (c) Disconnection of the plasma tail of comet Halley during its 1986 appearance. The image is a 3-min exposure taken by the Michigan Schmidt Telescope at the Cerro Tololo Inter-American Observatory (produced by Jet Propulsion Laboratory/NASA).

mechanism for the heating of solar and stellar coronae to extremely high temperatures [$>10^6$ K; Fig. 1.3(c)]. Even more fundamental is the role played by reconnection in the generation of solar, stellar, and planetary magnetic fields. The generation of magnetic fields in astrophysics and space physics is based almost exclusively on the concept of a self-excited magnetic dynamo. In a dynamo, complex motions of a plasma with a weak seed magnetic field can generate a stronger large-scale magnetic field. Magnetic reconnection is an essential part of this generation process, so, in this sense, the very existence of all magnetic phenomena on the Sun such as prominences, flares, the corona, and sunspots (Fig. 1.3) requires reconnection (Cowling, 1965).

The role of reconnection in plasma environments beyond the solar system remains both speculative and controversial. Because of the great distances involved, observations provide very few facts with which to constrain

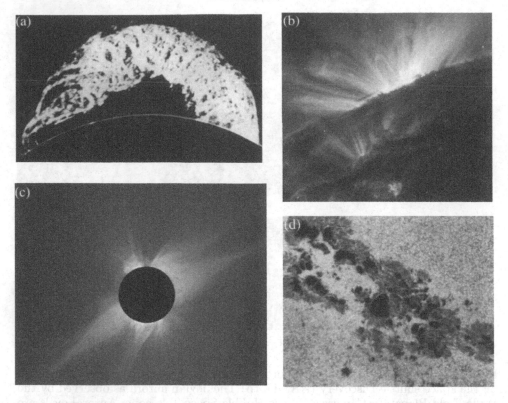

Fig. 1.3. The Sun. (a) Hα photograph of the extremely large prominence eruption (known as "Grandaddy") which occurred on 4 June 1946 (courtesy of the High Altitude Observatory). (b) X-ray image of the west limb of the Sun in Fe IX/X lines obtained by TRACE (Transition Region and Coronal Explorer) on 24 April 1998. On the limb is a system of "post"-flare loops, while in the centre foreground is an X-ray bright "point." Both features are thought to be produced by reconnection (courtesy of A. Title and L. Golub). (c) Solar corona photographed from Baja California, Mexico during the total eclipse on 11 July 1991 (courtesy of S. Albers). (d) Sunspots (courtesy of the National Solar Observatory at Sacramento Peak, New Mexico).

theories. One of the main reasons that reconnection is often invoked in astrophysical phenomena such as stellar flares [Fig. 1.4(a)] or galactic magnetotails is that these phenomena may be analogues of the same processes occurring in the solar corona and the Earth's magnetosphere. However, reconnection has also been invoked to account for viscous dissipation in accretion disks [Fig. 1.4(b)], for which no counterpart exists in the solar system. It may even be involved in the formation of chondritic inclusions in certain types of meteorite [Fig. 1.4(c)]. The most popular theory for the formation of these inclusions is that they were formed by reconnection in the

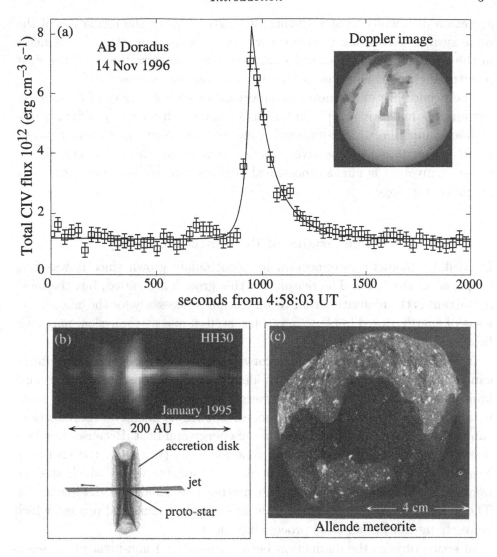

Fig. 1.4. Astrophysics. (a) Light-curve in the UV carbon IV line obtained by the Hubble Space Telescope of an impulsive flare on AB Doradus (Vilhu et al., 1998). The inset shows a Doppler image of AB Doradus constructed from photospheric spectral lines over a period of several hours on a different date. The dark areas are strongly magnetized regions which are the stellar equivalent of sunspots (courtesy of A.C. Cameron). (b) Hubble Space Telescope image of the accretion disk around the protostellar object called HH-30, which is 450 light years away in the constellation Taurus. The disk is seen edge on, and the light from the forming star illuminates both the top and bottom surfaces of the disk. The star itself is hidden behind the densest parts of the disk (courtesy of C. Burrows). (c) Part of the Allende meteorite, which struck the region of Chihuahua, Mexico, on 8 February 1969. It contains chondritic inclusions that are thought to have been created by nebular flares in the accretion disk which surrounded the Sun during its formation (courtesy of E.A. King).

accretion disk which existed during the early phase of the formation of the solar system. According to this theory, the reconnection of magnetic fields in the dust-filled disk produced solar nebular flares, which welded the dust into the inclusions found in many carbonaceous meteorites.

The occurrence of magnetic reconnection in a wide variety of plasma environments makes it a topic in which researchers from many different disciplines contribute to the understanding of a fundamental plasma process. From a theoretical perspective, reconnection is especially challenging because it involves nonlinear, non-ideal processes in an inherently complex magnetic topology.

1.1 The Origins of Reconnection Theory

Interest in magnetic reconnection has continually grown since it was first proposed in the 1940s. The reasons for this growth are varied, but the most important is the realisation that reconnection is necessary for the efficient release of energy stored in planetary, solar, stellar, and astrophysical magnetic fields.

In MHD theory, processes which convert magnetic energy into other forms can be distinguished as ideal or non-ideal. Ideal processes, such as the ideal kink instability, convert magnetic energy into kinetic energy without magnetic dissipation, while non-ideal processes, such as magnetic reconnection, can convert magnetic energy into kinetic energy and heat. Because they lack dissipation, ideal processes cannot generate heat (i.e., raise the entropy). Yet, in practice, they typically generate compressive waves which steepen nonlinearly to form shock waves no matter how small the dissipation is. The dissipation in such shocks is thus an important non-ideal process which converts motions of an ideal process into heat.

In space physics the distinction between ideal and non-ideal processes is important because simple estimates imply that magnetic dissipation acts on a time-scale which is many orders of magnitude slower than the observed time-scales of dynamic phenomena. For example, solar flares release stored magnetic energy in the corona within a period of 100 s. By comparison, the time-scale for magnetic dissipation based on a global scale-length of 10^5 km is of the order of 10^6 yr. Typically, phenomena like the solar flare and the substorm require a significant fraction of the stored magnetic energy to be converted within a few Alfvén time-scales. Such rapid time-scales are easily achieved in ideal MHD processes but not in non-ideal ones. Although ideal MHD processes can release energy quickly, they rarely release a significant amount because of the topological constraints which exist in the absence of dissipation. In contrast, magnetic reconnection is not topologically constrained, and therefore it can release a much greater amount of magnetic energy (Kivelson and Russell, 1995).

One of the principal goals of reconnection theory is to explain how reconnection can occur on short enough time-scales (although very recently there has been a shift in emphasis to explore the three-dimensional aspects, as described in Chapter 8). Historically, two approaches have been adopted. The first, based on the kinetic theory of plasmas, has been to find an anomalous resistivity mechanism which will allow rapid dissipation. The second approach, based on MHD (the main focus of this book), has been to find a geometrical configuration which greatly reduces the effective dissipation scale-length. These two approaches are often combined in analyzing specific phenomena.

For example, the ion-tearing mode, which is a wave-particle theory of reconnection, has been proposed as a mechanism for the onset of reconnection in the geomagnetic tail at the start of an auroral substorm (§10.5). However, in order for the ion-tearing mode to occur, it is first necessary that the current sheet in the tail become very much thinner than normal. Usually, the width of the current sheet is of the order of 7000 km (approximately an Earth radius), but the ion-tearing mode becomes effective only when the width is of the order of 200 km (approximately an ion-gyro radius). Recent observations that such thinning of the current sheet actually occurs at substorm onset has motivated researchers to develop MHD models to explain how conditions in the solar wind upstream of the Earth initiate the thinning in the tail (§10.5).

The process of magnetic reconnection has its origins in suggestions by R.G. Giovanelli (1946) and F. Hoyle (1949) that magnetic X-type null points can serve as locations for plasma heating and acceleration in solar flares and auroral substorms. Their interest in X-points may have been due to the fact that a magnetic field tends to inhibit particle acceleration unless the electric field has a component parallel to the magnetic field. Early researchers usually assumed that electrons quickly short out any parallel electric field, so magnetic null regions (or points) were thought to offer the best locations for particle acceleration. Magnetic null points occur naturally when there are two or more sources of magnetic field, such as the Sun and the Earth, for example. If an electric field exists in the vicinity of such a null point, then charged particles there will undergo acceleration until they travel into a region where the magnetic field is no longer negligible. This view of acceleration is now somewhat outdated, because many mechanisms have since been proposed for particle acceleration which do not require a null point (see Lyons and

Williams, 1984, or Tandberg-Hanssen and Emslie, 1988, for example). These mechanisms use electric and magnetic fields which are nonuniform in space and sometimes in time (see Chapter 13).

Cowling (1953) pointed out that, if a solar flare is due to ohmic dissipation, a current sheet only a few metres thick is needed to power it. Then J.W. Dungey (1953), who was a student of Hoyle, showed that such a current sheet can indeed form by the collapse of the magnetic field near an X-type neutral point, and he was the first to suggest that "lines of force can be broken and rejoined." He considered the self-consistent behaviour of both field and particles at an X-point (see §1.3.1). This is a quite different approach from simply considering the motion of charged particles in a given set of electric and magnetic fields. Moving charged particles are themselves a source of electric and magnetic fields, and in a plasma these self-fields must be taken into account by combining Maxwell's equations with Newton's equations of motion. To analyze the effect of such fields, Dungey used the MHD equations for a plasma with negligible gas pressure, and he found that small perturbations in the vicinity of the null point lead to the explosive formation of a current sheet (§2.1). Near the null point, plasma motions induced by a small current perturbation cause the current to grow, and this current further enhances the plasma motions resulting in a positive feedback.

After Dungey's pioneering work on the formation of current sheets, P.A. Sweet (1958a,b) and E.N. Parker (1957) were the first to develop a simple MHD model for how steady-state reconnection might work in a current sheet formed at a null point (§4.2). At the 1956 symposium in Stockholm, Sweet (1958a) stressed that conditions far from an X-point, as well as plasma pressure, may play important roles in forming a current sheet. As a model for solar flares, he considered a magnetic field with an X-point produced by sources in the photosphere. If the sources approach one another and the magnetic field remains frozen to the plasma, a narrow "colliding layer" (Sweet's term) forms around the null point. The field flattens, and magnetohydrostatic equilibrium implies that the plasma pressure inside the layer, or current sheet, is the same as the external magnetic pressure. Plasma is squeezed out of the ends of the sheet by this excess pressure, just as a fluid would be squeezed out between approaching plates. Parker, while listening to Sweet's talk, realised how to model the process in terms of MHD and so eagerly went back to his room on the evening of Sweet's talk and worked out the details. Parker derived scaling laws for the process and coined the words "reconnection of field lines" and "merging of magnetic fields" (Parker, 1957). Later Parker (1963) also coined the phrase "annihilation of magnetic fields" and gave an in-depth development of the mechanism. He modelled the internal structure across the current sheet and, in passing, made the challenging comment that "it would be instructive if the exact equations could be integrated on a machine," a challenge that took nearly thirty years to be accomplished reasonably and which still needs a full treatment. In the same paper, he also included compressibility and other effects for enhancing reconnection, such as ambipolar diffusion and fluid instabilities. He then applied this quantitative model to solar flares, but he found that the rate at which magnetic energy is converted to kinetic energy and heat is far too slow (by a factor of at least a hundred) to account for flares. A conversion rate several orders of magnitude greater was needed to explain the energy release in solar flares, and thus the Sweet–Parker model is often referred to as a model for *slow* reconnection. Ever since Sweet and Parker presented their model, the search has continued for a reconnection process which would be fast enough to work for solar flares.

As we mentioned previously, there have been two separate approaches in the search for a theory of fast reconnection – one based on showing that the plasma resistivity is sufficiently high, the other based on showing that the dissipation scale is sufficiently small. In calculating the reconnection rate, Sweet and Parker assumed that Spitzer's (1962) formula for the resistivity can be applied to the corona. However, there is good reason to suspect that Spitzer's formula is not valid for the type of plasma that is produced in the corona during a flare, since it assumes that there are many more collisions between particles than is appropriate for the solar corona or many other space and astrophysical plasmas. Thus, Sweet and Parker's mechanism might be fast enough only when combined with an anomalous resistivity.

The length of the current sheet in Sweet and Parker's model is approximately the same as the global scale length of the flaring region. However, Petschek (1964) developed an alternative model with a current sheet whose length is many orders of magnitude smaller than the one assumed by Sweet and Parker (§4.3). Because of this much smaller current sheet, Petschek's model predicts a reconnection rate which is close to the rate needed in solar flares, even if Spitzer resistivity is assumed. Thus, Petschek's model was the first model of *fast* reconnection to be proposed. Since then, a new generation of more general almost-uniform (§5.1) and non-uniform (§5.2) models has been developed.

One of the most important developments in the theory of reconnection occurred at about the same time when Furth, Killeen, and Rosenbluth published their classic paper on the tearing mode (Furth et al., 1963). In this paper they analyzed the stability of a simple, static current sheet and found that a sheet whose length is at least 2π times greater than its width will spontaneously reconnect to form magnetic islands (see Chapter 6). Thus, although in some applications reconnection occurs as a steady or quasi-steady process (i.e., modulated on a time-scale longer than the reconnection time), in other applications the reconnection is inherently time-dependent with impulsive and bursty releases of energy.

1.2 Magnetohydrodynamic Equations

Since derivations of the magnetohydrodynamic (MHD) equations are readily available elsewhere (e.g., Roberts, 1967; Priest, 1982), we give here only a short summary of them. The limitations of the MHD equations, especially regarding their use in collisionless plasmas, are discussed briefly in Section 1.7. MKS units are adopted, but their relationship to cgs units is summarised in Appendix 2, so that, for example, magnetic fields are measured in tesla, where 1 tesla $= 10^4$ gauss. The shorthand (G) for gauss is used, but the usual shorthand (T) for tesla is not used here in order to avoid possible confusion with the symbol (T) for temperature.

1.2.1 Basic Equations

The MHD equations embody the following conservation principles derived from the equations of fluid mechanics and electromagnetism.

Mass conservation:

$$\frac{d\rho}{dt} \equiv \frac{\partial \rho}{\partial t} + \mathbf{v} \cdot \nabla \rho = -\rho \nabla \cdot \mathbf{v}, \tag{1.1}$$

where ρ is the mass density, $\mathbf{v}$ is the bulk flow velocity, t is the time, and d/dt is the convective derivative, which represents the time rate of change following a plasma element as it moves.

Momentum conservation:

$$\rho \frac{d\mathbf{v}}{dt} = -\nabla p + \mathbf{j} \times \mathbf{B} + \nabla \cdot \underline{\mathbf{S}} + \mathbf{F}_g, \tag{1.2}$$

where p is the plasma pressure, $\mathbf{j}$ is the current density, $\mathbf{B}$ is the magnetic induction, $\underline{\mathbf{S}}$ is the viscous stress tensor, and $\mathbf{F}_g$ is an external force such as gravity. The induction ($\mathbf{B}$) is usually referred to as the "magnetic field," although technically the magnetic field is $\mathbf{H} = \mathbf{B}/\mu$, where μ is the magnetic permeability of free space $(4\pi \times 10^{-7}\,\mathrm{H\,m^{-1}})$.

In very weak magnetic fields the stress-tensor components are

$$S_{ij} = \rho\left(\zeta - \frac{2\nu}{3}\right)\nabla \cdot \mathbf{v}\, \delta_{ij} + \rho\nu\left(\frac{\partial v_i}{\partial x_j} + \frac{\partial v_j}{\partial x_i}\right), \qquad (1.3)$$

where ν and ζ are the coefficients of kinematic shear and bulk viscosity, respectively. The divergence of this expression (when the dynamic viscosity $\rho\nu$ and bulk viscosity $\rho\zeta$ are assumed uniform) gives the viscous force

$$\nabla \cdot \underline{\mathbf{S}} = \rho\nu\nabla^2\mathbf{v} + \rho(\zeta + \nu/3)\nabla(\nabla \cdot \mathbf{v}). \qquad (1.4)$$

Substituting this form into Eq. (1.2) and ignoring gravity and the magnetic field gives the standard form of the Navier–Stokes equations for a viscous fluid. In plasmas with non-negligible magnetic fields, however, the stress tensor is considerably more complex (e.g., Braginsky, 1965; Hollweg, 1986).

Internal energy conservation:

$$\rho\frac{de}{dt} + p\nabla \cdot \mathbf{v} = \nabla\cdot(\underline{\boldsymbol{\kappa}} \cdot \nabla T) + (\underline{\boldsymbol{\eta}}_e \cdot \mathbf{j}) \cdot \mathbf{j} + Q_\nu - Q_r, \qquad (1.5)$$

where

$$e = \frac{p}{(\gamma - 1)\rho}$$

is the internal energy per unit mass, $\underline{\boldsymbol{\kappa}}$ is the thermal conductivity tensor, T is the temperature, $\underline{\boldsymbol{\eta}}_e$ is the electrical resistivity tensor, Q_ν is the heating by viscous dissipation, Q_r is the radiative energy loss, and γ is the ratio of specific heats. In terms of the stress tensor ($\underline{\mathbf{S}}$), Q_ν is

$$Q_\nu = \mathbf{v} \cdot (\nabla \cdot \underline{\mathbf{S}}) - \nabla \cdot (\underline{\mathbf{S}} \cdot \mathbf{v}). \qquad (1.6)$$

In many astrophysical and solar applications, the plasma is optically thin and so the radiative loss term in Eq. (1.5) can be expressed as

$$Q_r = \rho^2\, Q(T),$$

where $Q(T)$ is a function describing the temperature variation of the radiative loss (Fig. 11.9).

Faraday's equation:

$$\nabla \times \mathbf{E} = -\frac{\partial \mathbf{B}}{\partial t}, \qquad (1.7)$$

where $\mathbf{E}$ is the electric field.

In MHD the usual displacement term ($\epsilon_0\,\partial\mathbf{E}/\partial t$, where ϵ_0 is the permittivity of free space) in Maxwell's equations is negligible and so the current is related to the field ($\mathbf{B}$) simply by

Ampère's law:

$$\nabla \times \mathbf{B} = \mu\mathbf{j}. \tag{1.8}$$

Because the divergence of the curl is zero, Eq. (1.8) immediately gives us the result that $\nabla \cdot \mathbf{j} = 0$, so that the electric current lines have no monopolar sources: they form closed paths unless they are ergodic or go off to infinity. Also, substitution of Eq. (1.8) for $\mathbf{j}$ in Eq. (1.2) enables the $\mathbf{j} \times \mathbf{B}$ force term to be replaced by $-\nabla[B^2/(2\mu)] + (\mathbf{B}\cdot\nabla)\mathbf{B}/\mu$. This form shows that the magnetic force in Eq. (1.2) can be divided into a magnetic pressure force $-\nabla[B^2/(2\mu)]$ and a magnetic tension force $(\mathbf{B}\cdot\nabla)\mathbf{B}/\mu$.

Gauss's law:

$$\nabla \cdot \mathbf{B} = 0. \tag{1.9}$$

Ohm's law:

$$\mathbf{E}' = \mathbf{E} + \mathbf{v} \times \mathbf{B} = \underline{\eta}_e \cdot \mathbf{j}. \tag{1.10}$$

In many applications it is sufficient to write the electrical resistivity tensor as $\underline{\eta}_e = \eta_e\,\delta_{ij}$, where η_e (the scalar electrical resistivity) is the inverse of the electrical conductivity (σ) so that

$$\mathbf{E}' = \mathbf{E} + \mathbf{v} \times \mathbf{B} = \frac{\mathbf{j}}{\sigma}. \tag{1.10a}$$

Here $\mathbf{E}' = \mathbf{E} + \mathbf{v} \times \mathbf{B}$ gives the Lorentz transformation from the electric field ($\mathbf{E}$) in a laboratory frame of reference to the electric field ($\mathbf{E}'$) in a frame moving with the plasma. It is important to note that Ohm's Law states that it is the electric field ($\mathbf{E}'$) in the moving frame (rather than in the laboratory frame) that is proportional to the current.

Equation of state:

$$p = \mathcal{R}\rho T = n k_B T, \tag{1.11}$$

where $\mathcal{R}$ is the universal gas constant ($8300\,\mathrm{m^2\,s^{-2}\,deg^{-1}}$), n is the total number of particles per unit volume, and k_B is Boltzmann's constant. The density can be written in terms of n as

$$\rho = n\overline{m},$$

where $\overline{m}$ is the mean particle mass, so that $k_B/\mathcal{R} = \overline{m}$. For a hydrogen plasma with electron number density n_e, the pressure becomes

$$p = 2n_e k_B T$$

and the plasma density may be written

$$\rho \approx n_e m_p,$$

where m_p is the proton mass.

The above system of time-dependent equations constitutes a set of 16 coupled equations for 15 unknowns ($\mathbf{v}, \mathbf{B}, \mathbf{j}, \mathbf{E}$, ρ, p, and T). Thus, it may seem that the system is over-determined. However, Gauss's law, Eq. (1.9), for the divergence of the magnetic field is a relatively weak constraint which has the status only of an initial condition. Taking the divergence of Faraday's equation, Eq. (1.7), and recalling that the divergence of the curl is always zero, we find that $\partial(\nabla \cdot \mathbf{B})/\partial t = 0$, so that, if $\nabla \cdot \mathbf{B}$ is zero initially, it will remain zero for all time. Therefore, given a divergence-free initial state, Gauss's law follows from Faraday's equation. It is interesting to note that the steady-state version of these equations has a different mathematical structure, with $\nabla \cdot \mathbf{B} = 0$ becoming a genuine equation. Instead, Faraday's equation (1.7) essentially represents only two rather than three equations, since the three equations $\nabla \times \mathbf{E} = \mathbf{0}$ may be replaced by the three equations $\mathbf{E} = -\nabla\Phi$ together with the introduction of an extra (sixteenth) variable Φ.

1.2.2 Other Useful Forms

Other useful forms can be obtained by combining the above equations. For example, the variables $\mathbf{E}$, $\mathbf{j}$, and T in the above system can easily be eliminated by substitution. Consequently, the eight remaining unknowns ($\mathbf{B}, \mathbf{v}, p$, and ρ) are usually thought of as the primary variables of MHD. For example, Faraday's equation can be combined with Ampère's law and Ohm's law to obtain the

Induction equation:

$$\frac{\partial \mathbf{B}}{\partial t} = \nabla \times (\mathbf{v} \times \mathbf{B}) - \nabla \times (\eta \nabla \times \mathbf{B}),$$

where $\eta = (\mu\sigma)^{-1} = \eta_e/\mu$ is the magnetic diffusivity. If η is uniform, then

$$\frac{\partial \mathbf{B}}{\partial t} = \nabla \times (\mathbf{v} \times \mathbf{B}) + \eta\nabla^2\mathbf{B}. \tag{1.12}$$

This is the basic equation for magnetic behaviour in MHD, and it determines **B** once **v** is known. In the electromagnetic theory of fixed conductors, the electric current and electric field are primary variables with the current driven by electric fields. In such a fixed system the magnetic field is a secondary variable derived from the currents. However, in MHD the basic physics is quite different, since the plasma velocity (**v**) and magnetic field (**B**) are the primary variables, determined by the induction equation and the equation of motion, while the resulting current density (**j**) and electric field (**E**) are secondary and may be deduced from Eqs. (1.8) and (1.10a) if required (Parker, 1996).

For a collisional plasma with a strong magnetic field but a relatively weak electric field, the magnetic diffusivity is approximately given by Spitzer's formula (Spitzer, 1962; Schmidt, 1966), namely

$$\eta_\perp = \frac{c^2 e^2 m_e^{1/2}}{3(2\pi)^{3/2}\epsilon_0} \ln\Lambda\,(k_B T_e)^{-3/2} = 1.05 \times 10^8\, T_e^{-3/2} \ln\Lambda\,\mathrm{m^2\,s^{-1}}, \quad (1.13)$$

where m_e is the electron mass, T_e the electron temperature, and $\ln\Lambda$ is the coulomb logarithm (Holt and Haskell, 1965), given approximately by

$$\ln\Lambda = \begin{cases} 16.3 + \frac{3}{2}\ln T - \frac{1}{2}\ln n, & T < 4.2 \times 10^5\,\mathrm{K}, \\ 22.8 + \ln T - \frac{1}{2}\ln n, & T > 4.2 \times 10^5\,\mathrm{K}. \end{cases} \quad (1.14)$$

For the final expression in Eq. (1.13) we have assumed a hydrogen plasma. The $\perp$ subscript indicates that Eq. (1.13) is the resistivity in a direction perpendicular to the magnetic field. If the current flows along the field, or if the plasma is unmagnetized, the parallel magnetic diffusivity is $\eta_\parallel \approx \eta_\perp/2$.

Since the collisional theory used to calculate Eq. (1.13) is a first-order expansion in powers of $1/\ln\Lambda$, the theory is only accurate when $\ln\Lambda \gg 1$. In laboratory plasmas, $\ln\Lambda$ is typically about 10, in the solar corona it is about 20, and in the magnetosphere it is about 30. Thus, transport coefficients such as Eq. (1.13) are only accurate to a few percent. The requirement that the electric field be weak means that the electric field must be less than the Dreicer field (E_D) for runaway electrons [see Eq. (1.66)], while the condition that the magnetic field be strong means that the electron gyro-radius must be much smaller than the mean-free path (λ_{mfp}) for electron-ion collisions.

If V_0, L_0 are typical velocity and length-scales, the ratio of the first to the second term on the right-hand side of Eq. (1.12) is, in order of magnitude, the magnetic Reynolds number

$$R_m = \frac{L_0 V_0}{\eta}.$$

Thus, for example, in the solar corona above an active region, where $T \approx 10^6$ K, $\eta \approx 1\,\mathrm{m^2\,s^{-1}}$, $L_0 \approx 10^5\,\mathrm{m}$, $V_0 \approx 10^4\,\mathrm{m\,s^{-1}}$, we find $R_m \approx 10^9$, and so the second term on the right of Eq. (1.12) is completely negligible. In turn, Eq. (1.10a) reduces to $\mathbf{E} = -\mathbf{v} \times \mathbf{B}$ to a very high degree of approximation. This is the case in almost all of the solar atmosphere, indeed in almost all of the plasma universe. The only exception is in regions (such as current sheets) where the length-scale is extremely small – so small that $R_m \lesssim 1$ and the second terms on the right-hand sides of Eqs. (1.10a) and (1.12) become important.

If $R_m \ll 1$, the induction equation reduces to

$$\frac{\partial \mathbf{B}}{\partial t} = \eta \nabla^2 \mathbf{B}, \tag{1.15}$$

and so $\mathbf{B}$ is governed by a diffusion equation, which implies that field variations (irregularities) on a scale L_0 diffuse away on a time-scale of

$$\tau_d = \frac{L_0^2}{\eta}, \tag{1.16}$$

which is obtained simply by equating the orders of magnitude of both sides of Eq. (1.15). The corresponding magnetic diffusion speed at which they slip through the plasma is

$$v_d = \frac{L_0}{\tau_d} = \frac{\eta}{L_0}. \tag{1.17}$$

With $\eta \approx 1\,\mathrm{m^2\,s^{-1}}$ and $L_0 = 10^8\,\mathrm{m}$, the decay-time for a sunspot is 30,000 yr, so that the process whereby sunspots disappear in a few weeks cannot just be diffusion. Note that the corresponding viscous diffusion speed, which arises from the equation of motion when viscous forces are included, is $v_D = \nu/L_0$.

The main reason for variations in R_m from one phenomenon to another is variations in the appropriate length-scale (L_0). If $R_m \gg 1$, the induction equation becomes

$$\frac{\partial \mathbf{B}}{\partial t} = \nabla \times (\mathbf{v} \times \mathbf{B}).$$

Although the electric field and current density are usually ignored during the process of solving the MHD equations, valuable insight can often be gained by determining the electric field and current density afterwards. In this regard, it is often helpful to introduce the concepts of the electric scalar potential (Φ) and magnetic vector potential ($\mathbf{A}$). Defining $\mathbf{A}$ from $\mathbf{B} = \nabla \times \mathbf{A}$ (to within a gauge), one then writes Faraday's equation and Ampère's law as

$$\mathbf{E} = -\frac{\partial \mathbf{A}}{\partial t} - \nabla\Phi, \tag{1.18}$$

$$\mathbf{j} = -\nabla^2\mathbf{A}/\mu, \tag{1.19}$$

where the coulomb gauge ($\nabla \cdot \mathbf{A} = 0$) has been assumed.

It is also sometimes useful to rewrite the momentum equation in terms of the fluid vorticity ($\boldsymbol{\omega}$), where $\boldsymbol{\omega} \equiv \nabla \times \mathbf{v}$. Dividing Eq. (1.2) by the density (ρ), assuming $\rho\nu$ and $\rho\zeta$ are uniform and taking the curl, we obtain, with the help of some common vector identities, the

Vorticity equation:

$$\frac{d\boldsymbol{\omega}}{dt} = (\boldsymbol{\omega} \cdot \nabla)\mathbf{v} - \boldsymbol{\omega}(\nabla \cdot \mathbf{v}) + \rho^{-2}(\nabla\rho \times \nabla p) + \rho^{-1}[(\mathbf{B} \cdot \nabla)\mathbf{j} - (\mathbf{j} \cdot \nabla)\mathbf{B}]$$

$$+ \nu\nabla^2\boldsymbol{\omega} - \rho^{-2}\nabla\rho \times [\mathbf{j} \times \mathbf{B} + \rho\nu\nabla^2\mathbf{v} + \rho(\zeta + \nu/3)\nabla(\nabla \cdot \mathbf{v})], \tag{1.20}$$

where the curl of the $\mathbf{j} \times \mathbf{B}$ term has been simplified by taking advantage of the fact that the divergences of both $\mathbf{j}$ and $\mathbf{B}$ are zero. We have also assumed that the external force is such that $\nabla \times (\mathbf{F}_g/\rho) = \mathbf{0}$, which is true when, for instance, $\mathbf{F}_g = \rho\mathbf{g}$ and g is constant. The third term on the right-hand side of Eq. (1.20) is sometimes referred to as Crocco's term (Milne-Thomson, 1960); it disappears in a fluid with constant and uniform entropy so that p is only a function of ρ. The general form (1.20) of the vorticity equation is usually simplified by invoking one or more additional assumptions. For example, in an incompressible, two-dimensional system with uniform density (ρ), $\mathbf{B} = (B_x, B_y, 0)$ and $\mathbf{v} = (v_x, v_y, 0)$, the vorticity equation reduces to

$$\frac{d\omega}{dt} = \nu\nabla^2\omega + (\mathbf{B} \cdot \nabla)j, \tag{1.21}$$

where $\boldsymbol{\omega} = \omega\hat{\mathbf{z}}$ and $\mathbf{j} = j\hat{\mathbf{z}}$. Equation (1.21) shows that, with these assumptions, vorticity is generated by variations of the current density (j) along field lines, while viscosity acts to dissipate it.

The flow of energy in a system is one of the most important concepts for understanding its physics. One of the main reasons that MHD is so useful, even in plasma environments where it is only an approximation, is that it provides insight into the energetics of the system, especially when complex geometry is involved. A particularly useful equation in this regard can be obtained by combining the basic equations of Section 1.2.1 to obtain the

Total energy equation:

$$\frac{\partial w}{\partial t} = -\nabla \cdot \mathbf{q} + \mathbf{v} \cdot \mathbf{F}_g - Q_r, \qquad (1.22)$$

where the total energy density (w) is given by

$$w = \rho \left(e + \frac{v^2}{2} \right) + \frac{B^2}{2\mu}, \qquad (1.23)$$

while the energy flux $(\mathbf{q})$ is given by

$$\mathbf{q} = \left[\rho \left(e + \frac{v^2}{2} \right) + p \right] \mathbf{v} - \underline{\kappa} \cdot \nabla T - \mathbf{v} \cdot \underline{\mathbf{S}} + \frac{\mathbf{E} \times \mathbf{B}}{\mu}. \qquad (1.24)$$

A full derivation can be found in Roberts (1967).

Another useful energy equation can be obtained from Eq. (1.5) by using entropy (s) in place of internal energy to write the

Entropy equation:

$$\rho^\gamma \frac{d}{dt} \left(\frac{p}{\rho^\gamma} \right) = (\gamma - 1) \left[\nabla \cdot (\underline{\kappa} \cdot \nabla T) + \frac{j^2}{\sigma} - Q_r + Q_\nu \right], \qquad (1.25)$$

where p/ρ^γ is related to the entropy (s) by

$$s = c_v \ln \left(\frac{p}{\rho^\gamma} \right) + \text{constant}$$

and $c_v = \mathcal{R}/(\gamma - 1)$ is the specific heat at constant volume.

1.2.3 Common Approximations

In many applications, some of the terms in the full system of MHD equations are negligible. Whether a particular term can be ignored or not is often determined by making rough estimates of the various terms by replacing the time and space differentials with characteristic time and space scales. If one term is orders of magnitude smaller than the other terms, it is usually ignored. However, this procedure is not without hazard because even small terms can have important consequences. The famous D'Alembert paradox concerning flow in a boundary layer is a classic example of where the dropping of a small term leads to a wrong answer. D'Alembert showed that the solution of the fluid equations without viscosity terms (the Euler equations) does not necessarily give the same solution as obtained by solving the equations with the viscous terms (the Navier–Stokes equations) and then allowing

Table 1.1. *Commonly Used Dimensionless Numbers*

Reynolds Number	Magnetic Reynolds Number
$R_e \equiv V_0/v_D$, where $v_D = \nu/L_0$	$R_m \equiv V_0/v_d$, where $v_d = \eta/L_0$
Plasma Beta	Lundquist Number
$\beta \equiv 2\mu p/B^2$	$L_u \equiv v_A/v_d$, where $v_A = B/\sqrt{\mu\rho}$
Prandtl Number	Magnetic Prandtl Number
$P_r \equiv \nu/\kappa$	$P_m \equiv \nu/\eta$
Mach Number	Alfvén Mach Number
$\overline{M} \equiv V_0/c_s$, where $c_s \equiv \sqrt{\gamma RT}$	$M \equiv V_0/v_A$

the viscosity to tend to zero. This discrepancy occurs because the latter procedure allows the formation of discontinuities, while the former procedure does not. Nevertheless, comparing the sizes of various terms is a very useful exercise because of the physical understanding it provides of the importance of the various forces acting on the plasma.

The sizes of various terms are normally expressed in terms of dimensionless numbers formed by taking the ratios of the rough estimates for the various terms. Some of the more common dimensionless numbers are shown in Table 1.1. Here V_0 is a characteristic speed, L_0 is a characteristic length-scale, v_D is the characteristic speed for viscous diffusion, v_d is the characteristic speed for magnetic diffusion, v_A is the Alfvén speed, and c_s is the sound speed. Other dimensionless numbers in MHD and fluid dynamics can be found in Priest (1982) and Lighthill (1978).

There are at least a dozen different approximations that are commonly used for the MHD system of equations. Some of the well-known ones are: incompressibility (ρ constant or $\nabla \cdot \mathbf{v} = 0$); a steady state ($\partial/\partial t = 0$ for all variables); the Boussinesq approximation (filtering out sound waves by including density variations only in the gravitational term in the equation of motion); an isothermal state (T = constant or κ large); an ideal MHD state ($R_m \gg 1$); an inviscid state ($R_e \gg 1$); an irrotational (potential) flow ($\boldsymbol{\omega} = 0$); an isentropic state (s = constant); a force-free field ($\mathbf{j} \times \mathbf{B} = 0$); a potential field ($\mathbf{j} = 0$); a strong magnetic field ($\beta \ll 1$); a weak magnetic field ($\beta \gg 1$); a supersonic flow regime ($\overline{M} > 1$); a subsonic flow regime ($\overline{M} < 1$); one-dimensional flow; two-dimensional co-planar flow ($\mathbf{v}$ and $\mathbf{B}$ both in the plane of variation); two-dimensional non-co-planar flow (the so called $2\frac{1}{2}$-D equations); and the Strauss approximation (for quasi-two-dimensional motions of a magnetic flux tube with a strong axial magnetic field, see Strauss, 1977).

The assumption of incompressibility is often made on the grounds of simplicity; namely, in order to understand some process one first of all excludes the complications of compressibility and then later on includes them. The condition for validity of this assumption in a compressible MHD medium depends on which terms are dominant in the equation of motion. Thus, consider an adiabatic change for which the changes ($\delta\rho$ and δp) in density and pressure are related by

$$\frac{\delta\rho}{\rho} = \frac{1}{\gamma}\frac{\delta p}{p}.$$

Then, if the pressure gradient and inertial terms dominate (e.g., if $\beta \gg 1$), then $\delta p \approx \rho v\,\delta v$ and so the condition that $\delta\rho/\rho \ll \delta v/v$ becomes

$$v \ll c_s.$$

If, on the other hand, there is a balance between pressure gradients and Lorentz forces, so that $\delta p \approx \delta[B^2/(2\mu)]$, then the condition that $\delta\rho/\rho \ll \delta B/B$ becomes

$$\beta \gg \frac{2}{\gamma}.$$

In other words, compressibility is important in the first case for rapid flows and in the second case for a low-beta plasma.

1.3 Null Points and Current Sheets

Neutral points or null points are locations where the magnetic field vanishes, and, as mentioned in Section 1.1, they are a common feature of magnetic fields containing multiple sources (for example, the field formed by two bar magnets). A neutral point can occur whether the magnetized medium is a conducting plasma or a neutral gas. However, the same is not true for a current sheet. Generally, a current sheet is a thin current-carrying layer across which the magnetic field changes either in direction or magnitude or both. Since, by definition, the layer contains a current, a current sheet can only exist in a conducting medium such as a plasma. In plasmas, null points typically give rise to current sheets, and so we now proceed to introduce the elementary, two-dimensional aspects of both concepts. Complications which arise in three dimensions are discussed in Chapter 8.

1.3.1 Two-Dimensional Neutral Points

If we expand the two-dimensional field (B_X, B_Y) near a neutral point in a Taylor series and keep only the first-order, linear terms, then the field components have the form

$$B_X = bX + 2cY, \tag{1.26}$$

$$B_Y = -2aX + dY, \tag{1.27}$$

where a, b, c, and d are arbitrary constants (the factors of 2 are introduced in order to simplify the final results). The condition $\nabla \cdot \mathbf{B} = 0$ implies that

$$d = -b.$$

In terms of the magnetic flux function (A) the field components are

$$B_X = \frac{\partial A}{\partial Y}, \quad B_Y = -\frac{\partial A}{\partial X}, \tag{1.28}$$

which from Eqs. (1.26) and (1.27) give the corresponding flux function as

$$A = aX^2 + bXY + cY^2, \tag{1.29}$$

where we have chosen the arbitrary constant of integration such that A vanishes at the origin.

Further simplification is achieved by rotating the XY-axes through an angle θ to give new xy-axes, such that

$$X = x\cos\theta - y\sin\theta, \quad Y = x\sin\theta + y\cos\theta.$$

Substitution of these forms into Eq. (1.29) gives

$$\begin{aligned}
A = &\tfrac{1}{2}\left[(a+c) + b\sin(2\theta) + (a-c)\cos(2\theta)\right]x^2 \\
&+ \left[(c-a)\sin(2\theta) + b\cos(2\theta)\right]xy \\
&+ \tfrac{1}{2}\left[(a+c) - b\sin(2\theta) - (a-c)\cos(2\theta)\right]y^2.
\end{aligned} \tag{1.30}$$

Then, if we choose the angle (θ) such that

$$\tan(2\theta) = \frac{b}{a-c},$$

the xy term vanishes and we are left with a flux function of the form

$$A = \frac{B_0}{2L_0}(y^2 - \overline{\alpha}^2 x^2), \tag{1.31}$$

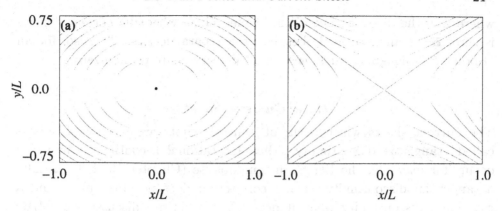

Fig. 1.5. Two-dimensional neutral points of (a) O-type, with $\bar{\alpha}^2 = -0.5$ and (b) X-type, with $\bar{\alpha}^2 = 0.5$ in Eq. (1.31).

where

$$\frac{B_0}{L_0} = (a+c) - \sqrt{b^2 + (a-c)^2}, \quad \bar{\alpha}^2 = \frac{\sqrt{b^2 + (a-c)^2} + (a+c)}{\sqrt{b^2 + (a-c)^2} - (a+c)}$$

and L_0 is the length-scale over which the field is varying. The corresponding field components are

$$B_x = B_0 \frac{y}{L_0}, \quad B_y = B_0 \bar{\alpha}^2 \frac{x}{L_0},$$

so that B_x vanishes on the x-axis and B_y on the y-axis.

The field lines are given by the curves

$$A = \text{constant}.$$

When $\bar{\alpha}^2 < 0$, Eq. (1.31) implies that the field lines are elliptical [Fig. 1.5(a)], and in this case the origin is referred to as an O-type neutral point. The particular case when $\bar{\alpha}^2 = -1$ produces circular field lines. When $\bar{\alpha}^2 > 0$, the field lines are hyperbolic [Fig. 1.5(b)] and we have an X-type neutral point (or X-point or X-line for short). The limiting field lines

$$y = \pm \bar{\alpha} x$$

through the origin are known as separatrices and are inclined at the angles $\pm \tan^{-1} \bar{\alpha}$ to the x-axis. The separatrices form an "X", from which the term "X-type" null point is derived. The value of $\bar{\alpha}$ (and therefore the angle between the separatrices) is related to the current density as follows. Taking the curl of $\mathbf{B}$ we find

$$\mathbf{j} = -\frac{1}{\mu} \nabla^2 A \,\hat{\mathbf{z}} = -\left(\frac{B_0}{\mu L_0}\right)(1 - \bar{\alpha}^2)\,\hat{\mathbf{z}}, \tag{1.32}$$

where $\hat{\mathbf{z}}$ is the direction out of the plane. Thus, when the current density is zero, $\bar{\alpha}=1$ and the angle between the separatrices is $90°$. Also, for an O-point $\bar{\alpha}$ is imaginary and the current density is always non-zero.

1.3.2 Current Sheets

The current sheets which occur at neutral points are usually of a type called a tangential discontinuity, although rotational discontinuities may also occur: for example, the Earth's magnetopause (Chapter 10) is essentially a tangential discontinuity when reconnection does not take place and a rotational discontinuity when it does. (For a general discussion of MHD discontinuities, see Landau and Lifshitz, 1984.) At a tangential discontinuity, the magnetic field is tangential and the plasma flow through the sheet is zero. In equilibrium, there is a total pressure balance

$$p_2 + \frac{B_2{}^2}{2\mu} = p_0 + \frac{B_0{}^2}{2\mu} = p_1 + \frac{B_1{}^2}{2\mu} \tag{1.33}$$

between the values on one side, in the centre of the sheet, and on the other side (denoted by subscripts 2, 0, 1, respectively). In particular, if the central field (B_0) vanishes, we have a neutral sheet. If also the ambient pressures $(p_1$ and $p_2)$ vanish, Eq. (1.33) reduces to

$$\frac{B_2{}^2}{2\mu} = p_0 = \frac{B_1{}^2}{2\mu},$$

and so, if the field changes sign across the sheet, there is an exact field reversal with $B_2 = -B_1$. If the magnetic field in the sheet is along the y-direction and varies with x,

$$\mathbf{B} = B_y(x)\,\hat{\mathbf{y}},$$

then Ampère's law (1.8) implies that

$$j_z = \frac{1}{\mu}\frac{dB_y}{dx},$$

so that a steep gradient in B_y with x produces a strong current along the sheet and perpendicular to the field lines. In general, in this book we shall denote the half-width and half-length of a current sheet by l and L, respectively (e.g., Fig. 4.2).

Current sheets formed from tangential discontinuities may be subject to several instabilities (Priest, 1982), such as the Kelvin–Helmholtz instability (Landau and Lifshitz, 1984). Suppose we have a simple current sheet with uniform fields $(\mathbf{B}_1)$ and $(\mathbf{B}_2)$ and uniform flows $(\mathbf{v}_1)$ and $(\mathbf{v}_2)$ on either side.

In an ideal, incompressible plasma, it transpires that the Kelvin–Helmholtz instability occurs in such a sheet if either of the following two conditions is met:

$$\tfrac{1}{2}\mu\rho\,(\mathbf{v}_2 - \mathbf{v}_1)^2 > B_1{}^2 + B_2{}^2, \tag{1.34}$$

$$\tfrac{1}{2}\mu\rho\left\{[\mathbf{B}_1 \times (\mathbf{v}_2 - \mathbf{v}_1)]^2 + [\mathbf{B}_2 \times (\mathbf{v}_2 - \mathbf{v}_1)]^2\right\} > (\mathbf{B}_1 \times \mathbf{B}_2)^2. \tag{1.35}$$

The first of these conditions (1.34) states essentially that the current sheet becomes unstable if the kinetic energy of the tangential velocity difference $(\mathbf{v}_2 - \mathbf{v}_1)$ exceeds the total magnetic energy density. A magnetic field acts to stabilize the current sheet because the fluid must do work to distort the field lines. The second condition (1.35), which involves the relative orientations of the flow and the field, arises because flows perpendicular to the field are more susceptible to instability. In a non-ideal plasma, a current sheet can still be unstable even if conditions (1.34) and (1.35) are not met. Another point to note is that for a simple current sheet with uniform fields and flows, both resistivity and viscosity cause the current sheet to thicken with time.

Almost all analyses of the Kelvin–Helmholtz instability have been undertaken for a one-dimensional current sheet with uniform flows and fields on either side. However, the current sheets which occur at neutral points have highly non-uniform flow and field on either side, and they are of finite length. Relatively little is known about the stability of such sheets at X-type neutral lines or points.

1.4 The Concepts of Frozen Flux and Field-Line Motion

The term "magnetic reconnection" is intimately linked to the concept of field-line motion, which was first introduced by Hannes Alfvén (1943). He showed that, in a fluid with a large magnetic Reynolds number $(R_m \gg 1)$, the field lines move as though they are "frozen" into the fluid. For such an ideal fluid Ohm's law (1.10a) just becomes $\mathbf{E} + \mathbf{v} \times \mathbf{B} = 0$, and the induction equation (1.12) reduces to

$$\frac{\partial \mathbf{B}}{\partial t} = \nabla \times (\mathbf{v} \times \mathbf{B}). \tag{1.36}$$

Then, if we consider a curve C (bounding a surface S) which is moving with the plasma, in a time dt an element $\mathbf{ds}$ of C sweeps out an element of area $\mathbf{v} \times \mathbf{ds}\,dt$. The rate of change of magnetic flux through C is

$$\frac{d}{dt}\int_s \mathbf{B} \cdot \mathbf{dS} = \int_s \frac{\partial \mathbf{B}}{\partial t} \cdot \mathbf{dS} + \int_c \mathbf{B} \cdot \mathbf{v} \times \mathbf{ds}. \tag{1.37}$$

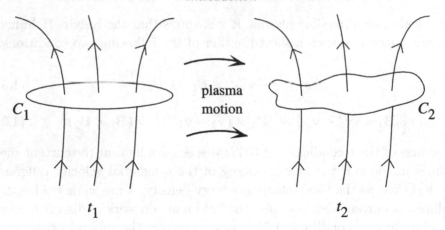

Fig. 1.6. Magnetic flux conservation: if a curve C_1 is distorted into C_2 by plasma motion, the flux through C_1 at t_1 equals the flux through C_2 at t_2.

As C moves, so the flux changes, both because the magnetic field changes with time [the first term on the right of Eq. (1.37)] and because the boundary moves in space [the second term on the right of Eq. (1.37)]. By setting $\mathbf{B} \cdot \mathbf{v} \times \mathbf{ds} = -\mathbf{v} \times \mathbf{B} \cdot \mathbf{ds}$ and applying Stokes' theorem to the second term, we obtain

$$\frac{d}{dt} \int_s \mathbf{B} \cdot \mathbf{dS} = \int_s \left[\frac{\partial \mathbf{B}}{\partial t} - \nabla \times (\mathbf{v} \times \mathbf{B}) \right] \cdot \mathbf{dS},$$

which vanishes in the ideal limit. Thus, the total magnetic flux through C remains constant as it moves with the plasma. In other words, we have proved *magnetic flux conservation*, namely the plasma elements that initially form a flux tube continue to do so at all later times (Fig. 1.6).

There is also *magnetic field line conservation*, namely that, if two plasma elements lie on a field line initially, then they will always do so (Fig. 1.7). Suppose that at $t = t_1$ the elements P_1 and P_2 lie on a field line, which may be defined as the intersection of two flux tubes. Then, at some later time $(t = t_2)$, according to magnetic flux conservation, P_1 and P_2 will still lie on both tubes, and so they will lie on the field line defined by their intersection. We have therefore established that, if Eq. (1.36) is true, then flux conservation holds and therefore line conservation holds.

Line conservation also may be proved directly from Eqs. (1.36) and (1.1) as follows. Applying the appropriate vector identity to the ideal induction equation gives

$$\frac{\partial \mathbf{B}}{\partial t} = (\mathbf{B} \cdot \nabla)\mathbf{v} - (\mathbf{v} \cdot \nabla)\mathbf{B} - \mathbf{B}(\nabla \cdot \mathbf{v}). \tag{1.38}$$

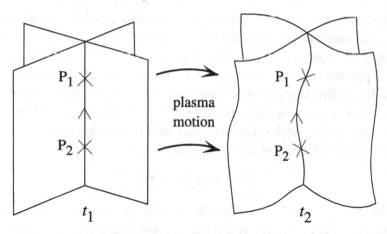

Fig. 1.7. Magnetic field-line conservation: if plasma elements P_1 and P_2 lie on a field line at time t_1, then they will lie on the same line at a later time t_2.

Using the mass continuity equation (1.1) to eliminate $\nabla \cdot \mathbf{v}$, we then obtain

$$\frac{d}{dt}\left(\frac{\mathbf{B}}{\rho}\right) = \left(\frac{\mathbf{B}}{\rho} \cdot \nabla\right)\mathbf{v}, \tag{1.39}$$

where $d/dt\,(= \partial/\partial t + \mathbf{v} \cdot \nabla)$ is the total or convective derivative. To see how this result leads to the conclusion that the field lines are "frozen" to the plasma, consider an elemental segment $\delta\mathbf{l}$ along a line moving with the plasma. If $\mathbf{v}$ is the plasma velocity at one end of the element and $\mathbf{v} + \delta\mathbf{v}$ is the velocity at the other end, then the differential velocity between the two ends is $\delta\mathbf{v} = (\delta\mathbf{l} \cdot \nabla)\mathbf{v}$. During the time interval dt, the segment $\delta\mathbf{l}$ changes at the rate

$$\frac{d\delta\mathbf{l}}{dt} = \delta\mathbf{v} = (\delta\mathbf{l} \cdot \nabla)\mathbf{v}.$$

Since this equation has exactly the same form as Eq. (1.39) for the vector $\mathbf{B}/\rho$, it necessarily follows that, if $\delta\mathbf{l}$ and $\mathbf{B}/\rho$ are initially parallel, then they will remain parallel for all time. Therefore, any two neighbouring plasma elements on a field line are always on the same field line, with the distance between them proportional to $\mathbf{B}/\rho$.

Defining $\mathbf{w}$ as the *magnetic field-line velocity*, we can state the *frozen-flux condition* for an ideal MHD plasma as

$$\mathbf{w} = \mathbf{v}.$$

The component of $\mathbf{w}$ in the direction of $\mathbf{B}$ is actually arbitrary because a one-to-one correspondence between lines of force does not require a one-to-one correspondence between the individual points lying on them. However, the usual convention is to assume that $\mathbf{w} \cdot \mathbf{B} = \mathbf{v} \cdot \mathbf{B}$.

Although the concept of moving field lines first appeared in connection with ideal plasmas, it can be extended to include non-ideal plasmas as well, although the definition is not unique (see §8.1.2). All that is required in order to define a field-line velocity in a resistive plasma is that we be able to find a velocity ($\mathbf{w}$) which has the same flux-preserving property as in the ideal MHD case. That is, $\mathbf{w}$ should be such that

$$\frac{\partial \mathbf{B}}{\partial t} = \nabla \times (\mathbf{w} \times \mathbf{B}). \tag{1.40}$$

(In Section 8.1.2 we shall consider the possibility of a more general definition of $\mathbf{w}$, which is field-line preserving but not necessarily flux preserving.) Comparing Eq. (1.40) with Faraday's equation (1.7), we see that $\mathbf{E}$ can be written in terms of $\mathbf{w}$ and a potential (Φ^*) such that

$$\mathbf{E} = -\mathbf{w} \times \mathbf{B} - \nabla \Phi^*. \tag{1.41}$$

[The asterisk after the symbol Φ is to remind the reader that Φ^* is not necessarily the same as the usual definition, Eq. (1.18), of the electrostatic potential.] The $\nabla \Phi^*$ term in Eq. (1.41) is necessary because in a resistive fluid $\mathbf{E}$ may have a component which is parallel to $\mathbf{B}$.

Substituting the resistive Ohm's law ($\mathbf{E} = -\mathbf{v} \times \mathbf{B} + \mathbf{j}/\sigma$) into Eq. (1.41) and assuming $(\mathbf{w} - \mathbf{v}) \cdot \mathbf{B} = 0$, we obtain the following general definition for the field-line velocity:

$$\mathbf{w} = \mathbf{v} + \frac{(\mathbf{j}/\sigma + \nabla \Phi^*) \times \mathbf{B}}{B^2}, \tag{1.42}$$

where integration of $\mathbf{E} \cdot \mathbf{B} = -\mathbf{B} \cdot \nabla \Phi^*$ from Eq. (1.41) implies that

$$\Phi^*(\mathbf{x}, t) = -\int_{\mathbf{x}_0}^{\mathbf{x}} \mathbf{E} \cdot \mathbf{dl} + \Phi_0^*(\mathbf{x}_0, t). \tag{1.43}$$

This integral is evaluated along a field line, and Φ_0^* is the value of Φ^* at the reference surface ($\mathbf{x}_0$). Since the reference potential (Φ_0^*) is an arbitrary function of $\mathbf{x}_0$ and t, it is not unique and consequently the field-line velocity ($\mathbf{w}$) is also not unique in general. We shall reconsider this aspect in detail in Section 8.1.2. The assumption that $(\mathbf{w} - \mathbf{v}) \cdot \mathbf{B} = 0$ is a matter of convenience, because the displacement of points along a field line is arbitrary in general. In cases in which there is no component of the electric field parallel to $\mathbf{B}$, the definition, Eq. (1.42), reduces to

$$\mathbf{w} = \mathbf{v} + \frac{\mathbf{j} \times \mathbf{B}}{\sigma B^2}, \tag{1.44}$$

since, according to Eq. (1.43), we can set $\nabla \Phi^* = 0$ if we choose $\Phi_0^* = 0$ at $\mathbf{x}_0$.

Note that both Eqs. (1.42) and (1.44) imply that the field-line velocity ($\mathbf{w}$) becomes infinite at a null point. A necessary condition for the validity of the definition (1.42) is that

$$\Phi^* = -\oint \mathbf{E} \cdot \mathbf{dl} = 0 \qquad (1.45)$$

along any closed field line in the plasma (see Roberts, 1967, pp. 53–57). This condition is necessary because without it Φ^* becomes multiple-valued as one integrates around the loop. This multiple-valuedness would in turn lead to the unphysical result that some of the components of $\mathbf{w}$ jump discontinuously as one reaches the starting point of the integration along the loop. A steady-state configuration always satisfies the necessary condition (1.45), but time-dependent configurations do not necessarily do so. A more extensive and rigorous discussion of field-line velocity can be found in Section 8.1.2.

Field-line velocity is also a useful concept for providing an intuitive way to look at the flow of electromagnetic energy in a plasma. In situations in which $\mathbf{E} \cdot \mathbf{B} = 0$, the Poynting flux can be expressed as

$$\mathbf{S} = \frac{\mathbf{E} \times \mathbf{B}}{\mu} = \frac{B^2}{\mu}[\mathbf{w} - (\mathbf{v} \cdot \hat{\mathbf{B}})\hat{\mathbf{B}}],$$

which shows that the field-line velocity perpendicular to $\mathbf{B}$ is the velocity at which magnetic enthalpy is conveyed. The magnetic enthalpy density is twice the magnetic energy density $[B^2/(2\mu)]$ because it also includes the work done by the magnetic pressure.

To complete the discussion of field-line velocity in a resistive plasma we present two examples in which the field-line motions are entirely due to their diffusion, that is, in plasmas at rest. The first example has an electric field which is everywhere perpendicular to $\mathbf{B}$, while the second has an electric field which is everywhere parallel to $\mathbf{B}$.

Example 1 ($\mathbf{E} \cdot \mathbf{B} = 0$)
Consider a cylinder with a uniform axial current such that

$$\mathbf{B} = B_0 \frac{r}{r_0}\hat{\boldsymbol{\theta}}$$

for $r < r_0$, as shown in Fig. 1.8a. For simplicity, let $\mathbf{v} = \mathbf{0}$, so that all field-line motions are diffusive. The electric field and field-line velocity are then

$$\mathbf{E} = \frac{\mathbf{j}}{\sigma} = 2\eta \frac{B_0}{r_0}\hat{\mathbf{z}},$$

$$\mathbf{w} = \frac{\mathbf{j} \times \mathbf{B}}{\sigma B^2} = -\frac{2\eta}{r}\hat{\mathbf{r}}.$$

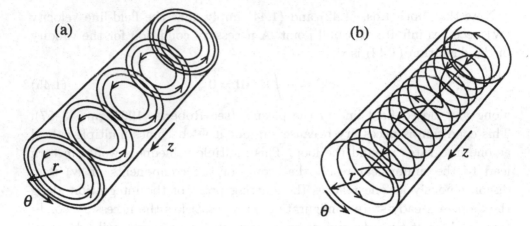

Fig. 1.8. (a) The circular field-line configuration of Example 1, and (b) the helical field-line configuration of Example 2.

The velocity corresponds to a radial motion of field lines towards the origin. This motion indicates that the null point at $r = 0$ is a sink of magnetic flux. The corresponding Poynting flux $[\mathbf{S} = -2\eta B_0^2 r/(\mu r_0^2)\hat{\mathbf{r}}]$ indicates that there is a flow of magnetic energy from the periphery of the cylinder towards the centre. Such a flow of energy is characteristic of a cylinder or wire attached to an external electrical generator and is an example of magnetic energy continually dissipated by ohmic heating in the cylinder.

Example 2 $(\mathbf{E} \cdot \mathbf{B} \neq 0)$

In this case we consider a force-free flux rope (i.e., with $\mathbf{j}$ parallel to $\mathbf{B}$) having a radius r_0 at which the field is purely azimuthal, such that

$$\mathbf{B} = B_0(t)[J_1(\alpha r)\hat{\boldsymbol{\theta}} + J_0(\alpha r)\hat{\mathbf{z}}]$$

for $r < r_0$, where J_1 and J_0 are Bessel functions, α is a constant, and $\alpha r_0 = 2.40$ is the first zero of $J_0(\alpha r)$. This field, which was first considered by Lundquist (1951), has the property that $\nabla \times \mathbf{B} = \mu \mathbf{j} = \alpha \mathbf{B}$ and is shown in Fig. 1.8(b). For simplicity, we again let $\mathbf{v} = 0$, so that the corresponding electric field is

$$\mathbf{E} = \frac{\mathbf{j}}{\sigma} = \eta \alpha \mathbf{B},$$

which upon substitution into Faraday's equation (1.7) leads to $\partial B_0/\partial t = -\eta \alpha^2 B_0$. Consequently,

$$B_0(t) = B_{00} e^{-\eta \alpha^2 t},$$

where B_{00} is a constant. From Eq. (1.42) the field-line velocity is just

$\mathbf{w} = \nabla \Phi^* \times \mathbf{B}/B^2$, where $\Phi^* = -\int_C E_{\parallel} dl = -jl/\sigma$ and l is the length along the field given by $l = \{1 + [J_1(\alpha r)/J_0(\alpha r)]^2\}^{1/2} z$. The potential Φ^* is thus

$$\Phi^* = -\eta \alpha B_0 \left(\frac{J_1^2 + J_0^2}{J_0} \right) z.$$

In obtaining this form of Φ^* we have arbitrarily assumed that Φ_0^* is zero, so that Φ^* vanishes at $z = 0$. Choosing Φ_0^* to be a constant (zero in this case) guarantees that the field-line velocity is continuous with respect to rotations in θ, but other choices are in fact possible (see §8.1.2). Taking the gradient of Φ^* leads to

$$\nabla \Phi^* = -\eta \alpha B_0 \left[\frac{J_1^2 + J_0^2}{J_0} \, \hat{\mathbf{z}} + \alpha z \left(J_1 - \frac{2J_1^2}{\alpha r J_0} + \frac{J_1^3}{J_0^2} \right) \hat{\mathbf{r}} \right],$$

and the field-line velocity is therefore

$$\mathbf{w} = \eta \alpha \left\{ \frac{J_1}{J_0} \, \hat{\mathbf{r}} - \alpha z \left[\frac{J_1}{J_0} - \frac{2J_1^2}{\alpha r (J_1^2 + J_0^2)} \right] \left(\frac{J_1}{J_0} \, \hat{\mathbf{z}} - \hat{\boldsymbol{\theta}} \right) \right\}.$$

The motion of field lines now consists of an inward radial motion plus a corkscrew-like motion along and around the axis of the cylinder. The fact that the z- and θ-components of the velocity increase with z is indicative of a general expansion of the field lines. Just as in the case of an expanding universe, field lines which are more distant from the local position ($z = 0$), where the coordinates are fixed, appear to be moving faster than nearby field lines. Unlike Example 1, the field-line velocity at $r = 0$ is now zero, instead of infinite, because there is no longer a null line there. However, the velocity does become infinite at $r = r_0$ where B_z is zero. At this surface the field lines are closed circles, and the necessary condition (1.45) for the existence of a well-defined field-line velocity is violated. This problem does not arise if one considers the motion of flux surfaces instead of field lines. In this example the field-line motion is not associated with a Poynting flux because $\mathbf{E} \times \mathbf{B} = 0$ everywhere. The motion simply represents the ohmic decay of a field which has no external power source to maintain it.

1.5 MHD Shock Waves

Magnetic shock waves play an important role in some reconnection mechanisms (§§4.3, 5.1, 5.2) and can efficiently accelerate fast particles (§13.3), so let us summarise their main properties.

Small-amplitude sound waves in a non-magnetic medium propagate without change of shape, but when the amplitude is finite the crest can move

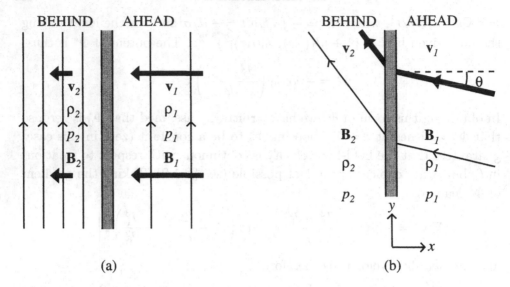

Fig. 1.9. Notation for (a) a perpendicular shock and (b) an oblique shock.

faster than its trough, causing a progressive steepening. Ultimately, the gradients become so large that dissipation becomes important, and a steady acoustic shock-wave shape may be attained, with a balance between the steepening effect of the nonlinear convective term and the broadening effect of dissipation. The dissipation inside the shock front converts the energy being carried by the wave gradually into heat. The effect of the passage of the shock is to compress and heat the gas.

1.5.1 Perpendicular Magnetic Shocks

In the presence of a magnetic field, there are three wave modes instead of just the sound mode, and when their amplitudes are large the Alfvén wave can propagate without steepening, whereas the slow and fast magnetoacoustic modes (i.e., slow- and fast-mode waves) may steepen to form shocks. Derivation of the jump relations is more complicated than for an acoustic shock, since: there is an extra variable ($\mathbf{B}$); $\mathbf{B}$ and $\mathbf{v}$ may be inclined away from the shock normal; and the condition that the entropy must increase for an acoustic shock is replaced by an evolutionary condition.

Let us for simplicity then first consider a perpendicular shock, in which $\mathbf{B}$ is parallel to the shock front and $\mathbf{v}$ is normal to it, and suppose that the variables ahead and behind the shock front in the frame of the shock front are denoted by 1 and 2, respectively [Fig. 1.9(a)]. The conservation (or jump) relations for mass, momentum, energy, and magnetic flux across the shock front are then

$$\rho_2 v_2 = \rho_1 v_1,$$

$$p_2 + \rho_2 v_2^2 + \frac{B_2^2}{2\mu} = p_1 + \rho_1 v_1^2 + \frac{B_1^2}{2\mu},$$

$$\left(p_2 + \frac{B_2^2}{2\mu}\right) v_2 + \left(\rho_2 e_2 + \tfrac{1}{2}\rho_2 v_2^2 + \frac{B_2^2}{2\mu}\right) v_2 =$$

$$\left(p_1 + \frac{B_1^2}{2\mu}\right) v_1 + \left(\rho_1 e_1 + \tfrac{1}{2}\rho_1 v_1^2 + \frac{B_1^2}{2\mu}\right) v_1,$$

$$B_2 v_2 = B_1 v_1, \qquad (1.46)$$

where $e = p/[(\gamma - 1)\rho]$ is the internal energy per unit mass. They imply that

$$\frac{v_2}{v_1} = \frac{1}{X}, \quad \frac{B_2}{B_1} = X, \quad \frac{p_2}{p_1} = \gamma \overline{M}_1^2 \left(1 - \frac{1}{X}\right) - \frac{1 - X^2}{\beta_1},$$

where $\beta_1 = 2\mu p_1 / B_1^2$, $\overline{M}_1 = v_1/c_{s1}$ is the Mach number (namely, the ratio of the shock speed (v_1) to the sound speed $[c_{s1} = (\gamma p_1/\rho_1)^{1/2}]$ and $X = \rho_2/\rho_1$ is the density ratio, which is the positive solution of

$$2(2 - \gamma)X^2 + [2\beta_1 + (\gamma - 1)\beta_1 \overline{M}_1^2 + 2]\gamma X - \gamma(\gamma + 1)\beta_1 \overline{M}_1^2 = 0. \quad (1.47)$$

Consequences of these jump relations together with the evolutionary condition (that the perturbation caused by a small disturbance be small and unique) are:

(i) Eq. (1.47) has only one positive real root since $1 < \gamma < 2$;
(ii) the effect of the magnetic field is to reduce X below its hydrodynamic value of $(\gamma + 1)\overline{M}_1^2/[2 + (\gamma - 1)\overline{M}_1^2]$;
(iii) the shock is compressive with $X \geq 1$;
(iv) the shock speed (v_1) must exceed the fast magnetoacoustic speed $\sqrt{(c_{s1}^2 + v_{A1}^2)}$ ahead of the shock;
(v) magnetic compression is limited to the range $1 < B_2/B_1 < (\gamma + 1)/(\gamma - 1)$, where for $\gamma = 5/3$ the upper limit is 4.

1.5.2 Oblique Magnetic Shocks

Suppose we set up axes in a frame moving with the shock, as before, and assume **v** and **B** lie in the *xy*-plane [Fig. 1.9(b)]. Then the jump relations for mass, x- and y-momentum, and magnetic flux may be written

$$\rho_2 v_{2x} = \rho_1 v_{1x},$$

$$p_2 + B_2^2/(2\mu) - B_{2x}^2/\mu + \rho_2 v_{2x}^2 = p_1 + B_1^2/(2\mu) - B_{1x}^2/\mu + \rho_1 v_{1x}^2,$$

$$\rho_2 v_{2x} v_{2y} - B_{2x} B_{2y}/\mu = \rho_1 v_{1x} v_{1y} - B_{1x} B_{1y}/\mu,$$

$$\left[p_2 + B_2^2/(2\mu) \right] v_{2x} - B_{2x}(\mathbf{B}_2 \cdot \mathbf{v}_2)/\mu + \left[\rho_2 e_2 + \tfrac{1}{2}\rho_2 v_2^2 + B_2^2/(2\mu) \right] v_{2x} =$$

$$\left[p_1 + B_1^2/(2\mu) \right] v_{1x} - B_{1x}(\mathbf{B}_1 \cdot \mathbf{v}_1)/\mu + \left[\rho_1 e_1 + \tfrac{1}{2}\rho_2 v_1^2 + B_1^2/(2\mu) \right] v_{1x},$$

$$B_{2x} = B_{1x}$$

$$v_{2x} B_{2y} - v_{2y} B_{2x} = v_{1x} B_{1y} - v_{1y} B_{1x}.$$

Now, if we choose axes moving parallel to the shock at such a speed that

$$v_{1y} = v_{1x} \frac{B_{1y}}{B_{1x}}, \tag{1.48}$$

so that $\mathbf{v}$ is parallel to $\mathbf{B}$, then the equations simplify greatly and may be solved to give

$$\frac{v_{2x}}{v_{1x}} = \frac{1}{X}, \quad \frac{v_{2y}}{v_{1y}} = \frac{v_1^2 - v_{A1}^2}{v_1^2 - X v_{A1}^2}, \quad \frac{B_{2x}}{B_{1x}} = 1,$$

$$\frac{B_{2y}}{B_{1y}} = \frac{\left(v_1^2 - v_{A1}^2\right) X}{v_1^2 - X v_{A1}^2}, \quad \frac{p_2}{p_1} = X + \frac{(\gamma - 1) X v_1^2}{2 c_{s1}^2}\left(1 - \frac{v_2^2}{v_1^2}\right), \tag{1.49}$$

where

$$X = \frac{\rho_2}{\rho_1}, \quad c_{s1}^2 = \frac{\gamma p_1}{\rho_1}, \quad v_{A1}^2 = \frac{B_1^2}{\mu \rho_1}, \quad \cos\theta = \frac{v_{1x}}{v_1}, \tag{1.50}$$

and the compression ratio (X) is a solution of

$$\left(v_1^2 - X v_{A1}^2\right)^2 \left\{ X c_{s1}^2 + \tfrac{1}{2} v_1^2 \cos^2\theta [X(\gamma - 1) - (\gamma + 1)] \right\} \tag{1.51}$$

$$+ \tfrac{1}{2} v_{A1}^2 v_1^2 \sin^2\theta \, X \{[\gamma + X(2 - \gamma)] v_1^2 - X v_{A1}^2 [(\gamma + 1) - X(\gamma - 1)]\} = 0.$$

Equation (1.51) has three solutions, which give the slow-mode shock, Alfvén wave, and fast-mode shock; the forms of the resulting field lines are shown in Fig. 1.10.

The slow and fast shocks have the following properties:

(i) they are compressive with $X > 1$ and $p_2 > p_1$;
(ii) they conserve the sign of B_y so that $B_{2y}/B_{1y} > 0$;
(iii) for the slow shock $B_2 < B_1$, so that $\mathbf{B}$ refracts towards the shock normal and B decreases as the shock passes;

BEHIND | AHEAD

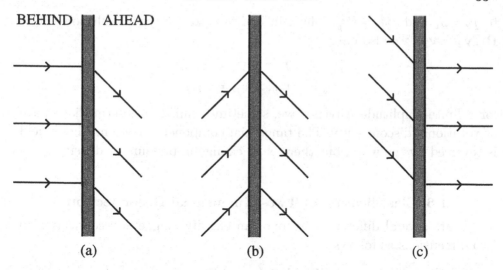

(a)   (b)   (c)

Fig. 1.10. Magnetic field lines for special oblique waves: (a) slow (switch-off) shock, (b) Alfvén wave, and (c) fast (switch-on) shock. In each case the wave is propagating into the region ahead of it to the right of the shaded shock front.

(iv) for the fast shock $B_2 > B_1$, so that the shock makes $\mathbf{B}$ refract away from the normal and B increase;

(v) the flow speed (v_{1x}) ahead of the shock exceeds the appropriate wave speed, while the speed (v_{2x}) behind it is smaller than the wave speed;

(vi) the flow normal to the shock is slowed down ($v_{2x} < v_{1x}$);

(vii) in the limit as $B_x \to 0$, the fast shock becomes a perpendicular shock, while the slow shock becomes a tangential discontinuity, for which v_x and B_x vanish and there are arbitrary jumps in v_y and B_y, subject only to total pressure balance $[p_2 + B_2^2/(2\mu) = p_1 + B_1^2/(2\mu)]$.

In the limit when $v_1 = v_{A1}$ and $X \neq 1$, Eq. (1.49) implies that $B_{2y} = 0$ and we have a switch-off shock [Fig. 1.10(a)]. Since $\mathbf{v}_1$ and $\mathbf{B}_1$ are parallel, this implies that

$$v_{1x} = \frac{B_{1x}}{\sqrt{(\mu\rho)}} \quad \text{and} \quad v_{1y} = \frac{B_{1y}}{\sqrt{(\mu\rho)}}, \tag{1.52}$$

so that the shock propagates at the Alfvén speed based on the normal component of the magnetic field and the frame of reference is moving along the shock at the Alfvén speed based on the parallel component. For a shock propagating along $\mathbf{B}_1$, the fast-shock solution is $X = v_1^2/v_{A1}^2$ for a switch-on shock [Fig. 1.10(c)].

When $v_1 = v_{A1}$, the conservation of flux implies that $v_{2y}/v_{1y} = B_{2y}/B_{1y}$, while the equations for conservation of x-momentum and energy reduce to $p_2 = p_1$ and $B_{2y}^2 = B_{1y}^2$. Thus, in addition to the trivial solution ($\mathbf{B}_2 = \mathbf{B}_1, \mathbf{v}_2 = \mathbf{v}_1$), we also have

$$B_{2y} = -B_{1y}, \quad B_{2x} = B_{1x},$$
$$v_{2y} = -v_{1y}, \quad v_{2x} = v_{1x},$$

for a finite-amplitude Alfvén wave, sometimes called an intermediate wave or rotational discontinuity. The tangential component of the magnetic field is reversed by the wave, but there is no change in pressure or density.

1.6 Classification of Two-Dimensional Reconnection

There are several different ways one can classify magnetic reconnection in two dimensions as follows:

(i) collisional versus collisionless (see §1.7);

(ii) spontaneous versus driven; in this book, we distinguish the latter from the former by defining spontaneous reconnection as occurring when there is no external forcing (i.e., for a bounded region, when there is no energy inflow at the boundary);

(iii) time-dependent (Chapters 6 and 7) versus steady (Chapters 4 and 5), depending on whether the configuration is varying on a time-scale that is faster or slower, respectively, than the typical time-scales (L_e/v_e, L_e/c_s, L_e/v_A) for the travel of plasma, sound waves, and Alfvén waves across a system of dimension L_e;

(iv) almost-uniform reconnection (when the overall geometry is mainly one-dimensional with a small amount of curvature) versus non-uniform (X-point) reconnection (when the basic geometry is that of an X-point so the inflow magnetic field is highly curved) – the former includes Petschek, Priest–Forbes (1986) and stagnation-point flow solutions (§3.2), while the latter includes the Priest–Lee (1990) model (§5.2), Biskamp's (1986) numerical experiments (§5.4), and the linear X-point model (§5.3).

However, since the rate of reconnection is the key issue on which most research concentrates, the primary way to classify reconnection is in terms of its rate. Then other aspects, such as whether the plasma is collisional or collisionless, can be viewed in terms of their effect on the reconnection rate. Moreover, when one goes into three dimensions there are other subtle issues to be considered (§8.1).

In two dimensions, the rate of reconnection is defined as the rate at which field lines move through the X-type neutral point. For a system in which the magnetic field and flow velocity are in the xy-plane, the field lines break and reconnect at the rate $\partial A/\partial t$, as prescribed by Faraday's equation in the form of Eq. (1.18). That is,

$$\frac{\partial A(x_0, y_0, t)}{\partial t} = -E(x_0, y_0, t)$$

for a neutral point located at (x_0, y_0). In other words, the rate of reconnection is directly given by the electric field at the neutral point.

For a steady-state two-dimensional configuration with flow and field components (v_x, v_y) and (B_x, B_y) depending on x and y, the electric field (E_0) has only a z-component from Eq. (1.10a). By Faraday's law (1.7), it is uniform in space and constant in time, and the rate of reconnection can be expressed in terms of the plasma flow at any point in the plane as

$$E_0 = v_y B_x - v_x B_y + \frac{j}{\sigma},$$

where E_0 is the uniform electric field.

The following three characteristic speeds are therefore important in steady two-dimensional reconnection: the *external flow speed* (v_e) at a fixed global distance (L_e) from the neutral point; the Alfvén speed $[v_{Ae} = B_e/(\mu\rho)^{1/2}]$ in terms of the corresponding magnetic field (B_e); and the global magnetic diffusion speed $(v_{de} = \eta/L_e)$. From these we may construct two independent dimensionless parameters which characterise steady, two-dimensional magnetic reconnection, namely the *external Alfvén Mach number*

$$M_e = \frac{v_e}{v_{Ae}}$$

and the *global magnetic Reynolds number*

$$R_{me} = \frac{v_{Ae}}{v_{de}}$$

based on the Alfvén speed (sometimes referred to as the Lundquist number).

Now for a simple X-line in which the current density is very small outside the diffusion region, the reconnection rate can be expressed in terms of the velocity (v_e) of the plasma flowing towards the neutral line as

$$E_0 = v_e B_e = M_e v_{Ae} B_e,$$

at the distance L_e from the X-point, where $v_y = v_e, B_x = B_e, v_x = 0, j = 0$, and v_{Ae} and B_e are regarded as fixed. Thus, for a steady-state system, the Alfvén Mach number (M_e) at the boundary of the system is a dimensionless measure of the rate of reconnection.

Whether the reconnection is super-slow, slow, or fast in a steady-state system depends on the size of the reconnection rate (M_e) relative to certain powers of the magnetic Reynolds number (R_{me}). Although there are potentially many other dimensionless numbers with which we could compare the Alfvén Mach number (M_e), the magnetic Reynolds number is the key parameter, because it measures the ability of magnetic field lines to diffuse through the plasma. As discussed in previous sections, reconnection cannot occur without such diffusion.

1.6.1 Super-slow Reconnection $(M_e < R_{me}^{-1})$

Now, if a magnetic field configuration only contains variations over the global scale (L_e), then the induction equation, Eq. (1.12), immediately tells us that the plasma velocity (v_e) is of order the global diffusion speed (v_{de}), so that the Alfvén Mach number (M_e) is of the same order as R_{me}^{-1}. If the scale of magnetic field variations is smaller than L_e, then M_e will be greater than R_{me}^{-1}. When the Alfvén Mach number is much lower than R_{me}^{-1}, the magnetic field diffuses in much the same way as it does in a solid conductor, because the advective term in Ohm's law, Eq. (1.10a), is unimportant. As we mentioned at the beginning of this chapter, this ordinary kind of magnetic diffusion is extremely slow in astrophysical plasmas. Therefore, we refer to it as super-slow reconnection, and we will discuss it in more detail in Section 5.3. An example is linear reconnection (Priest *et al.*, 1994c), for which one linearizes about a simple X-point field and finds a current spike at the origin and current sheets all along the separatrices. There is a balance between pressure-gradient and Lorentz forces, and the flow is everywhere much slower than the Alfvén speed.

1.6.2 Slow Reconnection $(R_{me}^{-1} < M_e < R_{me}^{-1/2})$

When the reconnection rate (M_e) exceeds R_{me}^{-1}, the $\mathbf{v} \times \mathbf{B}$ advective term in Ohm's law becomes important. The first model of how reconnection can occur with M_e exceeding R_{me}^{-1} was constructed by Sweet (1958a) and Parker (1957). They introduced the idea that the reconnection can occur at a thin current sheet whose half-thickness (l) is much less than the global scale-length (L_e). The plasma outside the current sheet is effectively frozen to the field; the breaking and reconnection of field lines takes place only within the current sheet, where inertia becomes important as the plasma is accelerated to Alfvénic speeds. In this current sheet model, the diffusion velocity is η/l, rather than η/L_e, so the rate of reconnection is faster than simple diffusion by the factor L_e/l. As will be shown in Section 4.2, the reconnection rate is $M_e = R_{me}^{-1/2}$ in their model. Although this is much faster than $M_e = R_{me}^{-1}$, it is still slow compared with the rate needed to explain energy release in many dynamic astrophysical processes such as flares.

1.6.3 Fast Reconnection $(M_e > R_{me}^{-1/2})$

We shall follow the standard notation and define fast reconnection as being reconnection that occurs much more rapidly than the Sweet–Parker rate $(R_{me}^{-1/2})$. The best-known example of fast reconnection is Petschek's mechanism, for which M_e scales as $(\log R_{me})^{-1}$. Even for the very large values of R_{me} typical of space and astrophysical plasmas $(R_{me} > 10^8)$, this rate is in the range 0.1 to 0.01.

Such a pragmatic definition of fast reconnection is to be preferred over a more mathematical one (such as M_e remaining bounded away from zero as R_{me} tends to infinity), since the real values are not in fact infinite. As we shall see, it is not possible in many situations to produce a totally analytical theory for such a highly complex process, so the limit as R_{me} tends to infinity is often not one that can be calculated in practice.

In Petschek's mechanism the half-length (L) of the diffusion region becomes much shorter than the global scale-length (L_e). When the Sweet–Parker reconnection rate is exceeded, the reconnection becomes highly nonlinear, and two pairs of slow-mode shock waves extend outwards from the tips of the current sheet. These shock waves play an important role in Petschek's model because they are the structures which accelerate the plasma to the Alfvén speed and convert most of the magnetic energy into kinetic energy and heat. Petschek's model is built around the Sweet–Parker model, which is still used to analyse the flow through the diffusion region, but the central current sheet is so small that very little plasma is processed through it. Since the half-length of the current sheet in Petschek's model is now much less than L_e, the width of the sheet is very much smaller than before, so the rate is much faster. A new generation of fast reconnection models, called almost-uniform and non-uniform reconnection, have now been developed and include Petschek's mechanism as a special case (§5.1 and §5.2).

In many applications the Reynolds numbers are enormous in size $(10^6 - 10^{12})$ and it is the fast reconnection mechanisms that are most relevant, but the slow mechanisms are important for reconnection theory as a whole and may be applicable when the effective Reynolds numbers are greatly reduced from their classical values by kinetic effects or turbulence.

1.7 Relevance of MHD to Collisionless Systems

Many of the environments in which reconnection occurs constitute collisionless plasmas; that is, plasmas where the mean-free path for binary collisions
between particles is much greater than the characteristic scale-length of
the system. The absence of binary collisions in such plasmas means that
individual particles can travel distances on the order of the global scale
of the system. Because MHD does not explicitly treat individual particle
motions, it may at first be thought that it is of little use in collisionless plasmas. However, MHD is always a correct description of the large-scale bulk
dynamics of a fluid, with or without internal collisions, so long as the fluid
cannot support a significant electric field in its own reference frame. The
equation of motion is just a statement of Newton's momentum equation,
and the induction equation is Faraday's equation applied to a fluid in whose
moving frame of reference there is no electric field ($\mathbf{E'}$): i.e., $\mathbf{E} \simeq -\mathbf{v} \times \mathbf{B}$ and
$\partial \mathbf{B}/\partial t \approx \nabla \times (\mathbf{v} \times \mathbf{B})$, so that the magnetic field is carried along by the fluid.
For example, beyond ten solar radii the solar wind is a completely collision-
less plasma. Yet MHD models describe its global velocity, temperature, and
density quite well, including such time-dependent aspects as shock disturbances, stream interactions, and turbulence. In principle, kinetic plasma theory derived from Vlasov's equation provides the most complete description
of a collisionless plasma. However, because of its mathematical intractability,
this theory has rarely been used to construct a global model of the solar wind.
Instead, its use has been confined primarily to calculating the effective transport coefficients (such as thermal conductivity) of the solar-wind plasma and
to modelling localised effects such as shock structure or particle acceleration.

There are several reasons why MHD is so successful in describing the
global properties of collisionless plasmas. Firstly, ideal (dissipationless) MHD
embodies conservation principles such as mass, momentum, and energy conservation, which are universal to both collisional and collisionless systems
(Parker, 1996). Secondly, if a magnetic field exists in the plasma, ionized
particles undergo a gyro-motion which prevents them from travelling unim-
peded in a direction perpendicular to the field. Thus, long-range interactions
occur only along the field, while short-range interactions across the field may
be described by MHD-like equations (Chew et al., 1956). Finally, most plasmas contain numerous wave-particle interactions, which tend to impede the
motion of charged particles even in the direction of the field, thereby local-
izing the interaction in all directions.

Cowley (1985) has pointed out that the frozen-flux condition of MHD is
a natural consequence of gyro-motion in a collisionless plasma whenever the

$\mathbf{E} \times \mathbf{B}$ drift of the gyro-centres dominates all other drifts. Unlike the other drifts, such as the gradient drift $\{$i.e., $[mv_g^2/(2qB^3)]\mathbf{B} \times \nabla \mathbf{B}\}$, the $\mathbf{E} \times \mathbf{B}$ drift remains large even as the particle's energy becomes small. [Here m is the particle mass, q its charge, and v_g is the gyro-speed (§13.1.4).] If a collisionless plasma contains particles undergoing gyro-motion and the $\mathbf{E} \times \mathbf{B}$ drift dominates, it will obey the ideal MHD Ohm's law and so the field lines will be frozen to the plasma.

The MHD force term ($\mathbf{j} \times \mathbf{B}$) is recovered in a collisionless plasma whenever the particle electric drift speeds ($v_{dr} \approx E/B$) are much smaller than the speed of light. Adding the Lorentz force on individual particles together gives a net force per unit volume of

$$\mathbf{F} = \rho_c \mathbf{E} + \mathbf{j} \times \mathbf{B}.$$

The electric charge density (ρ_c) is normally non-zero in a collisionless plasma whenever the bulk plasma has non-zero vorticity, but the electric force is negligible compared to the Lorentz force in each particle momentum equation (Goedbloed, 1983) when

$$v_{dr}^2 \ll c^2$$

and is negligible compared to the pressure gradient (acting over a scale L_0) when

$$\lambda_D^2 \ll L_0^2,$$

where λ_D is the Debye length defined below in Eq. (1.64).

Note that the above conditions may be established for, say, electrons as follows. The electron momentum equation has the form

$$m_e n_e \frac{d\mathbf{v}_e}{dt} = -\nabla \cdot \underline{\mathbf{p}}_e - en_e \mathbf{v}_e \times \mathbf{B} - en_e \mathbf{E},$$

where $\mathbf{v}_e$ is the electron velocity and $\underline{\mathbf{p}}_e$ is the electron stress tensor. Thus, if the inertial term (on the left-hand side) and the pressure gradient are negligible, this reduces to $0 = \mathbf{E} + \mathbf{v}_e \times \mathbf{B}$, for which

$$\mathbf{v}_e = \mathbf{v}_{dr} = \frac{\mathbf{E} \times \mathbf{B}}{B^2},$$

so that the electron fluid moves with the electric drift velocity ($\mathbf{v}_{dr}$). Now, Poisson's equation with ion density n_i and temperature T_i is

$$\nabla \cdot \mathbf{E} = \frac{\rho_c}{\epsilon_0} = \frac{e(n_i - n_e)}{\epsilon_0},$$

which implies that

$$\rho_c \approx \frac{\epsilon_0 E}{L_0}$$

in order of magnitude, while Ampère's Law $(\nabla \times \mathbf{B} = \mu \mathbf{j})$ implies that

$$j \approx \frac{B}{\mu L_0}.$$

Therefore the ratio of the net electric to the magnetic force is

$$\left| \frac{\rho_c \mathbf{E}}{\mathbf{j} \times \mathbf{B}} \right| \approx \frac{\epsilon_0 E^2 / L_0}{B^2 / (\mu L_0)} \approx \frac{v_{dr}^2}{c^2},$$

which is much less than unity if $v_{dr} \ll c$, as required. Furthermore, if the electric field and pressure terms have similar size in the electron momentum equation [so that $E \approx p_e / (e n_e L)$], then the ratio of the net electric to the pressure force is

$$\left| \frac{\rho_c \mathbf{E}}{\nabla p_e} \right| \approx \frac{\epsilon_0 E_0^2 / L_0}{p_0 / L_0} \approx \frac{\epsilon_0 p_0}{e^2 n_e^2 L_0^2} = \frac{\lambda_D^2}{L_0^2},$$

which is much less than unity if $\lambda_D \ll L_0$, as required.

The most important difference which typically arises in a collisionless system occurs in the pressure and viscous terms in the momentum equation. In a collisionless plasma, the gas pressure is not generally isotropic but is instead anisotropic relative to the magnetic field. Generally, the plasma exerts a pressure $(p_\parallel)$ along the field which is distinct from the pressure $(p_\perp)$ perpendicular to the field. Similarly, the viscous stress tensor is also anisotropic and dependent on the magnetic field. If $p_\parallel$ and $p_\perp$ can be expressed in terms of the local bulk properties of the plasma, then the MHD equations still apply, but the pressure and viscous terms now depend on $p_\parallel$ and $p_\perp$, and there is an additional term proportional to $(p_\parallel - p_\perp)$, (see Hollweg, 1986; Parker, 1996).

Despite the past successes of MHD in explaining the dynamics of collisionless plasmas, there is good reason to be cautious when applying it to reconnection. By definition, reconnection cannot occur in ideal MHD, because it depends on the diffusion of field lines through the plasma, often in small-scale structures such as current sheets. In this situation kinetic theory is essential to calculate the diffusivity.

To determine a realistic resistivity for a collisionless, or nearly collisionless, plasma requires consideration of the generalized Ohm's law. For a fully ionized plasma it can be written as

$$\mathbf{E}' \equiv \mathbf{E} + \mathbf{v} \times \mathbf{B} = \frac{\mathbf{j}}{\sigma} + \frac{m_e}{ne^2}\left[\frac{\partial \mathbf{j}}{\partial t} + \nabla \cdot (\mathbf{vj} + \mathbf{jv})\right] - \frac{\mathbf{j} \times \mathbf{B}}{ne} - \frac{\nabla \cdot \underline{\mathbf{p}}_e}{ne}, \quad (1.53)$$

where $\mathbf{vj}$ and $\mathbf{jv}$ are dyadic tensors and $\underline{\mathbf{p}}_e$ is the electron stress tensor (Braginsky, 1965; Rossi and Olbert, 1970). The $\mathbf{v} \times \mathbf{B}$ term is the convective electric field, while the first term on the right-hand side of this equation is the field associated with Ohmic dissipation caused by electron-ion collisions. The next three terms describe the effects of electron inertia, and the next to the last term is the Hall effect. Finally, the last term includes the electron gyro-viscosity, which is considered by many to be important at a magnetic null (e.g., Strauss, 1986; Dungey, 1994). For a partially ionized plasma, collisions between charged particles and neutrals lead to additional terms associated with ambipolar diffusion (§12.2.2.)

Although all of the terms on the right-hand side of Eq. (1.53), other than the first, allow field lines to slip through the plasma, they do not all produce dissipation. For example, the inertial terms do not directly cause the entropy of the plasma to increase but may lead to particle acceleration. Thus, even though one may speak of inertial effects as creating an effective resistivity, this resistivity does not necessarily lead to dissipation.

Which terms are important in a particular environment depends not only upon the plasma parameters, but also upon the length- and time-scales for variations of these parameters (Elliott, 1993; Sturrock, 1994). For magnetic reconnection, we normally want to know which non-ideal terms are likely to be significant within the current sheet where the frozen-flux condition is violated. Since each non-ideal (i.e., diffusion) term in the generalized Ohm's law contains either a spatial or temporal gradient, we can estimate the significance of any particular term by computing the gradient length-scale (L_0) required to make the term as large as the value of the convective electric field ($\mathbf{v} \times \mathbf{B}$) outside the diffusion region.

Consider, for example, the three inertial terms $[\partial \mathbf{j}/\partial t + \nabla \cdot (\mathbf{vj} + \mathbf{jv})]$ on the right-hand side of Eq. (1.53). If we assume $\nabla \approx 1/L_0$, $|\mathbf{j}| \approx B_0/(\mu L_0)$, and $\partial/\partial t \approx V_0/L_0$, say, where L_0 is a typical length-scale, then these three terms will be of the same order as the convective electric field if

$$\frac{m_e}{ne^2}\frac{V_0 B_0}{\mu L_0^2} \approx V_0 B_0.$$

In other words, in order for the inertial terms to be important in a current sheet, its thickness (L_{inertia}) should be

$$L_{\text{inertia}} \approx c \left(\frac{m_e \epsilon_0}{ne^2} \right)^{1/2} \approx \lambda_e, \tag{1.54}$$

where

$$\lambda_e = \frac{c}{\omega_{pe}} = \left(\frac{m_e}{ne^2 \mu} \right)^{1/2} = 5.30 \times 10^6 \, n^{-1/2} \tag{1.55}$$

is the electron-inertial length or skin-depth, $c = (\epsilon_0 \mu)^{-1/2}$ is the speed of light, and $\omega_{pe} = [(e^2 n_e)/(\epsilon_0 m_e)]^{1/2}$ is the electron plasma frequency.

Similarly, for the Hall term $[\mathbf{j} \times \mathbf{B}/(ne)]$,

$$\frac{B_0^2}{ne\mu L_0} \approx V_0 B_0$$

or

$$L_{\text{Hall}} \approx \frac{c}{M} \left(\frac{\tilde{\mu} m_p \epsilon_0}{ne^2} \right)^{1/2} \approx \frac{\lambda_i}{M}, \tag{1.56}$$

where

$$\lambda_i = \frac{c}{\omega_{pi}} = \left(\frac{\tilde{\mu} m_p}{ne^2 \mu} \right)^{1/2} = 2.27 \times 10^8 \left(\frac{\tilde{\mu}}{n} \right)^{1/2} \tag{1.57}$$

is the ion-inertial length or skin depth, $M = V_0/v_A = V_0 B_0^{-1} (\mu \tilde{\mu} m_p n)^{1/2}$, $\tilde{\mu} = \overline{m}/m_p$ is the mean atomic weight, and $\omega_{pi} = [(q_i^2 n_i)/(\epsilon_0 m_i)]^{1/2}$ is the ion plasma frequency.

For the electron-stress term $[\nabla \cdot \underline{\mathbf{p}}_e/(ne)]$ we can write

$$\frac{nk_B T}{ne L_0} \approx V_0 B_0$$

if we assume $|\underline{\mathbf{p}}_e| \approx nk_B T_e$ and $T_e \approx T_i \approx T$. Solving for L_0 leads to

$$L_{\text{stress}} \approx \frac{\beta^{1/2}}{M} R_{gi}, \tag{1.58}$$

where

$$\beta = nk_B T \left(\frac{2\mu}{B_0^2} \right) = 3.47 \times 10^{-29} \frac{nT}{B_0^2} \tag{1.59}$$

and

$$R_{gi} = \frac{(k_B T_i m_p \tilde{\mu})^{1/2}}{e B_0} = 9.49 \times 10^{-7} \frac{(T_i \tilde{\mu})^{1/2}}{B_0} \tag{1.60}$$

are the plasma-beta parameter and the ion-gyro radius, respectively.

Finally, for the collision term $(\mathbf{j}/\sigma)$

$$\frac{B_0}{\mu_0 \sigma L_0} \approx M v_A B_0.$$

Since the product $(\mu \sigma)^{-1}$ is just the magnetic diffusivity, substitution of Eq. (1.13) for $\eta_\perp$ produces

$$\frac{1}{\sigma} = \frac{(k_B m_e T_e)^{1/2}}{n e^2 \lambda_{\mathrm{mfp}}}, \tag{1.61}$$

where

$$\lambda_{\mathrm{mfp}} = 3(2\pi)^{3/2} \frac{(k_B T_e \epsilon_0)^2}{n e^4 \ln \Lambda} = 1.07 \times 10^9 \frac{T_e^2}{n \ln \Lambda} \tag{1.62}$$

is the mean-free path for electron-ion collisions (Schmidt, 1966). Combining Eq. (1.61) with the equation below (1.60) and the expressions (1.55) and (1.57) for the electron- and ion-inertial lengths, we obtain

$$L_{\mathrm{collision}} \approx \frac{\beta^{1/2}}{M} \frac{\lambda_e \lambda_i}{\lambda_{\mathrm{mfp}}}. \tag{1.63}$$

Note that the length-scale $(L_{\mathrm{collision}})$ of the spatial variations required to achieve significant field-line diffusion is inversely proportional to the mean-free path (λ_{mfp}). As λ_{mfp} increases, the diffusion caused by collisions becomes less effective, and increasingly sharper gradients are required to maintain the size of the dissipation term $(\mathbf{j}/\sigma)$.

Tables 1.2a and 1.2b list various plasma parameters along with the characteristic scale-lengths for four different regions where reconnection is thought to occur. The parameter L_e is the global (external) scale-size of the region, and the fundamental quantities from which all other parameters are derived are the density, temperature, and magnetic field. For convenience, we assume that the Alfvén Mach number (M) is unity and that the electron and ion temperatures are roughly equal. The most extreme plasma environments listed in Table 1.2 occur in the magnetosphere, which is completely collisionless, and in the solar interior, which is highly collisional.

In addition to the parameters discussed above, Table 1.2a also lists the value of the Debye length:

$$\lambda_D = \left(\frac{\epsilon_0 k_B T_e}{n e^2}\right)^{1/2} = 69.0 \left(\frac{T_e}{n}\right)^{1/2}. \tag{1.64}$$

The number of particles within a Debye sphere (i.e., $4\pi n \lambda_D^3 / 3$) must be larger than unity in order for the generalized Ohm's law (1.53) to hold. The number of particles in a Debye sphere for the environments shown in Table 1.2a ranges from 10^{14} for the magnetosphere to only about 4 for the

Table 1.2a. *Comparison of Plasma Parameters in Different Environments*

Parameter[1]	Laboratory Experiments[2]	Terrestrial Magnetosphere[3]	Solar Corona[4]	Solar Interior[5]
L_e	10^{-1}	10^7	10^8	10^7
n	10^{20}	10^5	10^{15}	10^{29}
T	10^5	10^7	10^6	10^6
B	10^{-1}	10^{-8}	10^{-2}	10^1
λ_D	10^{-6}	10^3	10^{-3}	10^{-10}
R_{gi}	10^{-3}	10^5	10^{-1}	10^{-4}
λ_i	10^{-2}	10^6	10^1	10^{-6}
$\ln\Lambda$	11	33	19	3
λ_{mfp}	10^{-2}	10^{16}	10^4	10^{-9}
β	10^{-2}	10^{-1}	10^{-4}	10^4
$L_u(\approx R_m)$	10^3	10^{14}	10^{14}	10^{10}
E_D	10^3	10^{-13}	10^{-2}	10^{11}
$E_A(= v_A B)$	10^4	10^{-2}	10^5	10^4
$E_{\mathrm{SP}}(= E_A/\sqrt{R_{me}})$	10^2	10^{-9}	10^{-3}	10^{-2}

Table 1.2b. *Characteristic Lengths in Metres from the Generalized Ohm's Law*

Characteristic Length	Laboratory Experiments[2]	Terrestrial Magnetosphere[3]	Solar Corona[4]	Solar Interior[5]
$L_{\mathrm{inertia}}(\lambda_e)$	10^{-4}	10^4	10^{-1}	10^{-8}
$L_{\mathrm{Hall}}(\lambda_i)$	10^{-2}	10^6	10^1	10^{-6}
L_{stress}	10^{-3}	10^5	10^{-3}	10^{-2}
$L_{\mathrm{collision}}$	10^{-4}	10^{-7}	10^{-7}	10^{-3}

[1] MKS units are used, i.e., length-scales in m, n in m^{-3}, T in K, B in tesla (1 tesla $= 10^4$ G), and electric fields in $V\,\mathrm{m}^{-1}$.
[2] MRX at Princeton Plasma Physics Laboratory.
[3] Plasma sheet.
[4] Above a solar active region.
[5] Base of the solar convection zone.

solar interior at the base of the convection zone. Also shown in the table is the Lundquist number (L_u), which is the same as the magnetic Reynolds number (R_m) when the flow and Alfvén speeds are the same. For a collisional plasma, the Lundquist number based on L_e becomes

$$L_u = \frac{v_A}{v_d} = \frac{L_e v_A}{\eta} = \frac{L_e T_e^{3/2} B_0}{(\tilde{\mu} n)^{1/2} \ln\Lambda} 2.07 \times 10^8 \qquad (1.65)$$

after substituting from Eq. (1.13) for η.

The characteristic scale-lengths in Table 1.2b provide an indication of which terms in the generalized Ohm's law are likely to be important for reconnecting current sheets. As with MHD shocks and turbulence, the large-scale dynamics of the flow cause the current sheet to thin until it reaches a length-scale where field-line diffusion is effective. Thus, in principle, the term with the largest characteristic length-scale in Table 1.2b is the one which will be most important. Since the Hall term has the largest length in every environment except the solar interior, one might conclude that it is generally the most important. However, this conclusion does not take into consideration the fact that the Hall term tends to zero in the region of a magnetic null point or sheet. The Hall term on its own does not contribute directly to reconnection, since it freezes the magnetic field to the electron flow. To know whether a particular term is really as important as suggested by its relative length-scale requires a complete analysis of the kinetic dynamics, which is a rather formidable task. An excessively small scale does indicate that any process associated with that term is unlikely to be important. Therefore, on this basis, we can conclude that collisional diffusion is not important in the terrestrial magnetosphere or the solar corona (above an active region), and that the electron-inertial terms and the Hall term are not important in the solar interior.

The rather large value of the length-scale (L_{stress}) associated with the electron-pressure tensor in the solar interior indicates that there are strong electrostatic fields due to charge separation arising from the electron mobility. In cases in which the electron pressure term has the form $\nabla p(n)/(ne)$, such as in an isentropic plasma, for example, the term has no effect on the diffusion of the magnetic field. When substituted into Faraday's equation, Eq. (1.7), the term disappears because $\nabla \times [\nabla p(n)/n]$ is zero, and the magnetic induction equation, Eq. (1.12), remains unchanged. Diffusion from collisions is, therefore, the dominant mechanism for unfreezing the field lines inside the Sun. Because the number of particles in a Debye sphere is only about 4 at the base of the convection zone, it is also necessary to apply electrostatic corrections to the standard equation of state (see Stix, 1989).

Although the collision length-scale ($L_{\mathrm{collision}}$) is equally small in both the magnetosphere and the corona, the general importance of collisions for these two regions is quite different. In the magnetosphere the collision mean-free path (λ_{mfp}) is nine orders of magnitude larger than the global scale-size (L_e), but in the corona it is four orders of magnitude smaller than the global scale. Thus, we can be confident that collisional transport theory applies to large-scale structures in the corona even though it is not applicable within thin current sheets or dissipation layers. By contrast, in the magnetosphere, collisions are so few that collisional transport theory does not apply at any scale.

Another important issue concerning the applicability of collisional theory is the strength of the electric field in a frame moving with the plasma. If this field exceeds the Dreicer electric field defined by

$$E_D = \frac{e \ln \Lambda}{4\pi\epsilon_0 \lambda_D^2} = \frac{e^3 \ln \Lambda \, n}{4\pi\epsilon_0^2 k_B T_e} = 3.02 \times 10^{-13} \frac{n \ln \Lambda}{T_e}, \quad (1.66)$$

runaway acceleration of electrons will occur (§13.1.2). The most likely location for the production of runaway electrons in a reconnection process is in a thin current sheet that forms at the null point. This field could be as large as the convective electric field based on the Alfvén speed, that is,

$$E_A = v_A B_0 = 2.18 \times 10^{16} \frac{B_0^2}{(\tilde{\mu}n)^{1/2}},$$

or as low as the Sweet–Parker electric field (§4.2),

$$E_{SP} = \frac{E_A}{R_m^{1/2}},$$

where R_m is the magnetic Reynolds number based on the inflow Alfvén speed (i.e., the inflow Lundquist number). As shown in Table 1.2a, on the one hand, the Dreicer field in the magnetosphere is much smaller than E_A or E_{SP}, so runaway electrons will always be generated by reconnection there. On the other hand, in the solar interior the Dreicer field is so large that runaway electrons never occur. In the intermediate regimes of the laboratory and the solar corona, the Dreicer field lies between E_A and E_{SP}, so perhaps runaway electrons are only produced when very fast reconnection occurs.

Even in completely collisionless environments such as the Earth's magnetosphere, it is still sometimes possible to express the relation between electric field and current density in terms of an anomalous resistivity. For example, Lyons and Speiser (1985) have shown that the electron-inertial terms in Eq. (1.53) lead to an anomalous resistivity

$$\frac{1}{\sigma^*} = \frac{\pi B_z}{2ne},$$

where B_z is the field normal to the current sheet. This resistivity is derived solely from a consideration of the particle orbits, and in the magneto-tail current sheet it may be larger than any anomalous resistivity due to wave-particle interactions. A typical example of the latter is the anomalous resistivity due to ion-acoustic waves (see §13.1.3).

Even if collisional MHD theory cannot be used to calculate the electrical resistivity in a thin current sheet, it can still be used in many circumstances to calculate the reconnection rate and the rate at which energy is dissipated. This statement may seem paradoxical, but, as we shall see in §4.3, MHD reconnection solutions exist where the rate of reconnection is largely independent of the magnitude of the electrical resistivity. These solutions often describe configurations where the reconnection is strongly driven by an external flow. The situation is analogous to the way dissipation works in MHD shock waves or in turbulent situations, where the rate of dissipation is independent of the value of the diffusion coefficient and depends only on the rate at which the flow is driven or energy is fed in. What the magnitude of the diffusion determines mainly is the width of the dissipating structure or the scale at which dissipation occurs.

In the linear phase of spontaneous reconnection (such as the simple tearing mode discussed in Section 6.2), the rate of energy dissipation depends strongly on the value of the magnetic diffusivity. However, in the nonlinear development to a Petschek (§4.3) or almost-uniform regime (§5.1), the dependence on diffusivity is weak. Nevertheless, the value and form of the diffusion still play an important role. Thus, one must always keep in mind that MHD provides a limited description of how reconnection works in a collisionless plasma. The principal strength of MHD in such systems is that it provides a global picture of how the reconnection occurs in response to various boundary conditions (§§5.1, 5.2). The appropriate boundary conditions depend strongly on the particular application being considered.

2

Current-Sheet Formation

We introduced briefly in Section 1.3.2 the idea of a current sheet as a narrow region across which the magnetic field changes rapidly. In this chapter we consider in detail the formation of such sheets in a medium where the magnetic field is frozen to the plasma (§1.4), and then in Chapter 4 we describe how they diffuse through the plasma.

There are several ways in which current sheets may form. One is by the collapse of an X-type neutral point (§2.1). Such a formation in two dimensions through a series of static potential field states may be described by complex variable theory, in which the sheet is treated as a branch cut in the complex plane (§2.2). Other techniques are required for three-dimensional axisymmetric fields (§2.2.5), force-free fields (§2.3.1) or more general magnetostatic fields (§2.3.2). The concept of magnetic relaxation as developed by Moffatt is described in Section 2.4, and a self-consistent theory for slow time-dependent formation is discussed in Section 2.5. Finally, two other ways of forming current sheets are described, namely by shearing a field with separatrices (§2.6) and by braiding (§2.7).

2.1 X-Point Collapse

As we shall discuss in detail in Section 7.1, an X-type neutral point in a magnetic configuration tends to be locally unstable, provided the sources of the field are free to move (Dungey, 1953). This may be demonstrated by considering the equilibrium current-free field,

$$B_x = y, \quad B_y = x, \tag{2.1}$$

whose field lines are the rectangular hyperbolae

$$y^2 - x^2 = \text{constant}, \tag{2.2}$$

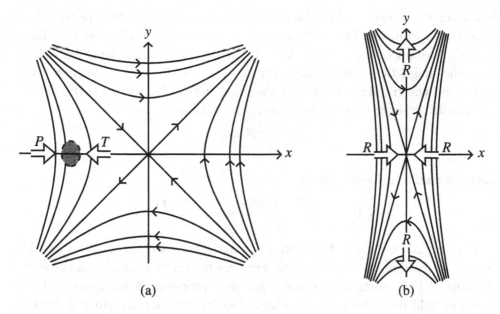

(a)   (b)

Fig. 2.1. The magnetic field lines near an X-type neutral point: (a) in equilibrium with no current, showing a plasma element (shaded) acted on by a magnetic pressure force (P) and a magnetic tension force (T); (b) away from equilibrium with a uniform current and a resultant force R.

as shown in Fig. 2.1(a). In sketching these field lines, we have taken account of the fact that the field strength increases with distance from the origin, so that the hyperbolae are situated closer together as one moves outwards. Any element of plasma, such as the one shown on the negative x-axis, experiences a magnetic tension force (T) due to the outwardly curving field lines. It acts outwards from the origin and is exactly balanced by a magnetic pressure force (P), which acts inwards because the magnetic field strength weakens as one approaches the origin.

Now suppose the magnetic field, Eq. (2.1), is distorted to the form

$$B_x = y, \quad B_y = \bar{\alpha}^2 x, \tag{2.3}$$

where $\bar{\alpha}^2 > 1$. The corresponding field lines are given by

$$y^2 - \bar{\alpha}^2 x^2 = \text{constant} \tag{2.4}$$

and are sketched in Fig. 2.1(b). The limiting field lines $(y = \pm\bar{\alpha}x)$ through the origin are no longer inclined at $\frac{1}{2}\pi$, but have closed up a little like a pair of scissors. On the x-axis, the field lines are more closely spaced than those in Fig. 2.1(a), so the magnetic pressure force has increased. They also have a smaller curvature than before, so the magnetic tension force has increased less than the pressure. The dominance of the magnetic pressure therefore produces a resultant force (R) acting inward. On the y-axis, the field lines have the same spacing as in Fig. 2.1(a), but they are more sharply curved, so that the magnetic pressure force remains the same while the tension force increases; the resultant force (R) therefore acts outward as shown. These comments are borne out by evaluating the current density

$$j_z = \frac{\overline{\alpha}^2 - 1}{\mu} \tag{2.5}$$

and the resulting Lorentz force

$$\mathbf{j} \times \mathbf{B} = -\frac{(\overline{\alpha}^2 - 1)\,\overline{\alpha}^2 x}{\mu}\,\hat{\mathbf{x}} + \frac{(\overline{\alpha}^2 - 1)\,y}{\mu}\,\hat{\mathbf{y}}. \tag{2.6}$$

The magnetic force is therefore such as to increase the original perturbation by closing up the "scissors" even more, which means that the equilibrium (2.1) is unstable. As the instability proceeds, $\overline{\alpha}$ increases and the limiting field lines through the origin close up, so that the current density and ohmic heating (j^2/σ) also increase. This process takes place only if conditions at distant boundaries permit the displacement, but, if the magnetic configuration contains energy stored in excess of potential, it is likely to be an efficient means of dissipating that energy.

The instability may be demonstrated formally by a linear stability analysis (§7.1.2), which gives rapid reconnection during an X-type collapse that depends only logarithmically on resistivity. Although self-similar nonlinear solutions exist (§7.1.1), the full nonlinear evolution is still unknown.

2.2 Current Sheets in Potential Fields

Here we model current sheets in two-dimensional potential fields by an elegant method, namely regarding them as cuts in the complex plane. Although we consider static current sheets in this section, the results can be applied to the quasi-static formation or evolution of such sheets through a series of equilibria. A different technique for analysing three-dimensional axisymmetric fields is outlined in Section 2.2.5.

2.2.1 Simple Current Sheet with Y-type Null Points at the Ends

Suppose that initially, as before, we start with a field of the form of Eq. (2.1), and that the sources of the magnetic field move slowly together and drive the formation of a series of equilibria containing a current sheet (Fig. 2.2). For example, for a steady two-dimensional motion, the electric field is uniform

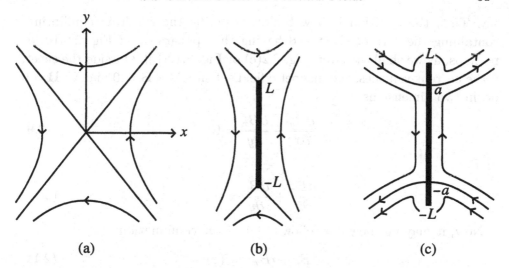

(a)   (b)   (c)

Fig. 2.2. (a) Sketch of the magnetic field near an X-type neutral point which evolves to a field with a current sheet having either (b) Y-points or (c) reversed currents and singularities at its ends.

and perpendicular to the xy-plane and is associated with a motion ($\mathbf{v}_\perp$) normal to the magnetic field. If the plasma is regarded as perfectly conducting, the field lines are carried with the motion at speed $v_\perp$, and Ohm's law may be written

$$E + v_\perp B = 0. \tag{2.7}$$

This is essentially an integral of the induction equation, with E being a constant of integration. It is clear that the instantaneous plasma speed ($v_\perp$) is determined by Eq. (2.7) everywhere except at the neutral point where B vanishes. Furthermore, if we suppose that the flow speed is much slower than the Alfvén speed ($M_\mathrm{A} \ll 1$) and the plasma pressure is much smaller than the magnetic pressure ($\beta \ll 1$), the configuration passes through a series of equilibria with vanishing Lorentz force so that, in this two-dimensional situation,

$$jB = 0. \tag{2.8}$$

The peculiar role of the neutral point is evident in both Eqs. (2.7) and (2.8), which imply that a continuous deformation of the original field [Eq. (2.1)] through a series of potential configurations with $j = 0$ is possible everywhere save in the vicinity of the neutral point.

At the neutral point itself, Eq. (2.8) allows non-zero currents. Indeed, a solution to the problem is that there develops a current sheet of length $2L$, say. Then the question is how best to describe the resulting equilibrium containing the current sheet and having the appearance of Fig. 2.2(b) [or perhaps, as we shall see later, Fig. 2.2(c) or Fig. 2.12(b)]. Outside the sheet, if the current vanishes, the magnetic field satisfies $\nabla \times \mathbf{B} = \mathbf{0}$ and $\nabla \cdot \mathbf{B} = 0$ or, in two dimensions,

$$\frac{\partial B_y}{\partial x} - \frac{\partial B_x}{\partial y} = 0 \tag{2.9}$$

and

$$\frac{\partial B_x}{\partial x} + \frac{\partial B_y}{\partial y} = 0. \tag{2.10}$$

Now, it may be shown as follows that, if the combination

$$B_y + iB_x = f(z) \tag{2.11}$$

is any analytic (i.e., differentiable) function of the complex variable $z = x + iy$, then Eqs. (2.9) and (2.10) are automatically satisfied. We are familiar with the fact in one dimension that, if the derivative $f'(x)$ of a function of x exists, then the gradient at x has the same value whether x is approached from the left or the right. In a similar way in two dimensions, if $f'(z)$ is analytic, then the gradient has the same value when z is approached from any direction, in particular keeping y zero (so that $z = x$) or keeping x zero (so that $z = iy$). In other words,

$$\frac{\partial}{\partial x}(B_y + iB_x) = \frac{\partial}{i\partial y}(B_y + iB_x),$$

or, by equating real and imaginary parts, we obtain Eqs. (2.9) and (2.10), as required. Thus, we can treat the current sheet as a cut in the complex plane, and the object is to find a function $f(z)$ which has such a cut.

The initial state, Eq. (2.1), has

$$B_y + iB_x = z, \tag{2.12}$$

and, when a sheet stretches from $z = -iL$ to $z = iL$, we may represent the field by (Green, 1965)

$$B_y + iB_x = (z^2 + L^2)^{1/2}, \tag{2.13}$$

which behaves like z when $z \gg L$ and reduces to z (as required) when $L = 0$. Thus, the evolution through a series of equilibria with a slowly growing sheet may simply be modelled by letting L slowly increase in value in Eq. (2.13). The field has limiting field lines (separatrices) through the ends of the sheet which are inclined to one another by $2\pi/3$. This may be shown by noting that, near the upper end of the sheet at $z = iL$, Eq. (2.13) becomes approximately

$$B_y + iB_x = (2iLZ)^{1/2}, \tag{2.14}$$

where $Z = z - iL$, and the corresponding flux function in terms of local polar coordinates $(\bar{r}, \bar{\theta})$ is

$$A = \frac{2}{3}(2L\bar{r}^3)^{1/2} \cos\left(\frac{3\bar{\theta}}{2} + \frac{\pi}{4}\right), \tag{2.15}$$

so that the particular field lines $A = 0$ are inclined at $2\pi/3$, as required.

2.2.2 Current Sheet with Singular End-Points

Somov and Syrovatsky (1976) suggested that, instead of evolving into the field (2.13), the initial field (2.1) may evolve into the more general field

$$B_y + iB_x = \frac{z^2 + a^2}{(z^2 + L^2)^{1/2}}, \tag{2.16}$$

where $a^2 < L^2$ [Fig. 2.2(c)]. The neutral points at the ends ($z = \pm iL$) of the sheet have now in general been replaced by singularities where the field becomes infinite at the rate $(z \mp iL)^{-1}$ as z tends to $\pm iL$. The special case when $a = L$ reduces to Eq. (2.13).

At the right of the current sheet ($x = 0+, y^2 < L^2$) the magnitude of the field is

$$B_y(0+, y) = \frac{a^2 - y^2}{(L^2 - y^2)^{1/2}},$$

which vanishes at $y = \pm a$. The current *per unit length* in the sheet is simply

$$J = \frac{1}{\mu}[B_y(0+, y) - B_y(0-, y)] = \frac{2(a^2 - y^2)}{\mu(L^2 - y^2)^{1/2}},$$

where 0+ signifies approaching 0 through positive values and 0− through negative values, and so the regions between the neutral points at $y = \pm a$ and the ends of the sheet contain reversed currents.

However, this new family of solutions was largely ignored outside Russia until numerical experiments on magnetic reconnection revealed in some circumstances reversed-current spikes near the ends of a reconnecting current sheet (Biskamp, 1986; Lee and Fu, 1986; Forbes and Priest, 1987). Also, Bajer (1990) has modelled numerically the collapse of an X-point in an isolated region and found that reversed and singular currents can indeed appear at the ends of the sheet.

Bungey and Priest (1995) have generalised the Somov–Syrovatsky solution (2.16) to give (Figs. 2.3 and 2.4)

$$B_y + iB_x = -B_0 \left[\frac{(b + \frac{1}{2})d^2 + 2dcz - z^2}{\sqrt{z^2 - d^2}} \right], \qquad (2.17)$$

where b, c, d, and B_0 are constants. The special case given by $c = 0$, $b = -\frac{1}{2} + a^2/L^2$, $B_0 = 1$, and $d^2 = -L^2$ is the solution Eq. (2.16).

Varying the value of the constant b simply moves the position of the null points, as can be seen from Fig. 2.3 (where $c = 0$ and $d = 1$). Figure 2.3(a) shows the configuration given by $b = -1.0$, which possesses two null points in the surrounding field, along the imaginary y-axis normal to the sheet. As b increases, these null points converge, meeting at $z = 0$ for $b = -0.5$ [Fig. 2.3(b)]. Further increases in b now cause a separation of the null points along the sheet, reaching the ends for $b = 0.5$, which gives Green's solution. Values of b greater than this create neutral points in the surrounding field on the x-axis [Fig. 2.3(f)].

Suppose instead the constant c is varied from zero. This has the effect of producing an asymmetry of the null points about the centre of the sheet. From Eq. (2.17) the null points of the field can be seen to occur at the roots of the numerator on the right-hand side. These roots lie at the points $z = cd \pm d\sqrt{c^2 + b + \frac{1}{2}}$. Thus, there are two null points which converge to the one point $z = cd$ when $c^2 = -(b + \frac{1}{2})$. Changing the value of c therefore changes the centre of convergence of the null points. Figures 2.4(a) and 2.4(b) show the configurations when $c = -0.5$, with $b = -0.75$ and $b = -0.5$, respectively. The asymmetry of the null points about the centre of the sheet can lead to sheets as in Fig. 2.4(b) with one singular end and one neutral end. Figures 2.4(c) and 2.4(d) are for the value $c = -1.0$, so that the centre of convergence is at one end of the sheet, with the values $b = -1.5$ and $b = -1.2$, respectively. The corresponding solutions for an ambient force-free field are described in §2.3.1 below.

In many of the above solutions the field behaves as

$$B \sim \frac{K}{(z - d)^{1/2}},$$

say, near an end ($z = d$) of the sheet, a behaviour which is also found in force-free fields (§2.3.1). Thus, the solutions fail at these singularities, where diffusive effects become important and slow-mode shocks may be generated (e.g., Petschek, 1964; Priest, 1985; Strachan and Priest, 1994). The failure is evidenced by the fact that the net magnetic force on the end of the sheet is non-zero (essentially because of the inward magnetic tension force) and so the end is not in equilibrium, which suggests that an extra feature such

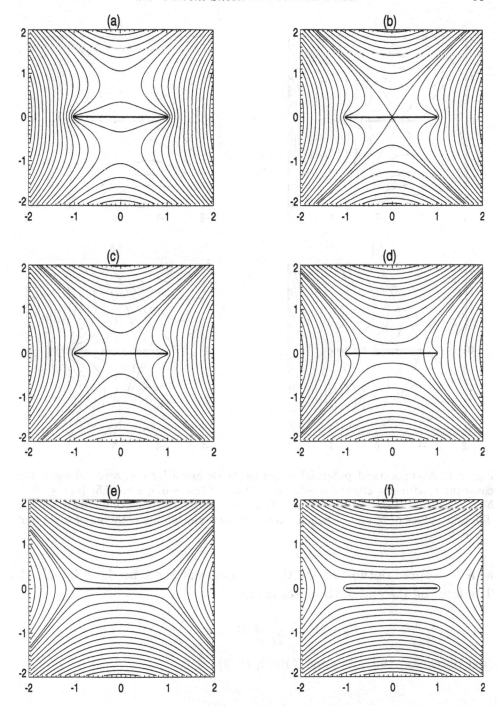

Fig. 2.3. Potential magnetic fields for different values of the constant b, with $c=0$ and $d=1$: (a) $b=-1.0$, (b) $b=-0.5$, (c) $b=-0.4$, (d) $b=0.0$, (e) $b=0.5$, (f) $b=1.0$. As b increases from -1.0 to 1.0 the null points converge from above and below on the centre of the sheet for $b=-0.5$, and then they separate symmetrically along it. (d) is a solution of the form of Syrovatsky, while (e) is Green's solution.

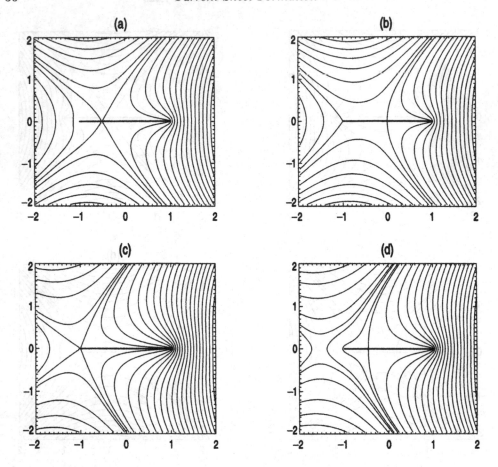

Fig. 2.4. Asymmetrical potential magnetic fields given by non-zero values of the constant c with $d=1$: (a) $c=-0.5$, $b=-0.75$, (b) $c=-0.5$, $b=-0.5$, (c) $c=-1.0$, $b=-1.5$, and (d) $c=-1.0$, $b=-1.2$. The upper and lower pairs of plots are for two different values of b, one of which makes the two null points combine at the point $z=c$.

as a high-speed outflow at the tip of the current sheet needs to be added. The force on the end of the sheet is given by

$$F_y + iF_x = \frac{1}{2\mu} \oint B^2 dz = \frac{\pi K^2 i}{\mu}$$

(see Batchelor, 1967; Aly and Amari, 1989).

2.2.3 Current Sheets in Other Configurations

The above complex variable technique (§2.2.1) for obtaining current sheets has been applied to a wide variety of different configurations. Priest and

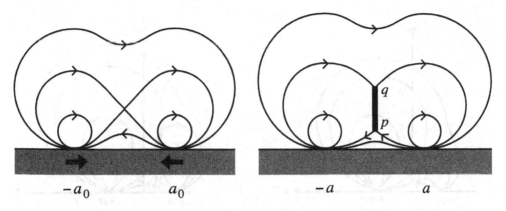

Fig. 2.5. The interactions of the magnetic field due to neighbouring bipolar magnetic fragments located in a line-tying surface such as the solar photosphere (shaded). Initially the separation of the magnetic sources is $2a_0$ and later it is $2a$.

Raadu (1975) have considered the formation of current sheets in the solar atmosphere. They supposed that the magnetic evolution is driven by the boundary conditions at the surface of the Sun, which are assumed to vary so slowly that the configuration passes through a series of equilibria. If the initial equilibrium contains an X-type neutral point, a current sheet forms and slowly grows in length (Fig. 2.5); the surrounding field is assumed to be two-dimensional and potential. The magnetic field sources are two line dipoles and the analysis is a natural extension of the previous work by Green (1965) for two line currents.

The dipoles have moment $(2\pi D/\mu)$ and are situated initially at $(a_0, 0)$ and $(-a_0, 0)$. The components (B_{0x}, B_{0y}) of the resulting potential field in the region $y > 0$ may be written in the combination

$$B_{0y} + iB_{0x} = \frac{iD}{(z + a_0)^2} + \frac{iD}{(z - a_0)^2}. \qquad (2.18)$$

Suppose the sources approach one another at a speed which is so fast that significant reconnection does not take place and that the field passes through a series of quasi-equilibrium states, with a vertical current sheet stretching from p to q along the y-axis when the dipole separation is $2a$. Outside the current sheet, the magnetic field is current-free and so potential; its components in the combination $(B_{0y} + iB_{0x})$ therefore form an analytic function of z, namely

$$B_y + iB_x = \frac{\overline{D}(z^2 + p^2)^{1/2}(z^2 + q^2)^{1/2}}{(z^2 - a^2)^2}, \qquad (2.19)$$

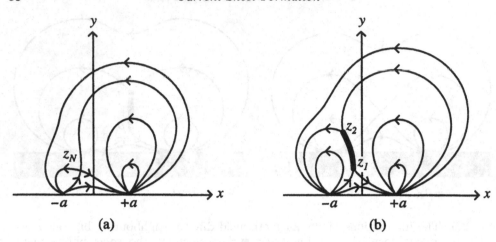

Fig. 2.6. Magnetic configuration due to two line dipoles: (a) purely potential situation; (b) curved current sheet stretching between z_1 and z_2 after an increase in the moment of the smaller dipole (from Tur and Priest, 1976).

where $\overline{D} = 4iDa^2(a^2 + p^2)^{-1/2}(a^2 + q^2)^{-1/2}$ is chosen to give the correct behaviour near the sources. The values of q and p are found as functions of a by using the conditions of frozen-in flux, which imply that the fluxes crossing the positive y-axis above and below the original neutral point on the y-axis must equal those crossing above q and below p, respectively. The fluxes can be found by performing the relevant integrals of Eqs. (2.18) and (2.19) with respect to y at $x = 0$. The dissipation of such a sheet may explain the two-thirds of X-ray bright points that are associated with cancelling magnetic features in the solar photosphere (see §11.3.1).

Tur and Priest (1976) have extended the analysis to include curved current sheets in more realistic configurations. For example, they modelled the current sheet which forms when a small emerging bipolar flux system presses up against a much larger bipolar field (Fig. 2.6) while the separation between the two bipoles remains constant (equal to $2a$). The initial potential field has components given by

$$B_{0y} + iB_{0x} = \frac{iD_0}{(z + a)^2} + \frac{iD_A}{(z - a)^2},$$

where $2\pi D_A/\mu$ and $2\pi D_0/\mu$ are the moments for the larger ambient dipole and the smaller evolving dipole, respectively. The emergence of additional flux is simulated by increasing the strength of the smaller dipole while

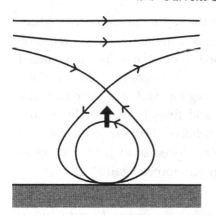 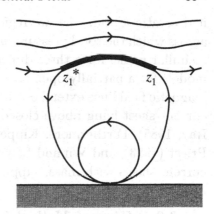

Fig. 2.7. The formation of a curved current sheet by the emergence of new magnetic flux through a boundary (shaded).

keeping its position fixed. The resulting magnetic field takes the form

$$B_y + iB_x = \frac{iK[(z - z_1)(z - z_2)(z - z_1^*)(z - z_2^*)]^{1/2}}{(z^2 - a^2)^2},$$

where there is a branch cut from z_1 to z_2 and the asterisks denote complex conjugates. The values of K, z_1, z_2 are determined from the behaviour near $z = \pm a$ and from the flux-conservation conditions.

Another model example considered by Tur and Priest is the emergence of new flux (represented by a dipole) into a uniform horizontal field (Fig. 2.7). The initial potential field is

$$B_{0x} - iB_{0y} = b + \frac{D_0}{z^2}.$$

When the dipole moment has increased from $2\pi D_0/\mu$ to $2\pi D/\mu$, the field becomes

$$B_x - iB_y = \frac{\overline{K}\left[(z^2 - z_1^2)(z^2 - z_1^{*2})\right]^{1/2}}{z^2}, \tag{2.20}$$

where $\overline{K}$ is a constant and the end-points of the current sheet are z_1 and $z_2 = -z_1^*$. The resulting sheet is again curved. Dissipation of this current sheet has been invoked to explain how the one-third of X-ray bright points that are associated with emerging flux can be produced (Heyvaerts, Priest and Rust, 1977; Low, 1987).

Low (1981, 1982, 1986) has used current sheets in a variety of equilibrium problems. For example, he modelled a solar coronal arcade as a two-dimensional bipolar potential field surrounded by an isothermal field-free hydrostatic region. The two regions are separated by a curved current sheet across which the total pressure $[p+B^2/(2\mu)]$ is continuous. He also presented a similar model for a three-dimensional coronal loop. Later, he constructed models for a partially open solar or stellar coronal magnetic field, in which magnetic field lines extend out from the polar regions and form an equatorial current sheet lying above closed equatorial field lines (see also Mestel and Ray, 1985). Furthermore, Kippenhahn and Schlüter (1957), Malherbe and Priest (1983), and Wu and Low (1987) have modelled solar prominences as current sheets with mass, supported by magnetic forces against gravity.

2.2.4 General Method for Constructing Current Sheets in Two-Dimensional Potential Configurations

Following on from the work for particular configurations of Priest and Raadu (1975), Tur and Priest (1976), and Low (1987), some general properties have been established by Aly and Amari (1989) and Amari and Aly (1990). These have in turn been used and extended by Titov (1992) to develop a general method for computing arbitrary two-dimensional potential fields with multiple curved current sheets and a given normal (B_y) component of magnetic field at a bounding plane (the x-axis) such as the base of a stellar corona.

A two-dimensional magnetic field may be written in terms of a flux function $[A(x,y,t)]$ as

$$B_x = \frac{\partial A}{\partial y}, \quad B_y = -\frac{\partial A}{\partial x}, \tag{2.21}$$

so as to satisfy $\nabla \cdot \mathbf{B} = 0$ automatically. Then for a potential field ($\nabla \times \mathbf{B} = 0$), A satisfies Laplace's equation

$$\nabla^2 A = 0, \tag{2.22}$$

and, if the magnetic field is frozen to the plasma, the induction equation may be uncurled to give

$$\frac{dA}{dt} \equiv \frac{\partial A}{\partial t} + \mathbf{v} \cdot \nabla A = 0. \tag{2.23}$$

Thus, A is frozen to the plasma, and the magnetic topology is preserved (Moffatt, 1985).

Starting with a potential field, our aim is to determine the field that results when $B_y(x,0,t)$ is prescribed in such a way that the normal component at the bounding plane ($y = 0$), such as the surface of a star, changes as a result of both horizontal motions of the footpoints and the appearance of new flux (such as emergence from below the photosphere). Thus, we seek a time-sequence of harmonic flux functions $A(x, y, t)$ in $y > 0$ for each t [satisfying Eq. (2.22)] whose topology is preserved, so that the values of A are constant in the course of time along the separatrices and the initial X-points bifurcate into current sheets. We suppose the distribution of B_y on the boundary is such that the corresponding potential field behaves like a dipole at infinity and contains N neutral points above the x-axis and none on it.

The field components of the potential field with current sheets may be combined as an analytic function,

$$\mathcal{B}(z, t) \equiv B_y(x, y, t) + i B_x(x, y, t),$$

of the complex variable $z = x + iy$. First of all, if $B_y(x, 0, t)$ is not prescribed and the coordinates of the two end-points of the kth current sheet are

$$z_{1k} = x_{1k} + iy_{1k}, \qquad z_{2k} = x_{2k} + iy_{2k},$$

then the field is

$$\mathcal{B}(z, t) = \mathcal{P}(z, t) \prod_{k=1}^{N} \sqrt{[(z - z_{1k})(z - z_{2k})]}, \qquad (2.24)$$

where $\mathcal{P}(z, t)$ is an analytic function without any null points. The shape of the kth current sheet is simply that of the field line passing through the branch points z_{1k} and z_{2k}.

Next, in order to solve our problem where $B_y(x, 0, t)$ is prescribed, we introduce extra factors in Eq. (2.24) corresponding to image current sheets below the x-axis with conjugate end-points z_{1k}^* and z_{2k}^*. The resulting field has the form

$$\mathcal{B}(z, t) = \mathcal{P}(z, t) \, \mathcal{Q}(z, \mathbf{r}_e), \qquad (2.25)$$

where

$$\mathcal{Q}(z, \mathbf{r}_e) = \prod_{k=1}^{N} \sqrt{\{[(z - x_{1k})^2 + y_{1k}^2][(z - x_{2k})^2 + y_{2k}^2]\}} \qquad (2.26)$$

and the set of end-points z_{1k}, z_{2k} (denoted by the vector $\mathbf{r}_e$) varies with time. The ingenious effect of this form is that on the x-axis $\mathcal{Q}$ is purely real, so that, when evaluated there, Eq. (2.25) implies that the real part of $\mathcal{P}$ on the x-axis is

$$\mathrm{Re}\, [\mathcal{P}(x, 0, t)] = \frac{B_y(x, 0, t)}{\mathcal{Q}(x, \mathbf{r}_e)}.$$

Then $\mathcal{P}$ is determined everywhere in terms of a Schwarz integral and Eq. (2.25) becomes

$$B(z,t) = -\frac{i\mathcal{Q}(z,\mathbf{r}_e)}{\pi} \int\limits_{-\infty}^{\infty} \frac{B_y(\xi,0,t)\,d\xi}{(\xi-z)\mathcal{Q}(\xi,\mathbf{r}_e)}. \tag{2.27}$$

At $t = 0$, the end-points of each sheet coincide ($z_{1k} = z_{2k}$) and each sheet becomes an X-point, so that our main result, Eq. (2.27), reduces to the classical result for a potential field, namely

$$B(z,0) = -\frac{i}{\pi} \int\limits_{-\infty}^{\infty} \frac{B_y(\xi,0,0)\,d\xi}{\xi-z}. \tag{2.28}$$

Finally, the $4N$ positions of the $2N$ current-sheet end-points are determined from the following conditions. First of all, the requirement that $\mathcal{B}$ behave like z^{-2} at infinity yields $2N+1$ equations of the form

$$\int\limits_{-\infty}^{\infty} \frac{\xi^m B_y(\xi,0,t)}{\mathcal{Q}(\xi,\mathbf{r}_e)}\,d\xi = 0, \quad m = 0,\dots,2N. \tag{2.29}$$

Secondly, the requirement of flux conservation at one end-point of each sheet gives N conditions, namely

$$\mathrm{Re} \int\limits_{-\infty}^{z_{1k}} B(\zeta,t)\,d\zeta = \mathrm{Re} \int\limits_{-\infty}^{z_{1k}} B(\zeta,0)\,d\zeta, \quad k = 1,\dots N. \tag{2.30}$$

Lastly, the conditions that each current sheet be a field line determines the remaining $N-1$ equations, namely

$$\mathrm{Re} \int\limits_{z_{1k}}^{z_{2k}} B(\zeta,t)\,d\zeta = 0, \quad k = 1,\dots N-1. \tag{2.31}$$

The Nth sheet then automatically lies along a field line.

The method may be generalised to allow infinitely long sheets or, for example, a uniform field at infinity. Other complex-variable techniques for solving Laplace's equation in two dimensions are given by Muskelishvilli (1953).

2.2.5 Axisymmetric Current Sheet in Three Dimensions

One limitation of the above models utilizing complex-variable techniques is that they are purely two-dimensional. This has been overcome by Tur (1977) for the case of the approach of symmetric dipoles (Fig. 2.5) when in three

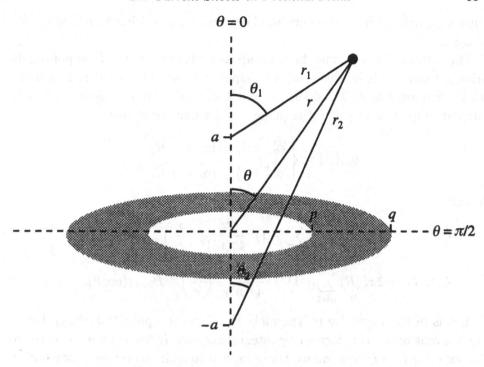

Fig. 2.8. An annular current sheet of radii p and q situated in the plane $\theta = \frac{1}{2}\pi$, with dipole sources at distances a above and below the plane.

dimensions the current sheet forms an annulus along a separator (Fig. 2.8). In spherical polar coordinates (r, θ, ϕ) the dipoles (of moment $2\pi D/\mu$) lie at a distance a from the origin in the directions $\theta = 0$ and $\theta = \pi$. The initial potential field ($\mathbf{B}_0$) is written in terms of a magnetic potential (Φ_{m0}) as

$$\mathbf{B}_0 = -\nabla\Phi_{m0},$$

where

$$\Phi_{m0} = \frac{D\cos\theta_1}{r_1} + \frac{D\cos\theta_2}{r_2} \tag{2.32}$$

and the values $r_1, r_2, \theta_1, \theta_2$ (which may be written in terms of r and θ) are indicated in Fig. 2.8. When the two dipoles approach one another and the current sheet forms, the magnetic field is

$$\mathbf{B} = -\nabla\Phi_m,$$

where

$$\Phi_m = \Phi_{ms} + \Phi_{mD},$$

with Φ_{ms} arising from the current sheet and Φ_{mD} arising from the dipole
sources.

The solution for Φ_{ms} may be built up as an integral sum of the potentials
arising from the individual current loops that may be considered to make
up the current sheet. Thus, the potential at a point (r, θ) due to a circular
current loop of radius R in the plane $\theta = \frac{1}{2}\pi$ may be written

$$\phi_m(R) = \begin{cases} \phi_m^{(-)}(R) & \text{for } r < R, \\ \phi_m^{(+)}(R) & \text{for } r > R, \end{cases}$$

where

$$\phi_m^{(-)}(R) = 2\pi I(R) \left[1 - \sum_{n=0}^{\infty} (-1)^n \frac{(2n)!}{4^n (n!)^2} \left(\frac{r}{R}\right)^{2n+1} P_{2n+1}(\cos\theta)\right],$$

$$\phi_m^{(+)}(R) = 2\pi I(R) \sum_{n=1}^{\infty} (-1)^{n+1} \frac{(2n)!}{4^n (n!)^2} \left(\frac{R}{r}\right)^{2n} P_{2n-1}(\cos\theta),$$

in terms of the Legendre polynomials P_n. Then the potential (Φ_{ms}) due to
the current annulus is formed by integrating over R from the interior (p) to
the exterior (q) of the annulus, being careful to split the range of integration
at $R = r$.

The function $I(R)$ is determined by the boundary conditions on Φ_m at
$\theta = \frac{1}{2}\pi$, which give an integral equation that can be solved numerically. One
finds a maximum in the current density function close to the inner boundary
$R = a$, because the field is stronger there. Just as for the two-dimensional
problem (Fig. 2.5), flux conservation may be used to determine numerically
the inner and outer radii $[p(a)$ and $q(a)]$ as the dipoles approach one another
with a separation $2a$. The resulting current sheet is somewhat larger than
in the two-dimensional case. Further properties of current-sheet formation
along nonaxisymmetric separators have been developed by Longcope and
Cowley (1996) and Longcope (1996, 1998).

2.3 Current Sheets in Force-Free and Magnetostatic Fields

2.3.1 Force-Free Fields

As we have seen, the technical advantage of considering potential fields is
that complex variable theory can be employed and, in particular, current
sheets may be treated as cuts in the complex plane. This advantage is no
longer directly available when we want to model force-free fields. However,
techniques are beginning to be developed to deal with current sheets in
force-free fields.

Bajer (1990) has considered an isolated plasmoid with a linear force-free field

$$(B_x, B_y, B_z) = \left(\frac{\partial A}{\partial y}, -\frac{\partial A}{\partial x}, B_z \right),$$ (2.33)

whose flux function (A) satisfies

$$\nabla^2 A + \alpha^2 A = 0,$$ (2.34)

where α is a constant. Initially, the field of the plasmoid is potential with

$$A = r^2(1 - r) \cos^2 \theta.$$

It contains an X-point at the origin and is contained within the cylinder $r = 1$. The collapse of the X-point is followed numerically and the resulting field contains a current sheet with reversed currents at its ends. It is calculated by mapping conformally the region outside the sheet to one outside a circle.

Bungey and Priest (1995) followed a similar approach when B_z^2 is of the form

$$B_z^2(A) = 2\mu J_0(A + A_0),$$

where J_0 is constant (with dimensionless value $J_0^* = \mu J_0 L_0/B_0$) and A_0 is a constant introduced to ensure the positivity of B_z^2. Then A satisfies

$$\nabla^2 A + \mu J_0 = 0,$$ (2.35)

and we have a constant-current force-free field (with current J_0). The conformal mapping

$$z = \frac{1}{2} \left(\tilde{z} + \frac{a^2}{\tilde{z}} \right)$$

maps the region outside a cut stretching from $z = -a$ to $z = +a$ onto the region outside a circle of radius $\tilde{r} = a$, centred at the origin of the $\tilde{z}$-plane. The Poisson equation, Eq. (2.35), can then be written as

$$\nabla^2 A(\tilde{r}, \tilde{\theta}) = -\frac{\mu J_0}{4} \left[1 - \frac{2a^2 \cos(2\tilde{\theta})}{\tilde{r}^2} + \frac{a^4}{\tilde{r}^4} \right].$$ (2.36)

The boundary conditions are that $A(a, \tilde{\theta}) = 0$ and that the flux function behave like that of an X-point at large distances [i.e., $B_0(x^2 - y^2)/(2L_0)$, say].

The resulting solution is the real part of the complex potential

$$
F(\tilde{r}, \tilde{\theta}) = \frac{B_0}{L_0} \left[a^2 b \log \frac{\tilde{z}}{a} + a(c - id)\tilde{z} - a^3 \frac{(c + id)}{\tilde{z}} - \frac{1}{8} \left(\tilde{z}^2 - \frac{a^4}{\tilde{z}^2} \right) \right.
$$
$$
\left. - \frac{J_0^*}{16} \left(|\tilde{z}|^2 + \frac{2a^2 \tilde{z}^2}{|\tilde{z}|^2} + \frac{a^4}{|\tilde{z}|^2} - 2a^2 - \frac{2a^4}{\tilde{z}^2} \right) \right],
\tag{2.37}
$$

where b, c, and d are constants and this may be mapped back onto the z-plane by using

$$
\tilde{z} = z + (z^2 - a^2)^{1/2}.
$$

When no current sheet is present, the X-point of the potential field becomes stretched out as $|J_0^*|$ increases, with the field lines closing over for $|J_0^*| > 2$ to give an elongated O-point field. The same feature is present when a current sheet of finite length is present. Figures 2.9(a)–2.9(f) show a sequence of force-free configurations for increasing values of the constant J_0^*. Starting with $J_0^* = -3.0$ in Fig. 2.9(a), where the field lines are closed over horizontally, the field lines can be seen to open up horizontally and become stretched out vertically. Eventually they reclose as shown in Fig. 2.9(f).

The force-free field configurations shown in Fig. 2.9 all have the constant values $b = c = 0$. Varying the values of these constants has exactly the same effect as in the potential case (§2.2.2), with b changing the position of the null points and c shifting the centre of symmetry away from the origin.

The question of the validity of these force-free solutions has to be raised. No difficulty arises in the potential case ($J_0 = 0$), where $B_z(A) = (2\mu J_0)^{\frac{1}{2}}$ $(A + A_0)^{\frac{1}{2}}$ is always zero. However, for non-zero α we must ensure that $2\mu J_0(A + A_0)$ be positive for meaningful solutions to exist. The particular boundary condition chosen at the sheet, namely $A = 0$, leads to a change in the sign of A as we cross a separatrix of the field, and hence both positive and negative values of the additional constant A_0 may be chosen so as to ensure the positivity of $B_z^2(A)$ in the considered domain, provided that domain is finite. Changing the value of the constant A_0 has the effect of increasing B_z^2 and therefore the shear of the magnetic field, but this does not alter the projection of the field lines in the xy-plane. Clearly, the larger the domain of validity that is required, the greater the size of the constant A_0 that must be incorporated.

The new features demonstrated in these solutions include the presence of an asymmetry along the length of the sheet, giving both potential and force-free fields with asymmetric current distributions. In particular, the

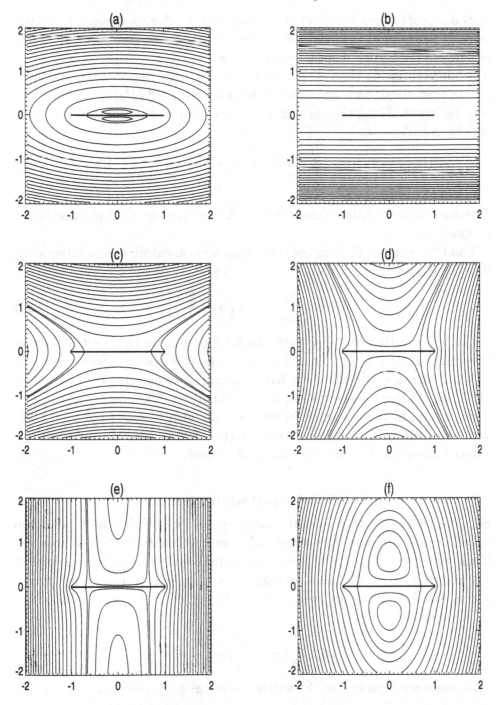

Fig. 2.9. Constant-current force-free field configurations for a range of values of the current J_0^*, with $b = c = 0$ held fixed: (a) $J_0^* = -3.0$, (b) $J_0^* = -2.0$, (c) $J_0^* = -1.0$, (d) $J_0^* = 1.0$, (e) $J_0^* = 2.0$, (f) $J_0^* = 3.0$. As J_0^* increases from -3 the field lines become open at $J_0^* = -2$ and then progressively stretched out in the y-direction, until beyond $J_0^* = 2$ they become closed again.

embedding of the current sheet inside a closed field-line configuration and the presence of magnetic islands close to the surfaces of the sheet are interesting features of these constant-current solutions.

By differentiating the real part of Eq. (2.37) with respect to y, and evaluating it at $y = 0$, an expression for the magnetic field (B_x) along the sheet may be found. Thus, the current per unit length at any point in the sheet may be calculated as

$$J = \frac{2B_x(y=0)}{\mu} = \frac{-2B_0}{\mu L_0} \left[\frac{a^2 b + 2acx - x^2(1 + \frac{1}{2}J_0^*) + \frac{1}{2}a^2}{\sqrt{a^2 - x^2}} \right]. \qquad (2.38)$$

The current distribution along the sheet is asymmetric when the constant c is non-zero.

The total current (J_T) in the sheet may also be calculated by integrating Eq. (2.38) along the length of the sheet to give

$$J_T = \frac{-2B_0}{\mu L_0} \left[a^2 \pi \left(b - \frac{J_0^*}{4} \right) \right]. \qquad (2.39)$$

When b is negative and J_0^* is zero, the total current in the sheet is positive (assuming $B_0/L_0 > 0$). When $b = J_0^*/4$, the reversed currents match the forward currents, and the sheet has an overall zero current. Positive values of $(b - J_0^*/4)$ give the sheet a negative total current. The effect of varying J_0^* from zero on the total current in the sheet can also be seen from Eq. (2.38), with negative values adding current in the reverse direction, and positive values increasing the size of the forward current.

2.3.2 Magnetostatic Fields

Suppose we start with a potential magnetic field containing an X-point and subject it to changes in the positions of its sources and to changes in pressure. Then we allow it to relax to a new equilibrium but preserve the topology by not allowing reconnection. In general, the X-point will bifurcate and current sheets will form along the separatrices (Fig. 2.10). If the field components are written as

$$(B_x, B_y) = \left(\frac{\partial A}{\partial y}, -\frac{\partial A}{\partial x} \right),$$

then the force balance may be written in terms of the flux function as

$$\nabla^2 A + \mu \frac{dp}{dA} = 0, \qquad (2.40)$$

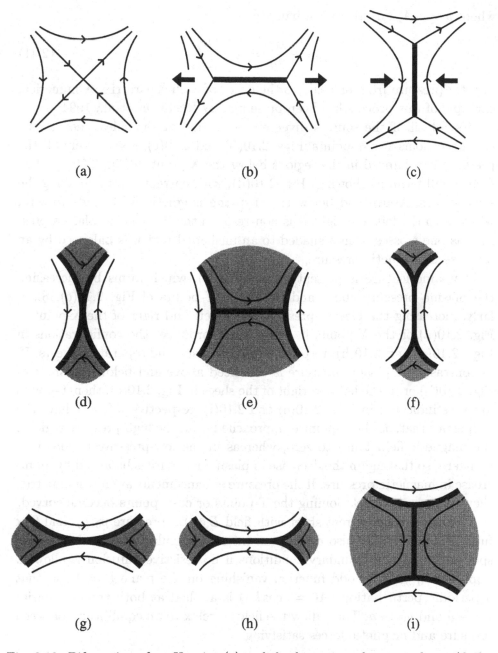

Fig. 2.10. Bifurcation of an X-point (a) and the formation of current sheets (dark curves) in response to (b) and (c) motions of sources and (d)–(i) increases in pressure in the shaded regions.

where $p = p(A)$ and there is a balance

$$p_1 + \frac{B_1^2}{2\mu} = p_2 + \frac{B_2^2}{2\mu} \qquad (2.41)$$

in total pressure from one side to the other of each separatrix. A numerical example of this process has been presented by Rastätter et al., 1994.

First of all, if the sources diverge or converge at the sides, then a current sheet joins two Y-points [Figs. 2.10(b) and 2.10(c), respectively]. If the pressure is enhanced in the regions below the X-point in Fig. 2.10(a), then a cusp will form, as shown in Fig. 2.10(d), with current sheets all along the separatrices. Above and below the cusp the magnetic field tends to zero, whereas to the left and right it is non-zero, so the decrease in plasma pressure as one crosses from a shaded to an unshaded region is balanced by an increase in magnetic pressure.

The same feature is present in Fig. 2.10(f), which forms by increasing the plasma pressure above and below the Y-points of Fig. 2.10(c). Similarly, increasing the plasma pressure to the left and right of the X-point in Fig. 2.10(a) or the Y-points in Fig. 2.10(b) produces the configurations in Figs. 2.10(g) and 2.10(h), respectively, with cusps and separatrix sheets. If, in contrast, the plasma pressure is enhanced above and below the sheet of Fig. 2.10(b) or to the left and right of the sheet in Fig. 2.10(c), then T-points form, as indicated in Figs. 2.10(e) and 2.10(i), respectively. These have the properties that, as the T-point is approached from the high-pressure regions, the magnetic field tends to zero, whereas in the low-pressure regions it is non-zero, so that again the decrease in plasma pressure is balanced by an increase in magnetic pressure. If the pressure enhancements are not symmetric, then the current sheets joining the T-points or cusp-points become curved.

A simple diffuse current sheet with field $B_0(y)\hat{\mathbf{x}}$, pressure $p_0(y)$, and flux function $A_0(y)\hat{\mathbf{z}}$ may also develop an internal singular current sheet in response to imposed boundary conditions if no resistive diffusion is allowed. Suppose $B_0(y)$ is an odd function vanishing on the plane $y = 0$ and that a periodic perturbation $(A_1 = \epsilon \cos kx)$ is applied at both the boundaries $y = +a$ and $y = -a$. Then allow the field to relax to an equilibrium between pressure and magnetic forces satisfying

$$\nabla^2 A + \mu \frac{dp}{dA} = 0,$$

whose linearized form is

$$\nabla^2 A_1 + \mu \frac{d^2 p_0}{dA_0^2} A_1 = 0. \qquad (2.42)$$

The general smooth solution consists of the sum of an odd function of y, which does not satisfy the boundary conditions, and an even function, which violates the condition ($A_1 = 0$ at $y = 0$) of frozen flux. There is thus no smooth solution to the problem, and instead the required solution is the odd function for $y > 0$ together with $A_1(x, y) = A_1(x, -y)$ for $y < 0$: this possesses a jump in $\partial A_1/\partial y$ at $y = 0$ and therefore produces a δ-function singularity of the current. The current sheet, whose formation is driven by the boundary conditions, may also reconnect; such a process is called forced magnetic reconnection (Kulsrud and Hahm, 1982; Hahm and Kulsrud, 1985). Indeed, the resulting energy release has been calculated and shown to be particularly large when the field is close to marginal tearing stability (Vekstein and Jain, 1998).

More generally, for the field $B_0(y) = \tanh y$, Schindler and Birn (1993) have imposed the boundary conditions

$$A_1(x, a) = A_1(x, -a) = \sum_{n=1}^{\infty} t_n \sin(k_n x)$$

at the top and bottom boundaries, where $k_n = n\pi/l$, and

$$A_1(0, y) = \sum_{n=1}^{\infty} r_n w_n(y)$$

at the left-hand boundary ($x = 0$), where

$$w_n(y) = \tanh(y) \cos(\kappa_n y) + k_n \sin(\kappa_n y)$$

and κ_n are the solutions of

$$\tanh(a) + \kappa_n \tan(\kappa_n a) = 0.$$

The resulting general solution of Eq. (2.42) is

$$A_1 = \sum_{n=1}^{\infty} \left[\frac{t_n}{v_n(a)} \sin(k_n x) v_n(y) + \frac{r_n}{u_n(0)} u_n(x) w_n(y) \right],$$

where

$$v_n(y) = \tanh(y) \cosh(k_n y) - k_n \sinh(k_n y),$$
$$u_n(y) = \tanh(\kappa_n l) \cosh(\kappa_n x) - \sinh(\kappa_n x).$$

The corresponding sheet current at $y = 0$ follows as

$$J(x) = -\frac{2}{\mu} \sum_{n=1}^{\infty} \left[\frac{t_n}{v_n(s)} \left(1 - k_n^2 \right) \sin(k_n x) + \frac{r_n}{u_n(0)} \left(1 + \kappa_n^2 \right) u_n(x) \right].$$

2.4 Magnetic Relaxation

Moffatt (1985, 1990) and his co-workers Bajer (1990) and Linardatos (1993)
have developed a method (originally suggested by Arnol'd, 1974) for determining magnetostatic equilibria containing current sheets by magnetic
relaxation. They considered a magnetic field threading an incompressible
fluid

$$\nabla \cdot \mathbf{v} = 0, \quad \nabla \cdot \mathbf{B} = 0 \tag{2.43}$$

contained in a domain $\mathcal{D}$ with boundary $\partial \mathcal{D}$ on which $\mathbf{v} \cdot \mathbf{n} = 0$ and $\mathbf{B} \cdot \mathbf{n} = 0$, so that neither fluid nor magnetic flux crosses the boundary.

The magnetic field is frozen to the plasma

$$\frac{\partial \mathbf{B}}{\partial t} = \nabla \times (\mathbf{v} \times \mathbf{B}), \tag{2.44}$$

and a simple model equation of motion is adopted in the form

$$\rho \frac{\partial \mathbf{v}}{\partial t} = -\nabla p + \mathbf{j} \times \mathbf{B} - K\mathbf{v}, \tag{2.45}$$

(where ρ = constant and $\mathbf{j} = \nabla \times \mathbf{B}/\mu$) such that nonlinearity in the acceleration term is neglected and the dissipation term on the right (with $K > 0$)
is linear. Starting from a smooth initial state ($\mathbf{B}_0$) at rest, an imbalance
of forces in general accelerates the plasma, but the dissipation eventually
slows down the flow. There is a relaxation towards an equilibrium state
($\mathbf{B}_E$), which is not necessarily unique and may be a saddle or a minimum.
During the evolution, the magnetic field is frozen to the fluid and so the
topology of the magnetic field is preserved, in the sense that magnetic field
lines may not cross each other or reconnect and so all knots and links are
conserved. The field therefore passes through a series of topologically equivalent states, but the final state may in general contain singularities. It is
topologically accessible from $\mathbf{B}_0$ in the sense that $\mathbf{B}_0$ may be convected to
$\mathbf{B}_E$ by a smooth incompressible flow with a finite amount of dissipation.

The justification for using the model equation of motion, Eq. (2.45), in
place of a full equation with a nonlinear inertial term and a full viscous term
is that it is merely a mathematical expedient to force the system towards an
equilibrium and so demonstrate the existence of the equilibrium. Of course,
once the equilibrium forms (or indeed is in the process of formation) it
may well be subject to instabilities and reconnection, as described in other
chapters. It is also of interest to note that magnetic relaxation as a numerical
technique for producing magnetic equilibria without current sheets has been
widely used in solar, astrophysical, and laboratory MHD (e.g., Sturrock and

Woodbury, 1967; Chodura and Schlüter, 1981; Klimchuk et al., 1988; Hesse and Birn, 1993; Rastätter and Neukirch, 1997).

If $\mathbf{X}(\mathbf{x})$ is the final position of a plasma element initially at $\mathbf{x}$, it is the non-continuity of the (homeomorphic) mapping from $\mathbf{x}$ to $\mathbf{X}$ which is associated with the singularities. The Lagrangian version of the frozen-field equation, Eq. (2.44), relates the i and j components of the final ($\mathbf{B}_E$) and initial ($\mathbf{B}_0$) fields by

$$B_{Ei}(\mathbf{X}) = B_{0j}(\mathbf{x}) \frac{\partial X_i}{\partial x_j},$$

and, although a finite stretching of field lines makes $\mathbf{x}$ differentiable in the direction of $\mathbf{B}_0$, it need not be differentiable in directions perpendicular to $\mathbf{B}_0$.

The magnetic energy $[W_m(t) = \int_\mathcal{D} B^2/(2\mu)dV]$ changes in time at a rate

$$\frac{dW_m}{dt} = \frac{1}{\mu} \int_\mathcal{D} \mathbf{B} \cdot \frac{\partial \mathbf{B}}{\partial t} dV = \frac{1}{\mu} \int_\mathcal{D} \mathbf{B} \cdot \nabla \times (\mathbf{v} \times \mathbf{B}) \, dV,$$

which may be transformed by using the divergence theorem and the boundary conditions on $\partial\mathcal{D}$ to give

$$\frac{dW_m}{dt} = -\int_\mathcal{D} \mathbf{v} \cdot \mathbf{j} \times \mathbf{B} \, dV, \tag{2.46}$$

so that the magnetic energy decreases when the Lorentz force does positive work on the plasma and vice versa. The kinetic energy $[W_k(t) = \frac{1}{2}\rho \int_\mathcal{D} v^2 dV]$ changes at a rate

$$\frac{dW_k}{dt} = \rho \int_\mathcal{D} \mathbf{v} \cdot \frac{\partial \mathbf{v}}{\partial t} dV = \int_\mathcal{D} -\mathbf{v} \cdot \nabla p + \mathbf{v} \cdot \mathbf{j} \times \mathbf{B} - K\mathbf{v} \cdot \mathbf{v} \, dV,$$

which may also be transformed by using the divergence theorem, the incompressibility condition, and the boundary conditions to

$$\frac{dW_k}{dt} = \int \mathbf{v} \cdot \mathbf{j} \times \mathbf{B} \, dV - 2KW_k, \tag{2.47}$$

so that the kinetic energy is increased by a (positive) Lorentz force and decreased by dissipation.

Combining Eqs. (2.46) and (2.47) we find

$$\frac{d}{dt}(W_m + W_k) = -2KW_k. \tag{2.48}$$

Since $W_k > 0$ and $K > 0$, the total energy decreases monotonically and, being positive, it must tend to a finite limit. Furthermore, if the topology of the initial magnetic field is nontrivial (in the sense that not all of the field lines can shrink to a point without cutting other field lines), then $W_m(t)$ (and also $W_m + W_k$) cannot decrease below some lower bound (> 0) determined by this topology.

The magnetic energy decreases as the magnetic tension causes the field lines to contract, but this is impeded by the linkage of the field lines, one measure of which is the magnetic helicity (see §8.5, §9.1.1, and Moffatt, 1969),

$$H = \int_D \mathbf{A} \cdot \mathbf{B} \, dV, \tag{2.49}$$

where $\mathbf{A}$ is a vector potential for $\mathbf{B}$ satisfying

$$\nabla \times \mathbf{A} = \mathbf{B}, \quad \nabla \cdot \mathbf{A} = 0. \tag{2.50}$$

The magnetic helicity is a measure of both the twist and the linkage of magnetic flux tubes: it is constant under frozen-field evolution (Woltjer, 1958) and is in general non-zero for fields of non-trivial topology. It is the invariance of H which places a lower bound on $W_m(t)$, since the Schwarz inequality

$$\int_D A^2 dV \int_D B^2 dV \geq H^2$$

and the standard result (Roberts, 1967)

$$\int_D A^2 dV \leq \frac{1}{K_0^2} \int_D B^2 dV$$

(where K_0 is a constant > 0) may be combined to give

$$W_m(t) \geq \frac{K_0}{2\mu} |H|,$$

as required.

In the asymptotic situation, the dissipation must vanish and so

$$\mathbf{v} \equiv 0 \text{ and } \mathbf{B} = \mathbf{B}_E,$$

where $\mathbf{B}_E$ is a magnetostatic field satisfying

$$\mathbf{j} \times \mathbf{B} = \nabla p.$$

In this equilibrium state $\mathbf{B}_E$ may have surfaces of tangential discontinuity (current sheets). Across such surfaces $\mathbf{n} \cdot \mathbf{B} = 0$ (otherwise there would be an infinite Lorentz force tangential to the sheet) and the total pressure $[p + B^2/(2\mu)]$ is continuous.

Several model problems have been considered which illustrate how the above ideas can be used to determine the evolution of a magnetic field. For example, Linardatos (1993) considered the relaxation of a two-dimensional field

$$(B_x, B_y) = \left(\frac{\partial A}{\partial y}, -\frac{\partial A}{\partial x} \right),$$

whose field lines are given by $A =$ constant and for which the frozen-field equation reduces to $dA/dt = 0$. Since the flow is incompressible, the area $[a(A)]$ inside a field line is conserved: it is known as the signature function and continues to characterise the field for all times. The field lines therefore tend to minimize their lengths while conserving the areas they enclose. The equilibrium flux function (A_E) is given by

$$\nabla^2 A_E = -\mu j(A_E),$$

where the current function $[j(A_E)]$ is determined partly by the signature function and partly by the shape of the boundary. Fields of finite energy that extend to infinity and have elliptical topology relax to a state in which all the field lines are concentric circles. If such a field is constrained within a finite boundary, the minimum-energy state is one in which the field lines close to the boundary tend to align themselves with it, while those deep in the interior tend to become circular. Fields containing X-points tend to collapse locally and form current sheets (see also Bajer, 1990).

A three-dimensional example was considered qualitatively by Moffatt (1985) in which two untwisted flux tubes are linked as shown in Fig. 2.11(a). The configuration has magnetic helicity

$$H = 2F_1F_2,$$

where F_1 and F_2 are the fluxes of the tubes. During the relaxation the field lines tend to shrink while the fluxes of each tube are conserved (and so are their volumes since the medium is incompressible). The shrinkage is halted when the two tubes come into contact [Fig. 2.11(b)], but the magnetic energy can decrease still further by a spreading of the outer field lines in azimuth around the inner tube to give a final axisymmetric state [Fig. 2.11(c)]. The two volumes are now separated by a thin current sheet on the toroidal surface of separation.

2.5 Self-Consistent Dynamic Time-Dependent Formation

Let us return now to the question of what happens when the X-point field

$$B_y + iB_x = z \tag{2.51}$$

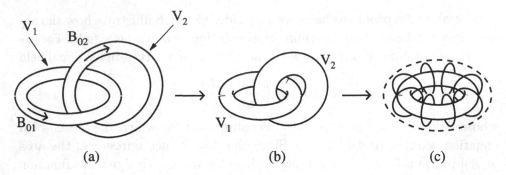

Fig. 2.11. The relaxation of a pair of linked flux tubes of volumes V_1 and V_2 and initial fields B_{01}, B_{02}, respectively (after Moffatt, 1985): (a) initial state, (b) during relaxation, (c) final relaxed state.

collapses. Does it evolve to the field

$$B_y + iB_x = (z^2 + L^2)^{1/2} \qquad (2.52)$$

with Y-points at the ends of a current sheet, as has commonly been supposed, or to the Syrovatsky field

$$B_y + iB_x = \frac{z^2 + a^2}{(z^2 + L^2)^{1/2}} \qquad (2.53)$$

with regions of reversed current and singularities at the ends? Or does it evolve to some other state?

The above solutions, Eqs. (2.52) and (2.53), represent a slow passive evolution through a series of equilibria, but Priest, Titov, and Rickard (1994b) have discovered some new nonlinear self-similar compressible solutions for the dynamic time-dependent formation of a current sheet. They assumed

$$c_s \ll v \ll v_A,$$

so that the flow speed (v) lies between the sound (c_s) and Alfvén speeds (v_A), and they solved the equation of motion

$$\rho \frac{d\mathbf{v}}{dt} = \mathbf{j} \times \mathbf{B} \qquad (2.54)$$

(with no pressure gradient) and the ideal induction equation

$$\frac{\partial \mathbf{B}}{\partial t} = \nabla \times (\mathbf{v} \times \mathbf{B}). \qquad (2.55)$$

To lowest order, they found a self-similar, quasi-static collapse through current-free states surrounding a growing current sheet. The motion of the field lines determines the flow speed perpendicular to the field lines, while the flow $(v_{\parallel})$ along the moving field lines is determined by a balance between the coriolis and centrifugal forces associated with the rotation of the field lines, such that the acceleration is perpendicular to the magnetic field. The resulting field lines are shown in Fig. 2.12. In dimensionless variables the current sheet stretches from $-\sqrt{t}$ to $+\sqrt{t}$ along the x-axis. As the sheet grows in length, the magnetic dissipation increases, and it swallows up half of the magnetic flux ahead of it, so creating a transverse y-component of field threading the sheet. The other half piles up ahead of the sheet and creates a region of reversed current near the ends. The individual plasma elements converge on the x-axis during the collapse.

The solution may be written in a concise form in terms of $z = x + iy$ as

$$B_y + iB_x = -\frac{[z + \sqrt{(z^2 - t)}]^2}{4\sqrt{(z^2 - t)}}. \tag{2.56}$$

It may be derived as follows. First of all, expand the variables in powers of the Alfvén Mach number $\epsilon(\ll 1)$, so that

$$\mathbf{v} = \epsilon \mathbf{v}_0 + \epsilon^2 \mathbf{v}_1 + \cdots, \quad \mathbf{B} = \mathbf{B}_0 + \epsilon \mathbf{B}_1 + \cdots.$$

To zeroth order Eq. (2.54) becomes

$$\mathbf{j}_0 = 0 \tag{2.57}$$

and to first order

$$\rho_0 \frac{d\mathbf{v}_0}{dt} = \mathbf{j}_1 \times \mathbf{B}_0. \tag{2.58}$$

Thus, Eq. (2.57) implies an evolution through potential states $(\mathbf{B}_0)$, while Eq. (2.58) implies that

$$\frac{d\mathbf{v}_0}{dt} \cdot \mathbf{B}_0 = 0, \tag{2.59}$$

so that the acceleration must be perpendicular to $\mathbf{B}_0$.

Next, write the field in terms of a flux function (A) as

$$(B_{0x}, B_{0y}) = \left(\frac{\partial A}{\partial y}, -\frac{\partial A}{\partial x} \right),$$

where Eqs. (2.57) and (2.55) imply that

$$\nabla^2 A = 0 \quad \text{and} \quad \frac{\partial A}{\partial t} + \mathbf{v}_0 \cdot \nabla A = 0,$$

so that (A) is potential and frozen to the plasma. Introduce then the conjugate harmonic function (Φ_m) and assume that it, too, is frozen to the plasma

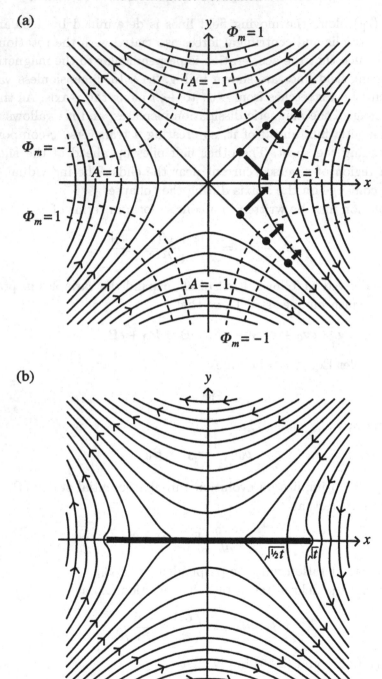

Fig. 2.12. Magnetic field lines (solid curves) at (a) $t = 0$ and (b) $t > 0$ when a self-consistent current sheet stretches from $z = -\sqrt{t}$ to $z = \sqrt{t}$.

so that

$$\frac{\partial \Phi_m}{\partial t} + \mathbf{v}_0 \cdot \nabla \Phi_m = 0.$$

Thus, a plasma element preserves its values of A and Φ_m as it moves. Then write

$$f = A + i\Phi_m$$

and

$$z = x + iy,$$

so that the goal is to determine $z(f, t)$ and its inverse $f(z, t)$. In this formalism Eq. (2.59) may be written

$$\frac{\partial^2 z}{\partial t^2} \bigg/ \frac{\partial z}{\partial f} - \frac{\partial^2 \bar{z}}{\partial t^2} \bigg/ \frac{\partial \bar{z}}{\partial f} = 0,$$

which implies that

$$\frac{\partial^2 z}{\partial t^2} = \chi(t)\frac{\partial z}{\partial f}, \tag{2.60}$$

where $\chi(t)$ is an arbitrary real function. For simplicity, now assume $\chi \equiv 0$ for acceleration-free flow, so the solution of Eq. (2.60) becomes simply

$$z = z_0(f) + v_0(f)t, \tag{2.61}$$

where $z_0(f)$ is the initial position of a plasma element and $v_0(f)$ its velocity. Suppose further that the initial flux function is $A_0 = \frac{1}{2}(x^2 - y^2)$, giving $f_0 = \frac{1}{2}z_0^2$ and so $z_0 = \sqrt{2f}$. If also the initial velocity is $v_0 = 1/(4\sqrt{2f})$ (i.e., the electric field equals $\frac{1}{4}$ initially), then Eq. (2.61) becomes

$$z = z_0 + \frac{t}{4z_0},$$

which may be rewritten to give z_0 as

$$2z_0 = z + (z^2 - t)^{\frac{1}{2}}.$$

However, the plasma preserves its value of $f(= f_0 = \frac{1}{2}z_0^2)$ as it moves and so

$$f(z, t) = \frac{1}{8}\left[z + (z^2 - t)^{\frac{1}{2}}\right]^2,$$

which is the required solution. The resulting magnetic field components are

$$B_y + iB_z = -\frac{\partial f}{\partial z} = -\frac{\left[z + (z^2 - t)^{\frac{1}{2}}\right]^2}{4\sqrt{(z^2 - t)}},$$

while the velocity components are

$$v_x + iv_y = \frac{\partial z}{\partial t} = \frac{\frac{1}{2}}{z + (z^2 - t)^{\frac{1}{2}}},$$

and the density and electric field are

$$\rho = \frac{\rho_0}{|\partial z/\partial z_0|^2} = \frac{\rho_0}{\left|1 - t\left[z + (z^2 - t)^{\frac{1}{2}}\right]^{-2}\right|^2}$$

and

$$E = -\mathrm{Re}\frac{\partial f}{\partial t} = \frac{1}{8}\mathrm{Re}\left[1 + z(z^2 - t)^{-\frac{1}{2}}\right].$$

Although this is an elegant formalism, the resulting expressions for, say, $B_x(x, y, t)$ are rather messy. In this simplest of a whole family of solutions, the individual plasma elements converge on the x-axis and form a current sheet of length $2\sqrt{t}$. As the magnetic field collapses, the sheet grows in length and the magnetic dissipation increases. The end of the sheet moves with speed $\frac{1}{2}/\sqrt{t}$: it swallows up half of the magnetic flux and causes the remainder to pile up in a region of reversed current.

The above particular solution can be generalised in several ways. First of all, the above analysis assumes an initial simple X-point field

$$f_0 = \tfrac{1}{2}z_0^2,$$

so instead different initial functions can be chosen, including those of one or two line current sources, two dipole sources, flux emergence, or four line currents. Secondly, the effect of initial velocity profiles that differ from

$$v_0 = \frac{1}{4\sqrt{2f}}$$

can be considered. Thirdly, in solving the basic equation (2.60) above, we assumed $\chi \equiv 0$, so that the flow is not accelerated. Another generalisation is, therefore, to consider accelerated flows with χ either constant or a given function of time. Fourthly, we assumed above that the magnetic potential (Φ_m) is frozen to the plasma. If this assumption is relaxed, it is more difficult to make analytical progress and in general a numerical approach is required. In principle, one may consider any topologically accessible set of potential solutions to the basic equations

$$\mathbf{j}_0 = \mathbf{0}, \tag{2.62}$$

$$\frac{\partial \mathbf{B}_0}{\partial t} = \nabla \times (\mathbf{v}_0 \times \mathbf{B}_0), \tag{2.63}$$

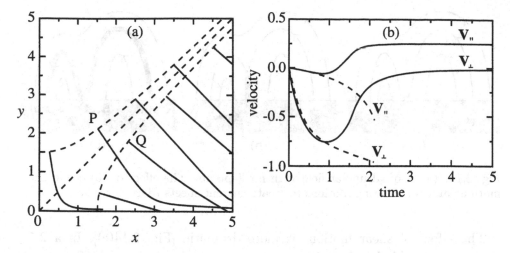

Fig. 2.13. Self-consistent, quasi-static formation of a current sheet at an X-point, when the current sheet has Y-points at its ends, its length is $2L = 2t$, and there is no reconnection. (a) Paths of plasma elements. (b) Velocity components of plasma elements initially at P (solid curves) and Q (dashed curves).

and

$$\frac{d\mathbf{v}_0}{dt} \cdot \mathbf{B}_0 = 0 \tag{2.64}$$

for the magnetic field $\mathbf{B}_0$. Then the flow speed $(v_\perp)$ normal to the magnetic field is determined explicitly by Eq. (2.63) and the difficult step is to solve the nonlinear equation (2.64) for the flow speed $(v_\parallel)$ parallel to the magnetic field. A general method for doing so has been developed and has been applied to the particular case of a growing current sheet of the form of Eq. (2.52) with Y-points at the ends (Fig. 2.13).

2.6 Creation of Current Sheets along Separatrices by Shearing

2.6.1 Two-Dimensional Fields without X-Points

Current sheets may also be produced in fields without X-points in the plasma volume (Low and Wolfson, 1988; Vekstein et al., 1990; Amari and Aly, 1990; Kliem and Seehafer, 1991). Consider, for example, a quadrupolar field which has a separatrix that intersects a surface (such as the solar photosphere) at a so-called *bald patch* (Titov et al., 1993), as shown in Fig. 2.14(a). The effect of converging motions is eventually to produce an X-point on the boundary, and from that point a current sheet extends upward from the surface in order to preserve the magnetic topology as the motion is continued [Fig. 2.14(b)].

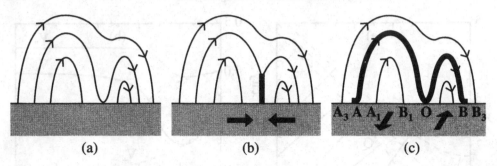

Fig. 2.14. (a) A quadrupolar field with no X-points. The effect of (b) converging motions and (c) shearing motions to create current sheets (dark curves).

The effect of shear motions is more dramatic [Fig. 2.14(c)]. In a 2.5-dimensional field the Cartesian components are, in terms of the flux function (A),

$$(B_x, B_y, B_z) = \left[\frac{\partial A}{\partial y}, -\frac{\partial A}{\partial x}, B_z(A)\right], \tag{2.65}$$

and the force-free equation $\mathbf{j} \times \mathbf{B} = \mathbf{0}$ reduces to the Grad–Shafranov equation

$$\nabla^2 A + B_z \frac{dB_z}{dA} = 0. \tag{2.66}$$

Starting with a potential field [Fig. 2.14(a)], suppose we impose a smooth perturbation $[\xi_z(x)]$ of the footpoints while keeping B_y a smooth fixed function at the base $y = 0$. The equation for a field line is

$$\frac{dz}{B_z} = \frac{ds}{B_p}, \tag{2.67}$$

where

$$\mathbf{B}_p = \left(\frac{\partial A}{\partial y}, -\frac{\partial A}{\partial x}\right) \tag{2.68}$$

is the poloidal field in the xy-plane, and so integrating Eq. (2.67) gives

$$B_z(A) = \frac{d(A)}{\overline{V}(A)}, \tag{2.69}$$

where

$$d(A) = [\xi_z] \tag{2.70}$$

is the difference in footpoint displacement between the ends of a field line

(a) 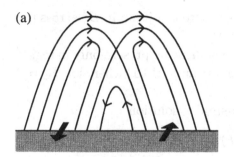   (b)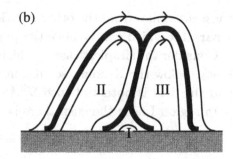

Fig. 2.15. (a) A quadrupolar field with an X-point. (b) The effect of shearing motions to create current sheets (dark curves).

and

$$\bar{V}(A) = \int \frac{ds}{B_p} \qquad (2.71)$$

is the differential flux volume of the poloidal field.

Thus, $B_z(A)$ is constant all along a given field line, but it may have quite a different value on the two field lines A_3B_3 and A_1B_1 that lie just above and just below the separatrix AOB, since the differences in the footpoint displacements are in general quite different. The whole separatrix therefore becomes a current sheet [Fig. 2.14(c)], and it touches the surface at the points A, O, and B.

2.6.2 Two-Dimensional Fields with X-Points

An X-point in a potential field is structurally unstable. Under the action of ideal converging motions it may split into a pair of Y-points joined by a current sheet. In contrast, shearing motions instead produce a pair of cusp points joined by a current sheet, together with current sheets all along the *separatrices*, namely, the field lines which link to the X-point (Zwingmann et al., 1985; Low and Wolfson, 1988; Vekstein et al., 1990; Vekstein and Priest, 1992; Platt and Neukirch, 1994), as indicated in Fig. 2.15.

At first sight, it may be expected that, in response to shearing, the X-point would just close up slightly, but in general this cannot remain in equilibrium since B_p tends to zero as one approaches the X-point from any direction, so the separatrix cannot support a jump in magnetic pressure across it associated with the jump in B_z. The answer to this problem is that the X-point becomes instead a current sheet with cusp-points at its ends (Vekstein and Priest, 1992, 1993). The separatrices connected to the cusp-point have the property that B_p tends to zero from one side and to constant values from the other two sides, so there is a jump in B_p^2 across the separatrix which can balance the jump in B_z^2.

Consider the simplest case in which there is shearing present only in region I, say, below the X-point, so that in regions II and III to either side $B_z = 0$ and the field is potential with $\nabla^2 A = 0$.

In region I near the cusp there is a self-similar solution

$$A = r^a f(\xi), \tag{2.72}$$

where

$$\xi = \frac{\theta}{r^b}, \tag{2.73}$$

so the separatrix ($A = 0$, say) is $\xi = 1$; in other words, it is not a straight line but a curve $\theta = r^b$, where $b > 0$. Then the field components are

$$B_r = \frac{1}{r}\frac{\partial A}{\partial \theta} = r^{a-1-b} f'(\xi),$$

$$B_\theta = -\frac{\partial A}{\partial r} = -ar^{a-1} f(\xi) + br^{a-1} f'(\xi)\xi.$$

The equilibrium equation

$$\nabla^2 A = -B_z \frac{dB_z}{dA}$$

must have the right-hand side of the form $-\epsilon A^{-n}$, where substitution of Eq. (2.72) gives

$$n = \frac{2b + 2 - a}{a}, \tag{2.74}$$

and to lowest order the function $f(\xi)$ is given by

$$f'' = -\epsilon f^{-n}. \tag{2.75}$$

In region II the field is potential and an appropriate form for A is

$$A = B_0 r \sin \theta + B_1 r^{K_1} \sin K_1(\theta - \pi), \tag{2.76}$$

where B_1 and K_1 are constants. This has $A = 0$ on $\theta = \pi$, the vertical arm of the separatrix, as required. As far as region II is concerned, the curved part of the separatrix is given by

$$\theta = \frac{B_1}{B_0} r^{K_1 - 1} \sin K_1 \pi, \tag{2.77}$$

so, by comparing with the form $\theta = r^b$ in region I, we see that

$$K_1 = 1 + b. \tag{2.78}$$

Finally, magnetic pressure balance across the separatrix dividing regions I and II gives

$$B_{z0}^2 + cr^{2(a-1-b)} = B_0^2 + 2KB_1B_0r^b \cos K_1\pi.$$

In order to match the variations in r across the separatrix, we need

$$a = 1 + \frac{3b}{2}. \tag{2.79}$$

Thus, our cusp solutions have one free parameter (namely, b) whose value can be determined by the global equilibrium solution. The parameters K_1 and a are given in terms of it by Eqs. (2.78) and (2.79), while the current parameter (n) follows from Eq. (2.74) as

$$n = \frac{2+b}{2+3b}.$$

2.7 Braiding by Random Footpoint Motions

Current sheets can be formed when a magnetostatic equilibrium ceases to exist at all, a situation known as *non-equilibrium*. If the footpoints of a simple bipolar magnetic field move slowly, a low-β plasma (such as the solar corona above the photosphere) responds by establishing a series of force-free configurations. In general, however, for a complex magnetic field which contains topologically distinct flux systems, a smooth transition from one equilibrium state to the next is not always possible. Parker (1972) and Syrovatsky (1978) have suggested that, as the footpoints of such a magnetic field move, the configuration cannot adjust to a new force-free equilibrium and current sheets are formed instead. These current sheets are themselves not in equilibrium, since they allow a rapid reconnection at some fraction of the Alfvén speed and the magnetic configuration reduces to the state of lowest potential energy. Parker referred to such a process as *topological dissipation*. Continual footpoint motion means that the field is all the time responding by reconnecting and converting magnetic energy into heat in the process. This may be the means whereby the solar corona is heated, especially inside active regions (see §11.3), and it may also be important for solar eruptions (§11.1) and magnetospheric substorms (§10.5).

Parker (1979, 1990) has pointed out that, if a series of flux tubes are closely packed together and each is twisted in the same direction, then current sheets will form at the boundaries of the tubes [Fig. 2.16(b)]. He considered the

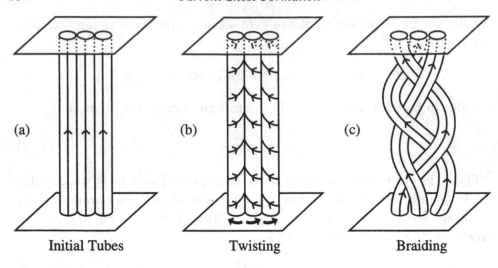

(a)	(b)	(c)
Initial Tubes	Twisting	Braiding

Fig. 2.16. The effect on (a) an initial field of (b) twisting and (c) braiding motions.

magnetic field

$$B_x = \frac{\partial A}{\partial y}, \quad B_y = -\frac{\partial A}{\partial x}, \quad B_z = \text{constant}$$

in equilibrium with the pressure $p = (k_x^2 + k_y^2)A^2/(2\mu)$ such that

$$A = K \sin k_x x \sin k_y y. \tag{2.80}$$

This represents a series of twisted flux tubes in rectangular cells with adjacent cells having opposite twist and the basic cell being $0 < x < \pi/k_x$, $0 < y < \pi/k_y$. Next, he demonstrated that, if all the cells have the same sense of twist, they are not in equilibrium, because they press together more strongly at the middle of each side than at the vertices and so squeeze out plasma from between the cells. Reconnection then leads to the configuration shown in Fig. 2.17, where closed field lines are created that circulate around the vertices of the initial cells with the opposite sense to the original circulation. One could imagine a pressure distribution $p(x,y)$ establishing equilibrium and preventing the reconnection by pumping plasma between the flux tubes until they are surrounded entirely by field-free regions and therefore isolated from one another, but this is extremely unlikely, especially in a low-β plasma such as the solar corona. Instead, nonequilibrium and reconnection will occur wherever neighbouring tubes have the same sense of twist. Even if a rectangular array such as Eq. (2.80) were formed with neighbouring tubes having opposite twist, they would tend to slip into a lower-energy hexagonal state of close packing, in which most of the tubes are squashed against at

Fig. 2.17. A section through a set of closely packed tubes with the same sense of twist after reconnection has set in (Parker, 1983).

least one neighbour that has the same sense of twist (Parker, 1983). Thus, the torsional energy of a set of close-packed twisted flux tubes will be rapidly dissipated until, according to Parker, a final state is reached in which the initial tubes have coalesced to give just two large tubes of opposite twist, which may live happily ever after in static equilibrium. Another possibility is that, at least on an intermediate time-scale, magnetic helicity is conserved in the reconnection process and so the tubes relax to an equilibrium that contains the same magnetic helicity (Taylor, 1974; Heyvaerts and Priest, 1983; Berger and Field, 1984; Browning et al., 1986).

Secondly, Parker (1972, 1979) has argued persuasively that, if instead the footpoints are braided around one another, a smooth solution exists when the footpoint motions are small, but finite-amplitude displacements lead to non-equilibrium and current-sheet formation at the boundaries of the braids (see also Tsinganos, 1982; Rosner and Knobloch, 1982; Zweibel and Li, 1987). In the original topological dissipation paper, Parker (1972) suggested that, if the pattern of small-scale variations is not uniform along a large-scale field, then the field cannot be in magnetostatic equilibrium. In other words, equilibrium exists only if the field variations consist of a simple twist extending from one footpoint to another. More complex topologies, such as braided flux tubes with several field lines wrapped around each other, are not in equilibrium.

Sakurai and Levine (1981) have since established that a *small* motion of the footpoints *does* determine the perturbed field uniquely. All the variations occur in the boundary layers near $z = \pm L$ with a uniform field elsewhere, and so the equilibrium is not of the form shown in Fig. 2.16(c) (see also Arendt and Schindler, 1988). Van Ballegooijen (1985) has shown that small random motions produce a cascade of magnetic energy to small scales with a mean-square current that grows exponentially in time and has magnetic and current spectra spreading to smaller and smaller scales. Nevertheless, the detailed consequences of Parker's idea remain to be worked out when finite-amplitude displacements at the boundary are imposed. A start has been made by Parker (1989 and subsequent papers in a series), using the optical and fluid dynamic analogies, and by Tsinganos et al. (1984). Mikić et al. (1990) have conducted a three-dimensional incompressible ideal MHD experiment with a 64^3 mesh in which they braid magnetic field lines with random footpoint motions. The field evolves through a series of smooth equilibria with a transfer to small scales and an exponential growth of filamentary currents towards curved current sheets. (See also Galsgaard and Nordlund, 1996, as described in Section 11.3.2).

Berger (1986, 1988, 1991a, b, 1993) has developed the ideas and techniques for understanding the braiding of magnetic field lines by means of second-order and third-order invariants. He has set up expressions for the linkage and magnetic helicity of field lines that have end-points on a surface such as the photosphere (§8.5.3). The resulting magnetic energy can be written in terms of linkage (when the field is force-free) or in terms of crossing numbers. He also investigated the twist acquired by solar coronal flux tubes as a consequence of random footpoint motions: after a day of such motions the typical twists are a few turns and the resulting current densities are extremely small, so that only in singular current sheets is there sufficient dissipation to heat the corona (§11.3).

On a different tack, Bobrova and Syrovatsky (1979) have shown that one may also form sheets by moving the boundary of a region. They considered the linear force-free field (Fig. 2.18)

$$B_{0x} = \cos \alpha z, \quad B_{0y} = \sin \alpha z, \tag{2.81}$$

and supposed that the top ($z = L$) of the region is moved up or down by a small distance $\xi_L(t)$, while the bottom ($z = 0$) remains at rest. The resulting equilibrium field ($\mathbf{B}_0 + \mathbf{B}_1$) satisfies

$$\mathbf{j}_1 \times \mathbf{B}_0 + \mathbf{j}_0 \times \mathbf{B}_1 = \mathbf{0}, \tag{2.82}$$

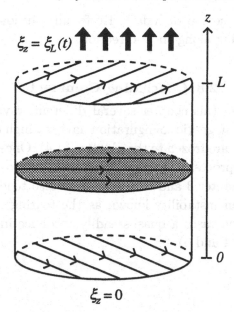

Fig. 2.18. A sheared cylindrical region of height L of a force-free magnetic field, subject to a perturbation $(\xi_z = \xi_L)$ on the top boundary.

where

$$\mathbf{j}_1 = \nabla_1 \times \mathbf{B}_1/\mu, \quad \mathbf{B}_1 = \nabla \times (\boldsymbol{\xi} \times \mathbf{B}_0),$$

the latter equation being a time-integral of the ideal induction equation since $\mathbf{v}_1 = \partial \boldsymbol{\xi}/\partial t$. The resulting solution is of the form

$$\xi_z = e^{(k_x x + k_y y)} \xi_L \frac{k_{\|}(L)}{k_{\|}(z)} f(z), \tag{2.83}$$

where

$$k_{\|} = k_x \cos \alpha z + k_y \sin \alpha z$$

and

$$f(z) = \begin{cases} \dfrac{\sin \sqrt{(\alpha^2 - k^2)z}}{\sin \sqrt{(\alpha^2 - k^2)L}}, & k < \alpha. \\[2ex] \dfrac{z}{L}, & k = \alpha. \end{cases}$$

Thus, the perturbation ξ_z and the resulting magnetic field have a singularity where $k_{\|}(z) = 0$, which is just the condition $(\mathbf{k} \cdot \mathbf{B} = 0)$ for a resonant surface. Parker (1990) has added line-tying and imposed a pressure perturbation in a small circle on the top boundary. He found solutions that represent a current sheet spiralling along the z-direction.

2.8 Concluding Comment

We have described in this chapter several different ways in which current sheets may form in a magnetic configuration having a high magnetic Reynolds number, so that they are frozen to the plasma (§1.4). Once they have formed, or indeed are in the process of forming, they will tend to diffuse away in the way described in Chapter 3 and they will often also tend to become unstable to a reconnection instability known as the tearing mode (Chapter 6). Furthermore, in some cases a quasi-steadily reconnecting state will be established (Chapters 4 and 5).

3

Magnetic Annihilation

We shall use the term *magnetic annihilation* to refer to the bringing in and cancelling of oppositely directed straight field lines in a one-dimensional current sheet. It is therefore an important ingredient of magnetic reconnection and is of interest in its own right. Section 3.1 discusses the two basic physical processes of magnetic diffusion and magnetic advection that occur in a current sheet and are described by the induction equation. Section 3.2 presents the classical solution for steady magnetic annihilation by a stagnation-point flow, with a balance between diffusion and advection. Then, in Section 3.3, more general steady and time-dependent solutions are discussed, where diffusion and advection are not necessarily in balance. Section 3.4 presents some more details of the complex time-dependent behaviour of one-dimensional flows in a current sheet, and, finally, solutions for reconnective annihilation are described (§3.5).

3.1 The Induction Equation

In Section 1.2.2 we introduced the induction equation

$$\frac{\partial \mathbf{B}}{\partial t} = \nabla \times (\mathbf{v} \times \mathbf{B}) + \eta \nabla^2 \mathbf{B}, \tag{3.1}$$

which describes for a given plasma velocity ($\mathbf{v}$) how the magnetic field ($\mathbf{B}$) changes in time as a result of advection of the magnetic field with the plasma (the first term on the right) and diffusion through the plasma (the second term). The ratio of these terms for a length-scale l_0 and velocity-scale V_0 is the magnetic Reynolds number

$$R_m = \frac{l_0 V_0}{\eta}, \tag{3.2}$$

and the resulting electric current and electric field are given by

$$\mathbf{j} = \nabla \times \mathbf{B}/\mu, \quad \mathbf{E} = -\mathbf{v} \times \mathbf{B} + \eta \nabla \times \mathbf{B}. \tag{3.3}$$

3.1.1 Diffusion

If $R_m \ll 1$, the first term on the right of Eq. (3.1) is negligible and we have a diffusion equation, Eq. (1.15), so that magnetic variations on a length-scale l_0 are smoothed out on a time-scale $\tau_d = l_0^2/\eta$ and with a speed

$$v_d = \frac{\eta}{l_0}. \tag{3.4}$$

Consider, for example, a one-dimensional magnetic field $[B(x,t)\hat{\mathbf{y}}]$ satisfying

$$\frac{\partial B}{\partial t} = \eta \frac{\partial^2 B}{\partial x^2}, \tag{3.5}$$

whose solution may be written in general as

$$B(x,t) = \int G(x - x', t) B(x', 0) \, dx'$$

in terms of some initial magnetic profile $[B(x',0)]$ and the Green's function

$$G(x - x', t) = \frac{1}{(4\pi\eta t)^{1/2}} \exp\left[-\frac{(x - x')^2}{4\eta t}\right].$$

As a particular example, suppose initially we have an infinitesimally thin current sheet with a piecewise-constant magnetic field ($B = B_0$ for $x > 0$ and $B = -B_0$ for $x < 0$), as shown in Fig. 3.1. Physically, what do we expect to happen? Since Eq. (3.5) has the form of a heat conduction equation and we know that heat tends to flow from a hot region to a cool one and to smooth out a temperature gradient, we expect the same diffusive process to occur for our magnetic field. Thus, the steep magnetic gradient at $x = 0$ spreads out, as shown in Fig. 3.1.

Mathematically, the solution of Eq. (3.5) for our example is given in terms of the error function $[\text{erf}\,(\zeta)]$ as

$$B(x,t) = \frac{2B_0}{\sqrt{\pi}} \, \text{erf}\left(\frac{x}{\sqrt{4\eta t}}\right) = \frac{2B_0}{\sqrt{\pi}} \int_0^{x/\sqrt{4\eta t}} e^{-u^2} \, du. \tag{3.6}$$

Solution (3.6) has the form shown in Fig. 3.1 and may be verified *a posteriori* by substituting it back into Eq. (3.5). The magnetic field diffuses away in time at a speed η/l, where the width (l) of the sheet is of the order of $(\eta t)^{1/2}$ and so increases in time. The resulting magnetic field strength at a

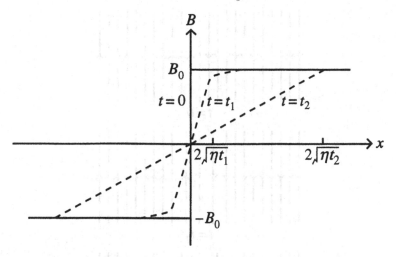

Fig. 3.1. The magnetic field (B) as a function of distance (x) in a one-dimensional sheet that is diffusing from one of zero thickness initially, for times $t = 0, t_1, t_2$, where $0 < t_1 < t_2$.

fixed position decreases with time, so the field is said, in a rather dramatic fashion, to be annihilated. The resulting field lines at subsequent times are sketched in Fig. 3.2, from which it can be seen that they are diffusing through the plasma and cancelling at $x = 0$. The total magnetic flux $(\int_{-\infty}^{\infty} B \, dx)$ remains constant (namely zero) and the total current

$$J = \int_{-\infty}^{\infty} j \, dx = \frac{1}{\mu} \int_{-\infty}^{\infty} \frac{\partial B}{\partial x} dx = \frac{2B_0}{\mu}$$

is conserved, since it simply spreads out in space. However, the magnetic energy $[\int_{-\infty}^{\infty} B^2/(2\mu) dx]$ decreases in time at a rate

$$\frac{\partial}{\partial t} \int_{-\infty}^{\infty} \frac{B^2}{2\mu} dx = \int_{-\infty}^{\infty} \frac{B}{\mu} \frac{\partial B}{\partial t} dx.$$

Substituting for $\partial B/\partial t$ from Eq. (3.5) and integrating by parts, we find that this becomes

$$\int_{-\infty}^{\infty} \frac{B\eta}{\mu} \frac{\partial^2 B}{\partial x^2} dx = \frac{1}{\mu^2 \sigma} \left\{ \left[B \frac{\partial B}{\partial x} \right]_{-\infty}^{\infty} - \int_{-\infty}^{\infty} \left(\frac{\partial B}{\partial x} \right)^2 dx \right\}.$$

Since $\partial B/\partial x$ remains equal to zero at infinity, the first term on the right vanishes, and, since the electric current is $j = \mu^{-1} \partial B/\partial x$, we finally have

$$\frac{\partial}{\partial t} \int_{-\infty}^{\infty} \frac{B^2}{2\mu} dx = - \int \frac{j^2}{\sigma} dx.$$

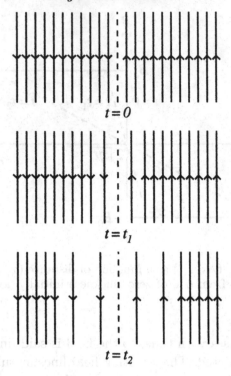

Fig. 3.2. Field lines at three different times $(0 < t_1 < t_2)$ in a diffusing current sheet.

In other words, not surprisingly, magnetic energy is converted entirely into heat by ohmic dissipation $(j^2/\sigma$ per unit volume).

3.1.2 Advection

If $R_m \gg 1$, the induction equation, Eq. (3.1), becomes

$$\frac{\partial \mathbf{B}}{\partial t} = \nabla \times (\mathbf{v} \times \mathbf{B}), \qquad (3.7)$$

and so the magnetic field lines are frozen into the plasma. Plasma can move freely along field lines, but in motion perpendicular to them they are dragged with the plasma or vice versa (§1.4).

As an example (Fig. 3.3), consider the effect of a flow

$$v_x = -\frac{V_0 x}{a}, \quad v_y = \frac{V_0 y}{a}$$

on a field that is initially

$$\mathbf{B} = B_0 \cos\frac{x}{a} \, \hat{\mathbf{y}} \quad \text{at} \quad t = 0$$

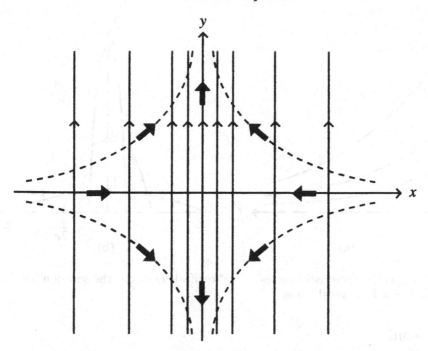

Fig. 3.3. A sketch of the magnetic lines (thin-headed arrows) and streamlines (thick-headed arrows) at $t = 0$ for $|x| < \pi a/2$ for the example in Section 3.1.2.

between $x = -\frac{1}{2}\pi a$ and $\frac{1}{2}\pi a$. The equations of the streamlines (namely, $xy = $ constant) are obtained from $dy/dx = v_y/v_x = -y/x$. These are rectangular hyperbolae (Fig. 3.3) with inflow along the x-axis and outflow along the y-axis when $V_0 > 0$.

The velocity field corresponds to a hydrodynamic *stagnation-point flow*. The effect of this flow on the magnetic field is to carry the field lines inwards from the sides and accumulate them near $x = 0$, increasing the field strength there. Since the component (v_x) of the velocity perpendicular to the field lines is constant along a particular field line ($x = $ constant), the field lines are not distorted but remain straight as they come in.

Now, the y-component of the induction equation, Eq. (3.7), when $R_m \gg 1$ is $\partial B/\partial t = -\partial(v_x B)/\partial x$ or

$$\frac{\partial B}{\partial t} - \frac{V_0 x}{a}\frac{\partial B}{\partial x} = \frac{V_0 B}{a}, \tag{3.8}$$

and this determines $B(x,t)$. In order to solve such a partial differential equation, we consider characteristic curves in the xt-plane, which are defined to be such that

$$\frac{dx}{dt} = -\frac{V_0 x}{a}, \tag{3.9}$$

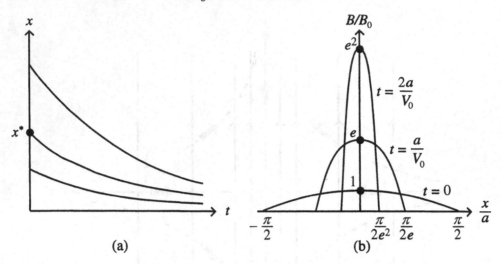

Fig. 3.4. (a) Characteristic curves $x = x^* \exp(-V_0 t/a)$; (b) the solution for B as a function of x for several times.

with solution

$$x = x^* e^{-V_0 t/a}, \tag{3.10}$$

where $x = x^*$, say, at $t = 0$. We wish to determine $B(x,t)$ at every point of the xt-plane, and the elegance of considering characteristic curves, $x = x(t)$ given by Eq. (3.10) [Fig. 3.4(a)], is that on such curves $B[x(t), t]$ has the derivative

$$\frac{dB}{dt} = \frac{\partial B}{\partial t} + \frac{dx}{dt}\frac{\partial B}{\partial x} = \frac{\partial B}{\partial t} - \frac{V_0 x}{a}\frac{\partial B}{\partial x},$$

by Eq. (3.9), or, from Eq. (3.8), $dB/dt = V_0 B/a$. In other words, on the characteristic curves we have a simple ordinary differential equation to solve in place of Eq. (3.8): the solution is $B = \text{constant } e^{V_0 t/a}$ or, since $x = x^*$ and $B = B_0 \cos(x^*/a)$ at $t = 0$, we have

$$B(x,t) = B_0 \cos(x^*/a)\, e^{V_0 t/a}.$$

However, in this solution x^* is a constant which we have introduced for convenience and which was not present in the initial statement of the problem, so we should eliminate it by using Eq. (3.10), with the final result

$$B(x,t) = B_0 \cos\left(\frac{x}{a}e^{V_0 t/a}\right) e^{V_0 t/a}.$$

This solution is plotted in Fig. 3.4(b) against x for several times. It can be seen that the field does indeed, as expected, concentrate near $x = 0$ as time proceeds. The field strength at the origin is $B(0,t) = B_0 e^{V_0 t/a}$, which grows exponentially in time (or decreases if the flow is reversed by taking $V_0 < 0$).

3.2 Stagnation-Point Flow Model

In the previous section we described how a current sheet naturally tends to diffuse away and convert magnetic energy into heat ohmically. The field lines diffuse inwards through the plasma and cancel, so that the region of diffused field spreads out (Fig. 3.2). Therefore, a steady state may be produced if magnetic flux is carried in (with the plasma) at the same rate as it is trying to diffuse. If the resistivity is small (in the sense that the global magnetic Reynolds number is much larger than unity), then an extremely small length-scale l (and therefore large magnetic gradient ∇B and current j) will be created. Furthermore, although the magnetic field may be destroyed by cancellation as it comes in, the plasma itself cannot be destroyed and has to flow out sideways, as illustrated in the following model (Parker, 1973; Sonnerup and Priest, 1975).

The standard equations for two-dimensional steady-state incompressible flow that we use for this model are

$$\mathbf{E} + \mathbf{v} \times \mathbf{B} = \eta \nabla \times \mathbf{B}, \tag{3.11}$$

$$\rho(\mathbf{v} \cdot \nabla)\mathbf{v} = -\nabla \left(p + \frac{B^2}{2\mu} \right) + (\mathbf{B} \cdot \nabla)\frac{\mathbf{B}}{\mu}, \tag{3.12}$$

where

$$\nabla \cdot \mathbf{B} = 0, \quad \nabla \cdot \mathbf{v} = 0 \tag{3.13}$$

and the components v_x, v_y, B_x, B_y depend on x and y alone. Faraday's Law ($\nabla \times \mathbf{E} = \mathbf{0}$) implies that $\partial E/\partial y = \partial E/\partial x = 0$, so that $\mathbf{E} = E\hat{\mathbf{z}}$ is uniform. Consider a steady-state flow

$$v_x = -\frac{V_0 x}{a}, \quad v_y = \frac{V_0 y}{a}, \tag{3.14}$$

for which $\nabla \cdot \mathbf{v} = 0$. The steady-state continuity equation $(\mathbf{v} \cdot \nabla)\rho + \rho(\nabla \cdot \mathbf{v}) = 0$ then reduces to $(\mathbf{v} \cdot \nabla)\rho = 0$, which implies that the density (ρ) is uniform if it is constant at the inflowing sides. The flow vanishes at the origin and therefore represents an incompressible, stagnation-point flow.

Suppose now that the magnetic field lines are straight with $\mathbf{B} = B(x)\hat{\mathbf{y}}$ and that they reverse sign at $x = 0$. Then in Ohm's Law, Eq. (3.11), both

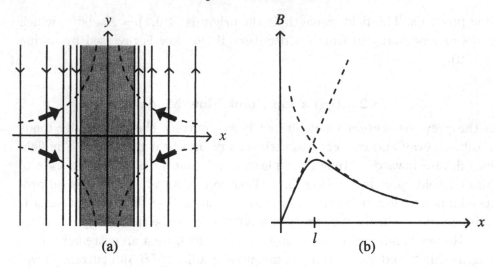

Fig. 3.5. (a) Stagnation-point flow creating a steady current sheet (shaded). (b) Magnetic field profile, with small-x and large-x approximations shown as dashed curves.

$\mathbf{v} \times \mathbf{B}$, $\nabla \times \mathbf{B}$, and therefore $\mathbf{E}$ are directed purely in the z-direction. Indeed, Eq. (3.11) is essentially an integral of the induction equation (3.1) and in the present case it reduces to

$$E - \frac{V_0 x}{a} B = \eta \frac{dB}{dx}. \tag{3.15}$$

Now, when x is sufficiently large, the right-hand side of Eq. (3.15) is negligible and $B \approx (Ea)/(V_0 x)$, whereas when x is very small the second term is negligible and $B \approx Ex/\eta$. These approximate solutions are indicated by dashed curves in Fig. 3.5(b). When x is large, the magnetic field lines are frozen to the plasma and are carried inwards, whereas when x is small the magnetic field diffuses through the plasma. The division between these two extremes, that is, the half-width of the resulting current sheet, occurs (by equating the two approximations for B) when $x = (a\eta/V_0)^{1/2}$. The full solution of Eq. (3.15) is shown in Fig. 3.5(b). It may be written formally in terms of the confluent hypergeometric Kummer function (M^*) as $B = \text{const} \times \exp(-X^2) M^*(\frac{1}{2}, \frac{3}{2}, X^2)$, where $X^2 = x^2 V_0/(2\eta a)$. Alternatively, simply by finding the integrating factor $\left[\exp(V_0 x^2/(2\eta a)\right]$ of the first-order linear equation (3.15) and integrating in the standard manner, the solution may be written in terms of the *Dawson integral function (daw)* as

$$B = \frac{2Ea}{V_0 l} \text{ daw} \left(\frac{x}{l}\right),$$

where $l^2 = 2\eta a/V_0$ and $\mathrm{daw}(X) = \exp(-X^2) \int_0^X \exp(t^2)dt$. A description of the properties of the Dawson integral function can be found in Spanier and Oldham (1987). For small argument $\mathrm{daw}(x/l) \approx x/l$, while for large argument $\mathrm{daw}(x/l) \approx l/(2x)$, as shown in Fig. 3.5(b).

The steady-state equation of motion can be shown to be satisfied also, and so the above solution represents an exact solution of the nonlinear MHD equations – one of the very few that exist. For straight magnetic field lines the equation of motion reduces to

$$\rho(\mathbf{v} \cdot \nabla)\mathbf{v} = -\nabla\left(p + \frac{B^2}{2\mu}\right)$$

or

$$\rho\left[-\mathbf{v} \times (\nabla \times \mathbf{v}) + \nabla\left(\tfrac{1}{2}v^2\right)\right] = -\nabla\left(p + \frac{B^2}{2\mu}\right).$$

However, the flow [Eq. (3.14)] has zero vorticity ($\nabla \times \mathbf{v} = 0$) and ρ is constant, so the first term in the above equation vanishes while the other terms imply that

$$\nabla\left(p + \frac{B^2}{2\mu} + \tfrac{1}{2}\rho v^2\right) = 0,$$

and thus

$$p = p_s - \tfrac{1}{2}\rho v^2 - \frac{B^2}{2\mu}, \tag{3.16}$$

where p_s is the pressure at the stagnation point situated at the origin.

The requirement that the pressure always be positive imposes a limit on how fast the field can be annihilated (Priest, 1996; Litvinenko et al., 1996) or on the plasma beta (Jardine et al., 1993). The pressure minimum and the maximum of B occur on the x-axis at the location $x = l$, which is the edge of the current sheet, and thus the requirement that $p > 0$ everywhere can be written as

$$M_e < \frac{1.7(1 + \beta_e)}{R_{me}},$$

where $M_e = E/(v_{Ae}B_e) = v_e/v_{Ae}$ is the Alfvén Mach number, $\beta_e = 2\mu p_e/B_e^2$, $R_{me} = v_{Ae}L_e/\eta$, and the subscript e denotes values at $x = L_e$, $y = 0$.

The above analysis has been generalised to include a three-dimensional stagnation-point flow with a spatially rotating field [$\mathbf{B}(x)$] by taking

$$v_x = -\frac{V_1}{a}x, \quad v_y = \frac{V_2}{a}y, \quad v_z = \frac{V_1 - V_2}{a}z,$$

$$B_x = 0, \quad B_y = B_y(x), \quad B_z = B_z(x).$$

In this case, the equation of motion again gives Eq. (3.16), but the y- and z-components of Eq. (3.1) reduce for a steady state to

$$a\eta\frac{d^2B_y}{dx^2} + V_1x\frac{dB_y}{dx} + V_2B_y = 0,$$

$$a\eta\frac{d^2B_z}{dx^2} + V_1x\frac{dB_z}{dx} + (V_1 - V_2)B_z = 0.$$

Sonnerup and Priest (1975) demonstrated the existence of solutions which join an arbitrary magnetic field vector at large positive values of x to another arbitrary one at large negative values of x. It is thus possible to bring together fields which are not just antiparallel but may be inclined at any angle. As the fields approach, their directions rotate and their magnitudes change, giving (in general) a magnetic field of non-zero strength at the centre of the current sheet and annihilating part of the magnetic flux in the process.

The above model, like much work on current sheets, deals only with the incompressible situation. The assumption of incompressibility is, of course, made for simplicity, in the hope that the essence of the annihilation process does not depend on compressible effects. But it would be valuable to extend many current sheet models to include such effects. The other limitation of the above exact solution is that it is relevant only when the flow speed into the current sheet is much less than the Alfvén speed, so that the length (L) of the sheet far exceeds its width (l). Furthermore, the solution fails near the ends of the sheet, where there are significant transverse components of magnetic field (in the x-direction). Relaxing these conditions, as we shall see in constructing reconnection solutions in Chapters 4 and 5, is a far from trivial undertaking.

3.3 More General Stagnation-Point Flow Solutions

3.3.1 Steady State

Consider more generally a two-dimensional incompressible flow with velocity components

$$v_x = \frac{\partial\psi}{\partial y}, \quad v_y = -\frac{\partial\psi}{\partial x}$$

written in terms of a *stream function* (ψ) so that $\nabla \cdot \mathbf{v} = 0$ automatically. The effect of this flow on a unidirectional field $B(x)\hat{\mathbf{y}}$ is determined by Ohm's Law [Eq. (3.11)], which becomes

$$E + \frac{\partial\psi}{\partial y}B = \eta\frac{dB}{dx}. \tag{3.17}$$

Since B and dB/dx are here functions of x alone and the electric field $(E\,\hat{\mathbf{z}})$ is uniform, this implies that $\partial\psi/\partial y$ is also a function of x alone, so that most generally

$$\psi = yf(x) + g(x) \tag{3.18}$$

and Eq. (3.17) becomes

$$E + fB = \eta\,\frac{dB}{dx}. \tag{3.19}$$

This equation determines the magnetic field $[B(x)]$ once $f(x)$ is known, but the function $g(x)$ has no effect on the field.

Now, if a viscous force $(\rho\nu\nabla^2\mathbf{v})$ is included in the equation of motion [Eq. (3.12)], its curl gives

$$(\mathbf{v}\cdot\nabla)\omega = \nu\,\nabla^2\omega$$

in terms of the vorticity $\omega\,\hat{\mathbf{z}} = \nabla\times\mathbf{v}$. Substituting the form of Eq. (3.18) into the forms for $\mathbf{v}$ and ω reduces this equation to

$$y\left[\nu f^{(4)} + ff''' - f'f''\right] + \left[\nu g^{(4)} + fg''' - g'f''\right] = 0. \tag{3.20}$$

Since the combinations in brackets are functions of x alone, we need separately

$$\nu f^{(4)} + ff''' - f'f'' = 0 \tag{3.21}$$

and

$$\nu g^{(4)} + fg''' - g'f'' = 0. \tag{3.22}$$

The function $f(x)$ is determined by Eq. (3.21) together with a set of four boundary conditions. Then $g(x)$ follows from Eq. (3.22): it is associated with an inflow that is skewed about the x-axis, and its effect has been studied by Besser et al. (1990) and Phan and Sonnerup (1990).

The simplest solution to Eq. (3.21) is $f(x) = x$, which is the case studied in Section 3.2. More general solutions have been considered by Gratton et al. (1988) and Jardine et al. (1992), allowing for vorticity in an inflow that carries straight, oppositely directed field lines towards a long, thin vortex-current sheet with both viscous and magnetic boundary layers. The widths of the boundary layers adjust themselves to accommodate different types and magnitudes of inflow, which may possess cellular patterns when the inflow vorticity is large enough. The effects of three-dimensional flows have been incorporated by Jardine et al. (1993). Also, Craig and Henton (1995) have discovered another solution, as described in Section 3.5.

3.3.2 Time-Dependent Stagnation-Point Flow

In general, a time-dependent flow with stream function

$$\psi = yf(x,t) + g(x,t)$$

acting on a field $B(x,t)\hat{\mathbf{y}}$ will satisfy the time-dependent equations of induction and motion, such that Eqs. (3.19), (3.21), and (3.22) generalise to

$$\frac{\partial B}{\partial t} = \frac{\partial f}{\partial x}B + f\frac{\partial B}{\partial x} + \eta\frac{\partial^2 B}{\partial x^2},$$

$$\frac{\partial^3 f}{\partial t \partial x^2} = f\frac{\partial^3 f}{\partial x^3} - \frac{\partial f}{\partial x}\frac{\partial^2 f}{\partial x^2} + \nu\frac{\partial^4 f}{\partial x^4},$$

$$\frac{\partial^3 g}{\partial t \partial x^2} = f\frac{\partial^3 g}{\partial x^3} - \frac{\partial^2 f}{\partial x^2}\frac{\partial g}{\partial x} + \nu\frac{\partial^4 g}{\partial x^4},$$

which determine the functions B, f, g, respectively (Anderson and Priest, 1993).

For simplicity, suppose $g \equiv 0$ and $f = -U(t)x/L_0$, so that the last two equations are satisfied trivially and the induction equation

$$\frac{\partial B}{\partial t} = \frac{UB}{L_0} + \frac{Ux}{L_0}\frac{\partial B}{\partial x} + \eta\frac{\partial^2 B}{\partial x^2} \qquad (3.23)$$

determines $B(x,t)$ for the given flow

$$(v_x, v_y) = -\frac{U(t)}{L_0}(x, -y). \qquad (3.24)$$

The equations $\nabla \cdot \mathbf{v} = \nabla \cdot \mathbf{B} = 0$ are satisfied trivially and the resulting pressure from the equation of motion is

$$p = \text{constant} - \frac{B^2}{2\mu} - \frac{\rho U^2}{2 L_0^2}(x^2 + y^2).$$

In order to solve Eq. (3.23) on an infinite interval, the constants L_0 and η may be scaled away and also the second term on the right may be transformed away by a change of independent variables from x and t to $X = x\bar{g}(t)$ and $\tau = \tau(t)$, where $\bar{g}$ and τ are determined for a given function $U(t)$ by

$$\frac{d\bar{g}}{dt} = U\bar{g}, \quad \frac{d\tau}{dt} = \bar{g}^2.$$

The result is that Eq. (3.23) simplifies to

$$\frac{\partial B}{\partial \tau} = \bar{f}(\tau)B + \frac{\partial^2 B}{\partial X^2}, \qquad (3.25)$$

where $\bar{f} = U/\bar{g}^2$. This may be solved by standard techniques to produce in general

$$B(X,\tau) = \frac{\bar{g}}{2\sqrt{\pi\tau}} \int_{-\infty}^{\infty} B_0(\xi) e^{-(\xi-X)^2/(4\tau)} d\xi, \qquad (3.26)$$

which is an analytical expression for the time-evolution of an initial magnetic field $B(X,0) = B_0(X)$. This solution was first obtained by Clark (1964) and it describes the effect on a unidirectional field $(B\hat{y})$ of any time-modulated stagnation-point flow of the form of Eq. (3.24).

Before describing some of the solutions of the form of Eq. (3.26), we find it instructive to consider the frozen-in solutions to the induction equation (3.23) when diffusion is negligible. Then, by the method of characteristics, we find simply

$$B(x,t) = \bar{g}(t)B_0[x\bar{g}(t)].$$

If the flow is constant with $U(t) \equiv 1$, we have $\bar{g}(t) = e^t$ and

$$B(x,t) = e^t B_0(xe^t). \qquad (3.27)$$

Suppose, for example, that the initial profile $B_0(x)$ behaves like x^{-n}, say, at large distances. Then Eq. (3.27) reduces to

$$B(x,t) \sim x^{-n}e^{(1-n)t},$$

and so the field at a particular point will decay with time if $n > 1$ but grow exponentially if $n < 1$. This striking behaviour produced by the frozen-in advection of flux may be understood as follows.

Consider what is happening to individual elements of plasma as they move along the x-axis towards $x = 0$ according to $dx/dt - -x$. This implies that the x-coordinate of any parcel of plasma is given by $x = Ke^{-t}$, where K is some constant determined by its initial position. So, the plasma element which at time t_0 lies between $x = a$ and $x = a + \delta$ will at the earlier time $t = 0$ have been lying between $x = ae^{t_0}$ and $x = (a + \delta)e^{t_0}$, as shown in Fig. 3.6. Now, if the magnetic field in the plasma element at time t_0 is B and the magnetic field at the earlier time $t = 0$ behaves like $1/x^n$, then, by conservation of magnetic flux, equating the products of the lengths of each plasma element and the magnetic field at times $t = 0$ and $t = t_0$ gives

$$B\delta = \frac{1}{(ae^{t_0})^n} \delta e^{t_0} \qquad (3.28)$$

or

$$B = \frac{1}{x^n} e^{(1-n)t}. \qquad (3.29)$$

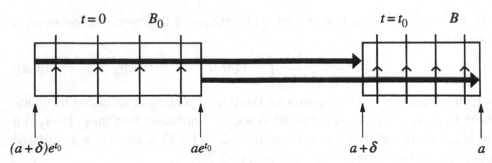

Fig. 3.6. Diagram showing a plasma element on the x-axis at times $t = 0$ and $t = t_0$.

This implies that the magnetic field will either decay, grow, or remain constant with time, depending on whether the value of n in the initial condition is greater, less than, or equal to unity.

Physically, Eq. (3.29) is showing that, although the magnetic field passing through $x = a$ is falling off as e^{-nt}, the flux element of which it is part is being compressed (the field lines are being piled together) by a factor of e^t due to the initial condition. Depending on n, these two effects will either balance or one will dominate the other. In other words, at large distances the magnetic flux is, respectively, piling up, steady, or spreading out as it is brought in by the flow, depending on whether $n < 1$, $n = 1$, or $n > 1$. In the first case, the flux between the two fixed positions is increasing because more flux is entering the region between the two positions than is leaving. In the second, the fluxes are equal, and in the third the flux is decreasing because more flux is leaving than entering.

From Eq. (3.26) the ideal asymptotic solution, Eq. (3.27), fails when $X^2 < 4\tau$ or, since $X = xe^t$ and $\tau = \frac{1}{2}(e^{2t} - 1)$,

$$|x| < \sqrt{2}(1 - e^{-2t})^{\frac{1}{2}}.$$

The right-hand side of this increases from zero towards a final value of $\sqrt{2}$ after a time of about 0.5. Thus, the effects of diffusion are localized in a diffusion layer of approximate dimensionless half-width $\sqrt{2}$ around $x = 0$. The inward plasma motion balances the outward flow of diffusing magnetic field, restricting the effects of diffusion to such a region. The diffusion region expands for approximately half a time-unit and extends to about $\sqrt{2}$ before being balanced by advection. After the diffusion layer is established, at approximately $t = 0.5$, its behaviour can be expected to be simply a response to the conditions imposed on it by the outer (ideal) solution. A time-scale

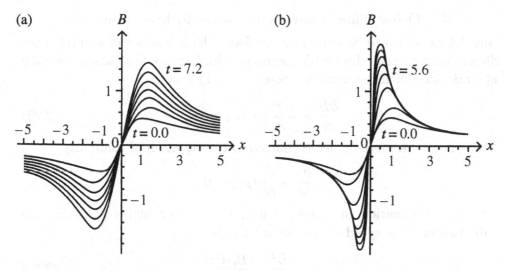

Fig. 3.7. The effect of (a) a steady flow on an initial profile $B_0(x) = x/(1 + x^{1.9})$ at times $t = 0$, 1.2, 2.4, ... ,7.2 and of (b) a linearly increasing flow on an initial profile $B_0(x) = x/(1 + x^2)$ at $t = 0$, 0.8, 1.6, ... ,5.6.

of 0.5 dimensionless units for the diffusion layer to become established corresponds to a dimensional time-scale of $0.5 L_0/V_0$.

Anderson and Priest (1993) next studied the solutions [Eq. (3.26)] with a variety of initial profiles of the form $B_0(x) = x/(1 + x^{1+n})$. First of all, the flow amplitude $U(t)$ was assumed constant (see also Clark, 1964). Near the origin, diffusion initially exceeds advection and so the field maximum moves outwards and the gradient at the origin decreases. If $n = 1$, on the one hand, a steady state is subsequently reached with the field strength increasing towards a final value. If $n < 1$ or $n > 1$, on the other hand, no steady state is reached and the field everywhere eventually grows or declines, respectively, with time [e.g., Fig. 3.7(a)]. Next, the magnitude of the magnetic diffusivity (η) was allowed to change rapidly: if η increases, the field diffuses towards a new steady state over the diffusion time, whereas, if η decreases, the field is advected towards a new steady balance. Finally, the flow amplitude $U(t)$ was assumed to increase linearly with time. When $n = 1$ [Fig. 3.7(b)] the field at large distances remains constant, whereas close to the origin after an initial diffusion the effect of advection dominates and makes the gradient increase indefinitely and the position of the maximum move inwards. Again, for $n < 1$ or $n > 1$ the field everywhere eventually grows or declines. An application of this type of solution to the dayside magnetopause has been made by Heyn and Pudovkin (1993), where they studied the response to sudden increases of solar-wind speed and magnetic diffusivity.

3.4 Other Time-Dependent Current-Sheet Solutions

Now let us return to consider in more detail the behaviour of a purely one-dimensional current sheet with magnetic field $B(x,t)\hat{y}$ and plasma velocity $v(x,t)\hat{x}$, for which the induction equation (3.1) becomes

$$\frac{\partial B}{\partial t} = -\frac{\partial}{\partial x}(vB) + \eta\frac{\partial^2 B}{\partial x^2}. \tag{3.30}$$

When flows are present, this is coupled to the equation of continuity

$$\frac{\partial \rho}{\partial t} + \frac{\partial}{\partial x}(\rho v) = 0 \tag{3.31}$$

and to the equation of motion, which reduces for highly subsonic and sub-Alfvénic flow speeds to the force balance

$$p + \frac{B^2}{2\mu} = \frac{B_e(t)^2}{2\mu}. \tag{3.32}$$

Let us assume that the edge of the neutral sheet $[x = \pm l(t)$, say] is at zero plasma pressure and at a magnetic pressure of $B_e(t)^2/(2\mu)$. In addition, there is an energy equation, but for simplicity we adopt the polytropic form

$$p = K\rho^a, \tag{3.33}$$

where K and a are constants.

3.4.1 Self-Similar Solutions

Compressive similarity solutions to Eqs. (3.31)–(3.33) have been discovered by Priest and Raadu (1975) in the form

$$B = B_e(t)\sin\frac{\pi x}{2l(t)}, \quad p = p_0(t)\cos^2\frac{\pi x}{2l(t)}, \quad v = \frac{x}{2t}, \quad \rho^a = \frac{p}{K} \tag{3.34}$$

for $t < 0$, where

$$l(t) = \pi\left(\frac{-\eta t}{2-a}\right)^{1/2},$$

$$B_e(t) = \text{constant}\times l^{-5/2}, \tag{3.35}$$

$$p_0(t) = \frac{B_e(t)^2}{2\mu}.$$

Similarity (or self-similar) solutions of a partial differential equation are particular solutions in which one of the variables (here x) appears only in a particular combination {here $x/[l(t)]$} with another variable (here t). In this solution, as t increases from $-\infty$ to 0, so l decreases from ∞ to 0, whereas

B_e, p_0, and $|v|$ increase from 0. As the width of the sheet decreases, so the magnetic field is compressed, but at the same time it diffuses into the neutral line. The magnetic flux in the sheet decreases like $l^{1-a/2}$, whereas the magnetic energy and ohmic dissipation increase like l^{1-a} and l^{-1-a}, respectively. There is no flow of plasma or electromagnetic energy into the sheet across its moving boundary, but work is done by the external magnetic pressure. Similarity solutions also exist with the width of the neutral sheet increasing in time. Kirkland and Sonnerup (1979) discovered some more general similarity solutions to the above equations but with Eq. (3.33) replaced by an energy balance between entropy advection, thermal conduction, and joule heating.

3.4.2 Numerical Solutions

When the pressure balance condition [Eq. (3.32)] is relaxed and replaced by the equation of motion,

$$\rho \left(\frac{\partial v}{\partial t} + v \frac{\partial v}{\partial x} \right) = -\frac{\partial}{\partial x} \left(p + \frac{B^2}{2\mu} \right), \tag{3.36}$$

one has to resort to numerical methods to study time-dependent current-sheet behaviour. Equation (3.36) is supplemented by the induction equation [Eq. (3.30)], the continuity equation [Eq. (3.31)], and some form of energy equation. One-dimensional computations have been carried out by Cheng (1979) and Forbes et al. (1982) to model the evolution of a current sheet in a plasma when the magnetic diffusivity is suddenly increased by the onset of a plasma micro-instability, such as might occur during a solar or stellar flare or in a geomagnetic substorm. They found that flow speeds several times greater than the local fast-mode speed can be produced if β_∞ (the plasma beta at large distances) and the initial sheet-width are small enough. Forbes et al. (1982) solved the above basic equations with a flux-corrected algorithm that is designed to treat shocks accurately, but they assumed the plasma is isothermal in order to avoid the complexities of a full energy equation. Initially, the magnetic field is $(2B_\infty/\sqrt{\pi})$ erf (x/l_0) and the plasma is stationary with a pressure gradient balancing the Lorentz force. The two independent parameters are the plasma beta (β_∞) and the Lundquist number ($l_0 v_{A\infty}/\eta$). In the limit $l_0 \to 0$ there is only one parameter, and the only independent natural scale-length is the acoustic-diffusion length (η/c_s), where c_s is the sound speed. For the numerical runs a variable grid was used and l_0 was taken to be very small, at most 0.01 of the total length (x_0) of the numerical grid, which was taken to be unity. The runs were stopped when a

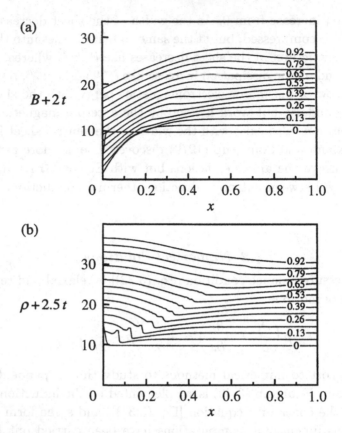

Fig. 3.8. Evolution of (a) the magnetic field (B) and (b) the plasma density (ρ) for a sheet undergoing resistive decay with $\beta_\infty = 2$. Curves at progressively later times are shifted upwards with the time (in units of $x_0/v_{A\infty}$), given at the right of each curve. The total length is $x_0 = 1$, and B and ρ are normalised with respect to their initial values at $x = 1$. The acoustic-diffusion length (η/c_s) is here 0.1 (from Forbes et al., 1982).

disturbance reached x_0, but different stages of the evolution were examined by adopting different values for η/c_s.

Figure 3.8 shows the results at high β_∞. The initially thin current sheet broadens and produces a magnetic pressure deficit in the region $x^2 < 4\eta t$, except near the origin where the magnetic pressure vanishes initially. The resulting inwards magnetic pressure gradient forces the plasma to flow primarily inward towards $x = 0$, so the density and pressure there increase. Eventually, the total pressure becomes very nearly uniform. On top of this overall picture, it can also be seen, especially from the density plots, that the evolution includes two other features. A small outflow near the origin is

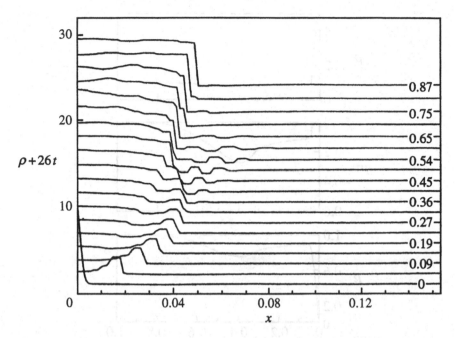

Fig. 3.9. Evolution of the density for $\beta_\infty = 0.1$ at small times when the shock is within the diffusion region. Here $\eta/c_s = 0.89$ (from Forbes et al., 1982).

driven by the total pressure maximum at $x = 0$ after the magnetic pressure in $x > 0$ has been eroded. Also, the initially sudden diffusion of the sheet causes a wave to propagate out from $x = 0$ and steepen into a shock. If the initial sheet width (l_0) is less than the acoustic-diffusion length (η/c_s), the shock is isomagnetic and is transformed into a fast magnetoacoustic wave when it reaches the edge of the sheet.

As the external plasma beta (β_∞) is decreased, the outgoing wave pulse is found to move more slowly (because of the reduction in sound speed), while the speed of the inflowing plasma is faster (because of the stronger magnetic forces). Figure 3.9 shows the density evolution for $\beta_\infty = 0.1$ close to $x = 0$ (i.e., well within the diffusion region). The main features here are the outward-propagation of the density pulse and the inflow of plasma that can be estimated by comparing the initial and final density profiles. It can also be seen that a second density jump appears behind the first one. It starts as a compressive wave, which steepens into a shock and overtakes the first shock at $t = 0.5$, which is itself being driven back towards $x = 0$ by a hyper-Alfvénic inflow with a fast-mode Mach number of about 4. The two shocks then coalesce to give a new, stronger shock that propagates in an outward direction. The development for small β_∞ at large distances from the

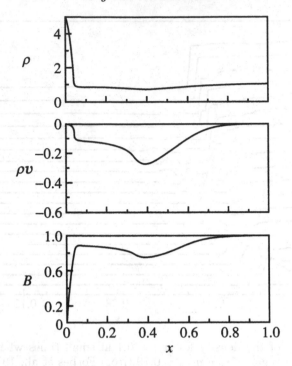

Fig. 3.10. Profiles of density (ρ), mass flux (ρv), and magnetic field (B) for $\beta = 0.2$ at large times when the waves are outside the diffusion region. Here $\eta/c_s = 0.016$ and $t = 0.66$ (from Forbes et al., 1982).

diffusion region ($x^2 \gg 4\eta t$) is shown in Fig. 3.10. A magnetoacoustic wave pulse is propagating outwards, consisting of a rarefaction at the leading edge followed by a compression at the trailing edge. The local fast-mode Mach number associated with the wave is here 0.3, but when $\beta_\infty \ll 0.1$ it can be much greater than unity.

3.5 Reconnective Annihilation

One of the few exact nonlinear solutions to the system, Eqs. (3.11)–(3.13), is the stagnation-point solution (§3.2) with inflow along the x-axis, namely

$$v_x = -x, \quad v_y = y, \quad \mathbf{B} = B_y(x)\,\hat{\mathbf{y}}. \qquad (3.37)$$

As we have seen in §3.2, it works because: Eq. (3.13) is satisfied identically; in Eq. (3.11) $\mathbf{v} \times \mathbf{B} = -x B_y(x)\,\hat{\mathbf{z}}$ and $\eta\nabla \times \mathbf{B} = \eta dB_y/dx\,\hat{\mathbf{z}}$ are functions of x alone; and in Eq. (3.12) the magnetic tension term $[(\mathbf{B} \cdot \nabla)\mathbf{B}/\mu]$ and the curl of $\rho(\mathbf{v} \cdot \nabla)\mathbf{v}$ both vanish.

Craig and Henton (1995) have discovered another exact solution that generalises the stagnation-point solution in a cunning way (Fig. 3.11). Consider

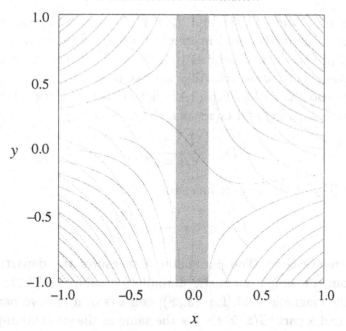

Fig. 3.11. Streamlines (dotted curves) and magnetic field lines (solid curves) for the Craig–Henton solution with $\lambda = 0.9$. There is a one-dimensional diffusion region extending to infinity in the positive and negative y-directions (shaded area).

a two-dimensional velocity $[v_x(x, y), v_y(x; y)]$ and a magnetic field of the form

$$\mathbf{B} = \lambda \mathbf{v} + G(x)\hat{\mathbf{y}}, \tag{3.38}$$

which is the superposition of a one-dimensional component $[G(x)\hat{\mathbf{y}}]$, as before, and a component $(\lambda \mathbf{v})$ parallel to the flow (where λ is a constant). Now suppose again that the terms in Eq. (3.11) depend only on x: this implies that both $v_x = v_x(x)$ and the vorticity $(\omega = \partial v_y/\partial x)$ depend only on x, and therefore the only form to within a constant multiplier that is also consistent with Eq. (3.13) is

$$v_x = -x, \quad v_y = y - F(x), \tag{3.39}$$

where $F(x)$ is an unknown function of x. In other words, both the vorticity $(\omega = -F')$ and electric current $[j = (G' - \lambda F')/\mu]$ depend only on x, and the stagnation-point flow is distorted by the addition of a simple shear flow $[v_y = -F(x)]$. In Eq. (3.11) $\mathbf{v} \times \mathbf{B}$ becomes $v_x B_y \hat{\mathbf{z}}$, so that the only advection of the magnetic field is of the part $G(x)$ of the y-component; furthermore, $\eta \nabla \times \mathbf{B}$ becomes $(\eta \, \partial B_y/\partial x \, \hat{\mathbf{z}})$, so the only diffusion of the magnetic field is of the y-component (B_y) purely in the x-direction.

Although the forms (3.38) and (3.39) give nonvanishing x- and y-components of the inertial and magnetic tension terms in the equation of motion, Eq. (3.12), the key point is that the curls of these terms are functions of x alone. This means that the curl-free part of Eq. (3.12) determines the pressure and the curl of Eq. (3.12) determines a relation between the two free functions, $F(x)$ and $G(x)$, namely

$$F(x) = -\frac{\lambda}{1 - \lambda^2} G(x).$$ (3.40)

Ohm's Law [Eq. (3.11)] then reduces to

$$E - x\,G = \frac{\eta}{1 - \lambda^2} \frac{dG}{dx},$$ (3.41)

which determines $G(x)$. The parameter λ measures the departure of the solution from the simple magnetic annihilation solution (§3.2). It can be seen that the magnetic field [Eq. (3.38)] consists of a passive part parallel to the flow and a part $[G(x)\hat{\mathbf{y}}]$ that is the same as the stagnation-point flow solution but with a scaling by a factor $(1-\lambda^2)$ of the x variable. In contrast, the plasma velocity [Eq. (3.39)] now consists of a stagnation-point flow plus a shear flow proportional to $G\hat{\mathbf{y}}$.

The solution is closer in spirit to standard magnetic annihilation than to classical reconnection, since the current sheet is one-dimensional and extends to infinity along the y-axis and the advection and diffusion are inherently one-dimensional. For these reasons we refer to the process as "reconnective annihilation". (In contrast, standard reconnection has a current sheet of finite length with two-dimensional advection and diffusion.) The separatrices of the magnetic field are located along the y-axis (across which there is no plasma flow) and a line inclined at a small angle $(2\lambda\eta/E)$ to it. As in magnetic annihilation, the width of the current sheet scales as $\eta^{1/2}$.

An alternative way of seeing why the Craig–Henton approach works is to recast Eqs. (3.11) and (3.12) in terms of Poisson brackets such as

$$[A, \psi] = \frac{\partial A}{\partial x}\frac{\partial \psi}{\partial y} - \frac{\partial A}{\partial y}\frac{\partial \psi}{\partial x}.$$ (3.42)

The flow and magnetic field

$$(v_x, v_y) = \left(\frac{\partial \psi}{\partial y}, -\frac{\partial \psi}{\partial x}\right), \quad (B_x, B_y) = \left(\frac{\partial A}{\partial y}, -\frac{\partial A}{\partial x}\right)$$

are written in terms of the flux function (A) and stream function (ψ), so that Eq. (3.13) is satisfied identically. Also, for simplicity in the rest of this section we work in units for which the constants μ and ρ are scaled out of the equations, so that we essentially set them equal to unity in (3.12) and Ampère's law. Then Eqs. (3.11) and (3.12) may be written concisely as

$$E - [A, \psi] = \frac{j}{\sigma}, \tag{3.43}$$

$$[\omega, \psi] = [j, A], \tag{3.44}$$

where $\omega = -\nabla^2 \psi$ and $j = -\nabla^2 A$. The Craig–Henton solution exploits the symmetry between A and ψ in these equations and is of the form

$$A = \lambda xy + g(x),$$
$$\psi = xy + \lambda g(x),$$

so that $j = -g''$, $\omega = -\lambda g''$, $[A, \psi] = x(1 - \lambda^2)g'$, and therefore Eq. (3.43) determines $g(x) = (1 - \lambda^2)^{-1} \int B_0(x)dx$. Furthermore, the Poisson brackets

$$-[\omega, \psi] = [\lambda g'', xy + \lambda g] = [\lambda g'', xy]$$

and

$$-[j, A] = [g'', \lambda xy + g] = [g'', \lambda xy]$$

in Eq. (3.44) are the same, which is therefore satisfied identically. Much more general exact solutions of the form

$$A = A_0(x) + A_1(x)y,$$
$$\psi = \psi_0(x) + \psi_1(x)y,$$

also exist (Priest, Titov, Grundy and Hood, in preparation).

The above solution has been extended naturally in several ways. Fabling and Craig (1996) included viscosity (which significantly affects the solution only inside the current sheet) and nonplanar components [$v_z(x)$ and $B_z(x)$, which are uniform except in the current sheet]. More significantly, Craig et al. (1995) extended it to a three-dimensional sheared stagnation-point flow of the form

$$v_x = -x, \quad v_y = Ky - F_y(x), \quad v_z = (1 - K)z - F_z(x)$$

with a magnetic field

$$\mathbf{B} = \lambda \mathbf{v} + G_y(x)\hat{\mathbf{y}} + G_z(x)\hat{\mathbf{z}},$$

and were thereby able to model the *fan reconnection* regime of Priest and Titov (1996) that is described in §8.6. This is the superposition of potential components $[-x, Ky, (1 - K)z]$ and one-and-a-half-dimensional components

(a)

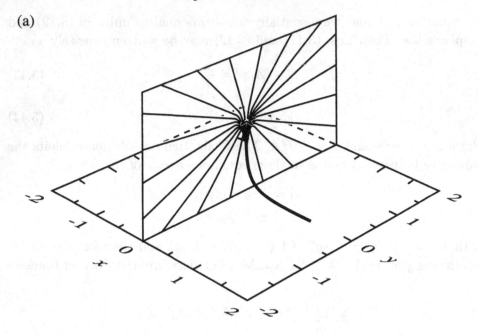

(b)

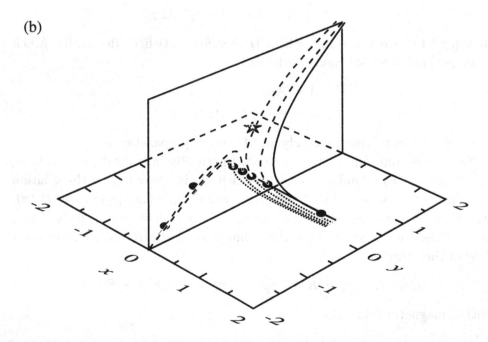

Fig. 3.12. (a) Spine curve and fan plane for fan reconnection with $\alpha = 2, K = 0.5, \lambda = 0.9, E_1 = E_2 = 0.1, \eta = 0.05$. (b) The reconnection of a curved field line carried in by a fluid element marked by dots (after Craig et al., 1995).

$[\mathbf{F}(x) = F_y(x)\hat{\mathbf{y}} + F_z(x)\hat{\mathbf{z}}$ and $\mathbf{G}(x) = G_y(x)\hat{\mathbf{y}} + G_z(x)\hat{\mathbf{z}}]$. The vorticity and current again depend on x alone, and the curl of Eq. (3.12) implies that

$$\mathbf{F} = -\frac{\lambda}{1-\lambda^2}\mathbf{G}.$$

Furthermore, the curl of Eq. (3.11) reduces to two ordinary differential equations in x, namely

$$xG_y' + KG_y = -\frac{\eta}{1-\lambda^2}G_y'',$$

$$xG_z' + (1-K)G_z = -\frac{\eta}{1-\lambda^2}G_z'',$$

which determine the unknown functions $G_y(x)$ and $G_z(x)$ separately.

The fan plane (see §8.2) is given by $x = 0$. Near the null point the spine (see §8.2) is inclined [Fig. 3.12(a)] and is given by

$$y = -\frac{E_1 x}{\lambda\eta(1+K)}, \qquad z = -\frac{E_2 x}{\lambda\eta(2-K)},$$

where E_1 and E_2 are integration constants. Figure 3.12(b) shows how curved field lines during fan reconnection are advected by the flow across the spine into the current sheet at $x = 0$, across which there is no flow.

In a similar way Craig and Fabling (1996) have discovered solutions for *spine reconnection* (§8.6.1) which may be written in cylindrical polars (R, ϕ, z) in the form

$$v_z = -2z + \lambda f(R)\cos m\phi, \qquad v_R = R,$$
$$B_z = -2\lambda z + f(R)\cos m\phi, \qquad B_R = R,$$

where $f(R)$ satisfies

$$f + \tfrac{1}{2}Rf' = \frac{\eta}{1-\lambda^2}\left(f'' + \frac{f'}{R} - \frac{f}{R^2}\right).$$

Flow across the fan produces a cylindrical diffusion region of radius $\sqrt{\eta}$ with a linearly increasing field surrounding the spine, outside which the field falls off as R^{-2}. More general exact solutions with flux function (A) and stream function (ψ) having the form

$$A = g(R)\sin\phi + G(R)z$$
$$\psi = f(R)\sin\phi + F(R)z$$

in cylindrical polars also exist (Titov and Priest, in preparation).

4

Steady Reconnection: The Classical Solutions

4.1 Introduction

In most of the universe the magnetic Reynolds number (R_m, §1.2.2) is very much larger than unity and so the magnetic field is attached very effectively to the plasma. It is only in extremely thin regions where the magnetic gradients are typically a million times or more stronger than normal that the magnetic field can slip through the plasma and reconnect. Thus, for example, a field line initially joining a plasma element at A to one at B in Fig. 4.1 may be carried towards another oppositely directed field line CD and a narrow region of very strong magnetic gradient (containing an X-type neutral point) may be formed between them. Then the field lines may diffuse, break, and reconnect, so that element A becomes linked instead to element C (Fig. 4.1).

There are several important effects of this local process:

(i) The global topology and connectivity of field lines change, affecting the paths of fast particles and heat conduction, since these are directed mainly along field lines;

(ii) Magnetic energy is converted to heat, kinetic energy, and fast particle energy;

(iii) Large electric currents and electric fields are created, as well as shock waves and filamentation, all of which may help to accelerate fast particles (Chapter 13).

As we discussed in Chapter 1, two questions that many of the early researchers tried to answer are: what is the nature of field-line breaking and reconnection when it takes place in a steady-state manner; and what is the rate at which it occurs – that is, what is the speed with which magnetic field lines can be carried in towards the reconnection site? Much of the focus

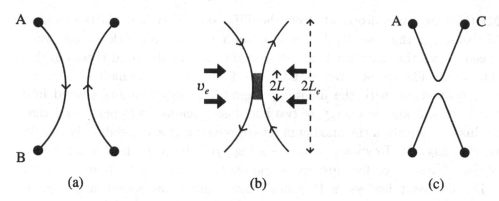

Fig. 4.1. Breaking and reconnection of magnetic field lines: (a) before reconnection the point A is joined to B; (b) during reconnection a narrow diffusion region (shaded) forms between the oppositely directed field lines; (c) after reconnection A is joined to C.

has been on the steady-state process (see Sonnerup, 1979), partly because it is simpler to analyse mathematically than time-varying reconnection, but also because the earliest motivation, namely the main phase of a large solar flare (§11.1.2), was realised to be essentially a steady-state energy release process for many hundreds of Alfvén travel-times (with a superimposed time-modulation).

Sweet (1958a, 1958b) and Parker (1957, 1963) put forward the original model (§4.2) in which a magnetic diffusion layer is present along the whole boundary between the opposing magnetic fields. Its length (L) is therefore equal to the global external length-scale (L_e), and the reconnection rate (v_e) is just equal to the speed with which field lines are entering the diffusion region. However, the resulting rate of reconnection is much too slow to explain energy release in a solar flare, and so the hunt was on for a faster mechanism.

Six years later, at a conference on solar flares, H.E. Petschek came up with a solution which was a stroke of genius and involved deep physical insight (Petschek, 1964). E.N. Parker, who was at the conference, immediately acclaimed the solution as the one people were seeking – or at least he did so after staying up late the previous evening discussing the details with Petschek privately! What Petschek did in this pioneering advance was to suggest that the Sweet–Parker diffusion region is only limited to a small segment (of length L) of the boundary between the opposing fields. Since the diffusion region is shorter, it is also thinner and so the diffusion (and therefore reconnection) process can take place faster. Petschek also considered the nature of the external flow outside the diffusion region and presented a clever mathematical model for it. He suggested that slow-mode

MHD shock waves propagate from the diffusion region, which acts as a kind of obstacle in the flow (the flow is supersonic relative to the slow-mode wave speed across the magnetic field, because the slow-mode speed is zero in this direction). The shock waves stand in the flow and are the main sites of energy conversion, with the inflowing magnetic energy being converted into the heat and kinetic energy of two hot, fast streams. Petschek found that for his configuration the maximum rate of reconnection depends only weakly on the magnetic Reynolds number – being typically a tenth or a hundredth of the Alfvén speed for most space and astrophysical applications.

For the next few years Petschek's mechanism was widely accepted as the answer to fast flare energy release, although few fully understood its complexities as it involves a combination of subtle physical effects, namely advection of flux, diffusion of flux, and magnetic shock-wave behaviour. However, this state of calm was shaken when Sonnerup (1970) came along with an alternative reconnection model that could operate at any rate up to the Alfvén speed. Furthermore, at about the same time Yeh and Axford (1970) sought self-similar solutions of the steady MHD equations and discovered a family of models, which includes Sonnerup's model as a special case. Then Vasyliunas (1975) clarified matters in a major review which highlighted various mathematical and physical difficulties with the Yeh and Axford and Sonnerup solutions. In particular, Sonnerup's model possesses an extra standing discontinuity in each quadrant in addition to the Petschek shock wave, and Vasyliunas realised that, whereas the Petschek shocks are generated by the diffusion region, the Sonnerup discontinuities have to be generated externally.

The result was that only Petschek's mechanism was accepted. Nevertheless, Yeh and Axford's idea of seeking self-similar solutions was a good one, and so this was later taken up by Soward and Priest (1977), who used a more sophisticated mathematical treatment to produce solutions which put Petschek's mechanism on a firm mathematical foundation when the central diffusion region can be regarded as a region of small dimensions as far as the external region is concerned.

Again a state of calm ensued – until, that is, the watershed year of 1986, when new generations of numerical experiments and theoretical models were to lead to a new state of ferment – but that is the story of the next chapter. For the present, let us go through the details of the classical models and summarise various analyses of the diffusion region.

The basic equations involved in this account include the induction equation

$$\frac{\partial \mathbf{B}}{\partial t} = \nabla \times (\mathbf{v} \times \mathbf{B}) + \eta \nabla^2 \mathbf{B}, \tag{4.1}$$

which for a two-dimensional steady state integrates to the Ohm's Law

$$\mathbf{E} + \mathbf{v} \times \mathbf{B} = \frac{\mathbf{j}}{\sigma}, \tag{4.2}$$

where $\mathbf{E} = E\hat{\mathbf{z}}$ is a constant uniform electric field normal to the xy-plane (in which the flow and magnetic field lie) and the current is given by

$$\mathbf{j} = \frac{\nabla \times \mathbf{B}}{\mu}, \tag{4.3}$$

so that Eq. (4.2) becomes

$$\mathbf{E} + \mathbf{v} \times \mathbf{B} = \eta \nabla \times \mathbf{B} \tag{4.4}$$

in terms of the magnetic diffusivity ($\eta = 1/(\mu\sigma)$). Equation (4.1) states that the magnetic field changes in time partly because of the transport of magnetic field with the plasma (the first term on the right-hand side) and partly the diffusion of magnetic field through the plasma (the second term). In most regions of the universe the magnetic field is frozen very effectively indeed to the plasma and the diffusion term is negligible, so that Eq. (4.4) correspondingly reduces to

$$\mathbf{E} + \mathbf{v} \times \mathbf{B} = 0, \tag{4.5}$$

to a very high accuracy. The other governing equations for a steady-state plasma are the equation of motion

$$\rho(\mathbf{v} \cdot \nabla)\mathbf{v} = -\nabla p + \mathbf{j} \times \mathbf{B} \tag{4.6}$$

when the dominant forces are a pressure gradient and a magnetic force, and the continuity equation

$$\nabla \cdot (\rho\mathbf{v}) = 0. \tag{4.7}$$

The time-scale for magnetic dissipation is given by equating the first and third terms in Eq. (4.1), namely

$$\tau_d = \frac{L^2}{\eta} = 10^{-9} L^2 T^{3/2}, \tag{4.8}$$

which is enormously large in most applications [e.g., 10^{14} s for a typical global solar coronal length-scale ($L = 10^7$ m) and coronal temperature ($T = 10^6$ K)]. Thus, in order to release magnetic energy (in, say, a stellar flare or a coronal heating event), one needs to create extremely small length-scales in sheets or filaments and therefore very large magnetic gradients and electric currents [Eq. (4.3)]. This may be accomplished in several ways: by the

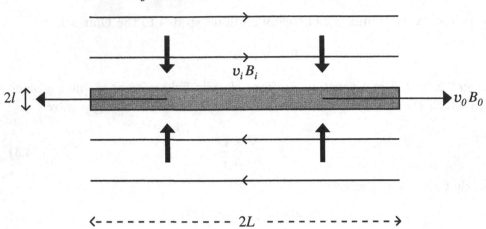

Fig. 4.2. Sweet–Parker reconnection. The diffusion region is shaded. The plasma velocity is indicated by thick-headed arrows and the magnetic field lines by thin-headed arrows.

creation of thin shock waves; by the development of instabilities such as the tearing mode (§6.2) or the coalescence instability (§6.5.3); by the growth of strong turbulence; and by the formation of current sheets at X-points. It is the latter possibility we are considering in this chapter.

4.2 Sweet–Parker Mechanism

Many misconceptions have appeared in the literature about this most fundamental model of reconnection, including incorrect factors of 2 in various places and the incorrect suggestion that the plasma is accelerated along the sheet solely by a pressure gradient. So let us start with the basics.

4.2.1 Basic Model

The Sweet–Parker model consists of a simple diffusion region of length $2L$ and width $2l$, say, lying between oppositely directed magnetic fields (Fig. 4.2), for which an order-of-magnitude analysis may be conducted as follows. First of all, let us ask the question: How fast can a magnetic field of strength B_i enter the diffusion layer at a speed v_i, say? For a steady state, the plasma must carry the field lines in at the same speed as they are trying to diffuse outward, so that

$$v_i = \frac{\eta}{l}. \tag{4.9}$$

This expression comes directly from Ohm's Law (4.2) as follows. Since the electric field is uniform for a steady two-dimensional state, its value may be found by evaluating Eq. (4.2) at the inflow to the diffusion region where the current vanishes, namely

$$E = v_i B_i.$$

However, at the centre (N) of the diffusion region, where the magnetic field vanishes, Eq. (4.2) becomes

$$E = \frac{j_N}{\sigma},$$

where Ampère's Law ($\mathbf{j} = \nabla \times \mathbf{B}/\mu$) implies that the current at N is roughly

$$j_N = \frac{B_i}{\mu l}. \tag{4.10}$$

Eliminating E and j_N between these three equations then gives Eq. (4.9), as required.

Conservation of mass implies that the rate ($4\rho L v_i$) at which mass is entering the sheet from both sides must equal the rate ($4\rho l v_o$) at which it is leaving at both ends, so that

$$\boxed{L\, v_i = l\, v_o,} \tag{4.11}$$

where v_o is the outflow speed. [Note that in a more detailed treatment including a non-rectangular shape such as a wedge (Parker, 1957) the variation of the diffusion region width with distance from the stagnation point may be included.] The width (l) may now be eliminated between our two basic Sweet–Parker equations, Eqs. (4.9) and (4.11), to give the square of the inflow speed as

$$v_i^2 = \frac{\eta\, v_o}{L}. \tag{4.12}$$

In dimensionless variables Eq. (4.12) may be rewritten

$$M_i = \frac{\sqrt{v_o/v_{Ai}}}{\sqrt{R_{mi}}}, \tag{4.13}$$

where

$$M_i = \frac{v_i}{v_{Ai}} \tag{4.14}$$

is the *inflow Alfvén Mach number* (or dimensionless *reconnection rate*) and

$$R_{mi} = \frac{L\, v_{Ai}}{\eta} \tag{4.15}$$

is the magnetic Reynolds number based on the inflow Alfvén speed.

Once v_o and therefore v_i from Eq. (4.12) are known for a given L, Eq. (4.11) in turn determines the sheet width as

$$l = L \frac{v_i}{v_o}, \tag{4.16}$$

and the outflow magnetic field strength (B_o) is determined from flux conservation

$$v_i B_i = v_o B_o$$

as

$$B_o = B_i \frac{v_i}{v_o}. \tag{4.17}$$

However, a key question is: What is the outflow speed v_o? This is determined as follows by the equation of motion (or equivalently the mechanical energy equation – see Section 4.2.3).

From Eq. (4.10) the order-of-magnitude electric current is $j \approx B_i/(\mu l)$ and so the Lorentz force along the sheet is $(\mathbf{j} \times \mathbf{B})_x \approx jB_o = B_iB_o/(\mu l)$. This force accelerates the plasma from rest at the neutral point to v_o over a distance L and so, by equating the magnitude of the inertial term $\rho(\mathbf{v} \cdot \nabla)v_x$ to the above Lorentz force and neglecting the plasma pressure gradient, we have

$$\rho \frac{v_o^2}{L} \approx \frac{B_i B_o}{\mu l}. \tag{4.18}$$

However, from $\nabla \cdot \mathbf{B} = 0$, or Eqs. (4.11) and (4.17),

$$\frac{B_o}{l} \approx \frac{B_i}{L}, \tag{4.19}$$

and so the right-hand side of Eq. (4.18) may be rewritten as $B_i^2/(\mu L)$ and we have

$$\boxed{v_o = \frac{B_i}{\sqrt{\mu\rho}} \equiv v_{Ai},} \tag{4.20}$$

where v_{Ai} is the Alfvén speed at the inflow. Not surprisingly, we have found that the magnetic force accelerates the plasma to the Alfvén speed.

The fields therefore reconnect for this basic model at a speed given by Eq. (4.13) as

$$\boxed{v_i = \frac{v_{Ai}}{R_{mi}^{1/2}}} \tag{4.21}$$

in terms of the (inflow) Alfvén speed (v_{Ai}) and magnetic Reynolds number ($R_{mi} = Lv_{Ai}/\eta$). The plasma is ejected from the sheet of width

$$l = \frac{L}{R_{mi}^{1/2}}$$

at a speed

$$v_o = v_{Ai}$$

and with a magnetic field strength

$$B_o = \frac{B_i}{R_{mi}^{1/2}}.$$

Since $R_{mi} \gg 1$, we therefore have $v_i \ll v_{Ai}$, $B_o \ll B_i$ and $l \ll L$.

Equations (4.9), (4.11), and (4.20) for flux diffusion, mass continuity, and momentum are thus the basic equations for the Sweet–Parker mechanism, and they together imply the reconnection rate [Eq. (4.21)]. In the Sweet–Parker mechanism, we identify the sheet length (L) with the global external length-scale (L_e) and R_{mi} therefore with the *global magnetic Reynolds number*

$$R_{me} = \frac{L_e v_{Ae}}{\eta}.$$

Since in practice $R_{me} \gg 1$, the reconnection rate is very small: for instance, in stellar coronae where R_{me} lies between, say, 10^6 and 10^{12}, the fields reconnect at between 10^{-3} and 10^{-6} of the Alfvén speed – much too slow for a solar or stellar flare.

4.2.2 Effect of Pressure Gradients

The basic version of the Sweet–Parker model presented in the previous section has an outflow speed (v_o) equal to the Alfvén speed (v_{Ai}). It assumes implicitly that the outflow pressure (p_o) is the same as, or close to, the pressure (p_N) at the neutral point (N) in the centre of the current sheet, so that plasma pressure gradients along the sheet play no role in accelerating the plasma from rest at the neutral point to v_o at the outflow. However, in principle the outflow pressure (p_o) may be imposed as any other (positive) value and so lead to a different outflow speed and therefore reconnection rate.

Let us set up x- and y-axes along and normal to the sheet (Fig. 4.3) and consider the steady equation of motion

$$\rho(\mathbf{v} \cdot \nabla)\mathbf{v} = \mathbf{j} \times \mathbf{B} - \nabla p. \tag{4.22}$$

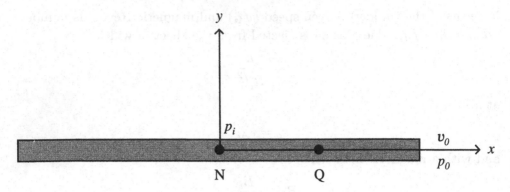

Fig. 4.3. Notation used for calculating the effect of pressure gradients in a Sweet–Parker current sheet.

When the inflow speed is much slower than the Alfvén speed and the sheet is long and thin, the inertial and magnetic tension forces may be neglected in the y-component of Eq. (4.22), which reduces to

$$0 = -\frac{\partial}{\partial y}\left(\frac{B^2}{2\mu} + p\right).$$

By integrating from the inflow point $(0, l)$ to the neutral point $(0, 0)$ where B vanishes, we find the neutral point pressure to be

$$p_N = p_i + \frac{B_i^2}{2\mu}. \tag{4.23}$$

Furthermore, the x-component of Eq. (4.22) along the x-axis is

$$\rho v_x \frac{\partial v_x}{\partial x} = jB_y - \frac{\partial p}{\partial x},$$

which may be evaluated at a point (Q) half-way along the sheet (Fig. 4.3) to give

$$\rho \frac{v_o}{2}\frac{v_o}{L} = \frac{B_i}{\mu l}\frac{B_o}{2} - \frac{p_o - p_N}{L}. \tag{4.24}$$

Here $j \approx \partial B_x/\partial y \approx 2B_i/(2l)$, since the field jumps from B_i to $-B_i$ across a width $2l$; also $\partial v_x/\partial x \approx v_o/L$ and the values of v_x and B_y at Q are $\frac{1}{2}v_o$ and $\frac{1}{2}B_o$, respectively. In this equation B_o may be eliminated by using Eq. (4.19), so that Eq. (4.24) becomes

$$v_o^2 = v_{Ai}^2 + \frac{2(p_N - p_o)}{\rho}, \tag{4.25}$$

or, after substituting for p_N from Eq. (4.23),

$$v_o^2 = 2v_{Ai}^2 + \frac{2(p_i - p_o)}{\rho}. \tag{4.26}$$

We notice several points about this result, which is the required generalisation of the standard Sweet–Parker result [Eq. (4.20)]. The outflow speed now depends on the imposed values of p_i and p_o and so affects the reconnection rate [Eq. (4.13)], which may be written after substituting for v_o/v_{Ai} from Eq. (4.26) in the form

$$M_i = \frac{2^{1/4}\left[1 + \tfrac{1}{2}\beta_i(1 - p_o/p_i)\right]^{1/4}}{\sqrt{R_{mi}}}; \tag{4.27}$$

thus, it is the values of the plasma beta ($\beta_i = 2\mu p_i/B_i^2$) and pressure ratio p_o/p_i which determine the departure from the Sweet–Parker rate ($M_i = 1/\sqrt{R_{mi}}$). In particular, it should be noted that the presence of reconnection outflows considerably slower than the Alfvén speed (as in, for example, explosive events (§11.2.3) or solar flares (§§11.1.2, 11.2.1)) may indicate that the outflows are being slowed by the presence of a large outflow pressure.

If $p_o = p_N$, on the one hand, we recover the Sweet–Parker result ($v_o = v_{Ai}$) from Eq. (4.25). If, on the other hand, $p_o = p_i$, then

$$v_o = \sqrt{2}\,v_{Ai}.$$

If the outflow pressure is so small that $p_o < p_N$, then $v_o > v_{Ai}$ and the reconnection rate is enhanced ($M_i > 1/\sqrt{R_{mi}}$). If the outflow pressure is large enough ($p_o > p_N$), the outflow is slowed ($v_o < v_{Ai}$) and the reconnection rate is lowered ($M_i < 1/\sqrt{R_{mi}}$). If the outflow pressure is smaller than the inflow pressure ($p_o < p_i$), then $v_o > \sqrt{2}v_{Ai}$, whereas if it is larger then $v_o < \sqrt{2}v_{Ai}$. (An additional possibility considered in Parker's original presentation is that the outflow is spread out into a fan; for example, if the flow spreads out into a 90° wedge ($\pm 45°$), the velocity component in the x-direction along the sheet is $v_{0x} = v_{Ai}$ when $p_0 = p_i$.)

It may also be noted that Eq. (4.25) may be recast in the form

$$p_o + \tfrac{1}{2}\rho v_o^2 - p_N = \tfrac{1}{2}\rho v_{Ai}^2,$$

so that the simple form of Bernouilli's law ($p + \tfrac{1}{2}\rho v^2 =$ constant) does not hold as the plasma moves along the sheet from the neutral point to the outflow: the reason lies in the extra effect of the magnetic force, as represented by the Alfvénic term on the right.

A final point is that the above equations are only order-of-magnitude estimates. Surprisingly, a more accurate treatment has not yet been accomplished properly. It would require a solution of the full equations throughout the diffusion layer, a determination of the shape of its boundary, and an integration along the boundary of the conservation equations, taking full account of the variation of the magnitude and inclinations of the velocity and magnetic field along the boundary.

4.2.3 Energy Considerations

It is interesting to consider the energetics of the reconnection process in a Sweet–Parker diffusion layer which is long and thin ($l \ll L$) and therefore has $v_i \ll v_{Ai}$, but there are again several misconceptions in the literature, as the alert reader will discover. The rate of inflow of electromagnetic energy is the Poynting flux ($\mathbf{E} \times \mathbf{H}$ per unit area), or, since $E = v_i B_i$ in magnitude,

$$E \frac{B_i}{\mu} L = v_i \frac{B_i^2}{\mu} L.$$

Therefore, the ratio of the inflow of kinetic to electromagnetic energy is

$$\frac{\text{inflow KE}}{\text{inflow EM}} = \frac{\frac{1}{2}\rho v_i^2}{B_i^2/\mu} = \frac{v_i^2}{2v_{Ai}^2}.$$

By Eq. (4.21) this is much smaller than unity, so most of the inflowing energy is magnetic.

Next, consider the energy outflow. By conservation of flux

$$v_o B_o = v_i B_i,$$

and so $B_o \ll B_i$. Also the outflow of electromagnetic energy is $EB_o l/\mu$, which is much less than the inflow of electromagnetic energy ($EB_i L/\mu$) since both $B_o \ll B_i$ and $l \ll L$. So what happens to the inflowing magnetic energy? Well, the ratio of outflowing kinetic to inflowing magnetic energy is

$$\frac{\text{outflow KE}}{\text{inflow EM}} = \frac{\frac{1}{2}\rho v_o^2 (v_o l)}{v_i B_i^2 L/\mu} = \frac{\frac{1}{2}v_o^2}{v_{Ai}^2} = \frac{1}{2}.$$

Thus, half of the inflowing magnetic energy is converted to kinetic energy, while the remaining half is converted to thermal energy. In other words, the effect of the reconnection is to create hot, fast streams of plasma with a rough equipartition between flow and thermal energy.

In a steady-state incompressible flow (with given density ρ), the equations of motion and continuity, namely

$$\rho(\mathbf{v} \cdot \nabla)\mathbf{v} = -\nabla p + \mathbf{j} \times \mathbf{B} \qquad (4.28)$$

and

$$\nabla \cdot \mathbf{v} = 0,$$

essentially determine p and $\mathbf{v}$, while the electromagnetic equations

$$\mathbf{E} + \mathbf{v} \times \mathbf{B} = \frac{\mathbf{j}}{\sigma} \tag{4.29}$$

and

$$\nabla \cdot \mathbf{B} = 0$$

essentially determine the three components of $\mathbf{B}$. Then Ampère's Law

$$\mathbf{j} = \frac{\nabla \times \mathbf{B}}{\mu} \tag{4.30}$$

determines the current and a heat energy equation would determine the temperature (T).

An important point to note is that the equations of electromagnetic energy and mechanical energy are secondary equations. They may be derived from the above set of primary equations and so provide no extra information about the values of the physical variables (although they do of course determine the energy partition). Thus, for example, a vector identity for the divergence of a vector product together with Eq. (4.30) and $\nabla \times \mathbf{E} = \mathbf{0}$ (for a steady state) imply that

$$-\nabla \cdot (\mathbf{E} \times \mathbf{H}) = \mathbf{E} \cdot \mathbf{j}.$$

However, the scalar product of Eq. (4.29) in turn with $\mathbf{j}$ gives

$$\mathbf{E} \cdot \mathbf{j} = \frac{j^2}{\sigma} + \mathbf{v} \cdot \mathbf{j} \times \mathbf{B}.$$

Combining these together, we obtain the electromagnetic energy equation

$$-\nabla \cdot (\mathbf{E} \times \mathbf{H}) = \frac{j^2}{\sigma} + \mathbf{v} \cdot \mathbf{j} \times \mathbf{B}, \tag{4.31}$$

which implies that the inflow of electromagnetic energy into a volume produces ohmic heat (j^2/σ) and work done by the Lorentz force $(\mathbf{j} \times \mathbf{B})$.

What can we deduce from Eq. (4.31)? Integrate it over the diffusion region with volume V and surface S and use the divergence theorem to give

$$-\int_S \mathbf{E} \times \mathbf{H} \cdot d\mathbf{S} = \int_V \left(\frac{j^2}{\sigma} + \mathbf{v} \cdot \mathbf{j} \times \mathbf{B} \right) dV, \tag{4.32}$$

in which $E \approx -v_i B_i$ along the inflow part of S. The outflow contribution $(E B_o l / \mu)$ is by Eq. (4.19) smaller than the inflow contribution $(E B_i L / \mu)$ by a factor $l^2/L^2 \approx R_{mi}^{-1}$ so that it may be neglected. Furthermore, the current in the centre of the sheet is roughly $B_i/(\mu l)$ and so we may approximate the mean value of j^2 by roughly $\frac{1}{2}B_i^2/(\mu l)^2$ and the mean value of $\mathbf{v} \cdot \mathbf{j} \times \mathbf{B}$ by $\frac{1}{2}v_o[B_i/(\mu l)]B_o$. Thus, Eq. (4.32) becomes approximately

$$v_i B_i \frac{B_i}{\mu} 4L = \left(\frac{B_i^2}{2\mu^2 l^2 \sigma} + \frac{v_o B_i B_o}{2\mu l} \right) 4Ll,$$

since the two opposing inflow sides of the current sheet have a total length of $4L$ and the volume is $4Ll$ (per unit length in the z-direction). Replacing $v_o B_o$ by $v_i B_i$ from Eq. (4.17) in the last term, we find

$$v_i = \frac{\eta}{2l} + \frac{v_i}{2} \tag{4.33}$$

or

$$v_i = \frac{\eta}{l}.$$

In other words, we recover the diffusion result [Eq. (4.9)], which is not surprising since both are essentially a consequence of Ohm's Law [Eq. (4.29)]. However, what we can deduce from Eqs. (4.32) and (4.33) is that half of the inflowing electromagnetic energy goes into ohmic heating and half into the work done by the magnetic force (which in turn, as we shall see, goes into kinetic energy when the pressure gradient along the sheet is negligible).

The equation of mechanical energy may be derived by taking the scalar product of Eq. (4.28) with $\mathbf{v}$ and using $\nabla \cdot \mathbf{v} = 0$, so that

$$\nabla \cdot \left(\tfrac{1}{2}\rho v^2 \mathbf{v} \right) = \mathbf{v} \cdot \mathbf{j} \times \mathbf{B} - \mathbf{v} \cdot \nabla p, \tag{4.34}$$

which implies that a change of kinetic energy is produced by the work done by $\mathbf{j} \times \mathbf{B}$ and $-\nabla p$. Next, rewriting $\mathbf{v} \cdot \nabla p$ as $\nabla \cdot (p\mathbf{v})$ since $\nabla \cdot \mathbf{v} = 0$ in the present model and integrating over the diffusion region, we find

$$\int_S \left(\tfrac{1}{2}\rho v^2 \mathbf{v} + p\mathbf{v} \right) \cdot d\mathbf{S} = \int_V (\mathbf{v} \cdot \mathbf{j} \times \mathbf{B}) \, dV, \tag{4.35}$$

and so the magnetic force term on the right of Eq. (4.32) is a combination of the change in kinetic energy and the net work done by pressure on the surface. Assuming that the inflow of kinetic energy is small and supposing that the mean values of the above integrands are half their maximum values, we find

$$\tfrac{1}{2}\left[\left(\tfrac{1}{2}\rho v_o^2 + p_o \right) 4v_o l - 4p_i v_i L \right] = \tfrac{1}{2}\left(v_o \frac{B_i}{\mu l} B_o \right) 4Ll,$$

which reduces to exactly the same equation, Eq. (4.26), for v_o as obtained

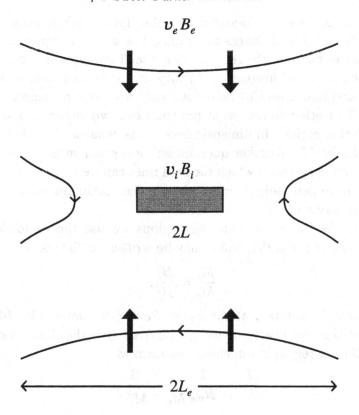

Fig. 4.4. The notation for fast reconnection regimes, with the diffusion region shaded.

directly from the equation of motion. Again, a more accurate evaluation of Eqs. (4.32) and (4.35) would require a full diffusion-region solution.

4.2.4 Fast Reconnection

The fast regimes of reconnection (see Fig. 4.4) that we shall consider next contain a tiny Sweet–Parker diffusion region around the X-point. Thus, to distinguish the outer region from the Sweet–Parker region, we shall denote the flow speed and magnetic field at large distances (L_e) from the X-point by v_e and B_e, as indicated in Fig. 4.4. The properties of fast reconnection models depend on two dimensionless parameters, namely the *external* reconnection rate ($M_e = v_e/v_{Ae}$) and the *external,* or global, magnetic Reynolds number ($R_{me} = L_e v_{Ae}/\eta$).

Reconnection is said to be fast when the reconnection rate (M_e) is much larger than the Sweet–Parker rate [Eq. (4.21)]. Properties at the inflow to the diffusion region (denoted by subscript i) are now related to the "external" values at large distances (denoted by subscript e). We suppose that the values v_e, B_e at large distances are imposed or given as typical ambient or external values far from the reconnection site. Two questions are then: how do the flow speed and magnetic field vary as the reconnection region is approached; and how does this behaviour vary from one reconnection regime to another? In other words, what are the unknown values v_i and B_i close to the diffusion region? In dimensionless terms, what is the value of M_i for a given value of M_e? Another question is: for a given value B_e, what is the maximum speed (v_e) with which the field lines can be brought in from large distances? In dimensionless terms, what is the maximum value of M_e and how does it vary with R_{me}?

To relate the inner and external regions we use the principle of flux conservation ($v_i B_i = v_e B_e$), which may be written in dimensionless form as

$$\frac{M_i}{M_e} = \frac{B_e^2}{B_i^2}. \tag{4.36}$$

Furthermore, the solution of the *Sweet–Parker relations* (4.9), (4.11) and (4.20) may then be rewritten using (4.36) to give the dimensions of the central diffusion region in dimensionless form as

$$\frac{L}{L_e} = \frac{1}{R_{me}} \frac{1}{M_e^{1/2}} \frac{1}{M_i^{3/2}}, \tag{4.37}$$

$$\frac{l}{L_e} = \frac{1}{R_{me}} \frac{1}{M_e^{1/2}} \frac{1}{M_i^{1/2}}. \tag{4.38}$$

Thus, once B_i/B_e has been determined from a model of the external region outside the diffusion region, Eq. (4.36) determines M_i/M_e and Eqs. (4.37) and (4.38) give the dimensions of the diffusion region in terms of M_e and R_{me}. In the following sections we consider different models for the external region, starting with Petschek's mechanism (§4.3), and then we move on to less familiar types. Later on, in Chapter 5, we shall show that Petschek's mechanism is part of a larger family of solutions, which we designate as *almost-uniform reconnection* (§5.1) in order to distinguish this family of solutions from another family that we call *non-uniform* (§5.2).

4.3 Petschek's Mechanism: Almost-Uniform, Potential Reconnection

Petschek (1964) realised that a slow magnetoacoustic shock (§1.5) provides another way (in addition to a diffusion region) of converting magnetic energy into heat and kinetic energy. In the switch-off limit the shock propagates at

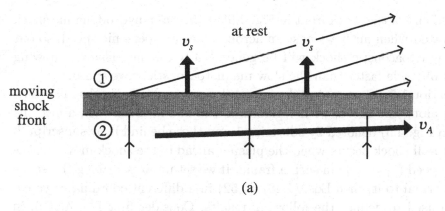

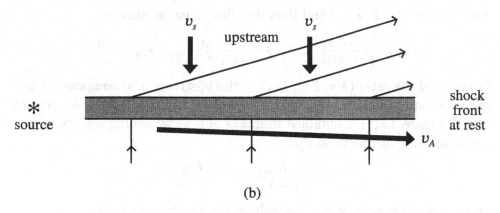

Fig. 4.5. The properties of a slow-mode shock (a) propagating upwards at speed v_s into a medium at rest and (b) in a frame of reference at rest.

a speed $v_s = B_N/\sqrt{\mu\rho}$ into a medium at rest (where B_N is the normal field component) and has the effect of turning the magnetic field towards the normal and therefore decreasing the downstream field strength [Fig. 4.5(a)]. At the same time the shock accelerates the plasma to the Alfvén speed (v_A) parallel to the shock front. Now, if the upstream plasma is moving downwards at the same speed (v_s) as the shock is trying to propagate upwards, then the shock front will remain stationary [Fig. 4.5(b)].

A further point to notice is that in Fig. 4.5(b), if you imagine a movie film in which a sequence of field lines comes in from above and passes down through the shock, the kink at the shock in an individual field line moves to the right along the shock front. In other words, the shock is a (finite-amplitude) disturbance which may be generated by some source at the left-hand end of the shock front in Fig. 4.5(b). Just as a hydrodynamic shock is generated when air flows past an aeroplane at a supersonic speed, so our slow magnetoacoustic shock will be generated when the plasma is flowing past an obstacle faster than the slow magnetoacoustic wave speed.

In Section 1.5, we presented the jump relations for MHD shocks in a frame moving along the shock, with variables ahead of the shock (i.e., above the shock in Fig. 4.5) denoted by subscript 1 and those behind it by subscript 2. A switch-off shock occurs when the plasma ahead of the shock moves at the Alfvén speed ($v_1 = v_{A1}$) in such a frame. If we set up axes $\bar{x}$ along the shock and $\bar{y}$ normal to it, then Eqs. (1.48)–(1.52) in a different coordinate system may be used to deduce the following results. Consider first Fig. 4.5(b), in which the reference frame is at rest in the shock front and the plasma is flowing in normal to the shock front from the region ahead (i.e., $v_{1\bar{x}} = 0$). Then $B_{2\bar{y}} = B_{1\bar{y}}, B_{2\bar{x}} = 0$ and the other flow components are

$$v_{1\bar{y}} = \frac{-\lambda v_{A1}}{(1+\lambda^2)^{1/2}}, \qquad v_{2\bar{x}} = \frac{v_{A1}}{(1+\lambda^2)^{1/2}}, \qquad v_{2\bar{y}} = \frac{v_{1\bar{y}}}{X}$$

in terms of the ratio ($\lambda = B_{1\bar{y}}/B_{1\bar{x}}$) of the upstream field components and the shock strength (X) given by the equivalent of Eq. (1.51) in the present frame. Figure 4.5(a) is obtained from Fig. 4.5(b) by adding a frame translation in the $\bar{y}$-direction, so that

$$v_s = \frac{\lambda v_{A1}}{(1+\lambda^2)^{1/2}} = \frac{B_{1\bar{y}}}{(\mu\rho)^{1/2}}$$

and the $\bar{y}$-component of the flow behind the shock is changed to

$$v_{2\bar{y}} = \left(1 - \frac{1}{X}\right) v_s.$$

Petschek appreciated that the Sweet–Parker diffusion region would act as a source for four slow magnetoacoustic shocks, which propagate in different directions from the diffusion region and which stand in the flow when a steady state is reached. The creation of standing shock waves without any solid obstacle in the flow has been confirmed by numerous numerical experiments (e.g., Sato, 1979; Ugai, 1984; Scholer, 1989; and §5.4). The generation of such shocks arises in part from the anisotropic nature of the slow-mode wave, which is not part of our everyday experience of sound waves.

Petschek's regime is *almost-uniform* in the sense that the field in the inflow region is a small perturbation to a uniform field (B_e). It is also potential in the sense that there is no current in the inflow region. Most of the energy conversion takes place at standing slow-mode shocks [Fig. 4.6(a)], which are

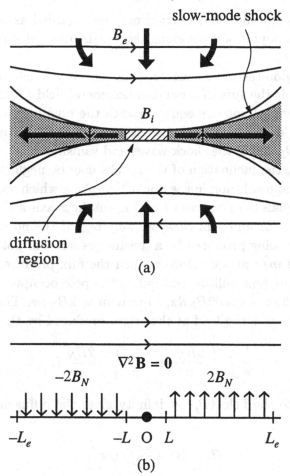

Fig. 4.6. (a) Petschek's model, in which the central shaded region is the diffusion region and the other two shaded regions represent plasma that is heated and accelerated by the shocks. (b) Notation for the analysis of the inflow region.

of the type known as *switch-off* (§1.5). These shocks accelerate and heat the plasma, with $\frac{2}{5}$ of the inflowing magnetic energy being changed to heat and $\frac{3}{5}$ to kinetic energy for a specific heat ratio (γ) of $\frac{5}{3}$.

The Petschek analysis is disarmingly simple. The magnetic field decreases substantially from a uniform value (B_e) at large distances to a value B_i at the entrance to the diffusion region, while the flow speed increases from v_e to v_i. The object is to determine for a given B_e the maximum value of v_e (in dimensionless form $M_e = v_e/v_{Ae}$). The effect of the shocks is to provide a normal field component (B_N) which is associated with the distortion in the inflow field from the uniform value (B_e) at large distances. Thus, if the inflow field is potential, the distortion may be regarded as being produced by a series of monopole sources along the x-axis between $-L_e$ and $-L$ and between L and L_e.

The inflow region therefore consists of slightly curved field lines. The magnetic field there is the sum of a uniform horizontal field $(B_e\hat{\mathbf{x}})$ and the field obtained by solving Laplace's equation with the boundary conditions that the magnetic field vanish at large distances and that the normal component of the field be B_N along the shock waves and vanish at the diffusion region. To lowest order, the inclination of the shocks may be neglected, and so the problem is to find a solution in the upper half-plane which vanishes at infinity and which equals $2B_N$ between L and L_e on the x-axis and, by symmetry, $-2B_N$ between $-L_e$ and $-L$. Now, we may regard the normal component on the x-axis as being produced by a continuous series of poles. If each pole produces a field m/r at a distance r, then the flux produced in the upper half-plane by that pole will be πm; but, if the pole occupies a length dx of the x-axis, the flux is also $2B_N dx$, so that $m = 2B_N/\pi$. Then, integrating along the x-axis gives the field at the origin produced by the poles as

$$\frac{1}{\pi}\int_{-L_e}^{-L}\frac{2B_N}{x}\,dx - \frac{1}{\pi}\int_{L}^{L_e}\frac{2B_N}{x}\,dx. \tag{4.39}$$

Adding this to the field (B_e) at infinity gives the diffusion-region inflow field as

$$B_i = B_e - \frac{4B_N}{\pi}\log\frac{L_e}{L}. \tag{4.40}$$

Remembering that in the switch-off limit slow shocks travel at the Alfvén speed based on the normal field, $B_N/\sqrt{\mu\rho} = v_e$, we can rewrite Eq. (4.40) as

$$B_i = B_e\left(1 - \frac{4M_e}{\pi}\log\frac{L_e}{L}\right), \tag{4.41}$$

which is the expression for B_i that we have been seeking.

Since $M_e \ll 1$ and $B_i \approx B_e$, the scalings (4.37) and (4.38) reduce to

$$\frac{L}{L_e} \approx \frac{1}{R_{me}M_e^2}, \quad \frac{l}{L_e} \approx \frac{1}{R_{me}M_e}, \tag{4.42}$$

which show that the dimensions of the central region decrease as the magnetic Reynolds number (R_{me}) or reconnection rate (M_e) increase. Petschek suggested that the mechanism chokes itself off when B_i becomes too small, and so he estimated a *maximum* reconnection rate (M_e^*) by putting

$B_i = \frac{1}{2}B_e$ in Eq. (4.41) to give

$$\boxed{M_e^* \approx \frac{\pi}{8 \log R_{me}}.}\qquad(4.43)$$

In value this lies typically in the range 0.1–0.01 since $\log R_{me}$ is slowly vary-ing, so we see that for typical R_{me} values the reconnection is much faster than the Sweet–Parker rate.

It should be noted that, although in a one-dimensional annihilation solution (§3.2) it is possible to have a formal matching of the external (ideal) and internal (diffusive) regions using matched asymptotic expansions, this does not appear to be feasible for Petschek's two-dimensional magnetic field solution. Instead, the matching is done in an integral or approximate sense, using the order-of-magnitude Sweet–Parker relations to link the internal and external regions. This is entirely reasonable and is shown by numerical experiments to work well when the resistivity is enhanced in the diffusion region, as is expected in many applications (§5.4).

Thus, for twenty years the problem of fast reconnection was thought to have been solved completely (by Petschek), until in the 1980s a new gener-ation of reconnection solutions were discovered (§§5.1, 5.2), with Petschek's mechanism as a special case, and high-resolution numerical experiments were undertaken (§5.4). It is now realised to be one member of a whole family of fast reconnection regimes (Chapter 5), which are likely to oc-cur in many applications when the magnetic diffusivity in the diffusion region is enhanced. In particular, the Petschek regime itself occurs when the ambient magnetic field is basically one-dimensional rather than hav-ing an X-type topology (§5.1) and when the distant boundary conditions are free so that the MHD characteristics propagate away from the diffusion region.

4.4 Early Attempts to Generalise and Analyse Petschek's Mechanism

Petschek's mechanism is not rigorous, but it shows great physical insight. Few understood it at the time that Petschek proposed it. Part of the difficulty was that it involves a combination of diffusion, potential theory, and MHD shock theory.

Parker had set the diffusion-region length as $L = L_e$, the external scale-length, and had used Eq. (4.11) to deduce the reconnection rate (v_i). A key to Petschek's mechanism, however, is that Eq. (4.11) merely deter-mines L once v_i has been prescribed from external conditions. Thus, the diffusion region responds to the driving, and the Sweet–Parker relations (4.9) and (4.11) simply determine the dimensions (l and L) of the diffusion region, which shrink as η becomes smaller. The Sweet–Parker relations imply that, as v_i varies between $v_A/R_{me}^{1/2}$ and v_A, so the length of the diffusion region decreases from a maximum value of L_e to a minimum value of l (Axford, 1967). Moreover, the reconnection rate can vary over a wide range, depending on the boundary conditions rather than the magnetic diffusivity (η): if the plasma is free to escape, reconnection can be rapid at the maximum rate, whereas, if there is a pressure build-up downstream, the reconnection may be throttled to an intermediate rate (formally demonstrated by Jardine and Priest, 1988c and by Priest and Lee, 1990).

One modification of Petschek's mechanism is to allow the external field to decrease along the shocks away from the diffusion region. This makes the field lines between pairs of shocks curve in a manner that is concave towards the diffusion region (Priest, 1972a; Vasyliunas, 1975). In turn this implies that the field lines reverse their tangential components as they are carried through the shocks (Petschek and Thorne, 1967). The shocks are therefore intermediate in character and split into pairs of intermediate waves and slow shocks in the compressible case. Intermediate shocks or waves are also required when the reconnection is asymmetric (Heyn et al., 1988). Another modification is a nonlinear version of Petschek's mechanism (Roberts and Priest, 1975) in which the shocks are inclined at angles that are not small.

Yeh and Axford (1970) sought self-similar solutions for the stream function (ψ) and magnetic flux function (A) of the form

$$\psi = rg(\theta), \quad A = rf(\theta), \tag{4.44}$$

so that the magnetic components are

$$B_r = \frac{1}{r}\frac{\partial A}{\partial \theta} = f'(\theta), \quad B_\theta = -\frac{\partial A}{\partial r} = -f(\theta),$$

i.e., they are functions of θ alone [and similarly for the velocity components in terms of $g(\theta)$]. The MHD equations then give ordinary differential equations for $f(\theta)$ and $g(\theta)$, which may be solved subject to the appropriate boundary conditions of symmetry, namely $g = f' = 0$ on $\theta = 0$ and $\theta = \frac{1}{2}\pi$. This was an excellent idea, but unfortunately the solutions contain a discontinuity at OL which Vasyliunas (1975) showed in general to be unacceptable physically or mathematically. A particular example of the Yeh–Axford solutions was discovered independently by Sonnerup (1970), in which each quadrant of the external region contains three uniform regions

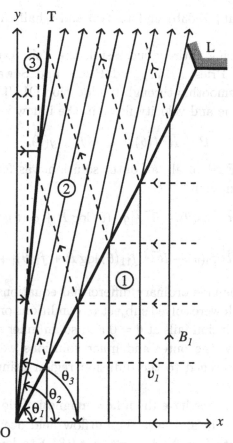

Fig. 4.7. Magnetic field lines (solid) and streamlines (dashed) for the first quadrant of Sonnerup's model with a Petschek shock (OT) and an extra discontinuity (OL).

linked by the discontinuity OL and the Petschek shock. Vasyliunas (1975) realised that, although the Sonnerup solution is acceptable mathematically, it is unlikely to occur in space or astrophysical plasmas since the discontinuity is not generated at the diffusion region at the origin; rather, it is generated by an external point or corner in the flow (this can be seen by imagining the field lines moving inwards and upwards from the right in Fig. 4.7, so that the kink in the field lines at OL is seen to propagate in towards O, while the one at OT propagates outwards from O). Furthermore, Sonnerup's discontinuity is an expansion fan rather than a shock, so it would broaden as it approaches the origin. Despite its shortcomings, Sonnerup's model has an elegance and simplicity which has made it feasible to generalise to asymmetric states (Cowley, 1974a; Mitchell and Kan,

1978; Semenov et al., 1983b) and to two-and-a-half dimensions (Cowley, 1974b).

In spite of not being entirely successful, the Yeh–Axford attempt stimulated Soward and Priest (1977) and Priest and Soward (1976) to seek solutions that pass smoothly through the location OL. These solutions have lowest-order magnetic and velocity fields in the inflow region of the form

$$B = \overline{R}^{\frac{1}{2}} f(\theta), \quad v = \overline{R}^{-\frac{1}{2}} g(\theta), \tag{4.45}$$

where $\overline{R} = \log_e r$. First of all, an outer solution was found having stream function and flux function

$$\psi = r \left\{ \overline{R}^{-\frac{1}{2}} g_0(\theta) + \overline{R}^{-\frac{3}{2}} [g_{11}(\theta) \log \overline{R} + g_1(\theta)] + \cdots \right\},$$

$$A = r \left\{ \overline{R}^{\frac{1}{2}} f_0(\theta) + \overline{R}^{-\frac{1}{2}} [f_{11}(\theta) \log \overline{R} + f_1(\theta)] + \cdots \right\}.$$

The MHD equations give ordinary differential equations (for the unknown functions of θ) which were solved subject to conditions on the inflow streamline $\theta = \frac{1}{2}\pi$. This solution fails at $\theta = 0$ and so an inner solution was sought near $\theta = 0$. Finally, the outer and inner solutions were matched to one another and then patched to the diffusion region using the Petschek approach.

The resulting solutions have the inflow magnetic field decreasing like $\overline{R}^{\frac{1}{2}}$ as the plasma comes in and have the outflow field decreasing like $\overline{R}^{-\frac{1}{2}}$ as the plasma goes out. The shock angle is $\pi/(8\overline{R})$ to lowest order, and the reconnection rate (M_e) is indeed found to possess a maximum value, as Petschek had suggested. The Petschek mechanism has therefore been put on a reasonably sound mathematical basis, at least as far as the analysis of the external region is concerned.

4.5 Compressibility

In the original Sweet–Parker model and in many subsequent analyses of reconnection, compressibility was neglected and the density was regarded as uniform. However, astrophysical plasmas are compressible, so a more sophisticated model is needed that includes density variations.

Suppose, for example, we have a simple Sweet–Parker sheet with inflow and outflow densities ρ_i and ρ_o, respectively, and acceleration by the magnetic force, as in §4.2.1. The diffusion equation, Eq. (4.9), is unchanged, but mass continuity, Eq. (4.11), is modified to

$$\rho_i L v_i = \rho_o l v_o,$$

so that the inflow speed, Eq. (4.12), becomes

$$v_i^2 = \frac{\rho_o}{\rho_i} \frac{\eta \, v_o}{L}.$$ (4.46)

Furthermore, the momentum balance, Eq. (4.18), is altered to

$$\rho_o \frac{v_o^2}{L} = \frac{B_i B_o}{\mu l},$$

so that the outflow speed, Eq. (4.20), becomes

$$v_o = \frac{B_i}{\sqrt{\mu \rho_o}} = v_{Ai} \left(\frac{\rho_i}{\rho_o}\right)^{\frac{1}{2}}.$$ (4.47)

In other words, the outflow speed is a hybrid Alfvén speed based on the inflow magnetic field and outflow density. The effect of compressibility is therefore to slow down the outflow by a factor $(\rho_i/\rho_o)^{\frac{1}{2}}$ when $\rho_o > \rho_i$.

The resulting effect on the reconnection rate (v_i), which from Eq. (4.46) becomes

$$v_i = \left(\frac{\rho_o}{\rho_i}\right)^{\frac{1}{4}} \left(\frac{\eta v_{Ai}}{L}\right)^{\frac{1}{2}},$$ (4.48)

is to enhance it by a factor $(\rho_o/\rho_i)^{\frac{1}{4}}$ when $\rho_o > \rho_i$. The basic reason for such an enhancement is that for given inflow conditions the width of the sheet is decreased by the compression.

Now, the question arises – what determines the ratio ρ_o/ρ_i ? We noted in Section 4.2.2 that the pressure ratio p_o/p_i is essentially a free parameter which may be imposed from the boundary conditions. Also, we remarked in Section 4.2.3 that in an incompressible plasma p and $\mathbf{v}$ are essentially determined from the equations of motion and continuity, while $\mathbf{B}$ follows from the equations of Ohm and Gauss, and a heat energy equation determines the temperature. When compressibility is added, there is an extra variable (the density), which may be regarded as being determined by an extra equation (namely the gas law $p = \mathcal{R}\rho T$). Alternatively, one may regard the force and continuity equations as determining ρ and $\mathbf{v}$, while the gas law determines the pressure. In both cases the net effect is the same. Thus, if the pressure ratio (p_o/p_i) is imposed, the density ratio $\rho_o/\rho_i = (p_o/p_i)(T_i/T_o)$ is given in terms of it and the temperature ratio, which is in turn determined by the heat energy equation. This procedure has been followed by Tur and Priest (1978) to deduce the properties of a reconnecting current sheet in the solar atmosphere.

A further point to note is that, in the presence of compressibility, the Petschek discontinuities become genuine slow-mode shocks, whereas in the incompressible case they are essentially Alfvénic discontinuities. Compressibility in turn means that the heat energy equation is important in determining the variations of density in the flow, and these in turn depend on the particular sources and sinks of energy that are important (such as heat conduction, viscous heating, radiation, and small-scale mechanical heating processes), which vary from one application to another.

4.6 Structure of the Diffusion Region

The diffusion region constitutes the whole structure of the Sweet–Parker mechanism. In turn, the Sweet–Parker mechanism also forms the central core of the fast reconnection mechanisms that we have discussed so far, although most of the energy conversion takes place in the external shocks. The internal structure of such a diffusion region has not yet been studied in detail, and to do so in a comprehensive manner would be a large undertaking since it depends on the particular form of heat energy equation that is appropriate in each application. Here we give a brief summary of some of the main approaches.

4.6.1 Local Behaviour Near the Null Point

Provided that a Taylor series expansion of the steady MHD equations about the neutral point (N) is valid, Priest and Cowley (1975) deduced that the magnetic field lines are locally cusp-like rather than X-type, with the separatrices in the diffusion region touching one another rather than intersecting at a non-zero angle. (When the flow velocity cannot be Taylor-expanded, however, solutions with X-type separatrices do exist; Uzdensky and Kulsrud, 1998.) The Priest–Cowley result may be proved as follows. In terms of a flux function (A) and stream function (ψ) such that $\mathbf{B} = \nabla \times (A\hat{\mathbf{z}})$ and $\mathbf{v} = \nabla \times (\psi\hat{\mathbf{z}})$, the equations of motion and Ohm for a steady, incompressible plasma may be written

$$\mu\rho\left(\frac{\partial\psi}{\partial y}\frac{\partial}{\partial x} - \frac{\partial\psi}{\partial x}\frac{\partial}{\partial y}\right)\nabla^2\psi = \left(\frac{\partial A}{\partial y}\frac{\partial}{\partial x} - \frac{\partial A}{\partial x}\frac{\partial}{\partial y}\right)\nabla^2 A, \qquad (4.49)$$

$$E + \frac{\partial\psi}{\partial x}\frac{\partial A}{\partial y} - \frac{\partial\psi}{\partial y}\frac{\partial A}{\partial x} = -\eta\nabla^2 A, \qquad (4.50)$$

where the electric field (E) is uniform and constant. Now assume the flow and magnetic field are analytic and symmetrical, with $B_x = v_y = 0$ on the x-axis and $B_y = v_x = 0$ on the y-axis. Then series expansions may be

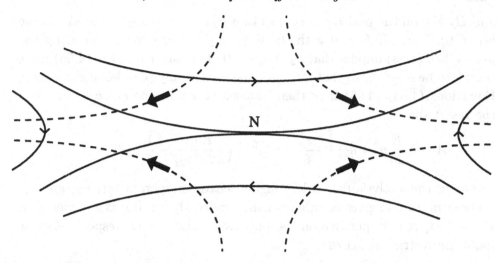

Fig. 4.8. The form of the magnetic field lines (solid) and streamlines (dashed) near the neutral point in a diffusion region when the variables can be Taylor-expanded.

assumed in the following forms:

$$A = (a_{20}x^2 + a_{02}y^2) + (a_{40}x^4 + a_{22}x^2y^2 + a_{04}y^4) + \cdots,$$
$$\psi = b_{11}xy + \cdots. \tag{4.51}$$

Substituting these forms into Eqs. (4.49) and (4.50) and equating the coefficients of powers of x and y up to second order yield

$$a_{02}(6a_{40} + a_{22}) - a_{20}(6a_{04} + a_{22}) = 0, \tag{4.52}$$
$$E + 2\eta(a_{20} + a_{02}) = 0, \tag{4.53}$$
$$b_{11}a_{20} - \eta(6a_{40} + a_{22}) = 0, \tag{4.54}$$
$$b_{11}a_{02} + \eta(6a_{04} + a_{22}) = 0. \tag{4.55}$$

However, these equations for the coefficients a_{ij} and b_{ij} are not linearly independent. The following consistency condition may be derived from the first and the last two equations:

$$b_{11}\, a_{02}\, a_{20} = 0. \tag{4.56}$$

In other words, either $b_{11} = 0$ or $a_{02}a_{20} = 0$. In the first case, the derivatives of the fluid velocity vanish at the neutral point as well as the velocity itself: thus the streamlines are not hyperbolic near N but the magnetic field lines are. In the second case, either a_{02} or a_{20} must vanish, so that the field lines are locally straight to lowest order, and to next order the field lines touch at N (Fig. 4.8). Suppose, for example, that $B_x > 0$ on the positive y-axis and $B_x > 0$ on the positive x-axis, so that $a_{20} \le 0$ and $a_{02} \ge 0$ (and assume $b_{11} \ne 0$). Then, if $E < 0$ with the fluid approaching the origin along the y-axis, Eq. (4.53) implies that $a_{20} + a_{02} > 0$. The solution of Eq. (4.56) must therefore be $a_{20} = 0$, so that the field lines are locally parallel to the x-axis. Equations (4.53)–(4.55) may then be used to determine a_{02}, a_{40}, a_{04}, with the result that

$$A = -\frac{E}{2\eta}y^2 + a_{22}\left[-\frac{x^4}{6} + x^2y^2 + \left(\frac{Eb_{11}}{12\eta^2a_{22}} - \frac{1}{6}\right)y^4\right] + \cdots.$$

Pursuing the analysis two orders higher leaves the parameters a_{22} and b_{11} undetermined and gives no other consistency condition. If it is assumed that B_x and B_y remain positive on the positive y- and x-axes, respectively, the following restriction arises

$$\frac{Eb_{11}}{\eta^2} > 2a_{22} > 0.$$

Instead of the form of Eq. (4.51), Sonnerup (1988) has considered expansions of the type

$$A = b_0(x) + y^2b_2(x) + y^4b_4(x) + \cdots,$$
$$\phi = y\,a_0(x) + y^3a_2(x) + \cdots. \tag{4.57}$$

Substituting them into the steady incompressible MHD equations and equating the two lowest powers of y yield three ordinary differential equations for five unknown functions of x (i.e., a_0, a_2, b_0, b_2, b_4). Each subsequent power of y gives two new equations for two extra coefficients (such as a_4, b_6). Thus, two of the original unknowns (such as a_0 and b_2) may be chosen at will and then the others follow, a procedure for which Sonnerup gave some examples. The fact that a_n is asymmetric in x while b_n is symmetric implies that $a_0'(0)\,b_0''(0)\,b_2(0) = 0$, which is essentially the same as Eq. (4.56).

4.6.2 Transverse Structure

Milne and Priest (1981) focussed on trying to determine the structure across the sheet (in the y-direction). Along the y-axis Ohm's Law becomes

$$E - v_yB_x = -\eta\frac{dB_x}{dy}, \tag{4.58}$$

and the y-component of the equation of motion reduces to

$$p + \frac{B_x^2}{2\mu} = \text{constant} \tag{4.59}$$

when $v_y^2 \ll v_A^2$. The equation of continuity

$$\frac{\partial}{\partial x}(\rho v_x) + \frac{\partial}{\partial y}(\rho v_y) = 0$$

presents a difficulty because of the need to evaluate the first term: the simplest rough way is to approximate it by $\rho v_x/L$ in terms of the sheet half-length (L) and to find v_x from the x-component of the equation of motion. If the magnetic force is neglected, the result is

$$\frac{d}{dy}(\rho v_y) = -\frac{[2\rho(p-p_o)]^{\frac{1}{2}}}{L}, \tag{4.60}$$

although, as we have seen in Section 4.2.2 the magnetic force should be included too. Equations (4.58)–(4.60) together with an energy equation and equation of state $(p = \mathcal{R}\rho T)$ have been used to determine $v_y(y), B_x(y)$, $p(y)$, $\rho(y)$, and $T(y)$ in current sheets, and from them a *beta-limitation* has been discovered such that steady reconnection is impossible when the plasma beta is too small.

4.6.3 Longitudinal Structure

Priest (1972b) calculated the structure along the sheet (in the x-direction) by assuming the sheet is thin and using a standard boundary-layer theory, known as Pohlhausen's method. The equations were scaled and small terms neglected and the resulting partial differential equations were integrated across the sheet. By assuming the forms of certain shape factors for the variations across the sheet, the MHD equations reduce to a set of ordinary differential equations for the variations along the sheet and for the variation of the boundary-layer thickness.

A simpler but similar order-of-magnitude approach was adopted by Somov (1992), who wrote Ohm's Law as

$$E + v_x B_y = -\eta \frac{B_i}{l}, \tag{4.61}$$

where v_x, B_i, B_y, l are functions of x, E is constant and the current $(j_z \approx -\mu^{-1}\partial B_x/\partial y)$ has been written roughly as $-B_i/(\mu l)$ in terms of the inflow external field $[B_i(x)]$ and the sheet width $[l(x)]$. The y-component of the equation of motion was for slow flow approximated by

$$p = p_i + \frac{B_i^2}{2\mu} \tag{4.62}$$

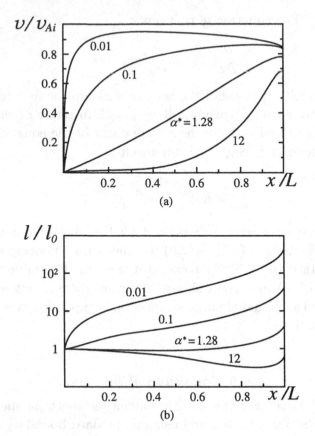

Fig. 4.9. Variation of (a) velocity $v(x)$ and (b) sheet width $l(x)$ along the sheet for $\beta_i = 0.1$ and various values of α^* ($\alpha^* < 1$ corresponds to faster inflow than the Sweet–Parker value). (From Somov, 1992).

and so determines the pressure $[p(x)]$ in the sheet. The equation of continuity was written as

$$\frac{d}{dx}[\rho\, v_x\, l(x)] = -\rho_i\, v_i. \tag{4.63}$$

The x-component of the equation of motion becomes

$$\frac{d}{dx}(\rho\, v_x^2\, l) = -l\frac{dp}{dx} + \frac{B_i B_y}{\mu}, \tag{4.64}$$

and finally an energy equation of the form

$$\frac{d}{dx}\left[\left(\frac{1}{2}\rho v_x^2 + \frac{5}{2}p\right)v_x l\right] = -\left(\frac{5}{2}p_i + \frac{B_i^2}{\mu}\right)v_i \tag{4.65}$$

was adopted, in which a thermal and kinetic energy outflux is provided by the thermal and magnetic enthalpy influx. (Cooling by radiation and thermal conduction was neglected.) The inflow density (ρ_i) and pressure (p_i) were assumed constant, while the inflow speed (v_i) was written as

$$v_i = \frac{E}{B_i},$$

assuming a frozen-in field outside the sheet with $B_x \gg B_y$. Finally, the inflow field (B_i) was assumed to take the form

$$B_i = -B_{i0}(1 - x^2/L^2)^{\frac{1}{2}},$$

appropriate for a simple current sheet, and so Eqs. (4.61)–(4.65) determine $v_x, B_y, l, p,$ and ρ as functions of x along the sheet, subject to the boundary conditions that v_x and B_y vanish at the middle $(x = 0)$ of the sheet. The resulting velocity and sheet-width vary along the sheet in a manner that depends on the parameters $\beta_i = 2\mu p_i/B_{i0}^2$ and $\alpha^* = (v_{sp}/v_{i0})^2$, which is the square of the ratio of the Sweet–Parker inflow speed [given by Eq. (4.12)] to the inflow speed at $x = 0$. The results are plotted for $\beta_i = 0.1$ in Fig. 4.9, in which $v_{Ai} = B_{i0}/(\mu\rho)^{\frac{1}{2}}$ is the inflow Alfvén speed at $x = 0$ and l_0 is the sheet width there.

Steady Reconnection: New Generation
of Fast Regimes

When solving partial differential equations, either analytically or numerically, the form, value, and number of the boundary conditions is of crucial importance. Indeed, often much physics is incorporated in the boundary conditions and, in the setting up of a numerical experiment with nonstandard boundary conditions, it is often the implementation of the boundary conditions that causes the most trouble. Petschek's mechanism, in which the boundary conditions at large distances are implicit, has been generalised in two distinct ways by adopting different boundary conditions to give regimes of *almost-uniform reconnection* (§5.1) and *non-uniform reconnection* (§5.2). Whereas Petschek's mechanism may be described as being almost-uniform and potential (§4.3), the first of these new families is in general nonpotential and the second is nonuniform. Also, surprisingly late in the day, a theory of *linear reconnection* was developed, which occurs when the reconnection rate is extremely slow (§5.3).

5.1 Almost-Uniform Non-Potential Reconnection

Vasyliunas (1975) clarified the physics of Petschek's mechanism by pointing out that the inflow region has the character of a diffuse fast-mode expansion, in which the pressure and field strength continuously decrease and the flow converges as the magnetic field is carried in. (This characterization of the inflow does not mean that a standing fast-mode wave is present in the inflow, since such a standing wave is not possible in a sub-fast flow.) A fast-mode disturbance has the plasma and magnetic pressure increasing or decreasing together, while a slow-mode disturbance has the plasma pressure changing in the opposite sense to the magnetic pressure. An expansion makes the pressure decrease, while a compression makes it increase, even in the incompressible limit. Sonnerup's model possesses discrete

146

slow-mode expansions that are unlikely to be found in astrophysics or space physics because they are generated externally rather than near the X-point. Vasyliunas suggested that a Sonnerup-like solution may, however, be possible with a diffuse slow-mode expansion spread throughout the inflow region, making the field strength increase, the pressure decrease, and the flow diverge as the field lines are carried in, although he was unable to find such a solution.

We wanted to understand Vasyliunas's distinction mathematically and were also puzzled at many strange features of some of the numerical reconnection experiments, such as much longer diffusion regions than in the standard Petschek mechanism, diverging flows, and large pressure gradients. Also, what is the relation to the stagnation-point flow solution? Can a Sonnerup-like solution be found without the extra discontinuities? And can a model in a finite region be produced, as in the numerical experiments?

During a summer collaboration we (Priest and Forbes, 1986) tried to answer such questions by seeking fast, steady, almost-uniform reconnection solutions to the equations for two-dimensional, incompressible flow, namely

$$\rho(\mathbf{v} \cdot \nabla)\mathbf{v} = -\nabla p + (\nabla \times \mathbf{B}) \times \mathbf{B}/\mu, \tag{5.1}$$

$$\mathbf{E} + \mathbf{v} \times \mathbf{B} = 0, \tag{5.2}$$

where $\nabla \cdot \mathbf{v} = 0$, $\nabla \cdot \mathbf{B} = 0$, $\mathbf{j} = \nabla \times \mathbf{B}/\mu$, and $\mathbf{E}$ is constant. In fast steady reconnection models (§4.2.4) the object is to analyse the inflow region and determine the relation between the inflow Alfvén Mach number ($M_e = v_e/v_{Ae}$) at large distances (i.e., the top of the box in Fig. 5.1) and the Alfvén Mach number ($M_i = v_i/v_{Ai}$) at the entrance to the Sweet–Parker diffusion region (at the bottom of the box in Fig. 5.1). Since the velocity in the inflow region is uniform to lowest order, we need only calculate the magnetic field strengths (B_e and B_i) at large distance and at the entrance to the diffusion region. We do this by seeking solutions in the form

$$\mathbf{B} = \mathbf{B}_e + M_e\mathbf{B}_1 + \cdots, \quad \mathbf{v} = M_e\mathbf{v}_1 + \cdots,$$

which correspond to a small perturbation about a uniform field ($\mathbf{B}_e = B_e\hat{\mathbf{x}}$), where the reconnection rate ($M_e \ll 1$) is the expansion parameter.

Now, if we arbitrarily assume at this stage that the pressure gradient is negligible to lowest order, then Eq. (5.1) becomes

$$j_1 B_e = 0. \tag{5.3}$$

Since the field components $(B_{1x}, B_{1y}) = (\partial A_1/\partial y, -\partial A_1/\partial x)$ can be written

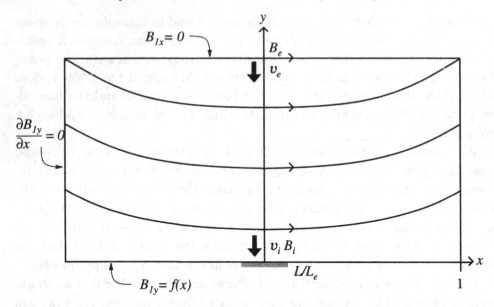

Fig. 5.1. Notation and boundary conditions for fast reconnection solutions. The subscript e refers to external quantities such as the inflow boundary parameters and the size of the box; subscript i refers to values at the inflow to the diffusion region (shaded).

in terms of a flux function (A_1), this reduces to Laplace's equation

$$\nabla^2 A_1 = 0. \tag{5.4}$$

Then, suppose we impose as boundary conditions that $B_{1x} = 0$ on the top boundary (so that $B_x = B_e$ is prescribed there), $\partial B_{1y}/\partial x = 0$ on the side boundaries (which is a rough form of free-floating condition), and on the lower boundary $(y = 0)$

$$B_{1y} = f(x) = \begin{cases} 2B_{\mathrm{N}}, & L < x < L_e, \\ 2B_{\mathrm{N}}\, x/L, & -L < x < L, \\ -2B_{\mathrm{N}}, & -L_e < x < -L, \end{cases}$$

in order to model the effect of the diffusion region and shocks. Here $B_{\mathrm{N}} \approx M_e B_e$ is the normal field component at the slow shocks. Note that the final solution is relatively insensitive to the assumed profile of B_{1y} in the diffusion region, since the diffusion region is often much shorter than the shocks when $R_{me} \gg 1$.

A separable solution of this system is

$$B_{1x} = -\sum_0^\infty a_n \cos\left[\left(n+\tfrac{1}{2}\right)\pi\frac{x}{L_e}\right]\sinh\left[\left(n+\tfrac{1}{2}\right)\pi\left(1-\frac{y}{L_e}\right)\right],$$

$$B_{1y} = \sum_0^\infty a_n \sin\left[\left(n+\tfrac{1}{2}\right)\pi\frac{x}{L_e}\right]\cosh\left[\left(n+\tfrac{1}{2}\right)\pi\left(1-\frac{y}{L_e}\right)\right],$$

where

$$A_1 = \sum_0^\infty \frac{a_n L_e}{(n+\tfrac{1}{2})\pi}\cos\left[\left(n+\tfrac{1}{2}\right)\pi\frac{x}{L_e}\right]\cosh\left[\left(n+\tfrac{1}{2}\right)\pi\left(1-\frac{y}{L_e}\right)\right] \quad (5.5)$$

and

$$a_n = \frac{4B_N\sin\left[\left(n+\tfrac{1}{2}\right)\pi L/L_e\right]}{L/L_e\left(n+\tfrac{1}{2}\right)^2\pi^2\cosh\left[\left(n+\tfrac{1}{2}\right)\pi\right]}. \quad (5.6)$$

This represents a Petschek-type solution with a weak fast-mode expansion. From Eq. (5.2) the first-order flow $[v_1 = (E/B_0)\hat{y}]$ is uniform, as expected, but the second-order flow is converging. These expressions enable us to calculate B_i/B_e as a substitute in Eq. (4.36) in place of Eq. (4.41). The resulting graphs of M_e against M_i for given $R_{me}(= L_e v_{Ae}/\eta)$ are shown in Fig. 5.2, from which it can be seen that for a given R_{me} there is indeed a maximum reconnection rate (M_e^*), as Petschek had suggested. Furthermore, the variation of M_e^* with R_{me} [the curve marked $b=0$ in Fig. 5.2(b)] is very close to Petschek's estimate (dashed curve). This analysis puts Petschek reconnection on a firmer foundation and models it in a finite box, such as one finds in numerical experiments.

It is always good to derive someone else's result in your own way, to make it real for yourself and understand in depth his or her assumptions. In the process of carrying out the above analysis, we realised that we could generalise it by relaxing one of the assumptions. What we decided to do is include pressure gradients, so that terms dp_1/dy and $-(\mu/B_e)dp_1/dy$ are added to the right-hand sides of Eqs. (5.3) and (5.4), which then becomes a form of Poisson's equation, namely,

$$\nabla^2 A_1 = -\frac{\mu}{B_e}\frac{dp_1}{dy}.$$

The effect on the solution [Eq. (5.5)] is simply to subtract a constant (b) from the x-dependent part of each term in the sum for A_1, which

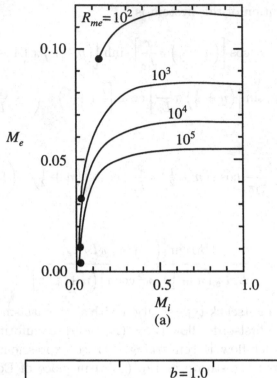

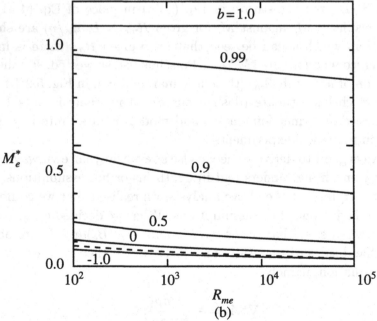

Fig. 5.2. (a) Reconnection rate (M_e) as a function of the inflow Alfvén Mach number (M_i) at the diffusion region for Petschek-like reconnection ($b = 0$). (b) Maximum reconnection rate (M_e^*) as a function of global magnetic Reynolds number (R_{me}) for different values of the parameter b.

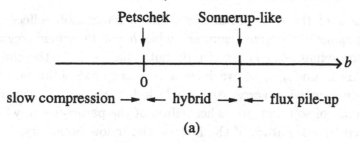

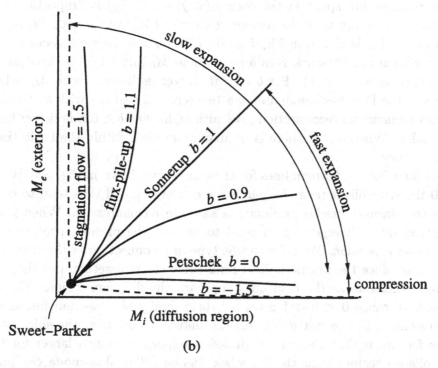

Fig. 5.3. (a) The different regimes of fast reconnection for different values of b. (b) Reconnection rate ($M_e = v_e/v_{Ae}$) as a function of $M_i = v_i/v_{Ai}$ for different b.

becomes

$$A_1 = \sum_0^\infty \frac{a_n L_e}{(n + \frac{1}{2})\pi} \left\{ \cos\left[\left(n + \frac{1}{2}\right)\pi \frac{x}{L}\right] - b \right\} \cosh\left[\left(n + \frac{1}{2}\right)\pi\left(1 - \frac{y}{L}\right)\right].$$

(5.7)

By evaluating the magnetic field at the entrance to the diffusion region in terms of B_e, we then showed that

$$\left(\frac{M_e}{M_i}\right)^2 \approx \frac{4M_e(1-b)}{\pi}\left[0.834 - \log_e \tan\left(\frac{4R_{me}M_e^{1/2}M_i^{1/2}}{\pi}\right)^{-1}\right].$$

The introduction of the new parameter b has a remarkable effect. It produces a whole range of different regimes: when $b = 0$, Petschek's regime (a weak fast-mode expansion) is recovered; but when $b = 1$, the inflow field on the y-axis is uniform, so we have a Sonnerup-like solution with a weak slow-mode expansion across the whole inflow region. In general, there is a continuum of solutions for other values of the parameter b, which can be determined by the nature of the flow on the inflow boundary, since the horizontal flow speed at the corner $(x, y) = (L_e, L_e)$ is proportional to $(b - 2/\pi)$. The way that the reconnection rate (M_e) varies with M_i and b for a given R_{me} is shown in Fig. 5.3(b). When $b > 0$ the reconnection rate is faster than the Petschek rate for the same M_i (although the analysis is only accurate for $M \ll 1$). For $b = 1$, M_e increases linearly with M_i, while for $b = 0$ the Petschek maximum can be seen. All regimes with $b < 1$ also possess a maximum reconnection rate, although when $b < 0$ it is slower than Petschek's. When $b > 1$, there is no maximum rate, within the limitations of the theory.

The field lines and streamlines for these cases are shown in Fig. 5.4. When $b < 0$ the streamlines near the y-axis are converging and thus tend to compress the plasma, thereby producing a *slow-mode compression*. When $b > 1$ the streamlines diverge and so tend to expand the plasma, producing a *slow-mode expansion*. We refer to this type of reconnection as the *flux pile-up regime*, since the magnetic field lines come closer together and the field strength increases as the field lines approach the diffusion region. The intermediate range $0 < b < 1$ gives a hybrid family of slow- and fast-mode expansions, with the fast-mode regions tending to occur at the sides. Another feature is that the central diffusion regions are much larger for the flux pile-up regime than the Petschek regime. The slow-mode compression regime also has central diffusion regions that are much larger than in Petschek's regime, but these solutions are rather slow – slower even than the Sweet–Parker regime.

The main results from the above analysis are that the type of reconnection regime and the rate of reconnection depend sensitively on the parameter b which characterises the inflow boundary conditions. Petschek $(b = 0)$ and Sonnerup-like $(b = 1)$ solutions are just particular members of a much wider class. As one progresses through the different regimes, the maximum reconnection rate may increase and the diffusion region lengthen – when it becomes too long it may become unstable to secondary tearing (§6.2) and a new regime of *impulsive bursty reconnection* results (Priest, 1986; Lee and Fu, 1986). The almost-uniform theory has been extended to include nonlinearity in the inflow (Jardine and Priest, 1988a), compressibility (Jardine

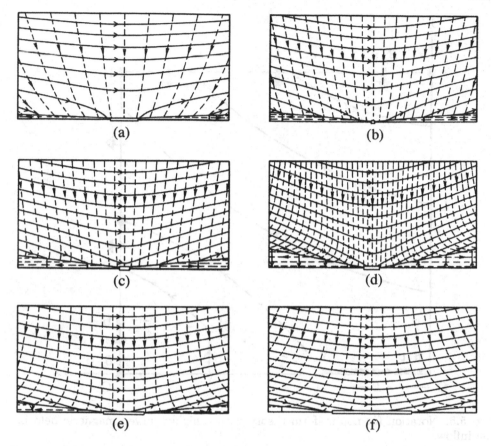

Fig. 5.4. Magnetic field lines (solid) and streamlines (dashed) in the upper half-plane $(y > 0)$ for several different regimes of almost-uniform reconnection: (a) slow compression $(b < 0)$, (b) Petschek $(b = 0)$, (c) hybrid expansion $(0 < b < 2/\pi)$, (d) hybrid expansion $(2/\pi < b < 1)$, (e) Sonnerup $(b = 1)$, (f) flux pile-up $(b > 1)$.

and Priest, 1989), energetics (Jardine and Priest, 1988a,b, 1990), and re-verse currents (Jardine and Priest, 1988c). Also, these solutions have been compared with a variety of numerical experiments (Forbes and Priest, 1987), as described in Section 5.4.

5.2 Non-Uniform Reconnection

Let us now consider what happens when the magnetic field lines in the inflow region are not approximately straight, but are instead highly curved. The existence of such curved field lines implies external field sources which are as important as the field produced by the diffusion region or slow-mode shocks. The terms "potential" and "nonpotential" reconnection are used

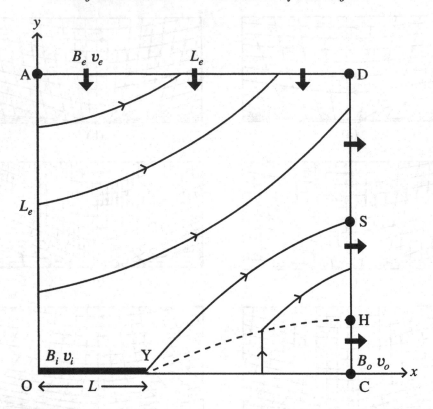

Fig. 5.5. Notation for non-uniform theory with a highly curved magnetic field in the inflow.

below to refer to the nature of the magnetic field in the inflow (external) region upstream of the diffusion region and slow-mode shocks.

5.2.1 Potential Reconnection

In order to avoid the assumption of small curvature inherent in the almost-uniform theory, Priest and Lee (1990) considered a strongly curved potential magnetic field in the inflow. They wrote the plasma velocity and magnetic field in terms of a stream function (ψ) and flux function (A) as

$$v_x = \frac{\partial \psi}{\partial y}, \quad v_y = -\frac{\partial \psi}{\partial x}, \quad B_x = \frac{\partial A}{\partial y}, \quad B_y = -\frac{\partial A}{\partial x}. \qquad (5.8)$$

Then the problem of non-uniform reconnection with an imposed inflow (v_e) and field strength (B_e) on the inflow boundary producing a diffusion region (OY) of half-length L with a separatrix YS and slow shock YH (Fig. 5.5) can be solved in three steps.

First of all, in the upstream region ahead of YH suppose, for simplicity, that both the plasma speed (v) and sound speed are much smaller than the Alfvén speed (v_A). Then the equation of motion, Eq. (5.1), implies that $j = 0$, and so for a potential field with the diffusion region treated as a current sheet we can use complex-variable theory to pick

$$B_y + iB_x = B_i \left(\frac{z^2}{L^2} - 1 \right)^{1/2},$$ (5.9)

where $z = x + iy$ and there is a cut (a current sheet) from $z = -L$ to $z = L$. Furthermore, Eq. (5.2) implies that the flow velocity may be deduced from

$$\psi = v_e B_e \int \frac{ds}{B},$$ (5.10)

where the integral is along a field line.

The second step is to calculate the position of the shock from the characteristic curve

$$\psi + A = \text{constant}$$

that passes through the end-point (Y) of the diffusion region. Then the shock relations are applied to deduce the conditions just downstream of YH. Finally, the MHD equations in the downstream region are solved subject to the appropriate boundary conditions at the shock and at the outflow boundary CH. In general these equations may be written

$$(\mathbf{v} \cdot \nabla) A = -v_e B_e,$$ (5.11)

$$(\mathbf{v} \cdot \nabla) \omega = \mathbf{B} \cdot \nabla j,$$ (5.12)

where $v_e B_e$ is the electric field and $\omega = -\nabla^2 \psi$ is the vorticity.

For example, the results of assuming $v \gg v_A$ so that

$$\omega = -\nabla^2 \psi = 0,$$

with ψ imposed along the boundary YHC and $\psi = $ constant on YC, are shown in Fig. 5.6. This exhibits many of the properties of Biskamp's experiment (§5.4). The field lines are highly curved because of the form of the inflow boundary conditions. Reversed current spikes at the ends of the diffusion region are revealed as regions of reversed magnetic curvature which slow down the inflowing plasma and divert it along the separatrices. Also, a separatrix jet occurs for long diffusion regions because the inflow to the diffusion region is diverted along the separatrix. The basic features of such a (vortex-current) separatrix layer have been discussed

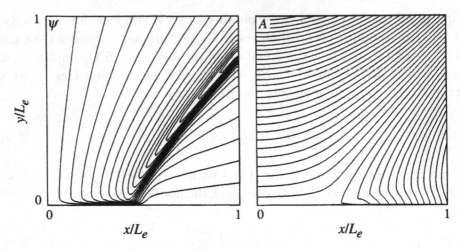

Fig. 5.6. Streamlines (left) and field lines (right) in one quadrant for non-uniform reconnection with a current-sheet half-length of $L = 0.4L_e$ (Priest and Lee, 1990).

by Soward and Priest (1986). Self-similar solutions have been presented by Sonnerup and Wang (1987) but not yet matched to the external flow. The shock (strictly speaking, an Alfvénic discontinuity of slow-mode compressional type in this incompressible model) is rather weak and has little effect on the magnetic field. As the current sheet decreases in length, so the inflow speed increases up to a value that depends on the inflow Alfvén Mach number. Results were also obtained by solving the full equations [Eqs. (5.11) and (5.12)] in the downstream region (Priest and Lee, 1990).

5.2.2 Non-Potential Reconnection

The scaling of diffusion region half-length (L) with magnetic Reynolds number in the Priest and Lee (1990) solution is not the same as in the Biskamp (1986) experiment. The reason is that in the Priest and Lee solution, like Petschek's potential solution, there is no extra degree of freedom which allows us to impose the extra particular boundary condition of Biskamp. Nevertheless, it is possible to generalise this potential solution as follows to include pressure gradients and so produce a nonpotential non-uniform solution (Strachan and Priest, 1994).

In the inflow region, the equation of magnetostatic balance ($\mathbf{j} \times \mathbf{B} = -\nabla p$) with $(B_x, B_y) = (\partial A / \partial y, -\partial A / \partial x)$ reduces to

$$\nabla^2 A = -\mu j(A). \tag{5.13}$$

If the flux function is written as $A = A_p + A_{np}$, where A_p is the potential solution that Priest and Lee (1990) used, and A_{np} is an extra nonpotential part, Eq. (5.13) becomes

$$\nabla^2 A_{np} = -\mu j(A_p + A_{np}),\qquad(5.14)$$

which in general is a tough equation to solve analytically because of the nonlinear function of $A_p + A_{np}$ on the right-hand side. However, there does exist a simple solution with uniform current, so that $\mu j(A) = \bar{c} B_e/L$, say, with $\bar{c}$ constant, for which the field becomes

$$B_y + iB_x = B_i \left(\frac{z^2}{L^2} - 1 \right)^{\frac{1}{2}} - i\bar{c}\frac{B_e}{L}y.\qquad(5.15)$$

Here $\bar{c}$ is a parameter analogous to the parameter b in the almost-uniform theory, since $\bar{c} < 0$ tends to produce converging flow and $\bar{c} > 0$ diverging flow. Also, $\bar{c}$ can be determined by imposing an appropriate boundary condition. Biskamp imposed $v_y = 0$, $B_x = 0$ on the outflow boundary and $v_y(x)$ on the inflow. His key condition, however, was also to prescribe the normal component of the magnetic field (namely, $B_y = B_e x/L$) on the inflow boundary. Although we do not have the freedom in our analytical theory to impose the functional form of $B_y(x)$ along this boundary, we can impose the same value as he did at the corner ($x = L_e$) where $B_y = B_e$ and use this to determine $\bar{c}$. The resulting scalings of L and l with M_e and R_{me} are shown in Fig. 5.7.

When $L/L_e \ll 1$, the relation between M_i and M_e reduces to

$$\frac{M_i}{M_e} \approx \frac{1}{M_e^2 R_{me}}.\qquad(5.16)$$

As Fig. 5.8 shows, the relation between M_i and M_e is quite different from that found previously for the almost-uniform theory, since now M_i is inversely proportional to M_e, whereas before it was proportional to M_e. (The closest similarity between the non-uniform and almost-uniform solutions occurs for the compressive branch, $b < 0$, when M_i is large.) Because of this change in the relation between M_i and M_e, the length L in the Strachan and Priest (1994) theory behaves very differently from the uniform theory. This different scaling is due to the fact that the orthogonal X-type background field now gives B_i tending to zero as M_e tends to zero, while the background field in the uniform theory gives B_i tending to B_e instead. As B_i approaches zero, the Alfvén speed at the entrance to the diffusion region vanishes, but the inflow velocity there approaches $R_{me}^{-1/2} v_{Ae}$. Thus, the Alfvén Mach number (M_i) there approaches infinity.

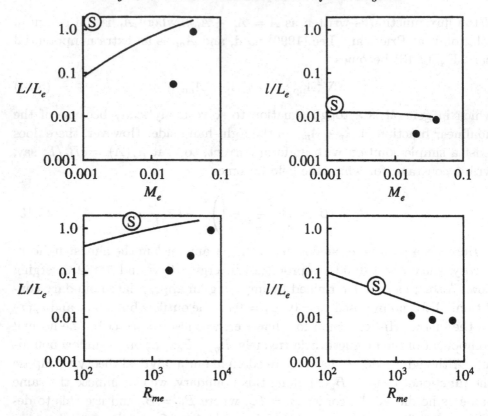

Fig. 5.7. The scaling of the half-length (L) and half-width (l) of the diffusion region with reconnection rate (M_e) and magnetic Reynolds number (R_{me}), showing a comparison between Biskamp's numerical experiments (dots) and non-uniform nonpotential theory (solid curves) when it includes the extra Biskamp boundary condition.

The similarity between the Strachan–Priest solution and the almost-uniform solution for the compressive case ($b = -10$) is not accidental. Along the y-axis the pressure (p) in the former solution is

$$p(0, y) = p_0 - \frac{B^2}{2\mu} - \frac{M_e^2 B_e^4}{2\mu B^2},$$

where p_0 is a constant. For $L/L_e \ll 1$ this pressure increases as one approaches the diffusion region, and, since at the same time the magnetic field decreases, such a behaviour is indicative of a slow-mode compression. However, for $L \approx L_e$, the Strachan–Priest solution gives a modest flux pile-up. For $L/L_e \ll 1$, both it and the Biskamp solutions are much slower than the Sweet–Parker rate (compare with Fig. 5.8), but when $L \approx L_e$ the reconnection becomes slightly faster than the Sweet–Parker rate because of the slight

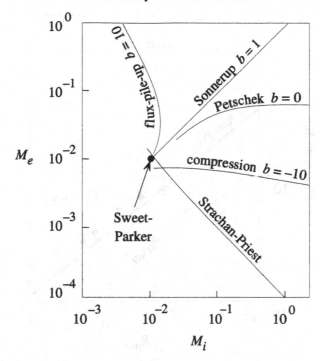

Fig. 5.8. External Alfvén Mach number (M_e) versus internal Alfvén Mach number (M_i) for various steady-state theories when the magnetic Reynolds number (R_{me}) is 10^4. In general, M_e is determined by the imposed boundary conditions, except for the Sweet–Parker theory, which has only one possible value (after Priest and Forbes, 1992b).

flux pile-up that occurs. This shows up in Fig. 5.8 as a slight extension of the Strachan–Priest curve beyond the Sweet–Parker point.

The maximum Alfvén Mach number, which occurs when $L = L_e$, is

$$M_e^* = \left[\frac{2}{(\sqrt{5} - 1)} \right]^{3/4} \frac{1}{R_{me}^{1/2}} \approx \frac{1.43}{R_{me}^{1/2}}. \tag{5.17}$$

The factor 1.43 is due to the modest flux pile-up that occurs when $L = L_e$. Thus, when the normal magnetic field component is fixed at the inflow boundary, the Strachan–Priest reconnection rate scales like the slow Sweet–Parker rate, as in Biskamp's experiment (Fig. 5.9).

The Strachan–Priest theory can adopt boundary conditions similar to those in Biskamp's simulations. As in the simulations, the normal magnetic field along the inflow may be fixed, so that the orthogonal magnetic field structure is always maintained. However, since the analytical theory does not have precisely the same boundary conditions as the experiments, we

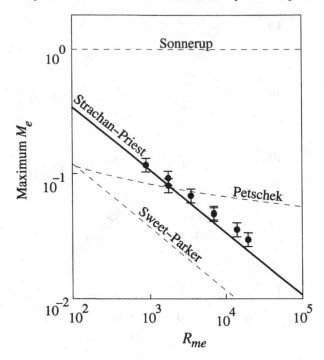

Fig. 5.9. The variation of the maximum reconnection rate with the magnetic Reynolds (Lundquist) number for various theories. When the theory by Strachan and Priest uses boundary conditions similar to those of Biskamp's simulations, it agrees with Biskamp's simulation results (dots with error bars). Note that there are no free parameters in the curves obtained from the analytical theories.

do not expect to be able to reproduce the experimental scalings exactly, but they do have very similar qualitative characteristics. Unlike the original Petschek scaling, we now have L increasing with both M_e and R_{me}, albeit not as rapidly as in the experiment. Furthermore, the maximum reconnection rates are in very close agreement with the theoretical and experimental values, scaling as $3.4\ R_{me}^{-1/2}$ and $3.5\ R_{me}^{-1/2}$, respectively (Fig. 5.9). The agreement between Biskamp's numerical results and the Strachan–Priest theory is really very good despite the slight differences in the functional form of the normal velocity component at the top of the box. Thus, we conclude that it is a combination of the X-type background and the fixed value of the normal magnetic field on the inflow boundary which produces these scalings and prevents fast reconnection in Biskamp's experiment.

Lee and Fu (1986) and Jin and Ip (1991) have also presented results on numerical reconnection experiments. Their scalings are different from Biskamp (1986) and have also been compared with the theory by Priest and Forbes

(1992b). Furthermore, Scholer (1989) has studied numerically the response to a local magnetic diffusivity enhancement. He imposes free-floating conditions (see Forbes and Priest, 1987) and finds Petschek reconnection and scaling, provided there is a small region of diffusivity enhancement. But, if the diffusivity is uniform, the diffusion-region current sheet grows with time. As we shall see in Section 7.2, the continual growth of the sheet is to be expected under these circumstances.

Lee and Fu (1986) have also undertaken a numerical experiment on steady reconnection, but they find different scaling results from Biskamp. The most noticeable of these is that the length (L) of the diffusion region decreases as R_{me} increases, whereas Biskamp found the opposite. Lee and Fu's scaling implies that the maximum reconnection rate increases with R_{me}, so for large magnetic Reynolds numbers there is no upper limit to the reconnection rate.

5.3 Linear (Super-Slow) Diffusion and Reconnection

Craig and Rickard (1994) have described a theory for steady linear X-point behaviour with no flow across the separatrices (§5.3.4). The magnetic field lines are carried towards the separatrices by a flow which turns sharply to become parallel to them. The field lines then diffuse across the separatrices, rather than being transported across them – we therefore refer to this process as *reconnective diffusion* to distinguish it from classical reconnection. As we shall see, the linear analysis used to obtain this type of solution requires $M_e < R_{me}^{-1}$, so that the process is super-slow in the standard usage (§1.6.1).

Let us consider here steady, two-dimensional incompressible flow with uniform density (ρ), magnetic diffusivity (η), and kinematic viscosity (ν). The plasma velocity $\mathbf{v}(x, y)$, magnetic field $\mathbf{B}(x, y)$, pressure $p(x, y)$, and electric current $\mathbf{j}(x, y)$ satisfy

$$\mathbf{E} + \mathbf{v} \times \mathbf{B} = \eta \nabla \times \mathbf{B}, \tag{5.18}$$

$$\rho(\mathbf{v} \cdot \nabla)\mathbf{v} = -\nabla p + \mathbf{j} \times \mathbf{B} + \rho \nu \nabla^2 \mathbf{v}, \tag{5.19}$$

where

$$\nabla \cdot \mathbf{v} = 0, \quad \nabla \cdot \mathbf{B} = 0, \tag{5.20}$$

$$\mathbf{j} = \frac{\nabla \times \mathbf{B}}{\mu}, \tag{5.21}$$

and Faraday's law $\nabla \times \mathbf{E} = \mathbf{0}$ implies that $\mathbf{E} = E\,\hat{\mathbf{z}}$ is uniform in space.

In the linear diffusion and reconnection models of this section, we linearise about an equilibrium X-point by assuming an extremely slow flow (slower than the global diffusion speed η/L_e). Therefore, in this section we are using the phrase "linear reconnection" to signify that the flow is *everywhere* very slow. This is unlike, say, the Petschek mechanism, where, although the flow is much slower than the Alfvén speed (v_A) in the external region, it does become comparable with v_A both at the exit from the diffusion region and after passing through the shocks. Before proceeding further, we first prove a general theorem which limits greatly the types of solution that exist for linear reconnection. Then we set up the equations (§5.3.2) and consider the ideal (§5.3.3) and reconnective diffusion solutions (§5.3.4), and, finally, we mention the much more complex reconnection solutions that do have flow across the separatrices (§5.3.5).

5.3.1 Anti-Reconnection Theorem

Theorem. Steady MHD reconnection in two dimensions with plasma flow across separatrices is impossible in an inviscid plasma with a highly sub-Alfvénic flow and a uniform magnetic diffusivity.

Proof. For an inviscid plasma whose flow speed is everywhere much smaller than the Alfvén speed ($v \ll v_A$), the equation of motion [Eq. (5.19)] reduces to the force balance

$$\mathbf{j} \times \mathbf{B} = \nabla p. \tag{5.22}$$

Taking the curl of this equation gives

$$\nabla \times (\mathbf{j} \times \mathbf{B}) = \mathbf{0},$$

or, expanding out the triple vector product,

$$\mathbf{j}(\nabla \cdot \mathbf{B}) + (\mathbf{B} \cdot \nabla)\mathbf{j} - \mathbf{B}(\nabla \cdot \mathbf{j}) - (\mathbf{j} \cdot \nabla)\mathbf{B} = \mathbf{0}. \tag{5.23}$$

However, by Eqs. (5.20) and (5.21), $\nabla \cdot \mathbf{B}$ and $\nabla \cdot \mathbf{j}$ vanish. Also, for a two-dimensional field [$B_x(x,y)$, $B_y(x,y)$] the current from Eq. (5.22) is directed purely in the z-direction, so that $(\mathbf{j} \cdot \nabla)\mathbf{B} = j_z\, \partial \mathbf{B}/\partial z = \mathbf{0}$. Thus Eq. (5.23) reduces to

$$(\mathbf{B} \cdot \nabla)j = 0, \tag{5.24}$$

which implies that j is constant along each magnetic field line, in particular along the separatrices (the field lines which pass through the X-point). This can be seen by introducing a coordinate (s) along a field line so that Eq. (5.24) becomes

$$B\frac{dj}{ds} = 0.$$

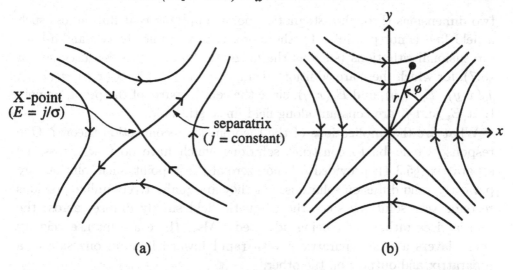

Fig. 5.10. (a) Magnetic configuration for linear two-dimensional reconnection. (b) Notation for a zeroth-order X-point field.

Thus, dj/ds vanishes and j is constant along a field line provided $B \neq 0$. For an isolated neutral point, if we make the additional assumption that j is continuous, the constant value of j along a separatrix will equal the value at the neutral point through which the separatrix passes.

Now, at a magnetic neutral point (such as the X-point at which we are considering reconnection to occur; Fig. 5.10(a)), $\mathbf{B}$ vanishes and so Ohm's law [Eq. (5.18)] reduces to

$$E = \frac{j}{\sigma},\qquad (5.25)$$

where σ is the electrical conductivity. However, everywhere along separatrices through such an X-point, both E and j are constant, so Ohm's law [Eq. (5.18)] implies that

$$\mathbf{v} \times \mathbf{B} = 0. \qquad (5.26)$$

In other words, there is no flow across the separatrices. Q.E.D.

Examples of solutions with no flow across the separatrices have been found by Craig and Rickard (1994) for a particular case of a linearised configuration about an X-type field with a uniform current. Note that the theorem applies not just for a linearised flow about a potential field (the case that we consider in the next section) but also for any general nonlinear magnetostatic field in two dimensions. In addition, it holds for any field line that passes in two dimensions through a stagnation point, implying that flow across such a field line is not possible. Furthermore, it also applies to two-and-a-half-dimensional situations (such as the magnetic flipping process discussed in §8.7), for which there are additional components $v_z(x,y)$, $B_z(x,y)$, $j_x(x,y)$, $j_y(x,y)$, $E_x(x,y)$, and $E_y(x,y)$, since the central parts of the proof, namely that E_z and j_z are constant along field lines, still hold.

What are the implications of the theorem for reconnection theory? One response would be to construct solutions which have no flow across the separatrix (§5.3.4): provided E is non-zero at the X-point, such solutions are perfectly valid diffusion solutions, but they are qualitatively unlike previous reconnection solutions since the magnetic field simply diffuses across the separatrices rather than being advected. Also, there are intense velocity shear layers at the separatrices with rapid inward flow on one side of a separatrix and outflow on the other.

Another response to the theorem is to try and build reconnection solutions with the usual flow across the separatrix which violate at least one of the assumptions of the theorem. There are three ways of trying to do so within the framework of steady, two-dimensional MHD reconnection. The first is to allow for a reduction in magnetic diffusivity as one moves outwards along the separatrix, so that j/σ no longer balances E, but this does not seem reasonable if the variation in diffusivity is produced by variations in microturbulence or temperature, since both are related to the current density which still remains constant along the separatrix. The second is to invoke the nonlinear effect of the inertial term by having Alfvénic flow in the central diffusion region and/or the separatrix layers, as in the classical slow and fast regimes of reconnection discussed in the previous chapter. The third way around the theorem, which is the only satisfactory one for a conventional super-slow regime, is to include the effect of viscosity in the central diffusion region and/or separatrix layers. This is the strategy we adopt in the next section.

5.3.2 Basic Equations for Linear (Super-Slow) Diffusion and Reconnection

By writing the velocity and magnetic field as

$$\mathbf{v} = \nabla \times (\psi \hat{\mathbf{z}}), \quad \mathbf{B} = \nabla \times (A\hat{\mathbf{z}})$$

in terms of a stream function (ψ) and flux function (A), we find that Ohm's law, Eq. (5.18), and the equation of motion, Eq. (5.19), reduce to

$$E + (\mathbf{B} \cdot \nabla)\psi = -\eta \, \nabla^2 A, \tag{5.27}$$

$$\rho(\mathbf{v} \cdot \nabla)\omega = (\mathbf{B} \cdot \nabla)j + \rho\nu \, \nabla^2 \omega, \tag{5.28}$$

where

$$\mu j = -\nabla^2 A, \qquad \omega = -\nabla^2 \psi.$$

We can nondimensionalise these equations in terms of the values $v_{Ae}, B_e, -v_e$ at distance L_e by writing $\bar{\mathbf{v}} = \mathbf{v}/v_{Ae}$, $\bar{\mathbf{B}} = \mathbf{B}/B_e$, $\bar{\mathbf{r}} = \mathbf{r}/L_e$, $\bar{\psi} = \psi/(L_e v_{Ae})$, $\bar{\omega} = \omega L_e/v_{Ae}$, $\bar{A} = A/(L_e B_e)$, $\bar{j} = j\mu L_e/B_e$, and putting $E = v_e B_e$. Then, after dropping the overbars, we find that Eqs. (5.27) and (5.28) become

$$M_e + (\mathbf{B} \cdot \nabla)\psi = R_{me}^{-1} \nabla^2 A, \tag{5.29}$$

$$(\mathbf{v} \cdot \nabla)\omega = (\mathbf{B} \cdot \nabla)j + R_e^{-1} \nabla^2 \omega, \tag{5.30}$$

where $R_e(= L_e v_{Ae}/\nu)$ and $R_{me}(= L_e v_{Ae}/\eta)$ are the Reynolds and magnetic Reynolds numbers, respectively. For the solutions below we shall assume without loss of generality that $v_e < 0$ and $B_e > 0$, so that $E < 0$ and $M_e < 0$.

In the limit when the reconnection rate (M_e) vanishes, Eqs. (5.29) and (5.30) have the simple solution

$$\mathbf{v}_0 \equiv \mathbf{0}, \quad j_0 \equiv 0,$$

and so, in particular, we may consider the zeroth-order X-type magnetic field with components

$$B_{0r} = r\cos 2\phi, \quad B_{0\phi} = -r\sin 2\phi, \tag{5.31}$$

such that ϕ is measured from the line $y = x$ [Fig. 5.10(b)], and the flux function is

$$A_0 = \tfrac{1}{2}r^2 \sin 2\phi.$$

In terms of x and y the corresponding field components are $B_{0x} = y$, $B_{0y} = x$, with $A_0 = \tfrac{1}{2}(y^2 - x^2)$.

We now seek a solution for small reconnection rates by expanding in powers of M_e ($|M_e| \ll 1$),

$$A = A_0 + M_e \, A_1 + \cdots, \quad \psi = M_e \, \psi_1 + \cdots,$$

so that the linearised versions of Eqs. (5.29) and (5.30) become two equations for ψ_1 and j_1, namely

$$1 + (\mathbf{B}_0 \cdot \nabla)\psi_1 = \frac{j_1}{R_{me}}, \tag{5.32}$$

$$(\mathbf{B}_0 \cdot \nabla)j_1 = R_e^{-1} \nabla^2 (\nabla^2 \psi_1), \tag{5.33}$$

where the dimensionless first-order current and vorticity are

$$j_1 = -\nabla^2 A_1/\mu \quad \text{and} \quad \omega_1 = -\nabla^2 \psi_1.$$

Then j_1 may be eliminated between Eqs. (5.32) and (5.33) to give our basic equation for ψ_1, namely

$$(\mathbf{B}_0 \cdot \nabla)^2 \psi_1 = \epsilon \nabla^2 \nabla^2 \psi_1, \tag{5.34}$$

which may be written in polar coordinates as

$$\left(r \cos 2\phi \frac{\partial}{\partial r} - \sin 2\phi \frac{\partial}{\partial \phi} \right)^2 \psi_1 = \epsilon \left[\frac{1}{r} \frac{\partial}{\partial r} \left(r \frac{\partial}{\partial r} \right) + \frac{1}{r^2} \frac{\partial^2}{\partial \phi^2} \right]^2 \psi_1, \tag{5.35}$$

where

$$\epsilon = \frac{1}{R_{me} R_e} = \frac{\nu \eta}{(L_e v_{Ae})^2}.$$

5.3.3 Ideal (External) Solution

If $R_{me} \gg 1$, the solution of Eq. (5.32) with $\psi_1 = 0$ on $\phi = \pi/4$ is

$$\psi_1 = \tfrac{1}{2} \log |\tan \phi|, \tag{5.36}$$

so that the plasma velocity has components

$$v_{1r} = \frac{1}{r \sin 2\phi}, \quad v_{1\phi} = 0.$$

Thus, the streamlines radially converge on or diverge from the origin and the velocity becomes singular both at the origin ($r = 0$) and at the separatrices $\phi = 0$, $\pi/2$, π, and $3\pi/2$ (Fig. 5.11).

An alternative way of writing Eq. (5.36) is

$$\psi_1 = -\tfrac{1}{2} \sinh^{-1} \cot 2\phi,$$

or in terms of Cartesian coordinates

$$\psi_1 = -\tfrac{1}{2} \log \left| \frac{x + y}{x - y} \right|,$$

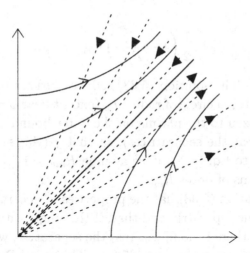

Fig. 5.11. A sketch of the external-region streamlines (dashed curves) and field lines (solid curves) in the first quadrant for linear (super-slow) reconnection.

with the velocity components

$$v_{1x} = -\frac{x}{x^2 - y^2}, \qquad v_{1y} = -\frac{y}{x^2 - y^2}.$$

5.3.4 Non-Ideal Reconnective Diffusion Solution

Now let us attempt to resolve the singularity in Eq. (5.36) by including the effects of resistivity. If $R_e \gg 1$, Eq. (5.33) implies that $j_1 = j_1(A_0)$, so that the current is constant along zeroth-order field lines. Then the general solution to Eq. (5.32) is of the form

$$\psi_1 = -\left(1 - \frac{j_1}{R_{me}}\right) \int \frac{ds}{B_0} = -\left(1 - \frac{j_1}{R_{me}}\right) \int \frac{r\, d\phi}{B_{0\phi}},$$

where $\mathbf{B}_0$ is given by Eq. (5.31) and the integration is carried out along a field line $A_0 = \text{const}$. Thus, by substituting $B_{0\phi}/r = -\sin 2\phi$ the integration may easily be carried out to yield

$$\psi_1 = \frac{1}{2}\left(1 - \frac{j_1}{R_{me}}\right) \log \left| \frac{\tan \phi}{\tan \phi_0} \right| + \psi_1^{(0)}(A_0), \qquad (5.37)$$

where $\psi_1 = \psi_1^{(0)}(A_0)$ on $\phi = \phi_0$.

Next, suppose that $\psi_1 = 0$ on $\phi = \pi/4$, so that the y-axis is a streamline and we have a symmetrical configuration; then $\psi_1^{(0)}(A_0) \equiv 0$ and Eq. (5.37) reduces to

$$\psi_1 = \frac{1}{2}\left(1 - \frac{j_1}{R_{me}}\right)\log|\tan\phi|, \tag{5.38}$$

where $j_1 = j_1(A_0)$. The solution (5.38) reduces further to Eq. (5.36) if we suppose that $j_1 = 0$ on a circle of radius L_e (except near $\phi = 0, \pi/2, \pi, 3\pi/2$): then the field is frozen to the plasma on the outer boundary $r = L_e$ (save in boundary layers near the separatrices), so that $j_1(A_0) \equiv 0$. Alternatively, Eq. (5.38) reduces to Eq. (5.36) if we assume $R_{me} \gg 1$, j_1 is of order unity, and we neglect terms of order R_{me}^{-1}.

The resistive solution (5.38) has the properties in general that the current is constant along the separatrix and that, if the singularity in ψ_1 is resolved by resistivity, then there is no flow across the separatrix, which is consistent with the anti-reconnection theorem (§5.3.1). Craig and Rickard (1994) considered electric currents that are highly concentrated near the separatrix, so that the function

$$f(A_0) \equiv 1 - \frac{j_1(A_0)}{R_{me}}$$

in Eq. (5.38) tends to zero when $A_0 \to 0$, and $f(A_0) \approx 1$ outside the separatrix boundary layer. However, there is no advection of magnetic flux across the separatrix, so that finite resistivity alone results only in a simple magnetic diffusion across the separatrices. Moreover, Priest et al. (1994c) pointed out that the arbitrariness in the choice of the form of $f(A_0)$ means that one can construct solutions with *any* prescribed thickness for the separatrix boundary layer (and therefore having any dissipation rate scaling with R_{me}) and not just the particular thickness adopted by Craig and Rickard.

The singularity at the separatrix is not caused by the singularity in the value of ψ_1 at the boundary $r = L_e$, say, but is intrinsic to the problem. Indeed, it may be shown more generally that the appearance of a separatrix singularity is an essential feature of linear X-point reconnection. For example, let us impose a continuous regular radial velocity distribution at the external boundary $r = L_e$ with inflow $[v_r(L_e, \phi) < 0]$ in one quadrant $(0 < \phi < \frac{1}{2}\pi)$ and outflow in another $(-\frac{1}{2}\pi < \phi < 0)$. Since $v_r = r^{-1}\partial\psi/\partial\phi$, this is equivalent to imposing the distribution of the flux function at this boundary $[\psi(L_e, \phi) = \Gamma(\phi)$, say]. Then, according to Eq. (5.32),

$$\frac{\partial\psi}{\partial s} = \frac{f(A_0)}{B_0},$$

and, after integrating this expression along a field line $A_0 = $ constant from footpoint 1 to footpoint 2 on the boundary $r = L_e$ $(0 < \phi < \frac{1}{2}\pi)$, we obtain

$$\psi(L_e, \phi_2) - \psi(L_e, \phi_1) \equiv \Delta\Gamma(A_0) = f(A_0) \int_1^2 \frac{ds}{B_0} \equiv f(A_0) \cdot V(A_0),$$

where

$$V(A_0) \equiv \int_1^2 \frac{ds}{B_0}$$

is the differential flux volume.

Thus, the given distribution of inflowing velocity determines $f(A_0)$, and so the electric current $j_1(A_0)$, from

$$f(A_0) = 1 - \frac{j_1(A_0)}{R_{me}} = \frac{\Delta\Gamma(A_0)}{V(A_0)}.$$

By inserting this $f(A_0)$ into expression (5.38), we obtain the stream function ψ_1 inside the quadrant $0 < \phi < \frac{1}{2}\pi$. Now, how does $f(A_0)$ behave near the separatrix? Since in this quadrant we have an inflow (or outflow), the function $\Delta\Gamma(A_0)$ tends to some nonzero positive (or negative) value when $A_0 \to 0$ (at the separatrix). At the same time, the function $V(A_0)$ becomes logarithmically divergent as A_0 approaches the separatrix value of zero, because B_0 behaves like r at the X-point. Thus, the function $f(A_0)$ behaves like

$$f(A_0) \sim \frac{1}{\ln A_0}$$

as $A_0 \to 0$. As a result, the components of the fluid velocity, which are determined by the derivatives of ψ_1 [i.e., the derivatives of $f(A_0)$], are singular near the separatrix inside the quadrant $0 < \phi < \frac{1}{2}\pi$ even when the imposed radial velocity distribution at the boundary $r = L_e$ is regular.

5.3.5 Reconnection Solutions

The natural next step is to try and include viscosity and so understand how a combination of viscous and resistive effects on the right-hand side of Eq. (5.34) can resolve the singularity in vorticity and current, so that the flow smoothly turns round and crosses the separatrix, connecting the regions above and below the separatrix in Fig. 5.11. Priest et al. (1994c) attempted to do so by constructing self-similar solutions in the resistive boundary layer with the stream function having a part that is a function of $r\phi/\epsilon^{1/4}$ alone. However, these still possess a weak discontinuity in the form of a jump in the third derivative of the stream function at the separatrix. Physically, a self-similar solution seems inappropriate because the separatrix boundary layer does not emanate from essentially a point at the origin but rather from a viscous region of size comparable with the boundary-layer width. This is borne out by a subsequent analysis of Titov and Priest (1997), who were successful in constructing reconnection solutions.

5.4 Related Numerical Experiments

Most numerical experiments of magnetic reconnection are highly time-dependent, but there are a few which have been able to achieve a nearly steady state, usually after an initial time-dependent phase. The first such experiments were those of Ugai and Tsuda (1977) and Sato (1979), which helped confirm the reality of the standing slow-mode shocks proposed by Petschek (1964). However, precise quantitative comparisons between numerical experiments and steady-state theory were not made until the mid-1980s, starting with the experiments by Biskamp (1986) and later Scholer (1989) and Yan et al. (1992, 1993). These latter studies were closely compared with Petschek's solution in particular, in order to determine whether the fast reconnection predicted by his solution occurs or not.

In his simulation Biskamp (1986) found that the reconnection rate was not fast, but slow with a scaling proportional to that predicted by the Sweet–Parker model (see Fig. 5.9). Both the length (L) and the thickness (l) of the diffusion region increased as M_e or R_{me} increases, which completely contradicts Petschek's result that L decreases with R_{me} (roughly as R_{me}^{-1}) and with M_e (roughly as M_e^{-2}). Thus, despite the presence of slow shocks, Biskamp realized that his numerical experiments had not found the scaling predicted by Petschek's model. Consequently, he concluded that Petschek's mechanism, and therefore fast reconnection, does not exist for large values of R_{me}. However, we have argued (Priest and Forbes, 1992b) that the boundary conditions used by Biskamp do not correspond to those required by Petschek's solution and that his particular scaling may be produced by adopting the appropriate boundary conditions (i.e., the Strachan–Priest solution, §5.2.2). We concluded, therefore, that fast reconnection can indeed exist provided the boundary conditions allow it.

Figure 5.12 shows a series of Biskamp's runs for $M_e = 0.042$ and $R_{me} = 1,746$ in the first column, $R_{me} = 3,492$ in the second column, and $R_{me} = 6,984$ in the third column. These show clearly an increase in the length of the diffusion region with R_{me}. They also possess several other interesting features:

(i) an inflow whose streamlines vary from converging to diverging;
(ii) inflow field lines which may be highly curved with a large shock angle;

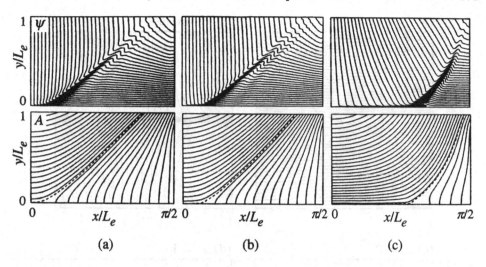

Fig. 5.12. Numerical experiment showing streamlines (top) and magnetic field lines (bottom) in the first quadrant (after Biskamp, 1986) for $M_e = 0.042$ and: (a) $R_{me} = 1,746$; (b) $R_{me} = 3,492$; (c) $R_{me} = 6,984$.

(iii) strong jets of plasma flowing out along the separatrices;
(iv) spikes of reversed current at the ends of the diffusion region.

The new generation of theoretical models in Sections 5.1 and 5.2 explains most of these properties. The family of almost-uniform models explains feature (i), and the family of non-uniform models is able to reproduce features (ii)–(iv).

In general, the number of boundary conditions that can be imposed equals the number of MHD characteristics that are propagating information into the region. For the case of ideal MHD flow that is two-dimensional, sub-Alfvénic, and incompressible, there are three imposed conditions (Forbes and Priest, 1987). If one prescribes, for example, the boundary conditions

$$v_x = 0, \quad v_y = \text{constant}, \quad p = \text{constant}$$

on the inflow boundary (AD in Fig. 5.5), then the MHD equations imply that

$$B_x = \text{constant}, \quad B_y = 0,$$

so that straight field lines are carried in by a uniform flow without curving and reconnection is impossible. If instead boundary conditions only slightly different from the above are imposed, then reconnection with weakly curved inflow field lines may be produced. Thus, it is entirely reasonable to expect that conditions greatly different from these could produce a highly curved inflow.

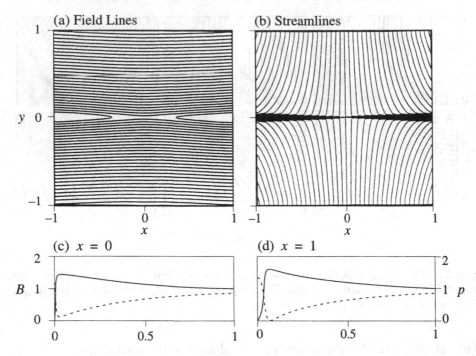

Fig. 5.13. (a) The magnetic field lines and (b) the streamlines for almost-uniform reconnection with $b = 2, M_e = 0.1, R_{me} = 2{,}000$. The variation of magnetic field (continuous curve) and pressure (dashed curve) along (c) $x = 0$ and (d) $x = 1$ (Yan et al., 1992).

In contrast to Biskamp, Yan et al. (1992) carried out numerical experiments with boundary conditions consistent with those required by Petschek's model, and they obtained fast reconnection with the scaling rate predicted by that model provided that a non-uniform resistivity is used. By proper choice of boundary conditions, Yan et al. (1992) recovered the entire family of almost-uniform solutions. On the inflow boundary they set $B_x = B_e$, $\partial j / \partial y = 0$, $v_y = -v_e$, and $\partial w / \partial y = 0$, and on the outflow boundary they put $\partial B_y / \partial x = 0$, $\partial w / \partial x = 0$, and imposed from the almost-uniform theory the value of v_y, which depends on b. 100×200 gridpoints were used in one quadrant and R_{me} was taken to be 2,000. When $b = 0$ a Petschek flow results; when $b = 1$ Sonnerup-like flow is found; when $b = 2$ they obtained an example of flux pile-up reconnection (Fig. 5.13), in which it can be seen that, as one approaches the diffusion region along the axis $x = 0$, the magnetic field increases and the plasma pressure decreases in a strong slow-mode expansion. When $b = -3$ they found an example of slow-mode compression reconnection with converging streamlines, in which the field decreases and the pressure increases as the diffusion region is approached.

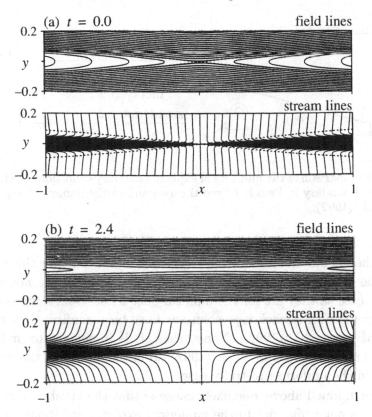

Fig. 5.14. Evolution in time of the numerical simulation of Yan et al. (1992) when the resistivity is uniform: (a) $t = 0.0$, (b) $t = 2.4$.

The configurations and scalings agree with almost-uniform theory, provided the magnetic diffusivity (η) is enhanced in the diffusion region.

Although this agreement between simulation and theory is conclusive, there is one aspect of their simulation that remains puzzling. When η is uniform throughout the numerical domain the steady-state solutions are no longer sustained, although the cause has not been identified. One possibility is that the slow-mode shocks become so thick that they are no longer resolved numerically. Another possibility is that the number of boundary conditions is no longer sufficient to specify a unique solution.

In order to persuade the simulation to match the theory, Yan et al. (1992) created a region of enhanced resistivity at the diffusion region. Specifically, they increased the resistivity in a strip whose width was chosen to match the length of the diffusion region predicted by the analytical solutions. However, when instead a uniform resistivity was used, the simulations developed a long current sheet extending the length of the box, as shown in Fig. 5.14.

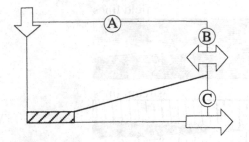

	inflow	outflow	
region	Ⓐ	Ⓑ	Ⓒ
theory	3	2	1
simulation	2	1	0

Fig. 5.15. Comparison of the numbers of explicit boundary conditions used on three parts of the boundary in Petschek-type theories and in the numerical experiments of Yan et al. (1992).

Since the ohmic electric field (j/σ) is negligible outside the diffusion region, the Petschek-type solutions are equally valid for the non-uniform resistivity case as they are for the uniform one. Thus, from a mathematical point of view, the Petschek-type solutions have been confirmed by the simulations of Yan et al. (1992). Nonetheless, it is important to understand why the simulations of Yan et al. (1992) do not converge to the analytical solutions in the uniform resistivity case.

As we mentioned above, one likely cause is that the number of boundary conditions is not sufficient. In the numerical experiment, about half of the boundary conditions required to obtain a unique analytical solution were replaced by open boundary conditions (sometimes referred to as free-floating or radiative boundary conditions). Figure 5.15 shows the number of boundary conditions which the analytical solutions require to be imposed at the boundary in order to specify a unique solution. In each boundary region the number required exceeds the number of explicit boundary conditions imposed in the simulation. In place of these explicit boundary conditions, the simulation imposes open boundary conditions that are specifically designed to make the boundary conditions as sensitive as possible to the initial conditions. This extensive use of open boundary conditions in Yan et al. (1992) allows the simulation boundary conditions to diverge away from the conditions required by Petschek's solution. As the numerical solution evolves away from the Petschek solution, so do the boundary conditions. In fact, the internal flow pattern and the boundary conditions track each other very much as predicted by the generalised solution of Priest and Forbes (1986). This type of difficulty, that is, the problem of persuading a time-dependent simulation to converge to a specific steady-state analytical solution, is well known in fluid dynamic simulations.

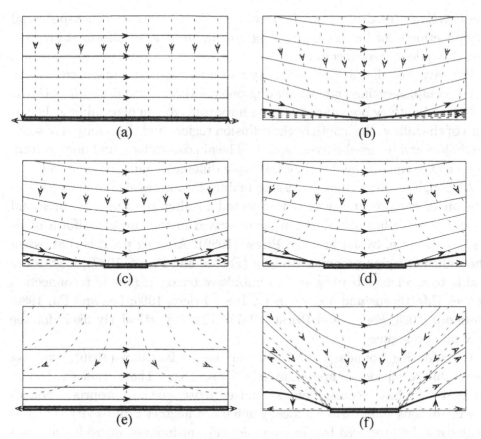

Fig. 5.16. Field lines and streamlines for various steady-state solutions with the X-line and diffusion region at the bottom centre of each box: (a) Sweet–Parker, (b) Petschek, (c) Sonnerup, (d) flux-pile-up, (e) stagnation-point flow, (f) Strachan–Priest. The shaded boxes show only the lengths of the diffusion regions (the thicknesses are not to scale).

5.5 Conclusions

The extensive differences in scaling laws of L and the maximum reconnection rate that are found in theories and experiments are clearly a direct consequence of the wide variety of boundary conditions that have been adopted (Fig. 5.16). The nature and value of the imposed boundary conditions are crucial, with much physics being hidden within them. The "correct" boundary conditions depend on the particular application of reconnection; spontaneous reconnection due to some localised instability and unaffected by distant magnetic fields would require free boundary conditions and would tend to produce potential reconnection, either almost-uniform or non-uniform, depending on the initial state; driven reconnection, in contrast, depends on the details of the driving but in general tends to give rise to non-potential reconnection – for instance, diverging flow in an almost-uniform state gives fast flux-pile-up reconnection.

We have seen that analytical theory and numerical experiment go hand in hand, complementing and stimulating one another. Conditions on both the inflow and outflow boundaries play an important role in determining the nature of the inflow, the length of the diffusion region, and the strength of separatrix jets and reversed-current spikes. The almost-uniform and non-uniform regimes are equally valid – they just have different boundary conditions.

Although we have been discussing mainly incompressible models, the inclusion of compressibility is not expected to alter the basic features and conclusions. It has already been incorporated in the almost-uniform models, for instance, by Jardine and Priest (1989). Another point is that, when the diffusion region becomes too long (Forbes and Priest, 1987), it goes unstable to secondary tearing and an impulsive bursty regime of reconnection ensues (Matthaeus and Montgomery, 1981; Priest, 1986; Lee and Fu, 1986; Biskamp, 1986; Forbes and Priest, 1987). This is particularly likely for the flux pile-up regime.

By adopting boundary conditions similar to Biskamp (1986), we have been able to reproduce approximately his scalings. These results therefore suggest that fast reconnection can indeed exist, given appropriate boundary conditions and possibly a locally enhanced magnetic diffusivity (§13.1.3) that could be produced by, for example, current-induced micro-instabilities in the diffusion region. The latter possibility is in fact highly likely in space and astrophysical applications because of the collisionless environment that typically occurs inside current sheets (§1.7). Therefore, we believe that fast reconnection, either almost-uniform or non-uniform, is a prime candidate for the rapid energy conversion that is often seen in solar, space, and astrophysical plasmas.

6

Unsteady Reconnection: The Tearing Mode

6.1 Introduction

In a conducting medium a typical current sheet tends to diffuse outward at a slow rate with a time-scale of $\tau_d = l^2/\eta$, where $2l$ is the width of the current sheet and $\eta = (\mu\sigma)^{-1}$ is the magnetic diffusivity. During the process of magnetic diffusion, magnetic energy is converted ohmically into heat at the same slow rate. However, in practice, the magnitude of τ_d is often far too large to explain the time-scale of dynamical cosmic processes. Nevertheless, Furth et al. (1963) showed how the diffusion can drive three distinct *resistive instabilities* at a rate which is often fast enough to be physically significant. These instabilities occur when the sheet is wide enough that $\tau_d \gg \tau_A$, where $\tau_A = l/v_A$ is the time it takes to traverse the sheet at the Alfvén speed $v_A = B_0(\mu\rho_0)^{-1/2}$. The instabilities occur on time-scales $\tau_d(\tau_A/\tau_d)^\lambda$, where $0 < \lambda < 1$, and they have the effect of creating in the sheet many small-scale magnetic loops. In other words, resistive instabilities produce current filaments in current sheets (or, indeed, in any sheared structure); subsequently, the filaments and associated magnetic loops diffuse away, releasing magnetic energy in the process.

The gravitational and rippling modes (§6.3) occur when the density $[\rho_0(x)]$ or resistivity $[\eta(x)]$ varies in the direction x across the sheet. They create a small-scale structure in the sheet (Fig. 6.1) and so are relatively harmless as far as the large-scale global stability of the configuration is concerned, although they may produce a turbulent diffusivity.

The third type of resistive instability is the tearing mode, which, in contrast to the other two, occurs with a wavelength greater than the width of the sheet ($kl < 1$). It has a growth-rate

$$\omega = [\tau_d^3 \tau_A^2 (kl)^2]^{-1/5}$$

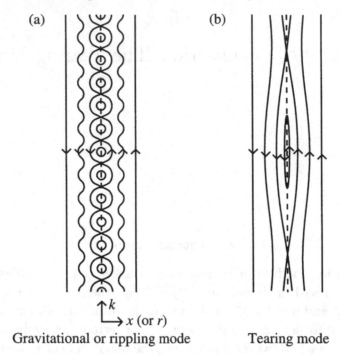

Gravitational or rippling mode   Tearing mode

Fig. 6.1. (a) Small-wavelength and (b) long-wavelength resistive instabilities in a current sheet or a sheared magnetic field, where x (or r in a cylindrical geometry) is the coordinate in a direction across the equilibrium magnetic field.

for wave-numbers (k) in the approximate range $(\tau_A/\tau_d)^{1/4} < kl < 1$. The smallest allowable wavelengths (l) grow in a time $\tau_d^{3/5}\tau_A^{2/5}$, whereas the longest wavelength, namely $(\tau_d/\tau_A)^{1/4}l$, is much larger than the width (l) of the sheet and has the fastest growth-rate, namely

$$\omega = \left(\frac{1}{\tau_d \, \tau_A}\right)^{1/2}.$$

Finite electrical conductivity effects are important only in a narrow layer of width $\epsilon l = (kl)^{-3/5}(\tau_A/\tau_d)^{-2/5}l$ about the centre of the sheet where the magnetic field vanishes. In this layer the magnetic field is able to slip rapidly through the plasma. The result is that, if one starts out with a one-dimensional sheet with straight field lines and then makes a perturbation such as the one in Fig. 6.1(b), the resulting forces are such as to make the perturbation grow. The magnetic tension tends to pull the new loops of field up and down away from the X-points, while the magnetic pressure gradient tends to push plasma in from the sides towards the X-points. Also, the field lines at the sides are curved and so possess a restoring magnetic tension force, which is minimised for long wavelengths.

The tearing mode is possibly the most important of the three resistive instabilities, since it has a long wavelength and requires neither a gravitational force nor a resistivity gradient to be excited. It may be important for magnetospheric substorms, laboratory plasmas, coronal heating, and solar and stellar flares (e.g., Shivamoggi, 1985).

The tearing mode also occurs often in a sheared field, not just a neutral sheet, since the analysis is largely unaffected by the addition of a constant magnetic field component normal to the plane of Fig. 6.1. Sheared fields are, in general, resistively unstable at many thin sheaths throughout the structure. At any particular location specified by the value of x (or r in a cylindrical flux tube), say, the instability has a vector wave-number ($\mathbf{k}$) in a direction normal to the equilibrium field ($\mathbf{B}_0$), that is,

$$\mathbf{k} \cdot \mathbf{B}_0 = 0,$$

so that the crests and troughs of the perturbation lie in the plane of Fig. 6.1. This contains the x-axis and the vector $\mathbf{k}$, since $\mathbf{B}_0$ is normal to that plane at $x = 0$.

In this chapter, we first present the basic stability analysis of the tearing mode (§6.2) and various modifications (§6.3), such as the influence of a steady flow or a normal field component and the double tearing mode. Then the effect of a flux tube geometry is considered (§6.4) in producing both ideal (§6.4.1) and resistive (§6.4.2) modes. Finally, various aspects of the nonlinear development are covered, including Rutherford saturation (§6.5.1), mode coupling (§6.5.2), and the coalescence instability (§6.5.3).

6.2 The Tearing-Mode Instability Analysis of Furth et al. (1963)

Consider a plasma at rest in a sheared equilibrium magnetic field

$$\mathbf{B}_0 = B_{0y}(x)\,\hat{\mathbf{y}} + B_{0z}(x)\,\hat{\mathbf{z}}, \tag{6.1}$$

whose field lines are confined to the yz-plane but rotate as one moves along the x-axis. Suppose departures from equilibrium satisfy the following equations of induction and vorticity for an incompressible plasma with a uniform magnetic diffusivity (η) and density:

$$\frac{\partial \mathbf{B}}{\partial t} = \nabla \times (\mathbf{v} \times \mathbf{B}) + \eta \nabla^2 \mathbf{B}, \tag{6.2}$$

$$\mu\rho \frac{D}{Dt}(\nabla \times \mathbf{v}) = \nabla \times [(\nabla \times \mathbf{B}) \times \mathbf{B}], \tag{6.3}$$

where

$$\nabla \cdot \mathbf{B} = 0, \quad \nabla \cdot \mathbf{v} = 0. \tag{6.4}$$

The vorticity equation, Eq. (6.3), is obtained by taking the curl of the equation of motion and so eliminating the plasma pressure. The assumption of incompressibility is reasonable since the resulting instability growth-time is much larger than the transit-time for MHD waves and the fluid speeds are subsonic.

Small perturbations of the initial state are assumed of the form

$$\mathbf{v}_1(x) \exp[i(k_y y + k_z z) + \omega t], \quad \mathbf{B}_1(x) \exp[i(k_y y + k_z z) + \omega t]. \tag{6.5}$$

Then the linearised x-component of the induction equation, Eq. (6.2), and the curl of the vorticity equation, Eq. (6.3), respectively, are

$$\omega B_{1x} = iv_{1x}(\mathbf{k} \cdot \mathbf{B}_0) + \eta(B_{1x}'' - k^2 B_{1x}), \tag{6.6}$$

$$\omega \left(v_{1x}'' - k^2 v_{1x}\right) = \frac{i(\mathbf{k} \cdot \mathbf{B}_0)}{\mu\rho} \left[-B_{1x} \frac{(\mathbf{k} \cdot \mathbf{B}_0)''}{\mathbf{k} \cdot \mathbf{B}_0} + \left(B_{1x}'' - k^2 B_{1x}\right)\right], \tag{6.7}$$

where $k^2 = k_y^2 + k_z^2$, a prime denotes a derivative with respect to x and Eq. (6.4) has been used.

Next, dimensionless variables are defined as

$$\overline{\mathbf{B}} = \frac{\mathbf{B}}{B_0}, \quad \overline{\mathbf{v}}_1 = -\mathbf{v}_1 \frac{ikl^2}{\eta}, \quad \overline{k} = kl, \quad \overline{\omega} = \frac{\omega l^2}{\eta}, \quad \overline{x} = \frac{x}{l}, \tag{6.8}$$

in terms of a typical field strength (B_0) and scale-length (l) with corresponding diffusion speed η/l and diffusion time l^2/η. Equations (6.6) and (6.7) then become

$$\overline{\omega}\overline{B}_{1x} = -\overline{v}_{1x} f + (\overline{B}_{1x}'' - \overline{k}^2 \overline{B}_{1x}), \tag{6.9}$$

$$\overline{\omega}(\overline{v}_{1x}'' - \overline{k}^2 \overline{v}_{1x}) = L_u^2 \overline{k}^2 f[-\overline{B}_{1x} f''/f + (\overline{B}_{1x}'' - \overline{k}^2 \overline{B}_{1x})], \tag{6.10}$$

where $f = \mathbf{k} \cdot \overline{\mathbf{B}}_0 / k$ and the Lundquist number

$$L_u = \frac{lv_A}{\eta} = \frac{\tau_d}{\tau_A} \tag{6.11}$$

is the ratio of the diffusion time (τ_d) to the Alfvén travel-time (τ_A) across the layer and is assumed to be much greater than unity. Often in the literature the notation S is used in place of L_u, but in this book S has already been given other meanings.

The assumption $L_u \gg 1$ means that the plasma is frozen to the magnetic field almost everywhere, and the diffusion term $(\eta\nabla^2 \mathbf{B})$ in Eq. (6.2) is negligible. The exception is in thin sheets where $\nabla \times (\mathbf{v} \times \mathbf{B})$ vanishes, that is, where f vanishes or $\mathbf{k} \cdot \mathbf{B}_0 = 0$. In such a sheet (of width $2\epsilon l$, say) the magnetic field lines diffuse through the plasma and reconnect. Let us suppose that the centre of the sheet is located at $x = 0$ and that $k_z = 0$, so that $\mathbf{k} \cdot \mathbf{B}_0 = 0$ reduces to $B_{0y} = 0$. Solutions to Eqs. (6.9) and (6.10) may then be obtained in an outer region ($|x| > \epsilon$) and an inner region ($|x| < \epsilon$) and can be patched (or, in a more precise analysis, mathematically matched) at the boundary between them.

In the *outer region*, terms of the order of L_u^{-1} are neglected by comparison with unity, so that Eqs. (6.9) and (6.10) become

$$\overline{\omega} \overline{B}_{1x} = - \overline{v}_{1x} \overline{B}_{0y} + (\overline{B}_{1x}'' - \overline{k}^2 \overline{B}_{1x}), \tag{6.12}$$

$$0 = - \overline{B}_{1x} \overline{B}_{0y}'' / \overline{B}_{0y} + (\overline{B}_{1x}'' - \overline{k}^2 \overline{B}_{1x}). \tag{6.13}$$

Given an equilibrium profile for $\overline{B}_{0y}(\overline{x})$, Eq. (6.13) can be solved for $\overline{B}_{1x}$. For instance, with the simple step-profile

$$\overline{B}_{0y} = \begin{cases} 1 & \text{for } \overline{x} > 1, \\ \overline{x} & \text{for } |\overline{x}| < 1, \\ -1 & \text{for } \overline{x} < -1, \end{cases} \tag{6.14}$$

the solution for $\overline{x} > 0$ which vanishes at large distances is

$$\overline{B}_{1x} = \begin{cases} a_1 \sinh \overline{k}\overline{x} + b_1 \cosh \overline{k}\overline{x} & \overline{x} < 1, \\ a_0 \exp(-\overline{k}\overline{x}) & \overline{x} > 1, \end{cases} \tag{6.15}$$

where the conditions that $\overline{B}_{1x}$ be continuous at $\overline{x} = 1$ and the integral of Eq. (6.13) across $\overline{x} = 1$ be satisfied imply

$$a_1 = a_0 \left[\exp(-\overline{k}) \left(\frac{\cosh \overline{k}}{\overline{k}} \right) - 1 \right],$$

$$b_1 = a_0 \left[1 - \exp(-\overline{k}) \left(\frac{\sinh \overline{k}}{\overline{k}} \right) \right], \tag{6.16}$$

for $\overline{x} > 0$. The corresponding conditions at $\overline{x} = -1$ give the same value of b_1 for $\overline{x} < 0$ but minus the above value of a_1. Thus, although $\overline{B}_{1x}$ is continuous at the origin, its slope is not. This means that the current density becomes large and the resistive term in the induction equation becomes important. Another reason why an inner region or singular layer is required near the origin is that the induction Eq. (6.12) implies that $\overline{v}_{1x} = -\overline{\omega} \overline{B}_{1x} / \overline{B}_{0y}$, so that the velocity becomes indefinitely large as $\overline{B}_{0y}$ approaches zero. The jump in $\overline{B}_{1x}' / \overline{B}_{1x}$ across the singular layer, denoted by Δ', is given from

Eqs. (6.15) and (6.16) as

$$\Delta' = \left[\frac{\overline{B}'_{1x}}{\overline{B}_{1x}}\right]_{0-}^{0+} = \frac{2a_1\overline{k}}{b_1}. \tag{6.17}$$

In particular, note that, when $\overline{k} \ll 1$, then $\Delta' \simeq 2/\overline{k}$.

In the *inner region* $\overline{B}_{0y}$ has become so small that diffusion is important in Eq. (6.9) and inertia is important in Eq. (6.10). The width of the region is of the order of ϵl, where

$$\epsilon^4 = \frac{\overline{\omega}}{4\overline{k}^2 L_u^2}. \tag{6.18}$$

In terms of new variables $X = \overline{x}/\epsilon$ and $V_{1x} = \overline{v}_{1x}(4\epsilon/\overline{\omega})$, Eqs. (6.6) and (6.7) become

$$\ddot{\overline{B}}_{1x} = \epsilon^2 \overline{k}^2 \overline{B}_{1x} + \epsilon^2 \overline{\omega}\left(\overline{B}_{1x} + \tfrac{1}{4}V_{1x}X\right), \tag{6.19}$$

$$\ddot{V}_{1x} = V_{1x}\left(\overline{k}^2\epsilon^2 + \tfrac{1}{4}X^2\right) + \overline{B}_{1x}X, \tag{6.20}$$

where overdots denote derivatives with respect to X. In order of magnitude, Eq. (6.19) implies that $\ddot{\overline{B}}_{1x} \sim \epsilon^2 \overline{\omega} \overline{B}_{1x}$ for long wavelengths ($\overline{k} \ll 1$). Thus,

$$\Delta' = 2\left(\frac{\overline{B}'_{1x}}{\overline{B}_{1x}}\right)_{x=\epsilon} = 2\left(\frac{\dot{\overline{B}}_{1x}}{\epsilon \overline{B}_{1x}}\right)_{X=1} \sim \epsilon \overline{\omega}.$$

If $\overline{B}_{1x}$ is constant in the inner region (the so-called *constant-psi approximation* in view of the original notation of ψ in place of $\overline{B}_{1x}$), Eq. (6.20) can be solved in terms of Hermite functions to give (when $\overline{k}\epsilon \ll 1$)

$$\Delta' = 3\,\epsilon\,\overline{\omega}. \tag{6.21}$$

The constant-psi approximation breaks down for the faster $m = 1$ resistive kink mode (§6.4.2) and the double tearing mode (§6.3.2) and for very large values of L_u and $\overline{k}^{-1}$. The outer and inner regions may be patched together by equating the two expressions (6.17) and (6.21) for Δ', so that (when $\overline{k} \ll 1$) $2/\overline{k} = 3\,\epsilon\overline{\omega}$, or, after substituting for ϵ from Eq. (6.18), $\overline{\omega} = [(8L_u)/(9\overline{k})]^{2/5}$. The solutions for the magnetic field and velocity perturbations across the singular layer are sketched in Fig. 6.2, and the corresponding field lines are shown in Fig. 6.3.

For the above analysis to be valid, the size of $\overline{k}$ must be restricted by two conditions. The first is that $\overline{B}'_{1x}/\overline{B}_{1x}$ be positive at the boundary between the inner and outer regions. From Eqs. (6.15) and (6.16) this implies that $\overline{k} < 0.64$. If the wavelength is so small that this inequality fails, the magnetic tension force in the outer region provides a stabilising force, which

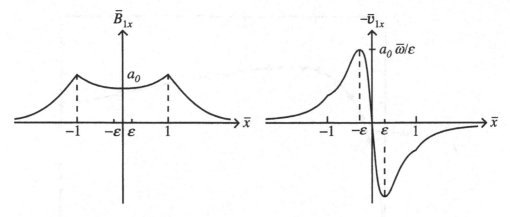

Fig. 6.2. A sketch of the perturbed magnetic field $(\overline{B}_{1x})$ and velocity $(\overline{v}_{1x})$ as functions of distance $(\overline{x} = x/l)$ normal to a current sheet for the tearing mode, where l is the half-width of the sheet and ϵl is the half-width of the inner diffusive layer.

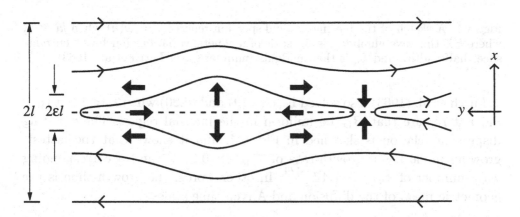

Fig. 6.3. Plasma velocity (thick arrows) and magnetic field lines (thin arrows) for the tearing-mode instability.

brings the fluid to rest before it reaches the inner region. The second restriction arises from the condition $(\epsilon|\overline{B}'_{1x}/\overline{B}_{1x}| < 1)$ that $\overline{B}_{1x}$ does not change significantly in the internal region. It gives a lower limit on $\overline{k}$, namely $\overline{k} > 2[8/(3L_u^2)]^{1/8}$. The corresponding bounds on the growth-rate are $(L_u/0.72)^{2/5} < \overline{\omega} < (2/27)^{1/4}L_u^{1/2}$. Since $L_u \gg 1$ by assumption, these bounds in turn imply that the instability grows at a rate (ω) that is intermediate between the very slow resistive diffusion rate $(\omega_d = \eta/l^2)$ and the much faster hydromagnetic rate $\omega_A = v_A/l$. Furthermore, for the constant-psi approximation to be valid, the inner-layer width must be much smaller than the resistive skin-depth $\sqrt{\eta/\omega}$.

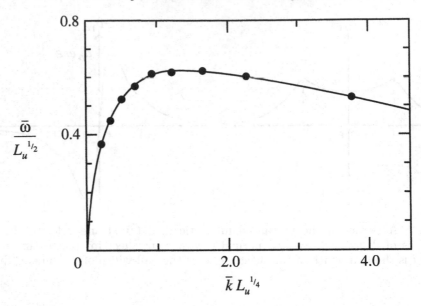

Fig. 6.4. A sketch of the tearing-mode dispersion relation $\omega = \omega(k)$ when $kl \ll 1$, where k is the wave-number, l is the scale of magnetic variations (such as a current-sheet half-width), and L_u is the Lundquist number (after Furth et al., 1963).

Furth et al. (1963) also solved Eqs. (6.19) and (6.20) when $\overline{k} \ll 1$ but they did not require that $\overline{B}_{1x}$ be uniform in the internal region. The resulting dispersion relation is sketched in Fig. 6.4, and it shows that the fastest-growing mode has a growth-rate of $\overline{\omega}_{\max} \simeq 0.6\,L_u^{1/2}$ and a corresponding wave-number of $\overline{k}_{\max} \simeq 1.4\,L_u^{-1/4}$. In other words, the growth-time is the geometric mean of the diffusion and Alfvén time-scales

$$\tau_m \simeq (\tau_d\,\tau_A)^{1/2}, \tag{6.22}$$

and $\overline{k} \ll 1$, so that the most unstable mode forms long narrow islands, very much longer than the width of the sheet. Most of the energy liberated in the outer region is dissipated ohmically in the inner region, typically less than 6% of it going into fluid motions there. The fourth-order equations may of course be solved numerically if L_u is not too large without making the above boundary-layer-type of approximation (e.g., Wesson, 1966).

6.3 Modifications of the Basic Tearing Analysis

The above analysis has been modified and extended by numerous authors. For example, Baldwin and Roberts (1972) have used the method of matched asymptotics to derive a uniformly valid treatment of the problem. Also,

Bobrova and Syrovatsky (1980) have studied the basic state

$$\mathbf{B}_0 = \sin \alpha x \, \hat{\mathbf{y}} + \cos \alpha x \, \hat{\mathbf{z}} \qquad (6.23)$$

in place of Eq. (6.14). Furthermore, diffusion of the initial state has been shown to increase the threshold for instability from $\Delta' > 0$ to $\Delta' > \pi |f''/f'| \tan(\pi/10)$ and to lower the growth-rate significantly.

Two other resistive modes with short wavelengths of the order of the sheet width ($kl \simeq 1$) were considered by Furth et al. (1963). The gravitational mode exists when a gravitational (or equivalent) force ($\rho g \, \hat{\mathbf{x}}$) acts transverse to the sheet to produce a density stratification [$\rho_0(x)$]. Its growth-rate is

$$\omega = \left[\frac{(kl)^2 \, \tau_A^2}{\tau_d \, \tau_G^4} \right]^{1/3}, \qquad (6.24)$$

where $\tau_G = (-g/\rho_0 \, d\rho_0/dx)^{-1/2}$ is the gravitational time-scale. The rippling mode occurs when there is a spatial variation across the sheet in magnetic diffusivity [$\eta_0(x)$], which may arise, for instance, from a temperature structure in the basic state. The growth-rate in this case is

$$\omega = \left[\left(\frac{d\eta_0}{dx} \frac{l}{\eta_0} \right)^4 \frac{(kl)^2}{\tau_d^3 \tau_A^2} \right]^{1/5}. \qquad (6.25)$$

Several other modifications are as follows. Although incompressibility is a good assumption at low plasma beta in a slab equilibrium, pressure gradients and curvature can make compressibility important in a cylinder or a torus (e.g., Finn and Manheimer, 1982). Also, Steinolfson and Van Hoven (1984a) have studied the coupling (caused by compressibility or a temperature-dependent resistivity) with the radiative mode. Furthermore, Steinolfson and Van Hoven (1983, 1984b) have considered linear tearing at very large L_u (up to 10^{12}) and small wave-number, where the constant-psi approximation fails, and have followed the nonlinear evolution for L_u values up to 10^6. More recently, Finn and Sovinec (1998) have given a comprehensive treatment of the nonlinear effect of both small and large islands, in both the constant-psi and non–constant-psi regimes.

6.3.1 Effect of Steady Flow

Bulanov et al. (1978) have shown how an equilibrium flow $\mathbf{v}_0 = (y/L_0)\hat{\mathbf{y}}$ along a sheet and away from the origin can provide extra stability against tearing. Such a flow may be present in a state of steady reconnection. They found instability when the length of the current sheet divided by its width exceeds $L_u^{3/7}$. Dobrowolny et al. (1983), in contrast, have included the effects of a viscous force $(\nu\nabla^2\mathbf{v})$ and of a velocity along the sheet of the form $V_0(x)\hat{\mathbf{v}}$, where $V_0'(0) \neq 0$. They obtained the startling result that the ordinary tearing mode represents a singular case, in the sense that its parity properties and scaling $(\omega \sim L_u^{-3/5})$ are destroyed as soon as the viscosity and/or shear are so large that $\nu/(\eta\rho) > L_u^{-1/(3n)}$ and/or $V_0'(0)/[V_0'(0) - B_0'(0)/\sqrt{\mu\rho}] > L_u^{-1/(3n)}$, where n (≥ 1) is an integer. The new growth-rate scales like $\omega \sim L_u^{-\bar{\lambda}}$, where $\bar{\lambda} = \frac{1}{3}(1 + n^{-1})$, so that $\bar{\lambda}$ lies between $\frac{1}{3}$ and $\frac{2}{3}$. Thus, all modes, except possibly those with $n = 1$, grow faster than the ordinary tearing mode. In particular, when there is no viscosity, the effect of the velocity shear is to make $n = 2$ and so drive an instability at a faster rate than ordinary tearing (namely $\omega \sim L_u^{-1/2}$) provided $|V_0'| < |v_A'|$. Flow and viscosity were also included later by Einaudi and Rubini (1989), while Ofman et al. (1991, 1993a) showed numerically that, for small shear flow, viscosity alters the growth-rate scaling from $L_u^{-3/5}$ to $L_u^{-2/3}(S_v/L_u)^{1/6}$, where S_v is the ratio of the viscous time to the Alfvén time; for large shear flow, Kelvin–Helmholtz instability effects are present. In the nonlinear regime with L_u between 10^2 and 10^5, the tearing saturates within one resistive time when the shear flow is small, but the energy release decreases and the saturation time increases as the shear flow increases.

Bulanov et al. (1979) have included the effect of an extra uniform magnetic field component $(B_n\hat{\mathbf{x}})$ across the sheet together with an extra gravitational (or equivalent) force $(-\rho g\hat{\mathbf{x}})$ that balances the Lorentz force $(j_z B_n)$ in the basic state. They found that without a flow the normal component does not suppress the instability, but a sufficiently strong flow can stabilize the mode. In contrast, Somov and Vernata (1994) have demonstrated a stabilizing effect of transverse and longitudinal components. Furthermore, Benedetti and Pegoraro (1995) have shown that resistive modes can be driven unstable at an X-point when the angle between the separatrices is smaller than $(\log L_u)^{-1}$.

6.3.2 Double Tearing

Rechester and Stix (1976) found that, when the equilibrium profile $[B_{0y}(x)]$ is such that two singular surfaces are close together, the linear growth-rate is greatly enhanced. In this *double* tearing mode the neighbouring islands are able to drive one another (Fig. 6.5). In particular, when B_{0y} is chosen to vanish at $x = \pm x_s$, say, Schnack and Killeen (1978) found numerically for $L_u = 100$ that the growth-rate is $0.26\, L_u^{-0.25}\tau_A^{-1}$ when $x_s = 0.25\,l$. When $L_u = 100$ the maximum growth-rate is found for $x_s = 0.5\,l$, but when $L_u = 10^5$ this occurs at a half-separation (x_s) of $0.2\,l$. Large values of v_x and

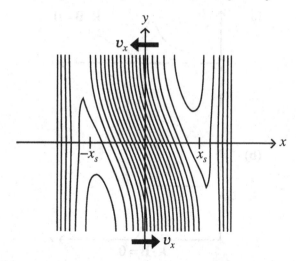

Fig. 6.5. Magnetic island structure for the double tearing mode, showing the driving influence of each island on the other (from Pritchett et al., 1980).

B_x are present over the entire region between $\pm x_s$, not just in the singular layers, so that more magnetic flux is carried in to be reconnected at $\pm x_s$. In the nonlinear regime, enhanced vortex flow and magnetic field distortion are found between the surfaces (see also Matthaeus and Montgomery, 1981).

In a particularly clear treatment of tearing, Pritchett et al. (1980) showed that, when the separation of the singular surfaces is small enough, the growth-rate scales like $L_u^{-1/3}$ with the mode structure being similar to that of the $m = 1$ cylindrical tearing mode. They also solved the nonlinear equations numerically for L_u-values up to 10^7. The difference between the radial perturbation (ξ_r) in different types of tearing is sketched in Fig. 6.6 for a cylindrical geometry. In standard tearing ξ_r is localised around the singular surface, whereas in $m = 1$ tearing there is a lateral displacement of the whole flux within the singular surface, and in double tearing the displacement is constant between the two surfaces.

6.3.3 Driving by Waves

Sakai (1983) suggested that fast magnetoacoustic waves can couple to and trigger the tearing mode with a faster growth-rate than normal. The waves of initial intensity $I_0 = (\delta B/B)^2$ drive a flow which is just such as to enhance the tearing. They are incident on a neutral sheet with a magnetic field $[B_{0y}(x) \simeq B_0 x/l]$ that vanishes at $x = 0$ and have a wavelength much smaller than the sheet half-width (l). The incident wave-amplitude is assumed so small

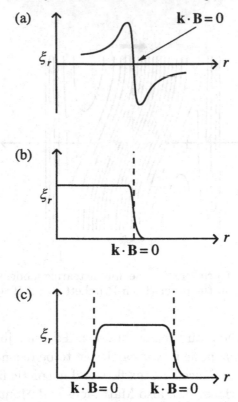

Fig. 6.6. Sketches of the radial displacement $\xi_r(r)$ in a cylinder for (a) standard tearing, (b) $m = 1$ tearing, and (c) double tearing (from Pritchett et al., 1980).

that the waves exert a significant force on the plasma only in the internal region, where their amplitude has grown substantially because of the fall in group speed (v_g). In particular, when the wave amplitude (I_0) is so large that $I_0 > (v_A/v_g)(kl)^{-9/5}(\tau_A/\tau_d)^{1/5}$, the wave-forcing term is important and tearing is driven by the waves.

In a similar analysis, Biskamp and Welter (1989) showed that the effect of 3D small-scale MHD turbulence on large-scale magnetic fields in a low-β plasma can be written in terms of an anomalous diffusivity $(\tilde{\eta})$ that is negative when the small-scale magnetic energy density exceeds the kinetic energy density (see also Pouquet, 1978). They investigated the effect of a negative resistivity on the $m = 2$ tearing mode by choosing the ansatz $\tilde{\eta}(k) = \tilde{\eta} \exp(-k^2/k_s^2)$, where k_s is the average turbulent wave-number. The results depend on the value of k_s relative to a critical wave-number $(k_c \sim |\tilde{\eta}|^{-2/5})$. When $k_s < k_c$ the negative resistivity is mainly stabilising, but

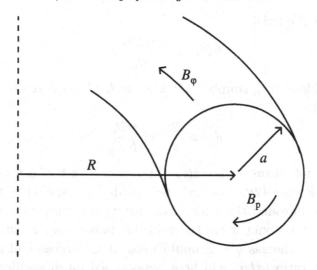

Fig. 6.7. Notation for a curved flux tube of minor radius $r = a$ and major radius R, having a toroidal field component B_ϕ and a poloidal component B_p.

when $k_s > k_c$ it is strongly destabilising such that for $k_s > 1.1\,k_c$ the mode is localised around the resonant surface (r_s) and the growth-rate approaches the value $k_s^2\,|\tilde\eta|$.

6.4 Instability of a Magnetic Flux Tube

The theory of magnetic instabilities in a cylindrical geometry is highly developed (e.g., Furth et al., 1973) and has been clearly summarised by, for instance, Bateman (1978) and Wesson (1981, 1997). It is of importance both for laboratory plasmas (Chapter 9) and for magnetic flux tubes in astrophysical environments (Chapter 12) such as the solar atmosphere (Chapter 11) or solar interior. The ideal modes grow fastest and have the magnetic field frozen to the plasma, whereas the resistive modes have lower thresholds for instability and allow the magnetic field to slip through the plasma in a narrow layer around a *resonant surface* where $\mathbf{k} \cdot \mathbf{B} = 0$.

Consider a magnetic flux tube of major radius R and minor radius a (both constant) with field components $B_p(r)$ (poloidal) and $B_\phi(r)$ (toroidal) that depend on the distance (r) from the magnetic axis (Fig. 6.7). Several useful quantities may be defined as follows. The plasma $\beta\,(= 2\mu p/B^2)$ is the ratio of plasma to magnetic pressure, and $\beta_p = 2\mu p/B_p^2$ is the corresponding ratio with the poloidal field component. For a toroidal flux tube, the amount by which a field line is twisted about the axis in going from one end of the tube to the other is the twist:

$$\Phi_T(r) = \frac{2\pi R B_p}{r B_\phi}.$$

(6.26)

(For a loop of length L, simply replace R by $L/(2\pi)$.) A related quantity is the safety factor:

$$q(r) = \frac{2\pi}{\Phi_T} = \frac{r\, B_\phi}{R\, B_p},$$

(6.27)

which for a whole torus is the ratio of the wavelength of a field line to the major circumference $(2\pi R)$ or, in other words, the number of turns that a field line makes around the major axis during one turn around the minor axis. Thus, $q = 1$ would mean the field line twists once around the major axis of a torus, whereas $q = 2$ would mean it undergoes half a twist. The inverse aspect ratio (a/R) will here be assumed much smaller than unity with $q \sim 1$ and $\beta \sim (a/R)^2$ (i.e., $\beta_p \sim 1$). The shear is $d(q^{-1})/dr$ and is a measure of the way the twist varies with radius. The electric current density along the tube is proportional to $\mu j_\phi(r) = r^{-1}\partial(r B_p)/\partial r$, which takes the value $2B_\phi/(Rq)$ on the axis $(r = 0)$. We shall consider a typical flux tube in which B_p increases with r from the axis while B_ϕ is roughly constant, j_ϕ decreases from a maximum, and q increases from a minimum. The effect of twisting up the tube is then to make j_ϕ increase and q fall in value.

A radial perturbation of the form $\xi = \xi(r)\exp[i(m\theta - n\phi)]$ produces a shape like a single helix if $m = 1$ or a double helix if $m = 2$ (Fig. 6.8). The resonant surface occurs at a radius (r_s) where $\mathbf{k} \cdot \mathbf{B} = 0$, namely $kB_\phi(r_s) + (m/r_s)B_p(r_s) = 0$. Since $k = n/R$, this implies that $q(r_s) = m/n$. On this rational surface the orientation of the perturbation matches that of the field, so that the crests and troughs of the helix follow the field lines.

6.4.1 Ideal Modes

To second order in a/R, there are no toroidal effects and the change in potential energy produced by the perturbation (assuming a vacuum outside the tube, which is often acceptable for laboratory plasmas) is

$$\delta W_2 = \frac{\pi^2 B_\phi^2}{\mu R}\left\{\int_0^a \left[\left(r\frac{d\xi}{dr}\right)^2 + (m^2 - 1)\xi^2\right]\left(\frac{n}{m} - \frac{1}{q}\right)^2 r dr + S_a\right\}$$

(6.28)

or

$$\delta W_2 = \frac{\pi^2 R}{\mu}\int_0^\infty \left[B_1^2 + B_p\left(1 - \frac{nq}{m}\right)\frac{dj_\phi}{dr}\xi^2\right] r dr,$$

(6.29)

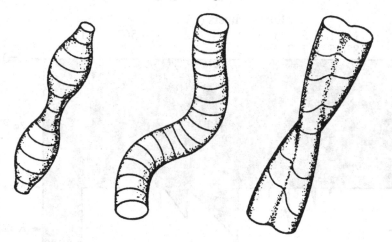

Fig. 6.8. The distortions produced by perturbations of a flux tube of type $m = 0$, $m = 1$, $m = 2$, respectively (after Bateman, 1978).

where

$$S_a = \left[\frac{2}{q_a} \left(\frac{n}{m} - \frac{1}{q_a} \right) + (1 + m) \left(\frac{n}{m} - \frac{1}{q_a} \right)^2 \right] a^2 \xi_a^2.$$

Here ξ_a^2 is the surface perturbation, q_a is the surface value of q, and $\mathbf{B}_1 = \nabla \times (\boldsymbol{\xi} \times \mathbf{B}_0)$ is the magnetic field perturbation. When $\delta W < 0$ the equilibrium is unstable, and otherwise it is stable.

Kink modes in a tokamak situation ($B_\phi \gg B_p$) are driven by the current gradient. They are the instabilities that arise at second order in a/R and are potentially the strongest. It can be seen from Eq. (6.28) that they need $q_a < m$, so that the resonant surface is outside the tube. Also, the second term in Eq. (6.29) shows that it is the torque arising from the current gradient (dj_ϕ/dr) that drives the instability and that the destabilising region is inside the resonant surface [i.e., $q(r) < m/n$]. Wesson (1978) has considered the current profile $j_\phi = j_{\phi 0}(1 - r^2/a^2)^K$ for which the total current is $I = \pi a^2 j_{\phi 0}(K + 1)^{-1}$ and the ratio of the q-values at the edge and axis of the tube is $q_a/q_0 = K + 1$. Figure 6.9 shows that when there is no shear ($q_a = q_0$) the tube is always kink unstable. At some value of K between 1 and 2.5 (depending on q_a), so that the current is sufficiently peaked, the mode becomes stabilised by shear. However, when $q_a < 1$ (the Kruskal–Shafranov boundary) the mode is always unstable (Shafranov, 1966). The effect of a potential or force-free magnetic field surrounding the tube is to provide some extra stability.

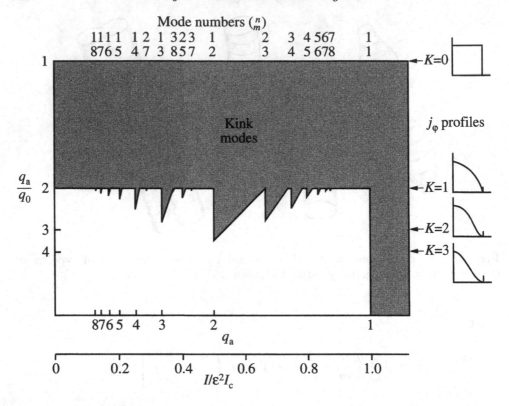

Fig. 6.9. Kink-instability diagram showing the ranges of values of q_a/q_0 and q_0 for which instability occurs (shaded), where q_a and q_0 are the values of the safety factor $q(r) = rB_\phi/(RB_p)$ at the edge and the axis, respectively, of the tube (Wesson, 1978).

Internal (interchange) modes are driven by a pressure gradient and do not require a surface perturbation (i.e., $\xi_a = 0$). A resonant surface now lies inside the tube and the potential energy is of fourth order in a/R with growth-rates smaller than those of the kinks by a factor a/R. The modes with $m > 1$ are localised around r_s (i.e., $\xi = 0$ except near $q = m/n$), so that $\delta W_2 \simeq 0$. In a cylindrical plasma they are unstable if

$$p' + r\frac{B_z^2}{8\mu}\left(\frac{q'}{q}\right)^2 < 0 \qquad \text{(Suydam's criterion)}, \qquad (6.30)$$

where a prime denotes a spatial derivative (d/dr). The first term in Eq. (6.30) is destabilising when $p' < 0$, and the second term represents the stabilising effect of shear. In a torus the curvature provides extra stability by multiplying p' by $(1 - q^2)$ (Mercier's criterion), so that a negative pressure gradient is only destabilising when $q_0 < 1$. Thus, the internal modes occur below a diagonal line $q_0 = 1$ in Fig. 6.9. For sufficiently high β, these modes balloon on the outer surface of a curved tube where the curvature is unfavourable. Such ballooning modes are driven by pressure gradients and have a large variation along the magnetic field.

6.4.2 Resistive Modes

The inclusion of resistivity removes a constraint by allowing field lines to break and rejoin in narrow layers around the resonant surfaces. The growth-times for the resulting instabilities lie between the diffusion time $(\tau_d = a^2/\eta)$ and the Alfvén time $(\tau_A = a/v_A)$, where $\tau_d \gg \tau_A$. The resistive form of the kink mode is a tearing mode. It is driven by the current gradient but now occurs when $q_a > m$, so that the resonant surface $(r = r_s)$ lies inside the tube. The Euler–Lagrange equation for Eq. (6.28) when $\eta = 0$ is

$$\frac{d}{dr}\left\{r\left[\frac{d}{dr}(rB_{r1})\right]\right\} - m^2 B_{r1} - \frac{\mu d j_\phi/dr}{[B_p/(mr^2)](m - nq)} B_{r1} = 0, \qquad (6.31)$$

where $B_{r1} = iB_p(m - nq)\xi/r$. This is also the equilibrium equation $[\nabla \times (\mathbf{j} \times \mathbf{B})_1 = 0]$ since the smallness of the growth-rate makes inertia negligible. The solutions to Eq. (6.31) starting at the axis and at infinity become singular at $r = r_s$, and so they need to be matched with those in the resistive layer. The result is that the mode is unstable when

$$\Delta' = \left[\frac{a}{B_{r1}}\frac{dB_{r1}}{dr}\right]_{r_s-\epsilon}^{r_s+\epsilon} > 0 \qquad (6.32)$$

and the growth-time behaves like $\tau_d^{3/5}\tau_A^{2/5}$. For $m = 2$ the effect of twisting up a flux tube is to move to the right in Fig. 6.10 and so to cross the threshold first for tearing $(q_0 = 2)$ and then, as the resonant surface crosses $r = a$, for kinking $(q_a = 2)$. The lower boundary in Fig. 6.10 appears because of shear stabilisation when q_a/q_0 is large enough. A similar figure is obtained for $m = 3$, but modes with $m > 3$ are stabilised because of the tension term $(-m^2 B_{r1})$ in Eq. (6.31).

The resistive interchange (or resistive-g) modes are the resistive form of the internal modes with $m > 1$ and have growth-times of the order of $\tau_d^{1/3}\tau_A^{2/3}$. The effect of increasing the shear is to localise the modes and reduce their growth-rate. They are unstable if

$$(-p')\left[q^2 - 1 + \frac{q^3 q'}{r^3}\int_0^r \frac{r^3}{q^2} + \frac{2R^2 r^2}{B_\phi^2}(-p')dr\right] < 0,$$

and so under normal conditions $(p' < 0, q' > 0)$ they require $q_0 < 1$. Increasing

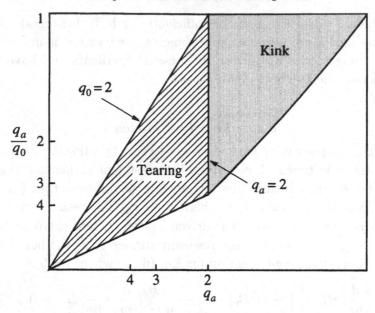

Fig. 6.10. Instability diagram for the $m = 2$ mode, where q_a and q_0 are the values of the safety factor at the edge of the tube and on its axis, respectively (Wesson, 1978).

β provides more stability by changing the solution in the resistive layer and modifying the instability criterion to $\Delta' > \Delta'_c$, where Δ'_c increases with β. The threshold for stability of $m = 2$ is moved to the right from $q_0 = 2$ in Fig. 6.10, and the mode is completely stabilised when $\overline{K} \gtrsim 60$ ($\overline{K} \gtrsim 10$ for $m = 3$), where $\overline{K} = \beta^{5/6}\epsilon^2(\tau_d/\tau_A)^{1/3}$. However, the increased pressure gradients may drive resistive ballooning modes (e.g., Strauss, 1981).

The $m = 1$ internal resistive kink mode becomes unstable when $q_0 < 1$. As the twist is increased, so the flux tube becomes tearing-mode unstable first with $\omega \sim \eta^{3/5}$; then it passes through a region where $\omega \sim \eta^{1/3}$, and finally it becomes unstable to the ideal mode. In a manner very similar to the Furth et al. treatment (§6.2), Coppi et al. (1976) analysed a linear force-free field [$B_p = B_0 J_1(\alpha r), B_\phi = B_0 J_0(\alpha r)$] that is stable to the ideal kink mode. They found that the fastest-growing perturbations have long wavelength ($k r_s \ll 1$) and $m = 1$, with a growth-time $\tau \sim \tau_d^{1/3}\tau_A^{2/3}$ that is somewhat shorter than the planar value [Eq. (6.22)].

For the solar and stellar flare applications, an important extra effect is the line-tying of the ends of coronal magnetic field lines in the dense stellar surface. In order of magnitude, it implies for the ideal kink mode that $2\pi/k \leq L$ so that the wavelength ($2\pi/k$) can fit into a loop of length L. But

k satisfies $\mathbf{k} \cdot \mathbf{B}_0 = 0$, and so this condition reduces to $\Phi_T \geq 2\pi$ in terms of the twist $[\Phi_T = LB_p/(rB_\phi)]$. A precise analysis gives instead a threshold for ideal instability of $\Phi_T = 2.5\pi$ for a field of uniform twist (Hood and Priest, 1981) or considerably larger for some other equilibria (Einaudi and Van Hoven, 1983). During the nonlinear development the flux tube may become highly kinked, creating current sheets where reconnection can take place.

6.5 Nonlinear Development of Tearing

The tearing-mode instability appears to be a rather delicate creature, which can easily be switched off by other effects (such as a flow, a normal magnetic field component, or a pressure anisotropy) and which often saturates at a low level. However, in a large tokamak it involves forces of tons and probably causes disruptions with forces of hundreds of tons! In its nonlinear development the tearing mode can follow several pathways, depending on the value of the magnetic Reynolds number, the wavelength, the boundary conditions, and the extra physics which may be added, such as the details of the energy equation. A full treatment of this nonlinear development at high magnetic Reynolds numbers has not yet been given. The emphasis in the theory of laboratory devices such as tokamaks (§9.1.2) has been on the slowing of reconnection by nonlinear saturation (§6.5.1), while space physicists and solar flare theorists have stressed a rapid nonlinear phase, which may either develop from linear tearing when the boundary conditions are free or be driven from outside by the motion of sources or an ideal instability (§6.4.1). Certainly, there is a need for a constructive comparison between these apparently contradictory applications of reconnection theory in order to understand the conditions that give rise to different forms of nonlinear development. In particular, numerical experiments, which we shall discuss below, are beginning to clarify the picture.

6.5.1 Saturation of a Single Mode

For standard slow tearing (§6.2), where the constant-psi approximation holds, Rutherford (1973) showed that in the nonlinear regime the instability quickly saturates. The unstable perturbation diffuses in from the external region and produces wider magnetic islands with higher inertia, which greatly slows the rate and magnitude of energy release. The linear theory breaks down at small amplitudes, such that $B_{1x}/(B'_{0y}l) \sim (\Delta'l)^{2/5}(kl)^{1/5}L_u^{-4/5}$. Then the island width exceeds the resistive layer width (ϵl), and the exponential growth is replaced by a slower algebraic growth such that the perturbed magnetic field grows like t^2 and the island width grows like t. In the linear phase the tearing flow pattern of Fig. 6.3 is driven by the linear force $j_{1z}B_{0y}$, but in the nonlinear phase this force is opposed by a third-order force $(j_{2z}B_{1x})$ from the second-order eddy current $(j_{2z} = -\sigma v_{1y}B_{1x})$. Nonlinear terms are included only in the singular layer, and the full width (w) of the island is found to grow with time at the rate

$$\frac{dw}{dt} \simeq \eta\,\Delta', \qquad (6.33)$$

where $w = 4(-A_1/B'_{0y})^{1/2}$ and A_1 is the perturbation of the flux function.

White et al. (1977) extended Rutherford's work by presenting an analytical model for island saturation by the quasi-linear development of a single mode. They wrote the magnetic field in terms of a flux function A (where $B_y = \partial A/\partial x$, $B_x = -\partial A/\partial y$) and supposed that the equilibrium profile is $A_0(x)$. A perturbation of the form $A_1(x)\cos ky$ produces islands of width $w = 4[-A_1(0)/A_0''(0)]^{1/2}$. Linear theory breaks down when the island width equals the tearing width $[\epsilon = (kA_0'')^{-1/2}(\omega\tau_A^2/\tau_d)^{1/4}]$, and they assumed at the same time that the inertia has become unimportant, so that $\mathbf{j}\times\mathbf{B} = 0$ or $\nabla^2 A = -\mu j_z(A)$. In the external region this becomes, to first order, $\nabla^2 A_1 = -\mu(dj_0/dA)A_1$, which, for an equilibrium profile $B_{0y} = B_0\tanh(x/l)$, has solution

$$A_1(x) = A_1\exp(\mp kx)\left[1 \pm \frac{\tanh(x/l)}{(kl)}\right] \qquad (6.34)$$

such that $A_0 \to 0$ as $|x| \to \infty$ and the upper (lower) sign corresponds to the region $x > \frac{1}{2}w$ $(x < -\frac{1}{2}w)$. This expression breaks down at the singular surface and gives a value for the jump $[\Delta_1'(w)]$ in A_1'/A across the surface. Inside the island the flux function and current are approximated by

$$A = A_0(x) + \epsilon[A_1(0) + A_1'(0)x]\cos ky, \quad j_z = a + bA.$$

Matching to the exterior solution determines the constants a and b. The time-evolution is determined by the y-average of the induction equation, namely $dA_1(0)/dt = -\eta(0)\,j_z(0) + E$ or, in terms of the island width,

$$\frac{dw}{dt} = \bar{c}\,\eta(0)\,[\Delta_1'(w) - K_0\,w], \qquad (6.35)$$

where $\bar{c}$ is constant and $K_0 = k^2 - 10/(9L^2)\{[1 - 124/(135\pi)]\}$. For small values of w, the function $\Delta_1'(w) - K_0 w$ is a positive decreasing function of

w which vanishes when $w = w_s$. This saturation width (w_s) is given by

$$\frac{w_s}{L} = \frac{9\Delta_1'(0)L}{2(1 + 31/27\pi)} \qquad (6.36)$$

for narrow islands ($kw \ll 1$, $w/L \ll 1$, $\Delta'(0)L \ll 1$). Since $w_s \sim \Delta_1'$, cases with a large Δ_1' would not necessarily be expected to saturate. The saturation is due to the island sampling a different part of the external solution, giving a quasi-linear decrease of the driving term Δ_1' modified by the finite island width. White et al. (1977) considered in detail the case of cylindrical geometry for which $x \to r$, $y \to r\theta$, $ky \to m\theta$, and the singular surface is located at $r = r_s$, rather than $x = 0$, with the constant of proportionality in Eq. (6.35) becoming $\bar{c} = 1.66$. Park et al. (1984) have confirmed the Rutherford scaling numerically for the $m = 2$ tearing mode, by finding that the linear growth ($\omega \sim \eta^{3/5}$) is quickly reduced in the nonlinear regime to an algebraic growth on the resistive time-scale ($\dot{A} \sim \eta$).

The nonlinear growth of the $m = 1$ tearing mode is quite different from that of modes with higher m-values, since f is no longer constant in the island and so Rutherford's analysis fails. Kadomtsev (1975) gave instead a heuristic argument that reconnection continues to grow at essentially the linear rate until the current density becomes flat inside the $q = 1$ surface, as described in Section 9.1.2. Schnack and Killeen (1979) followed numerically the instability of a linear force-free equilibrium [$B_p = B_0 J_1(\alpha r)$, $B_\phi = B_0 J_0(\alpha r)$] with $L_u = 100$ and $\beta = 0.6$. Exponential growth of tearing continues for three linear e-folding times (i.e., $30\tau_A$) and is followed by nonlinear saturation (Fig. 6.11). A magnetic island appears at the singular surface and grows rapidly to occupy a large portion of the plasma at saturation, after which the reconnection continues slowly until the old island is completely eaten away and a new set of nested surfaces with just one O-point has been produced (just as in the Kadomtsev scenario). The kinetic and magnetic energies saturate at about 1% of the total energy, with the maxima in magnetic and kinetic energies occurring at times $40\tau_A$ and $80\tau_A$, respectively. Park et al. (1984) found similar behaviour to Schnack and Killeen when the viscosity is important ($\nu/\rho \geq \eta$), but when viscosity is negligible ($\nu/\rho \ll \eta$) the current sheet becomes so long that secondary tearing takes place, since the sheet becomes unstable to $m > 1$ tearing modes and breaks up into multiple current sheets. Ultimately, a turbulent region may be produced with a higher effective resistivity and therefore a higher reconnection rate.

Dubois and Samain (1980) used a variational method to examine the nonlinear evolution of an $m = 1$ island. They found that the island width

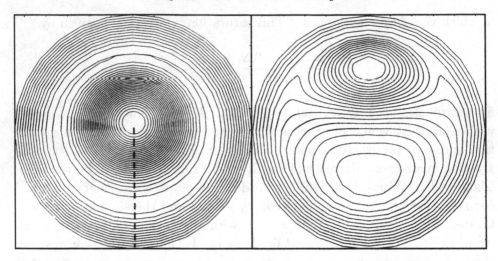

Fig. 6.11. Magnetic flux surfaces (flux contours of $\mathbf{k} \cdot \mathbf{B}/k$ and B_r) during the linear $(t = 10\tau_A)$ and nonlinear $(t = 60\tau_A)$ phases of the $m = 1$ resistive kink instability (after Schnack and Killeen, 1979).

grows in time according to the law

$$w^2 = 3\eta t,$$

which agrees with the numerical simulations up until the appearance of a current singularity. Thyagaraga (1981) has used singular perturbation theory to construct nonlinearly saturated island structures. By contrast with constant-psi saturated islands in the Rutherford regime, in the non-constant-psi regime the islands can tend to saturate to configurations with current sheets at the X-points and along the separatrices, as shown, for example, by Hahm and Kulsrud (1985) and Wang and Bhattacharjee (1992). A particularly clear plot of the resulting current density structures can be found in Fig. 8 of Finn and Sovinec (1998).

6.5.2 Mode Coupling

Waddell et al. (1976) developed a numerical code for studying the nonlinear coupling in three dimensions of tearing modes of different pitch in cylindrical geometry. In particular, they considered an initial equilibrium profile for which the $m = 2$, $n = 1$ (referred to as 2/1) and 3/2 tearing modes are linearly unstable. The results (for $L_u = 2 \times 10^4$ at the $q = 2$ surface) show that the $m = 2$ magnetic island does not slow down and saturate at its usual small size ($\simeq 0.1a$), but rather grows to a maximum width of $0.48a$ before rapidly contracting and saturating at a width of $0.38a$. Furthermore, the 3/2

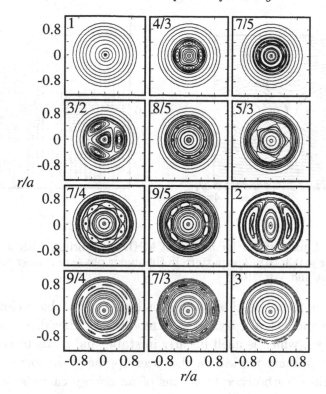

Fig. 6.12. Helical flux contours of different pitch at a time $0.01\tau_d$ for a Lundquist number $L_u = 2 \times 10^4$. Pitch values are given for each diagram (after Waddell et al., 1978).

mode is strongly destabilised by the presence of the 2/1 mode and grows to a width of $0.35a$, increasing the energy released from the magnetic field. Many other modes such as the 5/3 and 1/1 modes are also significantly destabilised because they overlap in radius. The destabilisation of the 3/2 and 5/3 modes is not simply due to the deformation of the equilibrium current by the 2/1 mode. Instead, the 2/1 and 3/2 modes drive the 1/1 and 5/3 modes, which in turn couple to the 2/1 mode and drive the 3/2 mode unstable. The flux contours for twelve different helicities are sketched in Fig. 6.12, which shows that the region of island activity extends across a significant fraction of the minor radius. This activity would increase greatly the heat transport across the toroidal field.

Carreras et al. (1981) considered analytically the effect of a static background of many modes with random phases and thick current layers on the growth of a tearing mode. Furthermore, Diamond et al. (1984) have shown how nonlinear mode coupling can produce enhanced growth of tearing. They

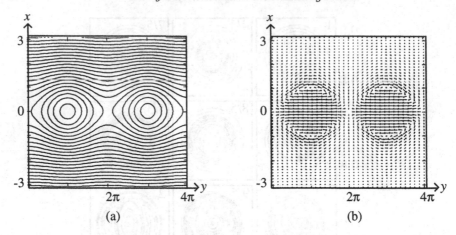

Fig. 6.13. (a) Flux surfaces for an initial MHD equilibrium with a perturbation magnitude of $\epsilon = 0.3$. (b) Flow pattern in the linear phase of coalescence instability (after Pritchett and Wu, 1979).

started with an initial equilibrium that is linearly unstable to low-m tearing modes (m, n) and (m', n'), say. As these modes grow they drive another mode $(m + m, n + n)$, which is itself linearly unstable and tends to accelerate the growth of the primary modes. Further overlapping of the low-m modes then produces high-m turbulence by means of an energy cascade to higher and higher m. The resulting turbulence causes a rapid tearing growth.

6.5.3 Coalescence Instability

Finn and Kaw (1977) considered the ideal stability of a chain of magnetic islands such as are created by the tearing mode. The magnetic structure is due to an array of current filaments superimposed on the original current sheet, with the O-points representing current maxima and the X-points representing current minima. Neighbouring islands were expected on physical grounds to coalesce because parallel currents (at the O-points) attract one another.

The assumed initial state, which is in equilibrium, possesses a flux function [Fig. 6.13(a)]

$$A_0 = \log_e(\cosh kx + \epsilon \cos ky) \tag{6.37}$$

satisfying

$$\nabla^2 A_0 = -\mu j_0(A_0) = (1 - \epsilon^2)\exp(-2A_0).$$

A perturbation A_1, say, causes a change in energy of

$$\delta W = \int |\nabla A_1|^2 + A_1^2 \, \mu \, \frac{dj_0}{dA_0} \, dV. \tag{6.38}$$

The first term is stabilising because of the increase in magnetic pressure as the flux piles up on both sides of the X-point, whereas the second term is destabilising because of the attraction of current filaments. A_1 is given by $\omega A_1 = -\mathbf{v}_1 \cdot \nabla A_0$ in terms of the perturbation velocity $\mathbf{v}_1 = -\hat{\mathbf{z}} \times \nabla \psi$ (assumed to be incompressible). In turn, the stream function was assumed for simplicity to take the form

$$\psi = \psi_0(x) \left(\sin \tfrac{1}{2}y + \lambda_1 \sin \tfrac{3}{2}y + \lambda_2 \sin \tfrac{5}{2}y + \cdots \right), \tag{6.39}$$

where

$$\psi_0(x) = \tanh^2(\lambda_0\, \epsilon^{-1/2}x) \exp\left(-\tfrac{1}{2}x\right) (2 + \cosh x),$$

which gives the correct behaviour for the single-mode solution to the linear equations at large distances ($|x| \gg \epsilon^{1/2}$). The trial function [Eq. (6.39)] possesses three parameters, which are determined by minimising δW to be $\lambda_0 \approx 1$, $\lambda_1 = -0.32$, and $\lambda_2 = 0.1$. Instability ($\delta W < 0$) was found with the minimising displacement peaked around the O-points and going to zero rapidly near the X-points. The same qualitative behaviour arises for cylindrical geometry. Since a non-zero dj_0/dA_0 within the island is required to drive the coalescence instability, the nonlinear saturated state of §6.5.1 would be immune if it were attained.

Subsequently, Pritchett and Wu (1979) and Biskamp (1982) performed detailed numerical computations of the instability. In the nonlinear evolution (Fig. 6.14) the neighbouring islands merge completely when $\eta \neq 0$ and the instability saturates, while the flow develops counter-cells and oscillates in direction on a period of 17 τ_A. With $\epsilon = 0.2$ and intermediate values of the Lundquist number ($200 < L_u < 10^4$), the reconnection rate measured by $\partial A_s/\partial t = \eta j_s - E_0 \approx \eta j_s$, namely, the change of magnetic flux enclosed by the separatrix, is practically independent of L_u, and the diffusion-region length (L) is constant at the time of maximum field compression. At large L_u-values ($10^4 < L_u < 5 \times 10^4$) a regime is reached with the sheet width scaling as $L_u^{-1/2}$ and L a constant. At still larger L_u-values ($5 \times 10^4 - 10^5$) one finds a new regime of "impulsive bursty reconnection" (Priest, 1986), when the reconnection region itself is disrupted and undergoes secondary tearing with the reconnection becoming significantly slower. The secondary islands grow and coalesce to produce one large island with one O- and two X-points replacing the original X. Secondary tearing (as distinct from the tearing that formed the initial islands) is initiated when the reconnection flow no longer stabilises tearing and B_x has grown to equal the field (B_∞) at large distances from the sheet. What appears to be happening in these simulations is that the ideal coalescence instability creates

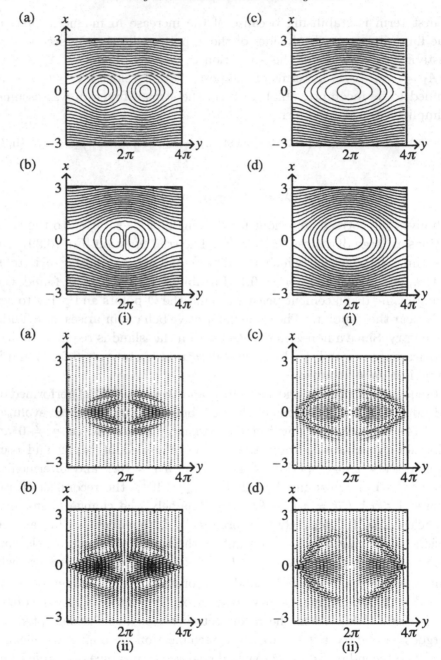

Fig. 6.14. (i) Nonlinear evolution of magnetic field lines during coalescence instability for a Lundquist number $L_u = 200$ and a perturbation amplitude $\epsilon = 0.3$ at times (a) 45 τ_A, (b) 55 τ_A, (c) 64 τ_A, and (d) 176 τ_A , where τ_A is the Alfvén travel time. (ii) Corresponding flow patterns at times (a) 61 τ_A, (b) 70 τ_A, (c) 80 τ_A, and (d) 141 τ_A (after Pritchett and Wu, 1979).

inflows that exceed the maximum reconnection rate, so that the flux piles up at the diffusion region. This in turn causes the region to grow in length until it becomes of the order of the global dimension (L_e) and goes unstable to tearing (cf. Forbes and Priest, 1987).

Bhattacharjee et al. (1983) have also investigated coalescence numerically. They found that the instability proceeds in three stages: the islands first approach and form a current sheet; the reconnection takes place; and the instability saturates after typically 30 τ_A, with the single island sitting there and oscillating in response to its jarring experience. The amount of reconnected flux grows linearly with time in the second stage like $L_u^{-1/2}t$ at a steady Sweet–Parker rate for L_u between 500 and 2,000. There is no evidence of shocks, presumably because the reconnection is too slow and the current sheet too long or perhaps because of insufficient grid-point resolution. Secondary tearing was observed but the secondary islands are always swept along the current sheet (cf. Forbes and Priest, 1982b), since no up–down symmetry conditions are imposed at the current sheet. The density contours reveal a highly compressed current sheet, which is larger than one expects from Petschek steady-state theory (probably because of flux pile-up) but which bifurcates at its ends to give pairs of structures that are suggestive of slow shocks.

Hayashi (1981) has followed numerically with $L_u = 10^3$ the formation of two magnetic islands (with a wave-number of 0.5 in the x-direction) due to linear tearing. Conditions of symmetry were adopted on $x = 0$, $y = 0$, and $x = 4\pi$, and fixed conditions were imposed on $y = 4\pi/3$, with an initial profile $B_x = B_0 \tanh y$. In the nonlinear development there is no sign of island coalescence, which casts some doubt on the reality of coalescence instability in a tearing situation. Hayashi suggested that the current density produced by tearing has a maximum at the X-point and a minimum at the O-point, which is the opposite of what is assumed in Finn and Kaw's initial state [Eq. (6.37)]. However, the relative values of the current density at the X- and O-points depend on the heating and heat transport at these locations and their effect on the resistivity. For example, in the numerical experiment of Finn and Sovinec (1998) with uniform resistivity in the bulk of the plasma, there are peaks in current at both the X- and O-points.

In practice, tearing generally favours the longest-wavelength mode that fits a given structure, since many structures are shorter than the fastest-growing mode in an infinite medium. Therefore, if such a mode grows and saturates, coalescence is unlikely. However, if an external disturbance or extra physical effects favour a shorter-wavelength mode that develops a peaked current-density profile in the island, then it can indeed coalesce.

Mikic et al. (1988) found coalescence in a periodic array of line-tied arcades, and later Finn et al. (1992) showed that this also occurs in a pair of arcades. Furthermore, Longcope and Strauss (1993) gave an interesting example of coalescence to form current sheets in a two-dimensional array.

An estimate of the role of coalescence may be made as follows. The coalescence time for islands of initial half-width $x_{i0} = 2l\epsilon^{1/2}$ is roughly

$$\tau_c = \tau_A \frac{x_{i0}}{2l}. \tag{6.40}$$

Primary coalescence of this initial island structure will be negligible if the tearing-mode time $(\tau_m = \tau_A R_m^{1/2})$ is much smaller than τ_c, that is, if

$$x_{i0} \ll \frac{2l}{R_m^{1/2}}. \tag{6.41}$$

For numerical experiments, where generally $R_m \approx 10^2 - 10^3$, primary coalescence is therefore likely to be negligible, but it may well be important in astrophysical plasmas where $R_m \approx 10^{10}$. It is also of interest to compare coalescence with saturation, which occurs after a time $\tau_s \approx \tau_m \log(l/x_{i0})$ such that the island half-width $[x_i = x_{i0} \exp(t/\tau_m)]$ has grown to equal l. Primary coalescence will therefore take place before saturation if $\tau_c < \tau_s$, or

$$2 \left(\frac{l}{x_{i0}} \right) < R_m^{1/2} \log \left(\frac{l}{x_{i0}} \right), \tag{6.42}$$

so that, for a given (l/x_{i0}), coalescence will occur first if R_m is large enough. In particular, many numerical experiments of tearing would not be expected to show primary coalescence, in contrast to most space and astrophysical plasmas.

The sequence of events that one may expect in an astrophysical or space-plasma current sheet that is longer than the fastest-growing mode and has a large magnetic Reynolds number is as follows. First of all, the sheet tears linearly at a wavelength $(4.5 \, l \, R_m^{1/4})$ of the fastest-growing mode. Then primary coalescence combines neighbouring islands. This may lead to a fast reconnection regime (such as Petschek or flux pile-up), probably disrupted by secondary tearing and secondary coalescence of the diffusion region (i.e., the regime of impulsive bursty reconnection). By contrast, in a current sheet that is line-tied at one end, say to the surface of a dense object such as a star, linear tearing produces dominant tearing at the neutral point closest to the surface. This neutral point does not coalesce with its neighbour but develops fast nonlinear reconnection, which subsequently undergoes secondary tearing and coalescence.

7

Unsteady Reconnection: Other Approaches

In this chapter we look at two theories for time-dependent reconnection that are not as well known as the tearing mode. The first of these is X-type collapse, which was first considered by Dungey (1953) and has been briefly described in Section 2.1. The second is the time-dependent, Petschek-type theory developed by Semenov et al. (1983a). Both theories provide new perspectives on reconnection because they describe behaviour which is not encompassed within the scope of either the steady-state or tearing-mode theories.

7.1 X-Type Collapse

Dungey's (1953) work on X-type collapse is the earliest analysis ever done on magnetic reconnection and predates both the tearing-mode (Furth et al., 1963) and the Sweet–Parker (1958) theories. Dungey considered what happens when a small, but uniform, current perturbation is imposed at a current-free X-line (i.e., an X-point in any intersecting plane). Before the current is imposed, the separatrices are at right angles to one another, but after the current is added the separatrices scissor, as shown in Figure 7.1. Assuming that the plasma pressure in a strongly magnetized plasma can be ignored, Dungey argued that the initial perturbation would grow with time and rapidly lead to the formation of a current sheet at the X-line. Cowling (1953) objected that the growth of the current density would violate Lenz's Law, but this was eventually resolved by Dungey (1958), who pointed out the role of the $\mathbf{v} \times \mathbf{B}$ term in the evolution of the plasma.

Dungey's (1953) qualitative analysis was later put on a firm mathematical foundation by Imshennik and Syrovatsky (1967), who found an exact, non-linear solution of the ideal MHD equations that describes the collapse. A remarkable aspect of this exact solution, which we discuss in the next section,

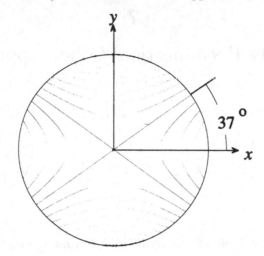

Fig. 7.1. Contours of the flux function (A) for the initial state of the X-type collapse considered by Dungey (1953). Here the separatrix half-angle is 37°.

is that it predicts an explosive growth of the current density as it becomes infinite in a finite time. In practice, there are two physical processes which limit the growth of the current density: one is gas pressure, the other is electrical resistivity, but neither process is included in the analyses of Dungey or Imshennik and Syrovatsky.

Chapman and Kendall (1963, 1966) and Uberoi (1963, 1966) attempted to assess the effects of pressure by using the incompressible MHD equations to follow the collapse. In an incompressible plasma, the pressure forces are strong, rather than weak, and so, if a collapse solution exists in the incompressible system, it suggests that pressure forces cannot prevent it. Chapman and Kendall (1963, 1966) did in fact find a solution in which the current density grows exponentially with time, and this led them to conclude that the collapse could not be stopped by pressure effects. However, their conclusion is not justified because the boundary conditions they used presuppose an external driver (Forbes and Speiser, 1979; Klapper, 1998). To maintain the exponential growth of the current density, an external generator must supply an electric potential which grows exponentially in time. Without this generator no growth occurs, so gas pressure can inhibit the collapse process.

7.1.1 Self-Similar Solution

It has been difficult to assess the physical significance of X-type collapse because, as originally formulated, the boundaries of the plasma domain are at infinity. Extending the boundaries to infinity eliminates all inherent scale-lengths and makes it possible to obtain solutions of the MHD equations, like those of Imshennik and Syrovatsky (1967), which are self-similar in time: that is, the time behaviour of the variables at any location is similar, so that the usual system of partial differential equations reduces to a much simpler system of ordinary differential equations.

Finding a self-similar solution is no guarantee that the solution is physically realizable or useful. However, some self-similar solutions, such as those for spherically symmetric explosions, are quite useful because they describe the long-term behaviour of the plasma. As we shall see, the self-similar solution describing compressible X-type collapse is more like an implosion. The reversal of the arrow of time in going from an explosion to an implosion means that the self-similar solution of the collapse is only valid for a short time, and instead of evolving naturally towards a self-similar regime, self-similarity must be introduced at the start by choosing a particular set of initial conditions. Because of these restrictions, the self-similar solution describing X-type collapse has been little more than a mathematical curiosity until quite recently. However, with the advent of numerical simulations (e.g., Ofman et al., 1993b; McClymont and Craig, 1996), it has now become possible to obtain solutions without the assumption of self-similarity and yet retaining the nonlinear dynamics inherent in an implosion.

The system of equations which describes Dungey's X-type collapse is:

$$\frac{\partial \rho}{\partial t} = -\left[\frac{\partial(\rho v_x)}{\partial x} + \frac{\partial(\rho v_y)}{\partial y} \right], \tag{7.1}$$

$$\frac{\partial B_x}{\partial t} = -\frac{\partial(v_y B_x - v_x B_y)}{\partial y} + \eta \left(\frac{\partial^2 B_x}{\partial x^2} + \frac{\partial^2 B_x}{\partial y^2} \right), \tag{7.2}$$

$$\frac{\partial B_y}{\partial t} = -\frac{\partial(v_x B_y - v_y B_x)}{\partial x} + \eta \left(\frac{\partial^2 B_y}{\partial x^2} + \frac{\partial^2 B_y}{\partial y^2} \right), \tag{7.3}$$

$$\rho \left(\frac{\partial v_x}{\partial t} + v_x \frac{\partial v_x}{\partial x} + v_y \frac{\partial v_x}{\partial y} \right) = -\frac{\partial p}{\partial x} - j B_y, \tag{7.4}$$

$$\rho \left(\frac{\partial v_y}{\partial t} + v_x \frac{\partial v_y}{\partial x} + v_y \frac{\partial v_y}{\partial y} \right) = -\frac{\partial p}{\partial y} + j B_x, \tag{7.5}$$

$$\frac{\partial s}{\partial t} + v_x \frac{\partial s}{\partial x} + v_y \frac{\partial s}{\partial y} = (\gamma - 1) \frac{j^2}{\sigma \rho^\gamma}, \tag{7.6}$$

$$j = \frac{1}{\mu}\left(\frac{\partial B_y}{\partial x} - \frac{\partial B_x}{\partial y}\right), \tag{7.7}$$

$$s = \frac{p}{\rho^\gamma}, \tag{7.8}$$

and the initial conditions are:

$$B_x = \frac{B_0\,y}{y_0}, \quad B_y = \frac{\lambda\,B_0\,x}{y_0}, \tag{7.9}$$

$$v_x = 0, \quad v_y = 0, \tag{7.10}$$

$$\rho = \rho_0, \quad p = p_0, \tag{7.11}$$

with $\lambda = y_0/x_0$, $\eta = 1/(\mu\sigma)$, and $\beta_0 = 2\mu p_0/B_0^2$. The corresponding magnetic flux function (A) at $t = 0$ is

$$A = \frac{B_0}{2y_0}(y^2 - \lambda x^2),$$

where $\mathbf{B} = \nabla \times (A\,\hat{\mathbf{z}})$ and the initial current density is

$$j = -\frac{\epsilon\,B_0}{\mu y_0},$$

where $\epsilon = 1 - \lambda$. Contours of A are shown in Fig. 7.1. In terms of the angle (ϕ_s) between the separatrices, the parameter λ is

$$\lambda = \tan(\tfrac{1}{2}\phi_s).$$

Ignoring gas pressure, Imshennik and Syrovatsky (1967) found a self-similar solution for the system of Eqs. (7.1)–(7.11). However, their solution can easily be extended to include gas pressure, as long as the pressure is initially uniform, so that the pressure gradient terms in Eqs. (7.4) and (7.5) disappear even though the gas pressure is not zero. Thus, the original solution of Imshennik and Syrovatsky (1967) can be written as

$$B_x = \frac{B_0\,y}{a(\bar{t})^2\,y_0}, \quad B_y = \frac{B_0\,x}{b(\bar{t})^2\,x_0}, \tag{7.12}$$

$$v_x = \chi_x(\bar{t})\frac{v_{A0}}{x_0}x, \quad v_y = \chi_y(\bar{t})\frac{v_{A0}}{y_0}y, \tag{7.13}$$

$$\rho = \frac{\rho_0}{a\,b}, \tag{7.14}$$

$$p = \frac{s\,p_0}{(\rho_0 a\,b)^\gamma}. \tag{7.15}$$

Here $\bar{t} = t v_{A0}/y_0$, $v_{A0} = B_0/(\mu\rho_0)^{1/2}$, while a and b satisfy

$$\ddot{a} = \left(\frac{\lambda}{b} - \frac{b}{a^2}\right), \tag{7.16}$$

$$\ddot{b} = \left(\frac{\lambda}{a} - \frac{\lambda^2 a}{b^2}\right), \tag{7.17}$$

with overdots denoting dimensionless time derivatives.

At $t = 0$, $a = b = 1$ and $\dot{a} = \dot{b} = 0$. Subsequently, the velocity gradients (χ_x and χ_y) are determined by

$$\chi_y = \frac{\dot{a}}{a}, \tag{7.18}$$

$$\chi_x = \frac{\dot{b}}{b}, \tag{7.19}$$

and the entropy function (s) is obtained by integrating

$$\dot{s} = (\gamma - 1)\frac{j^2}{\sigma\rho^\gamma}, \tag{7.20}$$

where the current density (j) is

$$j = \frac{B_0}{\mu \, y_0} \left(\frac{\lambda}{b^2} - \frac{1}{a^2}\right).$$

Because the current density (j) is uniform, the diffusion terms in Eqs. (7.2) and (7.3) are also eliminated from the system of equations. Other self-similar solutions closely related to those of Imshennik and Syrovatsky are discussed in Bulanov and Olshanetsky (1984) and Sakai (1990).

When $\epsilon = 1$ (i.e., $\lambda = 0$), the coupled equations (7.16) and (7.17) describe the evolution of a sheet pinch, and they have the exact solution:

$$a(a^{-1} - 1)^{1/2} + \tan^{-1}(a^{-1} - 1) = \sqrt{2}\bar{t}, \quad b = 1. \tag{7.21}$$

The corresponding solutions for χ_x and χ_y are

$$\chi_x = 0, \quad \chi_y = -\sqrt{2}a^{-1}(a^{-1} - 1)^{1/2}.$$

At $\bar{t} = \pi/(2\sqrt{2})$ the magnetic field, current density, and mass density all become infinite. This singularity results from the absence of any force to counter the initial $\mathbf{j} \times \mathbf{B}$ force which drives the plasma toward the x-axis. Figure 7.2 shows the behaviour of the system for initial current perturbations of $\epsilon = 0.9$ and 10^{-5} obtained by numerically solving Eqs. (7.16) and (7.17). When ϵ is reduced in value the singularity still occurs, but it takes longer

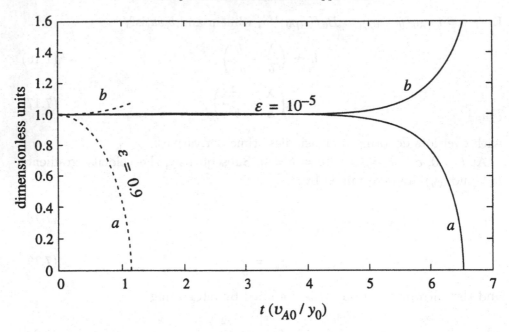

Fig. 7.2. The growth of the magnetic field parameters a and b as functions of time in the self-similar solution of Imshennik and Syrovatsky (1967) for two values of the initial current parameter ($\epsilon = 1 - \lambda$). When the variable a reaches zero, the current density becomes infinite.

to appear. Imshennik and Syrovatsky (1967) showed that the asymptotic behaviour for any value of ϵ is given by

$$a \propto (t_s - t)^{2/3}, \quad \chi_y \propto (t_s - t)^{-1},$$
$$\rho \propto (t_s - t)^{-2/3}, \quad j \propto (t_s - t)^{-4/3},$$

where t_s is the time at which the singularity occurs.

Even though the diffusion terms in the induction equations (7.2) and (7.3) are zero, reconnection still occurs as long as η is non-zero. As for any two-dimensional system, the reconnection rate is just the value of the electric field (E) at the X-line, namely

$$E = \frac{j}{\sigma} = \frac{B_0}{\mu \sigma y_0} \left[\left(\frac{1 - \epsilon}{b^2} \right) - \frac{1}{a^2} \right], \tag{7.22}$$

which is plotted in Fig. 7.3 as a function of time. Another indication that reconnection is occurring is the presence of ohmic heating in the energy equation [Eq. (7.6)].

The physical significance of the singularity is much clearer if we consider a system that is not of infinite extent in the two-dimensional plane, but is instead bounded. Suppose that the plasma is surrounded by a rigid, circular cylinder of radius r_0 located at $x^2 + y^2 = r_0^2$, as shown in Fig. 7.1. As long

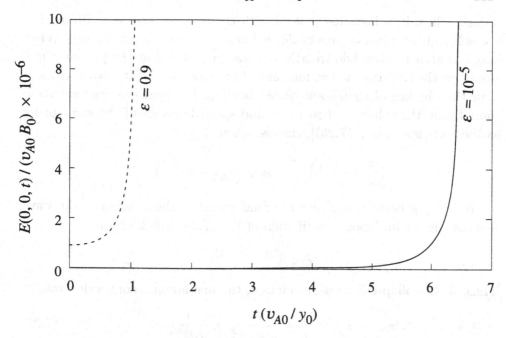

Fig. 7.3. The reconnection rate (E) as a function of time (t) during X-type collapse for initial current perturbations of $\epsilon = 0.9$ and 10^{-5}.

as the fluid is ideal and compressible, the boundary conditions imposed at the cylinder's surface cannot affect a particular location within the interior until there has been time for a fast-mode wave to travel from the boundary to that particular location. Along the y-axis the dimensionless location (y_w) of the wave carrying the boundary information is determined by

$$\dot{y}_w = \left\{ y_w \chi_y - \left[y_w^2 a^{-3} b + \tfrac{1}{2}\beta_0 (a\,b)^{\gamma-1} \right]^{1/2} \right\}, \qquad (7.23)$$

where the two terms on the right-hand side of Eq. (7.23) are the flow speed and fast-mode wave speed at y_w.

For the one-dimensional, ideal case with no initial gas pressure ($\epsilon = 1, \eta = 0$, and $\beta_0 = 0$), Eq. (7.23) has the solution (Forbes, 1982)

$$y_w = a[a^{-1/2} + (a^{-1} - 1)^{1/2}]^{-\sqrt{2}}, \qquad (7.24)$$

where a is prescribed by Eq. (7.21). From this result, we see that, in the limit of negligible pressure, the wave reaches the origin ($y_w = 0$) at precisely the moment when the singularity occurs ($a = 0$). Now, if either the initial pressure or the diffusivity is greater than zero, the wave will travel faster than predicted by Eq. (7.24) and reach the origin before the singularity occurs. Thus, in a bounded system with either gas pressure or diffusivity, the self-similar solution breaks down before the singularity occurs. What happens after the breakdown still remains unanswered at the present time. However, the situation at the moment of breakdown provides some clues.

In the absence of significant ohmic heating, the pressure gradient force should stop the collapse when the sound speed becomes of the same order as the wave speed [Eq. (7.23)], that is, when

$$\left(\frac{\gamma}{2}\beta_0 a_f^{1-\gamma}\right)^{1/2} \approx y_{wf}\left(\chi_{yf} - a_f^{-3/2}\right),$$

where the f subscripts indicate the final values at the time when the wave reaches the origin. Upon substitution of Eq. (7.24) this leads to

$$a_f \propto \beta_0^{1/(\gamma+\sqrt{2}-2)}.$$

Thus, if the collapse is pressure-limited, the maximum reconnection rate is

$$E_f = \frac{j_f}{\sigma} = \frac{B_0}{\mu y_0\, \sigma a_f^2} \propto \frac{1}{\sigma \beta_0^{1.85}} \tag{7.25}$$

for $\gamma = 5/3$, and increasing the initial pressure lowers the maximum rate of reconnection.

Now, let us suppose that the initial pressure is zero, but the diffusivity of the plasma is not. In this circumstance, diffusivity becomes important when the wave speed $(\dot{y}_w)$ is of the same order as the diffusion speed $[\eta/(v_{A0}y_0y_w)]$. Setting $\beta_0 = 0$ and using Eq. (7.24) with Eq. (7.23) for the wave speed leads to

$$\frac{\eta}{v_{A0}\,y_0} = \left[a_f^{-1/2} + \left(a_f^{-1} - 1\right)^{1/2}\right]^{-2\sqrt{2}}\left[\sqrt{2}a_f\left(a_f^{-1} - 1\right)^{1/2} + a_f^{1/2}\right]. \tag{7.26}$$

Assuming $\eta/(v_{A0}y_0)$ small and solving for a_f in terms of η gives

$$a_f \propto \eta^{0.522}. \tag{7.27}$$

Thus, if the collapse is diffusion-limited, the maximum reconnection rate is now

$$E_f = \frac{B_0\eta}{y_0\,a_f^2} \propto \eta^{-0.045}. \tag{7.28}$$

This is a rather remarkable result because it implies that, as the magnetic diffusivity $\eta = 1/(\mu\sigma)$ tends to zero, the reconnection rate becomes infinite and not zero as we might expect for an ideal MHD system. The ohmic heating rate also tends to infinity as η tends to zero, and both of these inverse scaling results have been confirmed numerically by McClymont and Craig (1996).

Equation (7.28) is the instantaneous reconnection rate, but it does not indicate how much flux is actually reconnected or merged by the time the self-similar solution breaks down. To find this flux, we must integrate the electric field over time. For the one-dimensional case, the amount of reconnected flux (A_f) is

$$A_f = \int_0^{t_f} \frac{j}{\sigma}\, dt = \frac{B_0}{\mu\sigma y_0} \int_1^{a_f} \frac{1}{a^2} \frac{dt}{da}\, da,$$

which integrates to

$$A_f = \sqrt{2}(\eta B_0/v_{A0}) \left(a_f^{-1} - 1\right)^{1/2}. \tag{7.29}$$

Substituting for a_f from Eq. (7.27) and assuming $\eta/(v_{A0}y_0)$ small produces

$$A_f = 1.067 B_0\, y_0 \left(\frac{\eta}{v_{A0}\, y_0}\right)^{0.739}.$$

Thus, despite the fact that the instantaneous reconnection rate at time t_f tends to ∞ as η tends to zero, the amount of flux which is reconnected by this time tends to zero. In terms of the average reconnection rate, namely,

$$E_{\text{ave}} = \frac{A_f}{t_f} = 0.961 (v_{A0} B_0) \left(\frac{\eta}{v_{A0}\, y_0}\right)^{0.739}, \tag{7.30}$$

the reconnection is slow, since η scales at a rate slower than the Sweet–Parker rate of $\eta^{0.5}$.

The final reconnection rate (E_f) predicted by Eq. (7.25) assumes that the diffusivity is negligible, while the rate predicted by Eq. (7.28) assumes that the pressure is negligible. By comparing these two expressions, we obtain

$$\beta_0 \lesssim \left(\frac{\eta}{v_{A0}\, y_0}\right)^{0.565} \tag{7.31}$$

as the condition for neglecting pressure. McClymont and Craig (1996) have pointed out that, unless the condition (7.31) is satisfied, the current sheet never becomes thin enough for the inverse scaling predicted by Eq. (7.28) to be achieved.

7.1.2 Linear Analyses

The self-similar solution provides a rather limited view of the collapse process. The solutions say nothing about what happens outside the self-similar region or what happens after the disappearance of self-similarity.

Furthermore, there is no way of knowing to what extent the evolution is just a result of the special initial conditions needed to set up the self-similarity in the first place. For example, if the system were perturbed at its edges, rather than by a uniform current density, would a collapse still occur? Progress in answering these and other questions has been made by using both linear analyses and numerical simulations.

Oddly, the first thorough linear analyses of the collapse process were not carried out until 1990s (i.e., Bulanov et al., 1990; Craig and McClymont, 1991), and these led to several additional analyses (Hassam, 1992; Craig and Watson, 1992; Craig and McClymont, 1993; Fontenla, 1993; Titov and Priest, 1993), all of which use some type of perturbation technique. One of the most important results from these approaches is that they demonstrate that the collapse process is not unique to the self-similar initial conditions but occurs for a wide variety of initial and boundary conditions. Indeed, the collapse is now seen to be the dynamic counterpart to the quasi-static formation of a current sheet examined by Green (1965) and Syrovatsky (1971). In these quasi-static solutions (§2.2), on the one hand, a current sheet is formed by perturbing the configuration at a very slow rate, so that the system passes through a series of equilibria. In the X-type collapse, on the other hand, the perturbation rate is fast, comparable to the fast-mode time-scale, and therefore dynamic effects are important. Another important result of the linear analyses is that they show that the reconnection process in the linear regime is quite fast, scaling as a function of $(\ln \eta)$.

The basic linear results are obtained as follows: First, we linearise the MHD equations (7.1)–(7.8) and express the flux function as $A = A_0 + A_1$, where A_0 is the current-free state $B_0 r_0 (\bar{y}^2 - \bar{x}^2)$, where $\bar{y} = y/r_0$, $\bar{x} = x/r_0$, and A_1 is the linear perturbation. Setting the pressure to zero then leads to the linear, third-order differential equation

$$\ddot{A}_1 = (\bar{x}^2 + \bar{y}^2)\nabla^2 A_1 + \bar{\eta}\nabla^2 \dot{A}_1, \qquad (7.32)$$

where the overdots denote dimensionless time-derivatives $(\partial/\partial \bar{t})$, $\bar{t} = tv_{A0}/r_0$, and $\bar{\eta} = \eta/(v_{A0}r_0)$. There are several techniques one can use to solve this equation (see Hassam, 1992, for example), but here we follow Craig and McClymont (1991), who looked for separable solutions of the form

$$A_1 = B_0 \, r_0 \, \epsilon \, \mathrm{Re} \left[f(\bar{r})e^{im\theta + \omega t} \right] \qquad (7.33)$$

with

$$\omega = \omega_R + i\omega_I,$$

where ϵ is now the dimensionless magnitude of the perturbation, $\bar{r} = r/r_0$

and θ are polar coordinates, and $f(\bar{r})$ is a complex function which satisfies

$$\bar{r}(\bar{r}f')' = \left(\frac{\bar{r}^2 \, \bar{\omega}^2}{\bar{r}^2 + \bar{\eta}\,\bar{\omega}} + m^2 \right) f, \tag{7.34}$$

where $\bar{\omega} = \omega r_0 / v_{A0}$.

Frozen-flux conditions are imposed at the surface of the cylinder by setting $f = 0$ at $\bar{r} = 1$, the surface location normalized to the radius (r_0) of the cylinder. Only the $m = 0$ mode corresponds to reconnection, since the solutions for $m > 0$ leave A unchanged at the origin.

The solution of Eq. (7.34) for $m = 0$ can be written in the form of an eigenfunction expansion, which leads to

$$\bar{\omega}_R = \frac{-\bar{\omega}_I^2}{2(n+1)}, \quad \bar{\omega}_I = -\frac{(2n+1)\pi}{\ln \bar{\eta}}, \tag{7.35}$$

with n being the number of radial nodes of the eigenmode. The solution (7.33) with Eq. (7.35) describes radial oscillations of waves propagating between the boundary and the origin, as shown in Fig. 7.4. The travel-time depends on the diffusivity (η) because, in the absence of pressure, the wave's speed tends to zero as the origin is approached, and so it is only the addition of diffusion that allows a perturbation to reach the origin and reflect.

For the lowest-order mode ($m = 0$, $n = 0$) the solution for the total flux function (A) is

$$A(\bar{r}, \theta, t) = B_0 r_0 \left\{ -\tfrac{1}{2}\bar{r}^2 \cos(2\theta) \right.$$
$$\left. + \epsilon \exp(\omega_R t)[f_R(\bar{r}) \cos(\omega_I t) - f_I(\bar{r}) \sin(\omega_I t)] \right\}, \tag{7.36}$$

where f_R and f_I are the real and imaginary parts of f and have the form shown in Fig. 7.5. The solution [Eq. (7.36)] describes oscillations which periodically flatten the field configuration along the x- and y-axes. During the course of these oscillations, the amplitude of the disturbance exponentially decays at the dimensionless rate $\bar{\omega}_R = -\pi^2 / (2 \ln^2 \bar{\eta})$.

Since $f_R = 1$ and $f_I = 0$ at $\bar{r} = 0$, the flux function at the X-line is

$$A(0, 0, t) = \epsilon \, B_0 \, r_0 \, e^{\omega_R t} \cos(\omega_I t),$$

where $\bar{\omega}_R = -\bar{\omega}_I^2 / 2$ and $\bar{\omega}_I = -\pi / \ln \bar{\eta}$. Thus, the collapse-time ($\bar{t}_f$) is one quarter of the cycle time $[2\pi / \bar{\omega}_I = -(\ln \bar{\eta})/2]$. For small η, the amount of flux reconnected by a time t_f is just ϵ and is independent of $\bar{\eta} = \eta/(v_{A0}r_0)$. The weak dependence on $\bar{\eta}$ becomes apparent only over the long term when the exponential decay is significant. Therefore, for the first collapse cycle the

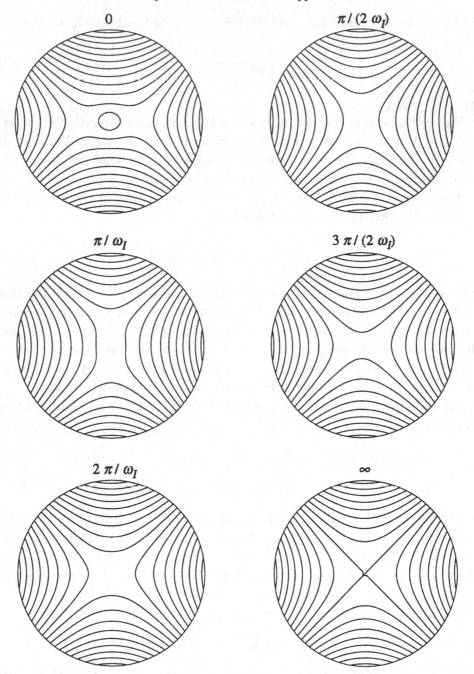

Fig. 7.4. Contours of the magnetic flux function at different times for the fundamental reconnection mode ($m = 0, n = 0$) in the linearised solution of Craig and McClymont (1991).

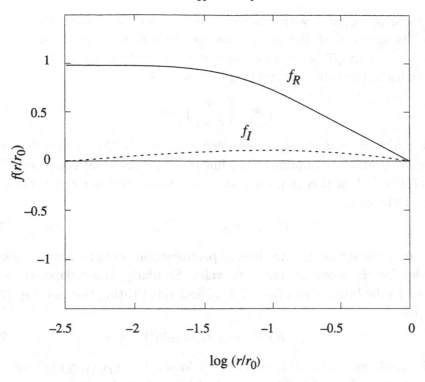

Fig. 7.5. Real (f_R) and imaginary (f_I) parts of the radial eigenmode function (f) for the $m = 0, n = 0$ mode as functions of r/r_0, where r_0 is the radius of the cylinder (after Craig and McClymont, 1991).

amount of flux reconnected (A_f) and the average electric field (E_{ave}) are

$$A_f \approx \epsilon B_0 r_0, \quad E_{\text{ave}} = \frac{A_f}{t_f} \approx -\frac{2\epsilon v_{A0} B_0}{\ln \bar{\eta}}.$$

Since this reconnection rate depends only weakly on $\bar{\eta}$ through the logarithm, the reconnection is fast.

The logarithmic dependence of the collapse-time (t_f) on the diffusivity (η) is a consequence of the decrease in wave speed $(\dot{r}_w)$ as the wave approaches the X-line. In the linear regime,

$$\dot{r}_w = -\frac{v_{A0} r_w}{r_0},$$

where r_w is the location of the wave at any instant of time. Therefore, for a wave starting at the boundary at $r = r_0$, its position as a function of time is just

$$r_w = r_0 \, e^{-v_{A0} t / r_0}.$$

The wave propagates with little dissipation until it reaches the location where its speed is of the same order as the diffusive speed (η/r_w). This occurs at $r_w = r_0\sqrt{\bar{\eta}}$, so, if the collapse stops roughly when the wave reaches this location, the collapse-time is approximately

$$t_f \approx -\left(\frac{r_0}{2v_{A0}}\right)\ln\bar{\eta}, \qquad (7.37)$$

which is the same as the value obtained from Eq. (7.35) for $m = 0, n = 0$.

Similar results are obtained if we linearise the nonlinear equations (7.16) and (7.17). When this is carried out, the reconnection electric field, Eq. (7.22), reduces to

$$E = \epsilon\bar{\eta}v_{A0}B_0\cosh(2\bar{t}), \qquad (7.38)$$

where now the initial current-density perturbation (ϵ) must always be small in order for the linearisation to be valid. Similarly, the collapse-time also reduces to the linear result [Eq. (7.37)], and substituting this into Eq. (7.38) leads to

$$E_f = \tfrac{1}{2}\epsilon v_{A0}B_0(1+\bar{\eta}^2) \qquad (7.39)$$

for the peak reconnection rate at $t = t_f$. When $\bar{\eta} = \eta/(v_{A0}r_0)$ is small, this rate is independent of $\bar{\eta}$. The amount of flux (A_f) which is reconnected by the collapse-time is

$$A_f = \tfrac{1}{4}\epsilon B_0 r_0(1-\bar{\eta}^2) \approx \tfrac{1}{4}\epsilon B_0 r_0, \qquad (7.40)$$

which is also independent of $\bar{\eta}$ when $\bar{\eta}$ is small. The average reconnection rate for small $\bar{\eta}$ is

$$E_{\text{ave}} \approx -\frac{\epsilon v_{A0}B_0}{2\ln\bar{\eta}}, \qquad (7.41)$$

which is the same scaling result as for the lowest-order eigenmode to within a constant factor.

Contrary to what we might expect, the average linear rate, Eq. (7.41), is faster than the average nonlinear rate, Eq. (7.30). The average linear rate scales as $(\ln\bar{\eta})^{-1}$, but the average nonlinear rate as $\bar{\eta}^{0.739}$ – slower even than that of Sweet–Parker. The nonlinear solution also predicts that the total amount of flux reconnected by the time t_f scales as $\bar{\eta}^{0.739}$, while the linear solution predicts that it is independent of $\bar{\eta}$. Numerical solution of the nonlinear, self-similar equations, Eqs. (7.16) and (7.17), confirms this behaviour. As shown in Fig. 7.6, the amount of flux which is reconnected by t_f is greater in the linear than in the nonlinear regime. The linear regime is valid throughout the evolution as long as $\bar{\eta}$ is greater than ϵ, but when $\bar{\eta}$

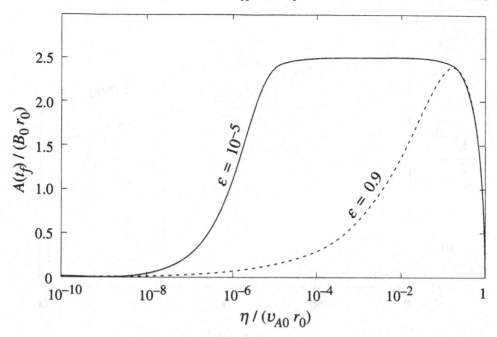

Fig. 7.6. The total amount of reconnected flux as a function of the diffusivity (η) for nonlinear, self-similar solutions. The results are obtained for a perturbation current $\epsilon = 10^{-5}$ by numerically integrating coupled equations (7.16) and (7.17). Nonlinear behaviour occurs when $\bar{\eta} = \eta/(v_{A0}r_0)$ is smaller than ϵ.

is less than ϵ, the nonlinear phase is achieved prior to t_f. In this latter case A_f decreases rapidly to zero as $\bar{\eta}$ tends to zero.

These results seem to contradict our intuition that the nonlinear process should be faster than the linear one unless some kind of saturation occurs. However, the effect of nonlinearity here is to increase the speed at which the incoming wave rushes towards the X-line. In the linear regime the time required for the incoming wave to reach the X-line tends to infinity as $\bar{\eta}$ tends to zero, but in the nonlinear regime this time remains on the order of the Alfvén time-scale no matter how small $\bar{\eta}$ becomes. Thus, as $\bar{\eta}$ is reduced, the amount of flux reconnected prior to the collapse at t_f tends to zero as $\bar{\eta}$ tends to zero.

Although the nonlinear reconnection process is slow prior to t_f, it is likely to become extremely fast after t_f because the electric field at t_f tends to infinity as $\bar{\eta}$ tends to zero. The only way at the present time to follow the nonlinear collapse beyond t_f is to use numerical simulations. Such simulations have been carried out by Brushlinskii et al. (1980), Ofman et al. (1993b), Roumeliotis and Moore (1993), and most recently by McClymont and Craig (1996). They confirm the description provided by the linear solutions and

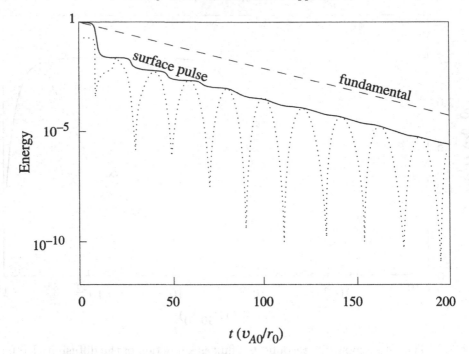

$$t\,(v_{A0}/r_0)$$

Fig. 7.7. The dimensionless magnetic plus kinetic energy (solid curve) and the magnetic energy alone (dotted curve) as functions of time obtained in a numerical solution of X-type collapse (Craig and Watson, 1992). The dashed line shows the slower decay of magnetic plus kinetic energy predicted by the linear solution.

suggest that the average reconnection rate is indeed faster in the nonlinear regime. However, a careful scaling analysis of the reconnection rate with $\bar{\eta}$ has yet to be done.

An important question that the linear solutions have been able to address is whether the collapse behaviour that occurs in the nonlinear, self-similar solution is universal or just a peculiarity of the special initial conditions required for self-similarity. As we discussed in the beginning of this section, the self-similar solution requires an initial perturbation corresponding to a uniform current density over a large area, but such a perturbation is physically rather implausible. Since the linear solutions are valid for a very wide range of boundary and initial conditions, they have no such requirement. Thus, when the linear solutions are used, it is possible to compare an initial global perturbation, like that occurring in the self-similar solution, with a perturbation which is localized at the boundary.

A comparison of this type is shown in Fig. 7.7, which plots magnetic plus kinetic energy as a function of time for two different initial conditions. The dashed line labelled "fundamental" is the $m = n = 0$ eigenmode of Craig and Watson (1992), and it corresponds to an initial perturbation over a large area, though in this case the current density is not uniform. The solid line labelled "surface pulse" is a sum over many eigenmodes and corresponds to an impulsive disturbance which originates at the boundary of the cylinder. For the fundamental mode solution the total (magnetic plus kinetic) energy decays exponentially with time at a rate which depends logarithmically on the magnetic diffusivity. However, for the surface-pulse solution the energy shows a step-like decay, which, on average, is faster than the fundamental mode. The step-like decay occurs because no significant dissipation takes place until the wave launched by the boundary disturbance reaches the centre of the cylinder. Once there, the wave creates a relatively thin current sheet, which dissipates energy more rapidly than the corresponding structure in the fundamental mode. These linear results imply that any disturbance, whether uniform or localized, will lead to rapid dissipation in the vicinity of the X-line in the absence of pressure.

Even though the collapse process was proposed by Dungey (1953) over 45 years ago, it still remains something of an enigma. Numerical solutions have established that the collapse occurs whenever an X-line configuration is perturbed in a low-resistivity, low-β plasma. If the perturbations are small ($\epsilon < \bar{\eta}$) the reconnection rate scales as $1/\ln\bar{\eta}$, while if they are large ($\epsilon > \bar{\eta}$) the average reconnection rate appears to be nearly independent of $\bar{\eta}$ as far as one can judge from the numerical simulations. One of the most important questions remaining about the collapse process is whether it is important in space or laboratory plasmas.

To obtain the rapid reconnection associated with the formation of a thin current sheet requires the plasma β_0 to be less than $\bar{\eta}^{0.565}$ (see Eqn. 7.31); otherwise, the collapse is choked off by the pressure. In solar and astrophysical applications the classical values of $\bar{\eta}$ are typically on the order of 10^{-10}, so this implies $\beta_0 \lesssim 10^{-6}$. This value of β_0 is smaller than the values usually inferred from observations, so it seems likely that the collapse in an astrophysical system would be limited by pressure rather than resistivity. However, if we consider laboratory plasmas or the possibility that the resistivity is anomalous, then a resistivity-limited collapse could still be important. For example, if $\bar{\eta} = 10^{-2}$, then $\beta_0 < 0.1$ is sufficient. In any case, the two-dimensional evolution after the initial collapse has not been fully explored. There is some indication in the simulations by McClymont and Craig (1996) that the nonlinear, two-dimensional effects mitigate the effects of pressure, and that a secondary phase of reconnection may occur which is not limited by pressure.

One application where the collapse process might be important is coronal mass ejections (CMEs), which we shall discuss in Section 11.1. In some models of CMEs, an X-type neutral line is formed during the eruption of a coronal magnetic structure. At the moment of its appearance, the X-line is current-free, but, because the global field is undergoing rapid evolution on the Alfvén time-scale, a current sheet immediately forms at the X-line. The formation of this current sheet somewhat resembles the formation of the current sheet in the X-type collapse. Therefore, it might be possible to predict the thickness of the current sheet created during a CME by comparing it to the pressure-limited ($\beta_0 > \overline{\eta}^{0.565}$) collapse process.

7.2 Time-Dependent Petschek-Type Reconnection

The X-type collapse that we discussed in the previous section presupposes an orthogonal X-type field generated by sources far from the neutral point. For this reason it is more closely related to the steady-state solution developed by Syrovatsky (1971) than the steady-state solution developed by Petschek (1964). However, time-dependent solutions of Petschek-type (§4.3) reconnection do exist. Semenov et al. (1983a, 1984), Heyn and Semenov (1996), and Kiendl et al. (1997) have considered the temporal evolution that results when reconnection is triggered by an increase in resistivity at a particular location in a current sheet. As in the tearing mode, the solution assumes that a simple current sheet exists initially, as shown in Fig. 7.8(a). Unlike the tearing mode, the effect of diffusion is neglected everywhere except in the region where the reconnection is initiated. The onset of reconnection in this localised region launches both fast and slow magnetoacoustic waves into the medium. In the incompressible version of the theory (Biernat et al., 1987, 1998; Rijnbeek and Semenov, 1993), the fast-mode waves propagate outwards instantaneously and set up an inflow of plasma towards the X-line. In contrast to the Petschek theory, the inflow is not uniform to lowest order but decreases with distance, becoming zero at infinity in the y-direction. After reconnection ceases at the reconnection point, there is a switch-off phase with the effect of the switch-off propagating outwards, which does not exist in the steady-state solutions.

As in the steady-state Petschek theory, the inflow near the X-line is supermagnetosonic with respect to the slow-mode wave speed, and so slow-mode shock pairs are formed. However, the shocks are curved and enclose the rear portion of an outflow region which is tear-drop in shape, as shown in Fig. 7.8(b). The front portion of this region is bounded by slow shocks, which connect to the slow-mode shocks in the rear portion. In the incompressible limit, the speed at which the outflow disturbance propagates along the x-axis is just v_{A0}, the ambient Alfvén speed, so the external scale-length (L_e)

(a)

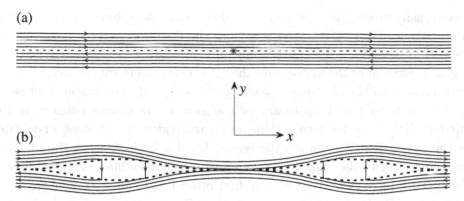

(b)

Fig. 7.8. (a) An initial current sheet. (b) The evolution of the magnetic field (solid curves) and shocks (dashed curves) in Semenov's time-dependent model of Petschek-type reconnection.

of the system continuously increases with time as

$$L_e = v_{A0} \, t. \tag{7.42}$$

For distances from the origin much smaller than this, the time-dependent solution tends to the steady-state Petschek solution.

Formally, the solution is set up in much the same manner as Petschek's. In the inflow region one expands the equations in terms of a smallness parameter (ϵ), now defined as

$$\epsilon(t) = \frac{E^*(t)}{v_{A0} B_0},$$

where $E^*(t)$ is the electric field at the X-line and B_0 is the ambient field outside the sheet. The parameter (ϵ) plays the same role as the Alfvén Mach number (M_e) of the steady-state theory, and the expansion assumes $\epsilon \ll 1$. Another restriction on E^* is that it must change slowly with respect to time, so that the evolution of the diffusion region at the X-line is quasi-steady. This restriction, combined with the fact that $\epsilon \ll 1$, makes the outflow region a thin layer and permits solutions of the MHD equations by using boundary-layer theory.

As in steady-state Petschek theory, the inflow in the incompressible case is current-free to first order (although this need not be so in the compressible case). However, this is no longer an arbitrary assumption, but a requirement which follows from the initial state. At $t = 0$, there is no current in the region outside the sheet, so any current which develops there must be set up by waves propagating outwards from the initial reconnection site. Fast-mode waves easily propagate upstream, but slow-mode waves have a more difficult time because their wave speed is zero in the direction perpendicular to the field. Once the inflow into the reconnection site becomes well developed, the region upstream of the slow-mode shocks is everywhere super-magnetosonic with respect to the slow-mode wave speed (cf. §5.2). The inability of slow-mode waves to travel upstream of the slow-mode shocks constrains the current density to be zero in the upstream region, to at least first order in the expansion. We can see the reason for this by looking at the momentum equation. In the region where a quasi-steady flow has been established, the inertial term $(\mathbf{v} \cdot \nabla)\mathbf{v}$ is zero to first order in ϵ since the inflow (v_y) is of first order. (In the case where v_y is also uniform to first order, the inertial term is zero to second order as well.) Therefore, to first order, the momentum equation is $\nabla p = \mathbf{j} \times \mathbf{B}$, or alternatively $j = dp/dA$. Since these are the equations describing a static equilibrium, variations in p and B across field lines must be in the opposite sense except for the special case when $j = 0$. Such variations can only be accomplished by slow-mode waves, and these waves must necessarily come from the exterior boundaries because the inflow is super-magnetosonic with respect to the slow-mode speed. Therefore, if we insist that the incoming slow waves carry no such variations (i.e., they come from a uniform state), then the only possible solution is $j = 0$ to first order in ϵ.

The low-order currents which are present in the inflow region of Petschek's solution (§5.1) are created by the fast-mode expansion launched into the inflow region by the initial reconnection event. In the steady-state theory we discussed in Chapter 5, Petschek's solution is the only one which is completely free of slow-mode effects in the inflow region (i.e., slow-mode expansions or compressions). Therefore, Petschek's solution can be associated with undriven reconnection, while other types of reconnection, such as flux pile-up, for example, may be generally associated with driven reconnection. Thinking in terms of the propagation properties of slow- and fast-mode waves also explains why there are no current layers along the separatrices in the Petschek configuration. As long as the separatrices lie upstream of the slow-mode shocks, they are necessarily current-free in the absence of external forcing. Only if the separatrices are downstream of the shocks or the shocks are absent can such layers exist.

Since j_1 is zero in the inflow region, to first order the flux function $A(x, y, t)$ is a solution of Laplace's equation ($\nabla^2 A = 0$), and the general solution for **B** in the inflow region is

$$B_x = B_0 + \frac{\partial A_1}{\partial y}, \quad B_y = -\frac{\partial A_1}{\partial x},$$

where

$$A_1 = \frac{y}{2\pi} \int_{-\infty}^{\infty} \frac{A_1(x',0,t)}{(x-x')^2 + y^2} dx' \tag{7.43}$$

and $A = B_0 y + A_1$. The function $A_1(x',0,t')$ is found by using the slow-mode jump conditions at the interface of the outflow region. The result is

$$A_1(x',0,t') = B_0 |x'| g(|x'| - v_{A0}t') - B_0 \int_0^{|x'|} g(\xi - v_{A0}t') d\xi, \tag{7.44}$$

where

$$g(x - v_{A0}t) = \epsilon(t - x/v_{A0}) = -\frac{E^*(t - x/v_{A0})}{B_0 v_{A0}}$$

relates the field structure to the time behaviour of the electric field at the X-line.

Since j is zero in the inflow region, the velocity there is irrotational to first order and the first-order stream function (ψ_1) satisfies $\nabla^2 \psi_1 = 0$. Thus, the general solution for the velocity in the inflow region is

$$v_{x1} = 0, \quad v_{y1} = -\frac{\partial \psi_1}{\partial x},$$

where

$$\psi_1 = \frac{y}{2\pi} \int_{-\infty}^{\infty} \frac{\psi_1(x',0,t)}{(x-x')^2 + y^2} dx' \tag{7.45}$$

and

$$\psi_1(x',0,t') = v_{A0} x' g(x' - v_{A0} t'). \tag{7.46}$$

In the outflow region the solution to zeroth order is

$$B_{x0} = 0, \quad B_{y0} = B_0 g(x - v_{A0} t),$$
$$v_{x0} = v_{A0}, \quad v_{y0} = 0,$$

and the location of the shock transition which separates the outflow from the inflow is prescribed by

$$f(x,t) = x g(x - v_{A0} t), \tag{7.47}$$

where $f(x,t)$ is the height of the shock at location x and time t. Expressions for the first-order corrections in the outflow region are not given here but may be found in Pudovkin and Semenov (1985). These corrections add curvature to the straight field lines shown in the outflow region of Fig. 7.8(b), such that the field lines no longer kink so sharply as they cross the shock transition.

To complete the solution it is necessary to relate the temporal behaviour of the normalized reconnection rate (ϵ) to the imposed variation in diffusivity (η). This is done by using the Sweet–Parker relations to match the average inflow and outflow in the diffusion region to the external region (Rijnbeek and Semenov, 1993). The procedure is similar to that used in steady-state Petschek theory to obtain the maximum rate of reconnection, but it does not give a complete description of the relation between $\eta(t)$ and $\epsilon(t)$. At the switch-on of the diffusivity there is no flow and convection is negligible. Therefore, at this time the description of the convection and waves provided by the Semenov-type solution is not yet in effect, but, once convection takes hold, $\eta(t)$ can be expressed in terms of $\epsilon(t)$ or vice versa. For η kept constant after the initial increase, the normalized electric field is

$$\epsilon(t) = \frac{\pi}{4 \ln[\epsilon^2(t) R_{me}(t)]} \approx \frac{\pi}{4 \ln R_{me}(t)}, \tag{7.48}$$

where

$$R_{me}(t) = \frac{v_{A0} L_e(t)}{\eta_{\text{ave}}} = \frac{v_{A0}^2}{\eta_{\text{ave}}} t.$$

Here η_{ave} is the average diffusivity of the diffusion region. Equation (7.48) is just Petschek's result but with the magnetic Reynolds number (R_{me}) now a function of time.

In the case in which the diffusivity is kept constant after its initial switch-on, the evolution of the diffusion region is simply obtained by rescaling the steady-state Petschek theory with the external, time-dependent scale-length (L_e) that is prescribed by Eq. (7.42). Doing this leads to

$$L = \frac{L_e(t)}{R_{me}(t)\,\epsilon^2(t)} = \frac{\eta_{\text{ave}}}{v_{A0}} \left[\frac{4 \ln(v_{A0}^2 t/\eta)}{\pi}\right]^2 \tag{7.49}$$

for the half-length of the diffusion region and

$$l = L_e(t)\epsilon(t) = \frac{\eta_{\text{ave}}}{v_{A0}} \frac{4 \ln(v_{A0}^2 t/\eta)}{\pi} \tag{7.50}$$

for its half-thickness. In contrast to the linear growth of the wave structures, the diffusion region grows logarithmically in time as long as η_{ave} is constant.

Numerical simulations which come closest to being like the time-dependent Petschek-type solutions are those of Ugai and Tsuda (1977, 1979), Ugai (1984, 1988, 1995a,b), Scholer (1989), and Schumacher and Kliem (1996). These simulations trigger reconnection in a simple current sheet by enhancing the diffusivity at a particular location, and they use open boundary conditions (i.e., the normal derivatives of most quantities are set to zero at the boundary). These open conditions make the boundaries very sensitive to the initial conditions and allow wave disturbances to exit the boundaries of the box without reflection. In some simulations an anomalous resistivity is also included with the resistivity a function of the current density. Figure 7.9 shows an example of the magnetic field and flow configuration that occurs in a simulation by Ugai (1995a). The overall tear-drop shape of the outflow region is qualitatively similar to that predicted by the analyses of Semenov et al. (1983a, 1992), but a quantitative comparison has yet to been made.

Both Ugai (1995b) and Scholer (1989) have found that a quasi-steady Petschek-like configuration is set up if the region of high resistivity is confined to a region which has the dimensions predicted by the steady-state theory. However, if the resistivity is made uniform and constant shortly after the initial triggering event, then the diffusion-region current sheet grows with time until it reaches the edges of the numerical box, at which time the reconnection becomes of Sweet–Parker type. This behaviour has sometimes been interpreted as a failure of Petschek theory (Jamitzky and Scholer, 1995), but in fact it agrees qualitatively with the expectation from the time-dependent theory that the diffusion region will grow in length until it reaches the length of the numerical domain.

Another feature seen in the numerical experiments that is not accounted for by the steady-state Petschek theory is the deflection of the inflowing plasma upstream of the separatrices. The deflection occurs in layers which lie along the upstream sides of the slow shocks and which start to turn the plasma in the direction of the outflow before the plasma crosses the separatrices (Scholer, 1989). Such layers only occur when the diffusion region is not forced to match the length required by steady-state Petschek theory, and this result has also led to doubt about the correctness of the Petschek solution. However, the compressible version of the time-dependent Petschek theory, as developed by Heyn and Semenov (1996) and Heyn (1997), implies that the layer observed in the simulations is caused by fast-mode waves generated by the evolving outflow region. Unlike the steady-state theory, the dimensions of the outflow region in the time-dependent theory continually increase, and this leads to a sustained fast-mode disturbance in the region upstream of the slow shocks.

Quantitative comparisons between numerical simulations and the time-dependent Petschek-type theory have yet to be done, but such simulations might actually be easier to implement than for the steady-state theory. The time-dependent theory is formulated as an initial-value problem in the infinite plane without boundaries, whereas the steady-state theory is formulated as a boundary-value problem. For MHD simulations initial conditions

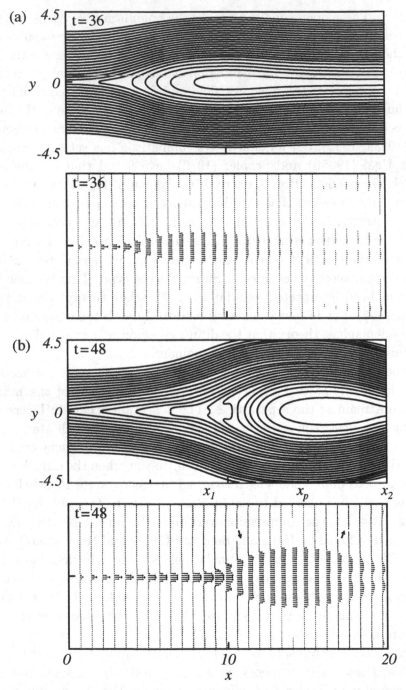

Fig. 7.9. Contours of the flux function for a numerical experiment with reconnection initiated by a region of locally enhanced resistivity. Several of the features predicted by the time-dependent Petschek-type model are present, including the tear-drop shaped outflow region and the shock discontinuity enclosing it (after Ugai, 1995a).

are much easier to impose than boundary conditions because of the variations in the number and type of boundary conditions that may be imposed (Forbes and Priest, 1987). Also, it is often more difficult numerically to obtain a steady-state solution than a time-dependent one, even if the steady-state solution is perfectly stable (Roache, 1982).

8

Reconnection in Three Dimensions

The theory of reconnection in two dimensions is now fairly well understood and is highly developed, and, as we have seen, the type of reconnection that is produced depends very much on the reconnection rate, the configuration, the boundary conditions, and the parameter values. Many questions do remain, however, such as: what are the properties of turbulent or impulsive bursty reconnection; why does the diffusion region in Petschek reconnection lengthen when the resistivity is uniform; what is the effect of outflow boundary conditions on fast reconnection; how does reconnection occur in a collisionless plasma; and how do the different terms in the energy equation such as radiation and conduction affect reconnection?

The theory of three-dimensional reconnection is much less developed. We have only just started a voyage of discovery that will last many years, but some important directions have already been indicated. Many features are quite different in three dimensions. For example, we discuss here the definition of reconnection (§8.1), the structure of null points (§8.2), the nature of the bifurcations (§8.3), the global magnetic topology (§8.4), and the nature of the reconnection itself (§§8.6, 8.7).

In this chapter we introduce several new concepts. At null points, magnetic reconnection can take place by *spine reconnection, fan reconnection,* or *separator reconnection* (§8.6). Regions where magnetic field lines touch a boundary and are concave towards the interior of the volume are referred to as *bald patches* (§8.4.1). When no null points or bald patches are present, the mapping of field lines from one boundary to another is continuous, so they all have the same topology and there are no separatrices. However, reconnection can still occur at *singular lines* (§8.1.3) and at special surfaces known as *quasi-separatrix layers* (§8.7).

We begin by describing the attempts to define what is meant by reconnection in three dimensions, including *general magnetic reconnection* (§8.1.1) and *singular field-line reconnection* (§8.1.3). We define what is meant by *magnetic topology* and show that in a non-ideal plasma the conservation of magnetic field lines and magnetic flux are not equivalent and that there are difficulties with defining field-line velocity (§8.1.2). Also, we describe the properties of 3D null points (§8.2) and the ways to define and calculate *magnetic helicity* (§8.5).

8.1 Definition of Reconnection

Null points (where the magnetic field vanishes) are said to be linear when the magnetic field components increase linearly with distance from the null point, and they are of higher order when one or more of the field components increases at a slower rate (quadratically or cubically, for instance). In *two dimensions*, linear null points fall into two types, namely X-points and O-points, where the neighbouring magnetic field lines have X-type or O-type topology, respectively. The notion of magnetic reconnection is then fairly straightforward and it has several properties [Fig. 8.1(a)]:

(i) Reconnection occurs at an X-point, where two pairs of separatrices meet; during the process of reconnection, pairs of magnetic field lines are brought in towards the X-point; they then lie along the separatrices and are broken and reconnected;

(ii) The electric field (**E**) is normal to the plane and so is directed along the X-line;

(iii) The mapping of magnetic field lines from one boundary to another is discontinuous; therefore, during the process of reconnection, there is a change of magnetic connectivity of plasma elements due to the presence of a localised diffusion region where ideal MHD breaks down; for instance, in Fig. 8.1(a), initially a field line joins plasma element A to B, whereas after reconnection when the plasma elements move to A$'$ and B$'$ they are no longer connected magnetically;

(iv) A flow of plasma is present across the separatrices.

The question arises: which of these properties are robust enough to be preserved to act as a definition in three dimensions? Property (ii) has been used as a definition of reconnection by Sonnerup (1984), property (iii) by Axford (1984), and property (iv) by Vasyliunas (1975). However, Schindler et al. (1988) have pointed out that reconnection can occur in the absence of nulls and separatrices, so that (ii) and (iv) cannot form the basis of a generalisation to three dimensions [Fig. 8.1(b)]. Instead they suggested that (iii) be

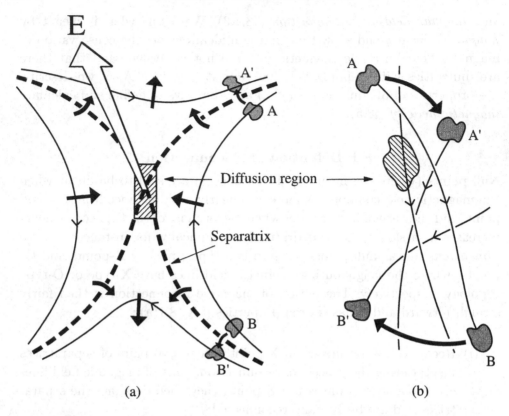

(a)   (b)

Fig. 8.1. General properties of (a) two-dimensional and (b) three-dimensional reconnection, in which plasma elements A and B that are initially joined by a magnetic
field line move to locations A′ and B′, where they are no longer joined magnetically.

used as the fundamental definition of so-called *general magnetic reconnection*
(§8.1.1). Their concept includes all effects of local nonidealness that produce
a component $(E_\parallel)$ of the electric field along a particular magnetic field line.
They therefore pointed out that a three-dimensional generalisation of (ii) is

$$\int E_\parallel \, ds \neq 0$$

as a necessary and sufficient condition for general magnetic reconnection,
where the integral is taken along that particular magnetic field. An equivalent condition is that the magnetic helicity (§8.5) change in time.

    This is a highly attractive definition, although in our view it is too gen-
eral, since it includes examples of magnetic diffusion or slippage (such as in
double layers or shock waves) that have not been traditionally included in
the concept of reconnection. We therefore prefer to restrict the definition of
reconnection to *singular field-line reconnection* (§§8.1.3, 8.1.4), in which the

presence of $E_\parallel$ along a field line is supplemented by the condition that the nearby field has an X-type topology in a plane perpendicular to the field line (Priest and Forbes, 1989). This concept of reconnection has been given an elegant covariant formulation by Hornig and Rastätter (1997, 1998).

8.1.1 General Magnetic Reconnection

Schindler et al. (1988) presented a physical discussion of their ideas, and in a companion paper Hesse and Schindler (1988) formulated general magnetic reconnection mathematically in terms of Euler potentials. They defined it as "a breakdown of magnetic connection due to non-idealness", and they wrote Ohm's law in a general form

$$\mathbf{E} + \mathbf{v} \times \mathbf{B} = \mathbf{R}, \tag{8.1}$$

where $\mathbf{R}$ is any general non-ideal term (due to, for instance, collisions, fluctuations, or particle inertia). $\mathbf{R}$ is assumed to be localised, so that it is nonzero inside a finite domain D_R (a diffusion region) and vanishes outside it. They therefore excluded global magnetic diffusion processes from their definition. In terms of resistive MHD, they included situations in which the global magnetic Reynolds number is large ($R_{me} \gg 1$) but excluded those where it is not ($R_{me} \leq 1$).

In order to model the formation of plasmoids in the Earth's magnetotail (see §10.5.2), they considered first a two-dimensional magnetic field of the form

$$B_x = Kz, \quad B_z = x^2 - 2ax - b,$$

with K, a, and b being constants such that $a^2 < b$. This field consists of a series of closed magnetic loops [Fig. 8.2(a)], but if a and b evolve in time to give $a^2 > b$, a pair of X- and O-points appears with a plasmoid [Fig. 8.2(b)]. A separatrix now divides the plane into three regions and reconnection at the X-point involves a transfer of flux from one region to another across the separatrix.

Then Schindler et al. (1988) added a constant B_y-component and supposed a and b become functions of y with the property that $a^2 > b$ only for a finite range of values of y, so that the plasmoid has a finite extent in the y direction. Now the startling feature is that, although the field in planes $y = $ constant still looks like Fig. 8.2, the separatrix surface completely disappears. There is no longer a topological difference between field lines inside and outside the plasmoid, since all field lines start at a left-hand (i.e., constant-x) boundary and end up there: a field line that is spiralling inside

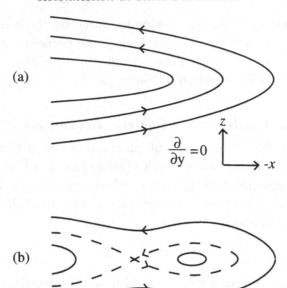

Fig. 8.2. Magnetic configuration (a) before and (b) after the formation of a plasmoid and a separatrix (dashed curve) in two dimensions (after Schindler et al., 1988).

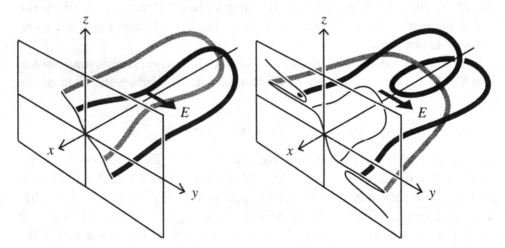

Fig. 8.3. Plasmoid formation in three dimensions in which two magnetic loops (light and dark) reconnect in the absence of a null point to create two new loops, one twisted around the other (after Schindler et al., 1988).

the plasmoid over a range of values of y will eventually leave the plasmoid at large values of $|y|$ (Figs. 8.3 and 10.14). The concept of a separatrix surface in a translationally invariant (e.g., y-independent) field such as Fig. 8.2 is therefore *structurally unstable*, since it may be destroyed by an arbitrarily small modification of the field. This is why Schindler et al. (1988) were led

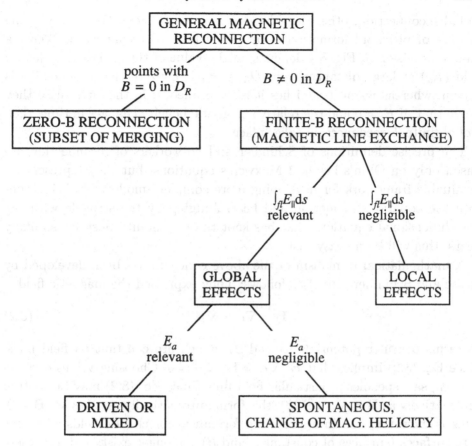

Fig. 8.4. A general classification of different types of reconnection in terms of a localised diffusion region (D_R), the electric field ($E_\parallel$) along the magnetic field, and the asymptotic electric field (E_a) at large distances from D_R (from Schindler et al., 1988).

to eschew the concept of separatrices in their formulation of a definition of reconnection.

They then proposed a classification of different kinds of general magnetic reconnection as shown in Fig. 8.4. On the one hand, zero-B (or null-point) reconnection takes place at a null point ($\mathbf{B} = \mathbf{0}$), for which separatrices do exist and so topological arguments can be applied. For finite-B reconnection, on the other hand, the magnetic field does not vanish. A necessary and sufficient condition for its occurrence is that the magnetic field be nonzero in a domain D_R (the diffusion region) and $\mathbf{B} \times (\nabla \times \mathbf{R}) \neq \mathbf{0}$ on some point-set in D_R. An important parameter here is the electric field ($E_\parallel$) parallel to the magnetic field: if $\int E_\parallel \, ds$ along a measurable set of field lines in D_R is nonzero, then the effect of reconnection is felt outside D_R and so we have global reconnection; otherwise, the reconnection is local. For example, the process of plasmoid formation is by global finite-B reconnection. Then, a final subdivision in Fig. 8.4 depends on the value of the asymptotic electric field (E_a) at large distances from D_R; if E_a is nonzero the reconnection is driven, whereas when it vanishes it is spontaneous; in the latter case they prove that, if $E_a = 0$ and the relative magnetic helicity changes in time, then global finite-B reconnection takes place.

The precise definitions of Schindler and co-workers of reconnection are based only on Ohm's law and Maxwell's equations, but they represent an invaluable framework for developing more complete models which incorporate the equations of motion and heat. Judging by the struggle with the two-dimensional equations that has kept the community busy for so many years, this will be no easy task.

A mathematical formalism of the above concepts has been developed by Hesse and Schindler (1988) as follows: They expressed the magnetic field

$$\mathbf{B} = \nabla\alpha \times \nabla\beta \qquad (8.2)$$

in terms of Euler potentials (α and β), which are constant on field lines, since Eq. (8.2) implies that $\mathbf{B} \cdot \nabla\alpha = \mathbf{B} \cdot \nabla\beta = 0$. Choosing values $\alpha = \alpha_0$, $\beta = \beta_0$, say, specifies a particular field line. Since Eq. (8.2) may be written alternatively as $\mathbf{B} = \nabla\times(\alpha\nabla\beta)$, this form automatically satisfies $\nabla \cdot \mathbf{B} = 0$. This is, however, not an appropriate formalism for chaotic fields when the flux surfaces (surfaces of constant α and β) no longer exist, and also there are difficulties associated with α and β becoming multi-valued in toroidal systems. In addition, care has to be taken near null points.

The evolution of the Euler potentials in time may be deduced as follows: If $\mathbf{A}$ is a vector potential (such that $\mathbf{B} = \nabla \times \mathbf{A}$) in a gauge with $\mathbf{A} \cdot \mathbf{B} = 0$, it can be expressed in terms of Euler potentials as

$$\mathbf{A} = \alpha\,\nabla\beta. \qquad (8.3)$$

Furthermore, for $\nabla \times \mathbf{E} = -\partial\mathbf{B}/\partial t$ to be satisfied identically, the electric field may be written in terms of potentials $\mathbf{A}$ and Φ as

$$\mathbf{E} = -\frac{\partial\mathbf{A}}{\partial t} - \nabla\Phi$$

$$= -\frac{\partial\alpha}{\partial t}\nabla\beta + \frac{\partial\beta}{\partial t}\nabla\alpha - \nabla\Psi, \qquad (8.4)$$

where, by use of Eq. (8.3), $\Psi = \Phi + \alpha\,\partial\beta/\partial t$. Then $\nabla\Psi$ and $\mathbf{R}$ in Eq. (8.1) may be written in terms of components parallel to $\nabla\alpha, \nabla\beta$, and ∇s, where

s measures the distance along field lines, so that, for instance,

$$\mathbf{R} = R_\alpha \nabla\alpha + R_\beta \nabla\beta + R_s \nabla s.$$

The corresponding components of Ohm's law [Eq. (8.1)] then become

$$\frac{d\alpha}{dt} = -\frac{\partial\Psi}{\partial\beta} - R_\beta, \tag{8.5}$$

$$\frac{d\beta}{dt} = \frac{\partial\Psi}{\partial\alpha} + R_\alpha, \tag{8.6}$$

$$\frac{\partial\Psi}{\partial s} = -R_s, \tag{8.7}$$

where $d/dt = \partial/\partial t + \mathbf{v} \cdot \nabla$ represents the time derivative following the plasma motion.

Outside the diffusion region the components of $\mathbf{R}$ all vanish, and so this set of equations implies that $\Psi = \Psi(\alpha, \beta, t)$ is independent of s and is a Hamiltonian for the dynamics of the coordinates α and β. Then line conservation is an immediate consequence, since the time-evolution of α and β is independent of s; that is, α and β change in the same way for all plasma elements on a given field line. Flux conservation also follows easily since Liouville's theorem (e.g., Arrowsmith and Place, 1990) implies that the area within any closed curve in $\alpha\beta$-space remains constant; that is, the magnetic flux through a closed curve moving in physical space with the plasma is constant in time.

Inside the diffusion region, where $\mathbf{R} \neq 0$, line conservation would still hold if the right-hand sides of Eqs. (8.5) and (8.6) were independent of s, that is,

$$\frac{\partial}{\partial s}\left(\frac{\partial\Psi}{\partial\beta} + R_\beta\right) = \frac{\partial}{\partial s}\left(\frac{\partial\Psi}{\partial\alpha} + R_\alpha\right) = 0, \tag{8.8}$$

so that the behaviour of α and β is independent of distance along a field line. The general solution of these equations, Eq. (8.8), is

$$R_\alpha = \frac{\partial f}{\partial\alpha} + \frac{\partial g}{\partial\beta}, \quad R_\beta = \frac{\partial f}{\partial\beta} - \frac{\partial g}{\partial\alpha}, \quad R_s = \frac{\partial f}{\partial s},$$

which in turn is equivalent to

$$\mathbf{B} \times (\nabla \times \mathbf{R}) = \mathbf{0}, \tag{8.9}$$

where $f(\alpha, \beta, s, t,)$ and $g(\alpha, \beta, t)$ are arbitrary functions. Furthermore, flux conservation would still hold if Liouville's theorem holds or, in other words, if α and β are of Hamiltonian form so that

$$\frac{d\alpha}{dt} = -\frac{\partial F}{\partial\beta}, \quad \frac{d\beta}{dt} = \frac{\partial F}{\partial\alpha}, \tag{8.10}$$

where $F = F(\alpha, \beta, t)$. Then, if we write the function Ψ in Eqs. (8.5)–(8.7) in the form $\Psi(\alpha, \beta, s, t) = F(\alpha, \beta, t) - G(\alpha, \beta, s, t)$, the pair (8.5) and (8.6) is the same as Eq. (8.10) if and only if

$$R_\alpha = \frac{\partial G}{\partial \alpha}, \quad R_\beta = \frac{\partial G}{\partial \beta},$$

or, in other words,

$$\nabla \times \mathbf{R} = \mathbf{0}. \tag{8.11}$$

We now proceed to prove that finite-B reconnection with global effects occurs if and only if

$$\int E_\parallel \, ds \neq 0 \tag{8.12}$$

on a measurable set of field lines in D_R. In the basic equations, Eqs. (8.5)–(8.7), Ψ is in general a function of α, β, s, and t, but outside the domain D_R $\partial \Psi / \partial s = 0$, so that Ψ is a function of α, β, and t alone. However, for a field line that passes through D_R, these may be different functions, Ψ_1 and Ψ_2, say, on either side of D_R. In other words,

$$\Psi_2(\alpha, \beta, t) - \Psi_1(\alpha, \beta, t) = \int_1^2 \frac{\partial \Psi}{\partial s} ds, \tag{8.13}$$

where the integration is carried out along a field line and the difference between Ψ_1 and Ψ_2 depends on the way Ψ varies in D_R. Thus, if the right-hand side of Eq. (8.13) is nonzero, so that by Eq. (8.4)

$$\int E_\parallel \, ds \neq 0,$$

the diffusion region creates a global effect along field lines that thread it: the change of α and β felt by plasma elements that are initially located on the same field line will be different on both sides of the non-ideal region.

Later, Birn et al. (1989) considered the three-dimensional structure of plasmoids, while Hesse et al. (1990) applied these ideas to flux-transfer events (§10.3) and Otto et al. (1990) applied them to geomagnetic tail reconnection. Also, Schindler et al. (1991) discussed the role of the resulting $E_\parallel$ in particle acceleration (§13.1.7). The structure of plasmoids, sometimes chaotic, was analysed further by Lau and Finn (1991, 1992), using a kinematic approach. Such an approach was also used by Wang and Bhattacharjee (1996) to model the magnetosphere. Van Hoven and Hendrix (1995) applied the general magnetic reconnection diagnostic to numerical simulations of tearing, while Greene (1993) compared reconnection with changes in topology of fluid streamlines. Furthermore, Jardine (1994) has presented a three-dimensional version of the almost-uniform family (§5.1) of reconnection models characterized by the vorticity with which plasma flows towards the reconnection site.

8.1.2 Conservation of Magnetic Topology in a Non-Ideal Plasma

In an ideal plasma Ohm's Law is

$$\mathbf{E} + \mathbf{v} \times \mathbf{B} = 0 \tag{8.14}$$

and the induction equation reduces to

$$\frac{\partial \mathbf{B}}{\partial t} = \nabla \times (\mathbf{v} \times \mathbf{B}). \tag{8.15}$$

We saw in §1.4 that Alfvèn's frozen-flux theorem implies then that:

  (i) ideal plasma flows conserve magnetic flux;
  (ii) they also conserve the magnetic field-line connection of plasma elements.

A consequence of (ii) is that such flows conserve the *magnetic topology*, which includes any property of the magnetic field that is preserved by an ideal displacement, such as the linkage and knottedness of magnetic field lines. The magnetic structure may be stretched and deformed by such a flow, but its topology is not changed. Furthermore, as we pointed out in Chapter 1, the components $(\mathbf{v}_\perp)$ and $(\mathbf{w}_\perp)$ of the plasma velocity and field-line velocity perpendicular to the magnetic field are identical and equal to

$$\mathbf{v}_\perp = \mathbf{w}_\perp = \frac{\mathbf{E} \times \mathbf{B}}{B^2}. \tag{8.16}$$

Now what happens to these properties in a non-ideal plasma flow, for which in general Eq. (8.1) holds, where $\mathbf{R}$ is any non-ideal term (such as $\mathbf{R} = \eta \nabla \times \mathbf{B}$ in resistive MHD)? It may be shown (e.g., Vasyliunas, 1972; Hornig and Schindler, 1996) that flux conservation (i) and line conservation (ii) are no longer equivalent and the field-line velocity is no longer unique! Specifically, the condition

$$\mathbf{B} \times (\nabla \times \mathbf{R}) = 0 \tag{8.17}$$

implies field-line conservation (ii). In contrast, the condition

$$\nabla \times \mathbf{R} = 0 \tag{8.18}$$

implies flux conservation (i) and therefore field-line conservation (ii). When $\mathbf{B} \neq 0$, Eq. (8.17) is equivalent to

$$\nabla \times \mathbf{R} = \lambda \mathbf{B},$$

where λ is a scalar function, but at a null point ($\mathbf{B} = \mathbf{0}$) this latter condition is more restrictive than Eq. (8.17) since it implies that $\nabla \times \mathbf{R}$ vanishes, which is not necessary for Eq. (8.17). However, Eq. (8.18) is equivalent to

$$\mathbf{R} = \nabla G,$$

where G is a scalar. Thus, the classical result (Newcomb, 1958) for an ideal plasma that flux conservation (i) implies line conservation (ii) also holds in a non-ideal plasma. However, in the non-ideal case, line conservation does not necessarily imply flux conservation.

Results (8.17) and (8.18) have already been proved above in §8.1.1 using Euler potentials. Result (8.17) may also be proved without Euler potentials as follows. The set of field lines for a magnetic field ($\mathbf{B}$) is given by

$$\frac{\partial x}{\partial s} = B_x, \quad \frac{\partial y}{\partial s} = B_y, \quad \frac{\partial z}{\partial s} = B_z, \tag{8.19}$$

where s is some parameter (not necessarily distance) that is a measure of the location of a point along a field line from an initial position ($\mathbf{r}_0$, say). In compact form, Eq. (8.19) is written

$$\frac{\partial \mathbf{r}}{\partial s} = \mathbf{B}, \tag{8.20}$$

and its solution

$$\mathbf{r} = \mathbf{F}_B(\mathbf{r}_0, s) \tag{8.21}$$

at some time t, say, is known as the *flow* of the magnetic field. Formally, the magnetic topology of the field is conserved if the field lines are deformed in a continuous manner in such a way that their mutual position, their sense (i.e., positive or negative direction), and their linkage stay the same and no field lines are cut or reconnected.

This is the case if a continuous mapping (a homeomorphism, $\mathbf{F}_w$) exists which maps the flow of the magnetic field at time t onto the flow at some later time $\bar{t}$, as sketched in Fig. 8.5. In order to avoid technical difficulties, Hornig and Schindler restricted their proof to *smooth* deformations of the field lines (diffeomorphisms), so that they excluded the formation of current sheets, for instance. The vector $\mathbf{F}_w$ is then associated with a velocity ($\mathbf{w}$) that satisfies

$$\mathbf{w} = \frac{\partial \mathbf{F}_w}{\partial \bar{t}} \tag{8.22}$$

and takes the points $\mathbf{r}_0$ and $\mathbf{r}$ to points $\bar{\mathbf{r}}_0$ and $\bar{\mathbf{r}}$, respectively, at time $\bar{t}$, so that

$$\bar{\mathbf{r}}_0 = \mathbf{F}_w(\mathbf{r}_0) \quad \text{and} \quad \bar{\mathbf{r}} = \mathbf{F}_w(\mathbf{r}). \tag{8.23}$$

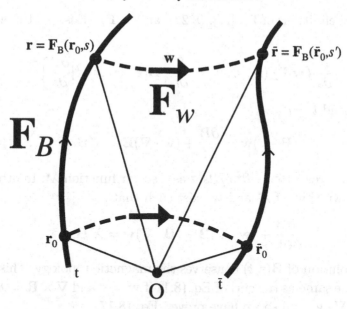

Fig. 8.5. Conservation of a magnetic field line mapped from time t to time $\bar{t}$ by a flow **w**. Points are mapped *along* field lines by the mapping $\mathbf{F}_B$ and *across* field lines by $\mathbf{F}_w$.

Then, if the topology is conserved, $\bar{\mathbf{r}}$ and $\bar{\mathbf{r}}_0$ still lie on the same field line at time $\bar{t}$ so that

$$\bar{\mathbf{r}} = \mathbf{F}_B(\bar{\mathbf{r}}_0, s'), \tag{8.24}$$

where the parametrisation (s') along the final field line may be different from the parametrisation (s) along the initial field line if there is motion along the field. By substituting for $\mathbf{r}, \mathbf{r}_0$, and $\bar{\mathbf{r}}$ from Eqs. (8.21) and (8.23), Eq. (8.24) becomes

$$\mathbf{F}_w[\mathbf{F}_B(\mathbf{r}_0)] = \mathbf{F}_B[\mathbf{F}_w(\mathbf{r}_0)],$$

which simply says that we may go from point $\mathbf{r}_0$ to $\bar{\mathbf{r}}$ in Fig. 8.5 either via point $\mathbf{r}$ or via point $\bar{\mathbf{r}}_0$. Including the distance parameters and times, this is expressed more fully as

$$\mathbf{F}_w[\mathbf{F}_B(\mathbf{r}_0, s, t), \bar{t}] = \mathbf{F}_B[\mathbf{F}_w(\mathbf{r}_0, t), s', \bar{t}].$$

Next, we differentiate this commuting condition with respect to s and $\bar{t}$ to give

$$\frac{\partial^2}{\partial s \partial \bar{t}} \{\mathbf{F}_w[\mathbf{F}_B(\mathbf{r}_0, s, t), \bar{t}]\} = \frac{\partial^2}{\partial \bar{t} \partial s} \{\mathbf{F}_B[\mathbf{F}_w(\mathbf{r}_0, t) s', \bar{t}]\},$$

and use the definitions of $\mathbf{F}_w$ [Eq. (8.22)] and of $\mathbf{F}_B$ [Eqs. (8.20) and (8.21)], so that

$$\frac{\partial}{\partial s}\{\mathbf{w}[\mathbf{F}_B(\mathbf{r}_0, s, t)]\} = \frac{\partial}{\partial t}\left\{\mathbf{B}[\mathbf{F}_w(\mathbf{r}_0, t), \bar{t}]\frac{\partial s'}{\partial s}\right\}$$

or at $s = 0$ and $\bar{t} = t$

$$(\mathbf{B}\cdot\nabla)\mathbf{w} = \frac{\partial\mathbf{B}}{\partial t} + (\mathbf{w}\cdot\nabla)\mathbf{B} - \lambda^*\mathbf{B},$$

where we have rewritten $-\partial^2 s'/\partial s\partial\bar{t}$ as a scalar function λ^*. In other words, we have shown that, if λ^* and $\mathbf{w}$ exist such that

$$\frac{\partial\mathbf{B}}{\partial t} + (\mathbf{w}\cdot\nabla)\mathbf{B} - (\mathbf{B}\cdot\nabla)\mathbf{w} = \lambda^*\mathbf{B},$$

then the evolution of $\mathbf{B}(r, t)$ conserves the magnetic topology. This equation is, finally, the same as the curl of Eq. (8.1) if $\mathbf{w} = \mathbf{v}$ and $\nabla\times\mathbf{R} = \lambda\mathbf{R}$, where $\lambda = -\lambda^* - \nabla\cdot\mathbf{v}$, and so we have proved Eq. (8.17).

In an ideal plasma we may define the magnetic field-line velocity (at least its component perpendicular to the field) by Eq. (8.16), but in a non-ideal plasma there is no unique definition of field-line velocity! The component $(\mathbf{w}_\perp)$ of a field-line velocity may be defined if and only if Ohm's Law, Eq. (8.1), can be transformed into the form

$$\mathbf{E} + \mathbf{w}\times\mathbf{B} = \mathbf{a}, \tag{8.25}$$

where

$$\nabla\times\mathbf{a} = \lambda_w\mathbf{B} \tag{8.26}$$

and λ_w is some scalar function. This implies that

$$\frac{\partial\mathbf{B}}{\partial t} = \nabla\times(\mathbf{w}\times\mathbf{B}) + \lambda_w\mathbf{B},$$

which is the most general form that conserves magnetic field lines (as we have seen above). This definition is field-line preserving but it is not flux preserving unless $\lambda_w = 0$. In the latter case $\nabla\times\mathbf{a} = \mathbf{0}$ and we recover the definitions (1.41) and (1.42) in Chapter 1 with $\mathbf{a} = -\nabla\Phi^*$. From Eq. (8.25) we can therefore deduce the expression

$$\mathbf{w}_\perp = \frac{(\mathbf{E} - \mathbf{a})\times\mathbf{B}}{B^2}$$

for the field-line velocity. However, this is not unique, since if we replace the vector $\mathbf{a}$ by $\mathbf{a}' = \mathbf{a} + \nabla\Psi^*$, where $(\mathbf{B}\cdot\nabla)\Psi^* = 0$, then the magnetic field from Eq. (8.26) is unaltered and $\mathbf{B} \cdot$ Eq. (8.25) is unchanged, whereas $\mathbf{B} \times$ Eq. (8.25) gives another flow

$$\mathbf{w}'_\perp = \frac{(\mathbf{E} - \mathbf{a}') \times \mathbf{B}}{B^2}$$

in which the magnetic field lines are frozen. The form of Eq. (8.25) is always possible for ideal flow [when Eq. (8.16) results], but it is also always possible when $\mathbf{E} \cdot \mathbf{B} = 0$.

Consider, for example, a two-dimensional flow and magnetic field in the xy-plane, say, satisfying the resistive Ohm's Law

$$\mathbf{E} + \mathbf{v} \times \mathbf{B} = \frac{\mathbf{j}}{\sigma},$$

where $\mathbf{E}$ and $\mathbf{j}$ are aligned in the z-direction such as we considered previously in Example 1 of Section 1.4. Then the plasma velocity normal to the field is

$$\mathbf{v}_\perp = \frac{(\mathbf{E} - \mathbf{j}/\sigma) \times \mathbf{B}}{B^2}$$

and a possible field-line velocity is

$$\mathbf{w}_\perp = \frac{\mathbf{E} \times \mathbf{B}}{B^2}, \tag{8.27}$$

so that the slippage velocity is the difference $(\mathbf{v}_\perp - \mathbf{w}_\perp)$. The form of Eq. (8.27) works because it implies that $\mathbf{E} + \mathbf{w}_\perp \times \mathbf{B} = 0$, which is of the required form, Eq. (8.25), with $\mathbf{a} = \mathbf{0}$ and $\lambda_w = 0$. However, another possible field-line velocity is

$$\mathbf{w}'_\perp = \frac{(\mathbf{E} - K^* \hat{\mathbf{z}}) \times \mathbf{B}}{B^2},$$

where K^* is a constant. This also works since it implies that

$$\mathbf{E} + \mathbf{w}_\perp \times \mathbf{B} = K^* \hat{\mathbf{z}},$$

which is again of the form (8.25) but now with $\mathbf{a} = K^* \hat{\mathbf{z}}$ and $\lambda_w = 0$. In other words, the field-line velocity is not unique.

Finally, let us reconsider Example 2 of Section 1.4, which describes the ohmic decay of a linear force-free field. Previously, assuming $\lambda_w = 0$, we obtained an expression for $\mathbf{w}$ which becomes singular at the surface $r = r_0$ where the helical field transforms into circular field lines. Suppose we instead set $\lambda_w = \eta \alpha^2$, where α is the force-free parameter. If we also put $\mathbf{a} = (\lambda_w/\alpha)\mathbf{B} - \nabla \overline{\Psi}$ with $\overline{\Psi} = 0$ on $z = 0$, so as to satisfy Eq. (8.26) and the force-free equation, then we find very simply that $\mathbf{w} \equiv \mathbf{0}$. Since $\mathbf{E}$ is parallel to $\mathbf{B}$ in this example, there is no flow of electromagnetic energy, and so the advantage of the particular boundary condition $\overline{\Psi} = 0$ on $z = 0$ is that it leads to a field-line velocity that is identical with the flow velocity of electromagnetic energy. Furthermore, with the above choice for λ_w, the field lines remain stationary even though the magnetic flux and field decay with time.

8.1.3 Singular Magnetic Field-Line Reconnection

It is possible to define reconnection in a less general way than Schindler et al. (1988) and thereby exclude many diffusive phenomena that are not normally associated with reconnection (Priest and Forbes, 1989). The idea here is to remain close in spirit to the normal two-dimensional approach. In two dimensions (x, y) the procedure has two parts:

(i) first of all, we seek X-points, where B_x and B_y both vanish and there is an X-type topology;

(ii) then reconnection occurs when there is an electric field E_z, or equivalently a hyperbolic flow (v_x, v_y), that carries flux in towards the X-point from two opposite quadrants and carries it out in the other two quadrants.

In three dimensions, by analogy, we may proceed as follows.

(i) Seek a set of *potential singular lines*, which are magnetic field lines near which the magnetic field has an X-type topology in a plane normal to that field line;

(ii) then reconnection takes place when there is an electric field $E_\parallel$ along a potential singular line; this is associated with a hyperbolic flow that brings magnetic flux in from two directions towards the potential singular line and carries it outwards in two other directions.

The potential singular line along which the electric field is directed is called a *singular field line*, and the resulting reconnection is known as *singular magnetic field-line reconnection*. In the plane perpendicular to the singular field line, the flow and field-line behaviour has a similar appearance to two-dimensional reconnection (Fig. 8.6). Thus, the emphasis in this definition of reconnection is that the magnetic geometry has to be appropriate (i.e., a potential singular line has to exist) and also that the plasma flow has to be of the correct form (i.e., corresponding to the existence of an $E_\parallel$). In the ideal region around a singular line, there are constraints on the electric field or flow: reconnection may occur at singularities of the electric field along the singular line where these constraints fail and diffusion becomes important.

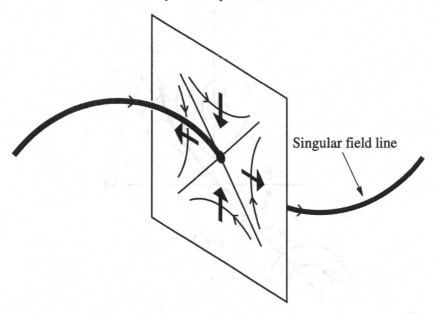

Fig. 8.6. Singular magnetic field-line reconnection.

In general, there is a continuum of neighbouring *potential* singular field lines. Consider, for example, the field

$$(B_x, B_y, B_z) = (y, x, 1). \qquad (8.28)$$

In the plane $z = \text{constant}$ there is an X-point at $x = y = 0$, but in the plane $z = \bar{a}y + z_0$ there is an X-point at $(-\bar{a}, 0, z_0)$, at which the magnetic field is directed normal to the plane. More generally, if one considers a translation of z_0 plus rotations of θ_1 and θ_2 about the x- and y-axes, respectively, there is an X-point at $(-\tan\theta_1, -\tan\theta_2/\cos\theta_1, z_0)$. Conversely, at any given point (x_0, y_0, z_0) there is an X-point in a plane with orientation $\tan\theta_1 = -x_0$, $\tan\theta_2 = -y_0\sqrt{1 + x_0^2}$. We conclude, therefore, that any field line of the configuration of Eq. (8.28) is a potential singular line.

Again consider, for example, the field with components

$$(B_x, B_y, B_z) = [\bar{f}(y, z), -x, 1],$$

where $\bar{f}(y, z) = (y-2)^2 - 1 + z^2$ (Fig. 8.7). In planes $z = z_0$ there are X-points at $(0, 2 - \sqrt{1 - z_0^2})$ and O-points at $(0, 2 + \sqrt{1 - z_0^2})$, provided that $z_0^2 < 1$. Also, at any point (x_0, y_0, z_0) there is a neutral point in a plane with orientations given by $\tan\theta_1 = x_0$, $\tan\theta_2 = -\bar{f}(y_0, z_0)/\sqrt{1 + x_0^2}$. It is an X-point below a certain surface, and so there is a continuum of potential singular lines passing through the X-points.

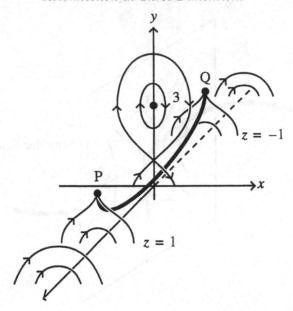

Fig. 8.7. Magnetic field lines in planes $z = $ constant for a model configuration around a solar prominence with a transition from normal ($|z| > 1$) to inverse ($|z| < 1$) polarity. There is a uniform field in the z-direction. The particular potential singular line that is located in the yz-plane is shown as a thick curve.

Now let us consider briefly the nature of the reconnection flow outside a diffusion region, where the ideal Ohm's Law holds

$$\mathbf{E} + \mathbf{v} \times \mathbf{B} = \mathbf{0}, \tag{8.29}$$

and for simplicity consider steady flow so that

$$\nabla \times \mathbf{E} = \mathbf{0}. \tag{8.30}$$

What is the nature of the flow in the ideal region around the singular line and what are the constraints on the electric field? Equation (8.29) implies that the electric field is perpendicular to the magnetic field almost everywhere, the exceptions being singular lines and surfaces where $\mathbf{E}$ becomes parallel to $\mathbf{B}$ and diffusion is important. The selection of a singular line from all the potential singular lines depends on the nature of the plasma flow or electric field. For example, consider again the field of Eq. (8.28), for which a flow $\mathbf{v}(x, y)$ that is independent of z produces a singular line along the z-axis. In this case, the components of Eq. (8.29) are

$$x v_x - y v_y = -E_z, \quad v_y - x v_z = -E_x, \quad v_x - y v_z = E_y, \tag{8.31}$$

where Eq. (8.30) implies that E_z is constant and $\partial E_y / \partial x = \partial E_x / \partial y$.

Eliminating E_y and the velocity components, we find an equation for E_x, namely

$$\frac{\partial E_x}{\partial y} + \frac{\partial}{\partial x}\left(\frac{E_z + yE_x}{x}\right) = 0,$$

whose solution by the method of characteristics is

$$E_x = \frac{yE_z}{x^2 - y^2} + x\,\overline{g}(x^2 - y^2), \tag{8.32}$$

where $\overline{g}(x^2 - y^2)$ is an arbitrary function of $x^2 - y^2$. There is therefore a singularity along the z-axis (i.e., the singular line) which is resolved by diffusion, while Eq. (8.31) implies that for $E_z > 0$ there is inflow ($v_x = -E_z/x$) along the x-axis and outflow ($v_y = E_z/y$) along the y-axis.

More generally, condition (8.30) implies $\mathbf{E} = -\nabla\Phi$ and then Eq. (8.29) implies that $\mathbf{E}\cdot\mathbf{B} = 0$ or $\mathbf{B}\cdot\nabla\Phi = 0$, so that Φ depends only on two coordinates normal to the field lines. For example, the field lines for Eq. (8.28) when $y \geq x$ may be described by

$$x = C^{1/2}\sinh s, \quad y = C^{1/2}\cosh s, \quad z = s + K,$$

so that the potential $\Phi = \Phi(C, K)$ depends only on C and K, and the electric field components are constrained to have the forms

$$E_x = 2C^{1/2}\sinh s\frac{\partial\Phi}{\partial C} + \frac{\cosh s}{C^{1/2}}\frac{\partial\Phi}{\partial K}, \quad E_y = \frac{yE_x + E_z}{-x}, \quad E_z = -\frac{\partial\Phi}{\partial K}.$$

$$\tag{8.33}$$

These forms ensure ideal flow satisfying Eqs. (8.29) and (8.30) except at singularities, where diffusion becomes important. For instance, the forms

$$\Phi = -E_z K \text{ and } \Phi = C\tan^{-1}\frac{K - K_0}{C - C_0}$$

lead to breakdown of ideal behaviour at the field lines $C = 0$ and $C = C_0$, $K = K_0$, respectively.

8.1.4 Covariant Formalism for Singular Field-Line Reconnection

Hornig and Rastätter (1998) have developed the above concept of singular field-line reconnection to include reconnection in three dimensions with or without null points. Their basic approach is again to start with a simple definition of two-dimensional reconnection and to expand it into three dimensions. In two dimensions consider a field-line velocity

$$\mathbf{w} = \frac{\mathbf{E}\times\mathbf{B}}{B^2}, \tag{8.34}$$

which transports the magnetic flux both in the ideal and non-ideal regions and assumes $\mathbf{a} = \mathbf{0}$ in Eq. (8.25). Reconnection then occurs when the field-line flow possesses an X-type velocity field with a singularity at the magnetic null line. The flow is therefore capable of carrying the magnetic flux in a finite time into the null line, where it splits and is transported outwards. For example, the magnetic field $\mathbf{B} = (y, x)$ gives rise to a field-line flow

$$\mathbf{w} = -\frac{E\,x}{x^2 + y^2}\hat{\mathbf{x}} + \frac{E\,y}{x^2 + y^2}\hat{\mathbf{y}},$$

which is singular on the z-axis.

What happens in three-dimensional reconnection? Well, for reconnection at a 3D null point or for any reconnection that satisfies $\mathbf{E} \cdot \mathbf{B} = 0$, we may define a field-line flow ($\mathbf{w}$) by Eq. (8.34) and so may use it to define reconnection exactly the same as in two dimensions as a process with a singular transporting hyperbolic flow. Such a singularity will exist along the singular line.

However, for finite-B reconnection in the absence of a null point, $\mathbf{E} \cdot \mathbf{B}$ does not vanish and no singularity in $\mathbf{w}$ occurs. The above definition then fails not because reconnection is fundamentally different but because the concept of magnetic flux conservation on which the above definition of $\mathbf{w}$ is based is too narrow a framework. Hornig and Rastätter (1998) realised that magnetic flux conservation is only a special case of a more general conservation of *electromagnetic flux* and that this provides an appropriate framework for both null-point and finite-B ($\mathbf{B} \neq \mathbf{0}$) reconnection. The equations for flux conservation from Eq. (8.18), namely

$$\frac{\partial \mathbf{B}}{\partial t} = \nabla \times (\mathbf{v} \times \mathbf{B}), \quad \mathbf{E} + \mathbf{v} \times \mathbf{B} = \nabla G,$$

may be rewritten in covariant form as

$$-\mathbf{E} \cdot \frac{\mathbf{W}}{c} = \frac{\partial G}{\partial x_0}, \quad \frac{W^0}{c}\mathbf{E} + \mathbf{W} \times \mathbf{B} = \nabla G, \tag{8.35}$$

where the index 0 denotes the time coordinate $x_0 = ct$. These equations imply conservation of electromagnetic flux

$$\int_C \mathbf{B} \cdot \mathbf{dA} + \frac{1}{c}\int_C \mathbf{E} \cdot \mathbf{dr}dx_0 = \text{constant} \tag{8.36}$$

for a surface comoving in four-dimensional Minkowski space with the four-velocity $\mathbf{W}^{(4)} = (W^0, \mathbf{W})$.

The usual magnetic flux threading a curve C in the first term of Eq. (8.36) is supplemented by the second term representing an extension of the curve along the time-axis. The four-velocity $\mathbf{W}^{(4)}$ is a transporting velocity for electromagnetic flux. When $W^0 \equiv 1$ we recover magnetic flux conservation, but when W^0 is a function of $\mathbf{r}$ and t we have a non–flux-conserving (and therefore non-ideal) evolution.

Magnetic reconnection may then be defined to occur when:

(a) there is no solution of the ideal covariant equations (8.35) with $W^0 > 0$ everywhere and

(b) there is a solution with $W^0 > 0$ with the exception of a line where $\mathbf{W}^{(4)}$ vanishes and $\nabla \mathbf{W}^{(4)}$ has a positive and negative eigenvalue.

This definition is structurally stable and applies to both null-point and finite-B reconnection. It is also covariant since the location $\mathbf{W}^{(4)} = \mathbf{0}$ is independent of the frame of reference. It may be noted that, although $\mathbf{W}^{(4)}$ is not unique, the location where $\mathbf{W}^{(4)}$ vanishes is unique. In general, reconnection occurs along a finite singular length of a moving line, and so, if it exists along such a line for a finite time, it shows up in space time as a two-dimensional surface. The site of such a singular line is determined by a global criterion. Also, the reconnection rate is measured by the quantity $\int \mathbf{E} \cdot \mathbf{dr}$ along the singular line.

Finally, it should be noted that the Hesse–Schindler definition of general reconnection is essentially equivalent to the violation of magnetic field-line conservation, but it is in our view too broad. However, the above definition of singular field-line reconnection is much more appealing, since it includes only those types of violation of field-line conservation that are traditionally regarded as being reconnection.

8.2 Three-Dimensional Null Points

Linear null points in three dimensions are rather different from those in two dimensions, so let us now describe them (Priest and Titov, 1996; Parnell et al., 1996), building on earlier work of, for instance, Cowley (1973b), Greene (1988), and Lau and Finn (1990, 1991). The simplest such null point (for which the magnetic field increases linearly from the origin) has field components

$$(B_x, B_y, B_z) = (x, y, -2z), \tag{8.37}$$

or in cylindrical polars

$$(B_r, B_\theta, B_z) = (r, 0, -2z), \tag{8.38}$$

so that the equation $\nabla \cdot \mathbf{B} = 0$ is satisfied identically. The resulting field

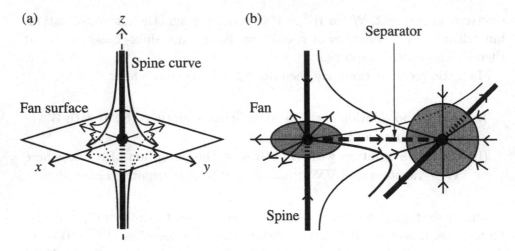

Fig. 8.8. The structure of (a) a 3D null point showing the spine and fan and (b) a structurally stable separator (see §8.4.1) joining two nulls and representing the intersection of two fans.

lines satisfy

$$\frac{dx}{B_x} = \frac{dy}{B_y} = \frac{dz}{B_z} \tag{8.39}$$

and are given by intersections of the two surfaces

$$y = Cx \text{ and } z = \frac{K}{x^2}. \tag{8.40}$$

Two quite distinct families of field lines pass through the null point. The *null spine* is the isolated field line in Fig. 8.8(a) which approaches or recedes from the null along the z-axis. Its neighbouring field lines form two bundles which spread out as they approach the xy-plane, which constitutes the *null fan*. This is a surface of field lines which spread out as one moves away from the null. When the fan field lines radiate from the null we refer to it as a *positive null point*, whereas when they converge on the null we call it a *negative null point*.

The null given by Eq. (8.37) is known as a *proper radial null*, since its fan field lines are straight lines. It is a particular member (with $a = 1$) of a class of null points having field components

$$(B_x, B_y, B_z) = [x, ay, -(a+1)z]. \tag{8.41}$$

When $a \neq 1$ the fan field lines are no longer straight and we have an *improper*

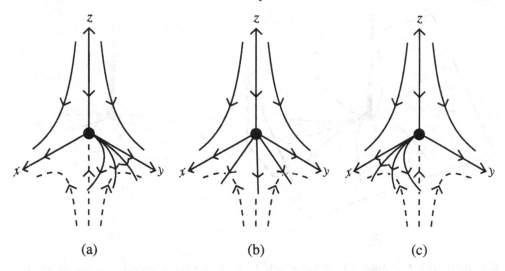

Fig. 8.9. Null points having a spine along the z-axis and a fan in the xy-plane for (a) $0 < a < 1$, (b) $a = 1$, and (c) $a > 1$.

radial null, such that when $0 < a < 1$ most of them touch the y-axis, while when $a > 1$ they touch the x-axis (Fig. 8.9).

More generally, each of the three field components of a linear null may be written in terms of three constants, making nine in all. However, by using $\nabla \cdot \mathbf{B} = 0$ and by normalising and rotating the coordinate axes, these may be reduced to four constants, namely a, b, $j_\parallel$, and $j_\perp$ such that

$$\begin{pmatrix} B_x \\ B_y \\ B_z \end{pmatrix} = \begin{pmatrix} 1 & \frac{1}{2}(b - j_\parallel) & 0 \\ \frac{1}{2}(b + j_\parallel) & a & 0 \\ 0 & j_\perp & -a - 1 \end{pmatrix} \begin{pmatrix} x \\ y \\ z \end{pmatrix}. \tag{8.42}$$

If the current $(j_\perp)$ normal to the spine is non-zero, the fan surface is inclined to the spine at an angle not equal to $\frac{1}{2}\pi$ and we have an *oblique null*. If the current $(j_\parallel)$ along the spine exceeds a critical value, the eigenvalues of the matrix in Eq. (8.42) are no longer all real and we have a *spiral null* with the field lines in the fan spiralling into or out of the null (Fig. 8.10).

8.3 Local Bifurcations

A magnetic field is said to be *structurally stable* if the essential features of its topology are unaffected by any small change in the field, that is, if the elements of its *skeleton* (which we define below in Section 8.4.1 to be its nulls, spines, and separatrices (Priest et al., 1996)) are preserved. Conversely, a magnetic field is structurally unstable if an arbitrary change in the field

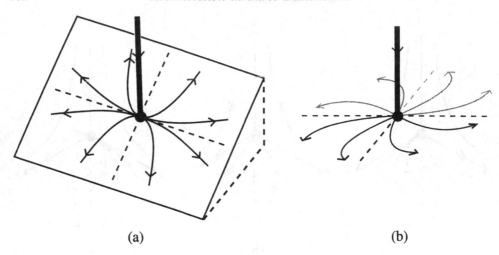

(a)   (b)

Fig. 8.10. (a) An oblique null for which $j_\perp \neq 0$ and (b) a spiral null for which $j_\parallel$ exceeds a critical value.

causes a change in the topology. Such a change in topology is known as a *bifurcation*.

A *local bifurcation* is one that involves a change in the number or nature of the null points. Thus, in three dimensions isolated linear nulls are structurally stable, but null lines or null sheets are structurally unstable, since they may break up into a series of null points. More generally, null points are structurally unstable if they are degenerate (i.e., when the Jacobian matrix ($D\mathbf{B}$) in Eq. (8.42) is singular at the null point) or if the null points are of second or higher order (i.e., when $D\mathbf{B}$ vanishes at the null point).

In two dimensions, a *global bifurcation* of a magnetic field (a Hamiltonian vector field whose divergence vanishes) is one that involves a change in connectivity of the separatrix field lines. At the moment of bifurcation there exists either a homoclinic separatrix (starting and ending at the same null point) or a heteroclinic separatrix (linking one null point to another null point). However, homoclinic and heteroclinic field lines are structurally unstable in two dimensions and so do not survive the bifurcation, since a general perturbation stops the field line from one null either coming back to the null or going to the other null.

In three dimensions, in contrast, a global bifurcation can involve the creation and continued existence of a separator; or it may involve the destruction of such a separator. Separators that are heteroclinic field lines linking two nulls and representing the intersection of two fan surfaces are structurally stable, whereas those that represent the intersection of the spine of one null with the spine or fan of another are structurally unstable, since a perturbation of the field in general destroys such a separator connection between two nulls.

Linear null points may coalesce at a second-order null (for which the field increases quadratically from zero), or a second-order null may split and give birth to linear nulls. Such local bifurcations may produce global changes of magnetic topology and stimulate release of magnetic energy.

In two dimensions the magnetic components may be written as

$$\frac{dx}{ds} = B_x = \frac{\partial A}{\partial y}, \quad \frac{dy}{ds} = B_y = -\frac{\partial A}{\partial x}, \tag{8.43}$$

where the flux function (A) is a Hamiltonian that is constant along field lines. The isolated linear nulls are either X-points or O-points with flux functions of the form $A = b^2 y^2 \pm a^2 x^2$. Thus, in order to find what kinds of bifurcation are possible, we may consider a general cubic function and suppose there are nulls at $(\pm\sqrt{K}, 0)$, say. The simplest, generic, nondegenerate type is a saddle-centre bifurcation, exemplified by the flux function

$$A = x^3 - 3Kx + y^2, \tag{8.44}$$

which represents a set of curved field lines when $K < 0$; they develop a null at the origin when $K = 0$, which then splits into an X and an O when $K > 0$. Pitchfork bifurcations are also possible when an X- or an O-point changes to one of opposite type and spawns two more of the same type. Examples of resonant and degenerate bifurcations may also be produced (Priest et al., 1996).

In three dimensions the system

$$\frac{dx}{ds} = B_x, \quad \frac{dy}{ds} = B_y, \quad \frac{dz}{ds} = B_z \tag{8.45}$$

is no longer Hamiltonian, but it is conservative since $\nabla \cdot \mathbf{B} = 0$. The null points and their bifurcations are much more complex. The first step has been to consider fields with cylindrical symmetry, for which a general categorisation has been completed (Priest et al., 1996), building on a particular example studied by Lau and Finn (1992). An interesting particular subclass in cylindrical polars (r, θ, z) is the field

$$B_r = rz, \quad B_\theta = 0, \quad B_z = Cr^2 + K - z^2, \tag{8.46}$$

with two parameters, C and K. If $C = 0$, a flux tube exists when $K < 0$. It undergoes a saddle-node bifurcation when $K = 0$ to produce a pair of radial nulls when $K > 0$ at points $(0, 0, \pm\sqrt{K})$ on the z-axis [Fig. 8.11(a)]. If $C < 0$, a ring of null points is also produced at $r = \sqrt{-K/C}, z = 0$,

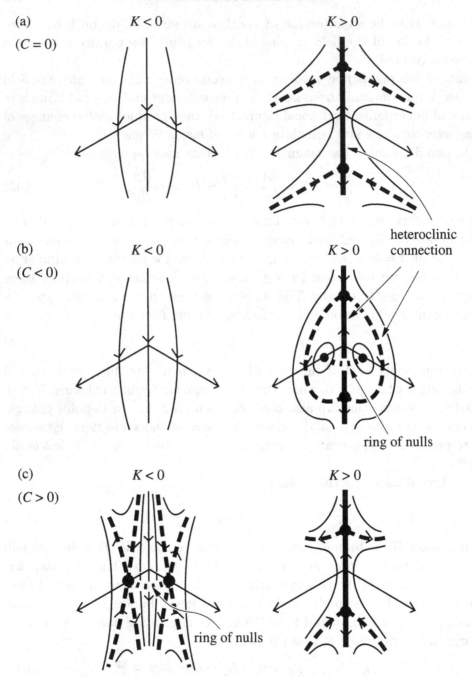

Fig. 8.11. Axisymmetric configurations when (a) $C = 0$, (b) $C < 0$, (c) $C > 0$, showing saddle-node (top) or saddle-node–Hopf (middle and bottom) bifurcations as K increases through zero.

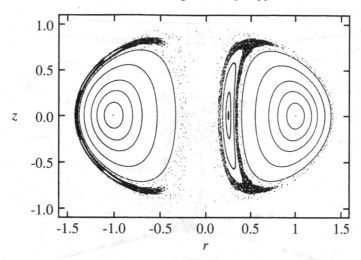

Fig. 8.12. The appearance of chaotic field lines when the axial symmetry of Fig. 8.11(b) with $K > 0$ is broken, as shown by a Poincaré plot of the intersections of field lines with the zr-plane.

surrounded by a set of nested toroidal surfaces, and so we have a saddle-node–Hopf bifurcation [Fig. 8.11(b)]. If $C > 0$, we start with a ring of nulls when $K < 0$ which disappears and becomes a pair of nulls on the z-axis when $K > 0$ [Fig. 8.11(c)]. If a component $B_\theta = r$ is added, the radial nulls on the z-axis become spiral nulls and the ring of nulls becomes a closed field line encircling the z-axis and surrounded by a set of nested flux surfaces.

A start has also been made at studying fully three-dimensional systems, a new feature being the possible onset of chaos. For example, the effect of breaking the symmetry of the example on the right of Fig. 8.11(b) may be considered by putting

$$(B_r, B_\theta, B_z) = (rz + \lambda r^2 \cos^2 \theta \sin \theta, r + \lambda r^2 \cos^3 \theta, 1 - r^2 - z^2). \qquad (8.47)$$

When $\lambda = 0$ we have the axisymmetric field with well-defined flux surfaces, but when λ is non-zero the field lines that pass close to the null points become chaotic; this can be seen in Fig. 8.12 for the case $\lambda = 0.01$, where a Poincaré surface of section monitors the points where field lines starting out on the r-axis intersect a plane $\theta =$ constant (Priest et al., 1996).

8.4 Global Magnetic Topology

A complex magnetic field configuration can have many different components (Fig. 8.13). Some regions will be laminar and consist of magnetic field lines that lie on flux surfaces, whereas others will be chaotic with space-filling

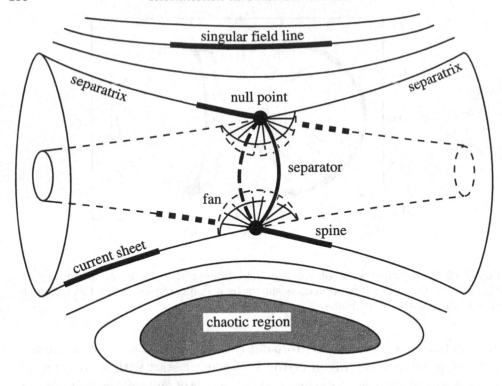

Fig. 8.13. The components of a complex magnetic field, containing both laminar and chaotic (shaded) regions. Separatrix surfaces intersect in separator lines which link null points. The null point regions contain spine curves (which lie on the separatrices) and fan surfaces (which is the name given to separatrices that converge on or radiate from a null). Current sheets and singular lines may also be present at other locations in the configuration.

field lines. The laminar regions will have a *skeleton* that consists of null points, spines, fans, and separatrix surfaces (§8.4.1), such that pairs of separatrices intersect in separator curves. There will also in general be current sheets (Chapter 2) and singular field lines (§8.1.3). Reconnection will occur both at the null points (§8.6) and at the singular field lines (§8.1.3) and strong dissipation will be present at current sheets along the spines, fans, and separators.

8.4.1 Skeleton of Laminar Fields

In two dimensions a general magnetic configuration [Fig. 8.14(a)] contains separatrix curves, which separate the plane into topologically distinct regions, in the sense that generally all the field lines in one region start at a

Separatrix curve

X-point

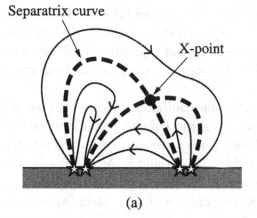

(a)

Separatrix (touching) curve

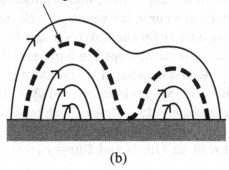

(b)

Separatrix surface

Separator

Null point

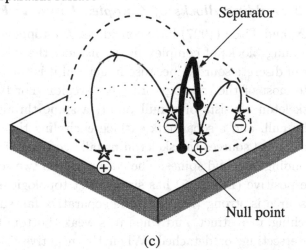

(c)

Fig. 8.14. Two-dimensional separatrix curves that (a) intersect at an X-point or (b) touch the boundary. (c) Three-dimensional separatrix surfaces that intersect in a separator (and that may instead touch the boundary).

particular source and end at a particular sink. The separatrices intersect at
an X-point where the field vanishes and is locally hyperbolic. Reconnection
then occurs by the breaking and rejoining of field lines at the X-point and
the transferring of flux across the separatrices from one topological region
to another.

In three dimensions, some configurations have similar properties, with separatrix surfaces separating the volume into topologically different regions.
They intersect each other in a *separator*, a field line which ends at null points
or on the boundary [Figs. 8.8(b) and 8.14(c)]. The *skeleton* or global structure of complex fields due to many sources then comprises the null points
and a network of spine curves and separatrix fan surfaces. The separatrix
surfaces are of two types, namely separatrix fan surfaces, which are extensions of the fan surfaces at null points, and separatrix touching surfaces,
which touch a boundary in a curve known as a touch curve or a bald patch.

The concept of a touch curve for the Earth's magnetic field was mentioned
by Hide (1979), but in the solar concept the importance of bald patches and
of complex topology was first developed by Seehafer (1985, 1986). Furthermore, Low and Wolfson (1988) and Amari and Aly (1990) showed how a
current sheet can grow from a two-dimensional bald patch (§2.6.1). More
recently, the development of bald patches in three dimensions has been modelled in detail by Titov et al. (1993) and Bungey et al. (1996).

8.4.2 Building Blocks of Complex Laminar Fields

Priest, Bungey, and Titov (1997) have considered the topological properties
of the key building blocks of complex fields, namely the field created by a
finite number of discrete sources, because in general it is the nearest sources
that dominate most of the topology of a given region. For two unbalanced
sources the skeleton consists of a null point, a spine through the weaker
source and the null, and a separatrix surface encircling the weaker source.
For three unbalanced sources the skeleton usually consists of two nulls with
their corresponding fans and spines. The case in which two sources are negative and one positive (Fig. 8.15) has six distinct topological states. When
the positive source is strong (top line) the separatrix fan surfaces may be
separate, touching, or distinct, and when it is weak (bottom line) they may
be nested, intersecting, or detached. When the negative sources are close
enough in location, there are *local separator bifurcations* (of saddle-node
type) along the top line from the separate state to the touching state (with
two extra nulls) and then to the enclosed state, in which two linear nulls and
a separator joining them are first created and then destroyed. When the

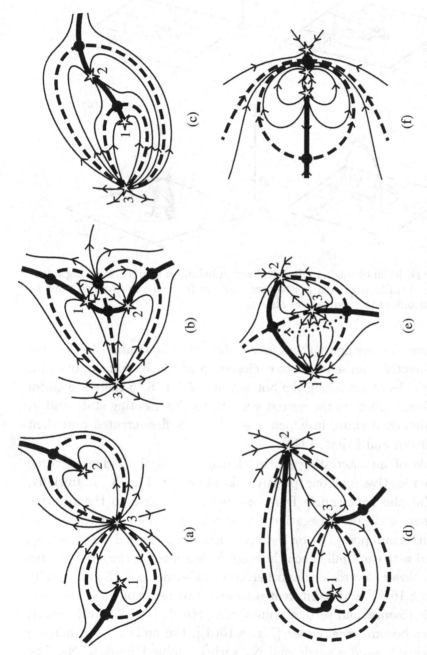

Fig. 8.15. The different types of topology due to two negative sources (labelled 1 and 2) and one positive source (labelled 3): (a) separate, (b) touching, (c) enclosed, (d) nested, (e) intersecting, (f) detached. The sources are indicated by stars, the nulls by dots, the spines by thick solid curves, and the fans by dashed curves. The three upper states occur when $f_1 + f_2 < f_3$ and the lower ones when $f_1 + f_2 > f_3$, where f_1, f_2, and f_3 are the magnitudes of the fluxes of the three sources. In (e) the dotted curve represents a separator.

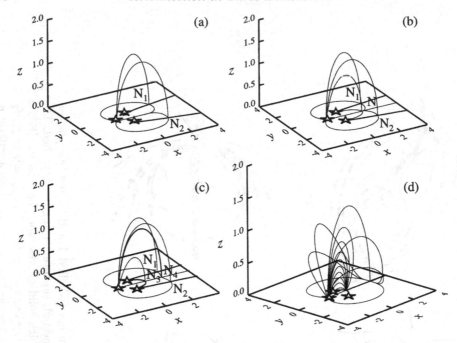

Fig. 8.16. The skeleton of the field due to three unbalanced sources, showing bifurcations from two nulls in (a) via (b) to four nulls in (c) and back to two nulls in (d) as the positions of the sources are changed.

negative sources are far apart there is a *global spine bifurcation* (of heteroclinic type) directly from separate to enclosed, in which the spine of one null passes through the other. Along the bottom line of Fig. 8.15 there are *global separator bifurcations* from the nested state to the intersecting state and in turn to the detached state, in which a separator is first created and then destroyed (Brown and Priest, 1999).

An example of an interesting set of changes of topology due to three sources (two negative and one positive) is shown in Fig. 8.16. Initially, in Fig. 8.16(a) the acute angle from one negative source to the positive source and then to the other negative source is large enough that each negative source has a null point (N_1 or N_2) and therefore two separate skeletons, one associated with each null. Then the angle is decreased as the two negative sources move closer towards one another: a second-order null (N) eventually appears [Fig. 8.16(b)] and it bifurcates to give two new nulls (N_3 and N_4) with locally horizontal and vertical spines, respectively [Fig. 8.16(c)]; finally, as the sources become even closer [Fig. 8.16(d)], the nulls N_1, N_4, and N_2 coalesce to leave behind a single null N_5 with its spine joining to N_3. The bifurcations with four sources are even richer: one interesting example shows a new process of emerging flux breakout, in which the effect of emerging flux is to create a separator that slowly rises to infinity and makes the field open in response to a finite change.

8.5 Magnetic Helicity

Magnetic helicity is a measure of the twisting and kinking of a flux tube (referred to as *self-helicity*) and also the linkage between different flux tubes (referred to as *mutual helicity*). It is a global topological invariant that cannot be changed in an ideal medium and that decays very slowly (over the global magnetic diffusion time τ_d) in a resistive medium. Magnetic reconnection on times much smaller than τ_d cannot destroy magnetic helicity but only convert it from one form to another. Thus, when a complex configuration undergoes reconnection, the conservation of magnetic helicity provides an important constraint on the nature of the final state.

Magnetic helicity was first discussed in an astrophysical connection by Woltjer (1958), who suggested that the lowest-energy state that conserves the total magnetic helicity but allows arbitrary reconnections is a linear force-free field (see §9.1.1). Most of the interest in the subject was then for many years on laboratory plasmas such as tokamaks that are bounded by a magnetic surface. Taylor (1974), for example, put forward the hypothesis that reversed-field pinches are so turbulent that the magnetic surfaces can break down easily by reconnection and can spread the magnetic helicity uniformly throughout the configuration and reduce it to a linear force-free state (§9.1.1). Then Heyvaerts and Priest (1984) were the first to realise the importance of magnetic helicity for the solar corona, and they developed Taylor's theory to allow magnetic flux to thread the coronal boundary (the solar surface): they suggested that coronal magnetic fields may be in a state of MHD turbulence with the energy continually being fed into the corona by photospheric motions, so that the corona is continually heated by turbulent reconnection as it evolves through a series of linear force-free states. They also suggested that if the magnetic helicity becomes too great it may be expelled by magnetic eruptions as eruptive flares, erupting prominences, and coronal mass ejections. These ideas have since been developed much further, and the usefulness of the concept of magnetic helicity in the solar corona is now widely accepted (e.g., Brown et al., 1999).

In solar and stellar atmospheres the continual motion of photospheric magnetic footpoints tends to build up the magnetic helicity until it is ejected by magnetic eruptions. Thus, the magnetic helicity in the atmosphere can change either by the emergence of twisted or linked structures from the

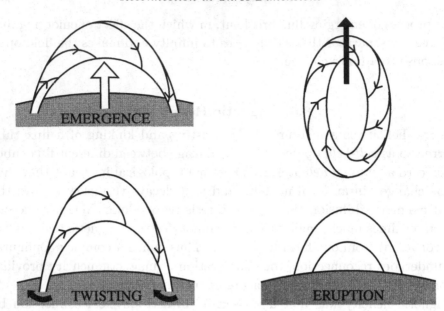

Fig. 8.17. Magnetic helicity changes associated with emergence of flux through a boundary or twisting motions at the boundary or a magnetic eruption.

interior of a star or by the twisting (or untwisting) of the footpoint motions of magnetic loops. Magnetic helicity of a coronal loop can also be decreased by the detachment and ejection of magnetic structures from the surface of a star (as occurs in solar prominence eruptions) or of a planet (as occurs in geomagnetic substorms; Fig. 8.17).

In this section we follow Moffatt (1978), Berger and Field (1984), and Berger (1984, 1998) in giving first the formal equations for magnetic helicity (§8.5.1) and its evolution (§8.5.2) and then practical rules for calculating the magnetic helicity of different types of structure (§8.5.3). We shall often drop the adjective "magnetic" when it is clear we are referring to magnetic helicity rather than, say, fluid or current helicity.

8.5.1 Definition of Magnetic Helicity

Let us begin with the expression

$$H_0 = \int_V \mathbf{A} \cdot \mathbf{B} \, dV \qquad (8.48)$$

for magnetic helicity, where $\mathbf{A}$ is the vector potential such that $\mathbf{B} = \nabla \times \mathbf{A}$. This definition is meaningful only if the boundary (S) of the volume (V) of integration is a magnetic surface (or the integration is over all space with a vanishing magnetic field at infinity). Otherwise, the field lines that close outside V will have linkage with those inside V, and such a linkage is ill-defined if only the field inside V is prescribed, since there are many ways of continuing it outside V. This is related to the question of gauge-invariance, since, if $\mathbf{A}$ is replaced by $\mathbf{A} + \nabla \Phi_A$, say, (which does not change $\mathbf{B}$) then H_0 is changed by

$$\Delta H_0 = \int_V \nabla \Phi_A \cdot \mathbf{B} \, dV = \int_V \nabla \cdot (\Phi_A \mathbf{B}) \, dV,$$

using $\nabla \cdot \mathbf{B} = 0$. Then invoking the divergence theorem for a simply connected volume gives

$$\Delta H_0 = \int_S \Phi_A \mathbf{B} \cdot \mathbf{n} \, dS,$$

where $\mathbf{n}$ is the unit normal to S and pointing out of V_0. We therefore have gauge invariance ($\Delta H_0 = 0$) if $\mathbf{B} \cdot \mathbf{n} = 0$ on the surface S.

Berger and Field (1984) realised that the difference in the magnetic helicities integrated over all space (V_∞) of any two fields that differ only inside V is independent of the field outside V, and that a particularly useful reference field inside V is a potential field, since it is completely determined by $\mathbf{B} \cdot \mathbf{n}$ on S. This led them to propose the *relative magnetic helicity*

$$H = \int_{V_\infty} \mathbf{A} \cdot \mathbf{B} - \mathbf{A}_0 \cdot \mathbf{B}_0 \, dV \qquad (8.49)$$

as a gauge-invariant alternative to H_0, where $\mathbf{B}_0 = \nabla \times \mathbf{A}_0$ is a field that is potential inside V with the same $\mathbf{B}$ outside V and has $\mathbf{A} \times \mathbf{n} = \mathbf{A}_0 \times \mathbf{n}$ on S. In other words, a generalisation of the definition of helicity requires a ground state that sets the zero value and the potential field is the best one for this purpose, since it minimises the magnetic energy when the normal field at the boundary is imposed. Furthermore, with the helicity of the potential field set to zero in this way, the helicity is now uniquely specified. An alternative form of relative helicity that is sometimes useful and does not require the boundary constraint on $\mathbf{A}$ is (Finn and Antonsen, 1985)

$$H = \int_V (\mathbf{A} + \mathbf{A}_0) \cdot (\mathbf{B} - \mathbf{B}_0) \, dV, \qquad (8.50)$$

where again $\mathbf{B}_0 = \nabla \times \mathbf{A}_0$ is potential inside V and has the same normal component as $\mathbf{B}$ on S.

In a multiply connected domain, there is the added complication that you can impose the magnetic fluxes through each of the holes. For example, in a torus you can impose the toroidal flux. However, the above definitions [Eq. (8.49) or (8.50)] of magnetic helicity are still valid, provided the potential field ($\mathbf{B}_0$) also possesses those fluxes through the holes. One convenient way to ensure the fluxes are correct is to replace the condition ($\mathbf{B}_0 \cdot \mathbf{n} = \mathbf{B} \cdot \mathbf{n}$) on the normal fields at the boundary in Eq. (8.50) with $\mathbf{A} \times \mathbf{n} = \mathbf{A}_0 \times \mathbf{n}$, so that the tangential components of the vector potentials are the same at the boundary. This implies $\mathbf{B} \cdot \mathbf{n} = \mathbf{B}_0 \cdot \mathbf{n}$ on S and also implies that $\int \mathbf{A} \cdot \mathbf{ds} = \int \mathbf{A}_0 \cdot \mathbf{ds}$ along any closed curve drawn on the boundary.

8.5.2 Rate of Change of Magnetic Helicity

Let us for convenience choose the gauge ($\mathbf{A}_p$) such that $\nabla \cdot \mathbf{A}_p = 0$ and $\mathbf{A}_p \cdot \mathbf{n} = 0$ on S. Then the time rate of change of the magnetic helicity is

$$\frac{dH}{dt} = -2 \int_V \mathbf{E} \cdot \mathbf{B} \, dV + 2 \int_S \mathbf{A}_p \times \mathbf{E} \cdot \mathbf{n} \, dS. \tag{8.51}$$

If the resistive Ohm's Law ($\mathbf{E} = -\mathbf{v} \times \mathbf{B} + \mathbf{j}/\sigma$) holds and there is no slippage on the boundary, this becomes

$$\frac{dH}{dt} = -2 \int_V \mathbf{j} \cdot \mathbf{B}/\sigma \, dV + 2 \int_S (\mathbf{B} \cdot \mathbf{A}_p)(\mathbf{v} \cdot \mathbf{n}) - (\mathbf{v} \cdot \mathbf{A}_p)(\mathbf{B} \cdot \mathbf{n}) \, dS. \tag{8.52}$$

The first term represents the internal helicity dissipation, and the surface integral measures the flow of helicity across the boundary.

From Eq. (8.52) we may make several useful deductions. First of all, by equating dH/dt and the first term in order of magnitude with $j \sim A/(\mu L^2)$, we see that the time-scale for magnetic helicity dissipation is the global diffusion time ($\tau_d = L^2/\eta$) if the magnetic field is varying on the global length-scale (L). Secondly, on time-scales much shorter than τ_d, changes of magnetic helicity are given by the surface term. Thus, for example, if the volume is closed, with $\mathbf{B} \cdot \mathbf{n} = \mathbf{v} \cdot \mathbf{n} = 0$, the magnetic helicity is conserved. But, if instead footpoint motions along the surface are prescribed, we may deduce the resulting injection or extraction of helicity as follows.

Consider for simplicity a plane surface S that is threaded by the magnetic field of the footpoints of N thin flux tubes of magnetic flux F_i. Suppose the motion of each footpoint consists of a translation plus a uniform rotation at a rate ω_i. Then we find that helicity is changed at the rate

$$\frac{dH}{dt} = -\frac{1}{2\pi} \left(\sum_{i=1}^{N} \omega_i F_i^2 + \sum_{i=1}^{N} \sum_{j=1}^{N} \dot{\theta}_{ij} F_i F_j \right), \tag{8.53}$$

where $\dot{\theta}_{ij}$ is the time derivative of the relative angle (θ_{ij}) between footpoints i and j (Berger, 1984). The first term in Eq. (8.53) arises because the footpoint rotation injects twist into a flux tube at a rate of one unit of twist every $2\pi/\omega$ seconds. The second term measures the rate at which footpoints circle each other and cause braiding of the flux tubes.

The proof of Eq. (8.53) from the surface term of Eq. (8.52) is as follows: Using Cartesian coordinates with the z-axis directed into V and the xy-plane as the surface S, the condition $\mathbf{A}_p \cdot \mathbf{n} = 0$ becomes $A_{pz} = 0$, which implies that $\partial A_{pz}/\partial x = \partial A_{pz}/\partial y = 0$. Thus, on the xy-plane the x- and y-coordinates of $\mathbf{A}_p$ satisfy

$$\frac{\partial A_{px}}{\partial x} + \frac{\partial A_{py}}{\partial y} + \frac{\partial A_{pz}}{\partial z} = 0, \qquad \frac{\partial A_{py}}{\partial x} - \frac{\partial A_{px}}{\partial y} = b,$$

where $b(x, y) = B_z(x, y, 0)$, which implies that

$$\frac{\partial^2 A_{px}}{\partial x^2} + \frac{\partial^2 A_{px}}{\partial y^2} = -\frac{\partial b}{\partial y}.$$

The usual Poisson solution is

$$A_{px} = -\int \frac{b(x', y')(y - y')}{2\pi[(x - x')^2 + (y - y')^2]} dx' dy',$$

$$\tag{8.54}$$

$$A_{py} = \int \frac{b(x', y')(x - x')}{2\pi[(x - x')^2 + (y - y')^2]} dx' dy'.$$

Now Eq. (8.52) for helicity evolution becomes in the ideal limit

$$\frac{dH}{dt} = 2 \int \int (\mathbf{B} \cdot \mathbf{A}_p)v_z - (\mathbf{v} \cdot \mathbf{A}_p)B_z \, dx dy, \tag{8.55}$$

where the integration is over the xy-plane. The first term ($F_\perp$, say) involves only motions in the z-direction and thus represents the effect of the emergence of structures carrying helicity through the surface. The second term ($F_\parallel$, say) involves only motions in the xy-plane and represents the way the shuffling of footpoints can inject helicity into fields already present in the volume.

Thus, if we set $v_z = 0$ on the surface and focus on the latter effect, Eq. (8.55) becomes, after using Eq. (8.54),

$$\frac{dH}{dt} = -2 \int \int b(x, y)b(x', y') \frac{\mathbf{r} \times \mathbf{v}(x, y) \cdot \hat{\mathbf{z}}}{2\pi r^2} dS dS', \tag{8.56}$$

where $\mathbf{r} = (x - x')\hat{\mathbf{x}} + (y - y')\hat{\mathbf{y}}$.

Consider, for example, two thin flux-tubes of flux F_1 and F_2 which have a translational motion plus uniform rotations at angular speeds ω_1 and ω_2

about points (x_1, y_1) and (x_2, y_2), respectively. Thus, near (x_i, y_i)

$$\mathbf{v}(x, y) = \mathbf{v}(x_i, y_i) + \omega_i(x - x_i)\hat{\mathbf{y}} - \omega_i(y - y_i)\hat{\mathbf{x}}. \tag{8.57}$$

Now write $b = b_1 + b_2$, where b_1 and b_2 are the fields of the flux tube at $\mathbf{r}_1 = (x_1, y_1)$ and $\mathbf{r}_2 = (x_2, y_2)$, respectively. The integrals in Eq. (8.56) therefore have three parts: integrating the first over $b_1 b_1'$ gives a contribution of $-F_1^2 \omega_1/(2\pi)$, the second over $b_2 b_2'$ gives $-F_2^2 \omega_2/2\pi$, and the third over $(b_1 b_2' + b_2 b_1')$ gives the cross-term $(-F_1 F_2 \dot{\theta}_{12}/\pi)$, as required in Eq. (8.53), where

$$\dot{\theta}_{12} = \frac{|(\mathbf{r}_2 - \mathbf{r}_1) \times (\mathbf{v}_2 - \mathbf{v}_1)|}{|\mathbf{r}_2 - \mathbf{r}_1|^2}.$$

8.5.3 A Primer on Calculating the Magnetic Helicity of Simple Structures

Equation (8.53) may be used to deduce the magnetic helicity of a variety of different magnetic configurations when the twisting and linking is built up from an initially potential field that has separate untwisted tubes at large distances from one another. The resulting total magnetic helicity consists of the self-helicity (H_s) of each tube due to its own internal twist and the mutual helicity (H_m) due to the linking of one tube with another. Thus, from Eq. (8.53)

$$H = \sum_{i=1}^{N} H_{si} + \sum_{\substack{i,j=1 \\ i<j}}^{N} H_{mij}, \tag{8.58}$$

where the self-helicity of the ith flux tube is

$$H_{si} = \frac{\Phi_{T_i}}{2\pi} F_i^2 \tag{8.59}$$

in terms of its twist Φ_{T_i} (or number of turns times 2π) and magnetic flux F_i, and the mutual helicity is

$$H_{mij} = 2L_{ij} F_i F_j \tag{8.60}$$

in terms of the *linking number* L_{ij}, which is described as follows.

Using the linking number is an efficient way to calculate magnetic helicity in a variety of situations. It was first introduced by Gauss in 1833. This number is a topological parameter that describes two curves, and it does not change as the curves are distorted provided they do not cross through each other during the distortion. Consider first of all a set of closed curves such as those shown in Fig. 8.18. Each curve is given a direction, and reversing

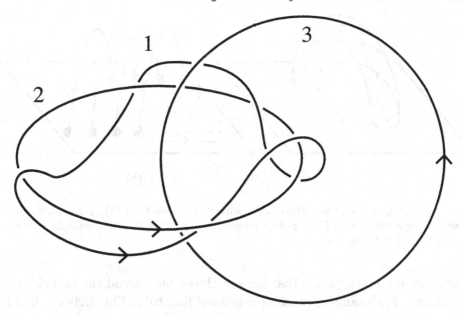

Fig. 8.18. Three linked tubes with linking numbers $L_{12} = 3$, $L_{13} = -1$, and $L_{23} = -1$ (after Berger, 1998).

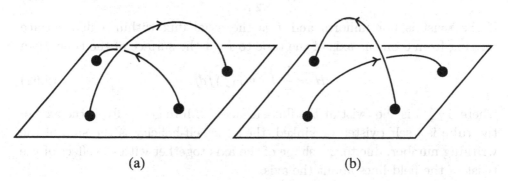

(a)   (b)

Fig. 8.19. A positive crossing (left) and a negative crossing (right) of two field lines.

one of the directions changes the sign of L. There are special points where two curves cross over each other which provide a convenient way to calculate linking numbers. Each crossing has a sign $+1$ or -1, and the linking number is just half the sum of the signed crossings, which is independent of the viewing angle and of a deformation of the curves. The sign is indicated in Fig. 8.19 and may be calculated by a right-hand rule: curl the fingers of your right hand along curve 1 in the direction of the arrow; then, at places where that curve crosses over another curve, curve 2, the crossing is positive if your thumb points in the direction of curve 2.

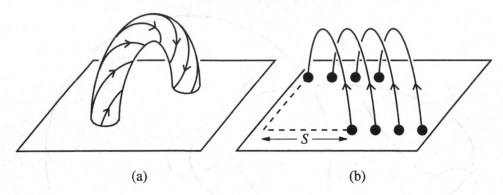

(a)   (b)

Fig. 8.20. (a) A twisted flux tube of self-helicity $H = \Phi_T F^2/(2\pi)$ and (b) a sheared arcade of helicity $H = SF^2$ per unit length, where F is the magnetic flux, Φ_T is the twist, and S is the shear.

Now let us consider fields that are not closed but thread the boundary of the volume. For example, consider a twisted flux tube [Fig. 8.20(a)]. If it is uniformly twisted, its helicity is, from Eq. (8.59),

$$H = \frac{\Phi_T}{2\pi}F^2. \tag{8.61}$$

If the twist is non-uniform and f is the axial flux within a flux surface ranging from 0 on the axis of the tube to F at the surface of the tube, then

$$H = \frac{1}{\pi}\int_0^F \Phi_T(f)f\,df, \tag{8.62}$$

where $\Phi_T(f)$ is the twist at the flux surface f. More generally, if the axis of the tube is itself twisted or kinked, then the self-helicity is the sum of the writhing number, due to the shape of the axis together with the effect of the twist of the field lines about the axis.

Next, consider an arcade of infinite extent that has flux F per unit length and is uniformly sheared by an amount S per unit length [Fig. 8.20(b)]. Suppose the arcade consists of a cylindrical shell of small thickness w, radius r, and has magnetic field components B_θ and B_z in cylindrical coordinates (r, θ, z). Then by Eq. (8.61) the magnetic helicity of the arcade (half a flux tube) is $H = \Phi_T F_0^2/(4\pi)$, where $\Phi_T = B_\theta L/(B_z r)$ and $F_0 = 2\pi r w B_z$ is twice the axial flux along the arcade. By putting $L = 1$ for a unit length of the arcade and noting that the axial field component is $B_z = B_\theta S/(\pi r)$, we find that the magnetic helicity per unit length of the arcade reduces to

$$H = SF^2, \tag{8.63}$$

where $F = wB_\theta$ is the (azimuthal) flux of the arcade per unit length. If

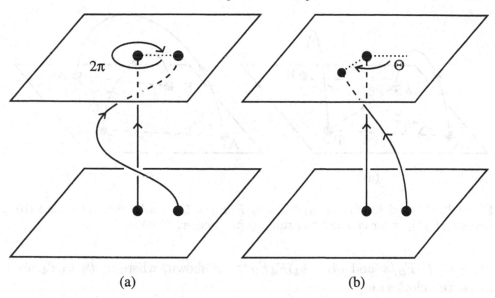

Fig. 8.21. Two braided tubes of fluxes F_1 and F_2 and (a) mutual helicity $2F_1F_2$ when 2π is the braiding angle, and (b) mutual helicity $(\Theta/\pi)F_1F_2$ when Θ is the braiding angle.

instead the shear $[S(f)]$ varies with flux from the arcade axis, the helicity is

$$H = 2 \int_0^F S(f)f\,df. \tag{8.64}$$

We have seen from Eq. (8.60) that, if two closed loops of flux F_1 and F_2 are linked once, they have two crossings, and their linking number is 1, while their mutual helicity is $2F_1F_2$. Correspondingly, if two tubes are stretched between two planes and one is braided by an angle 2π about the other, then their mutual helicity will also be $2F_1F_2$ [Fig. 8.21(a)]. If instead the braiding angle is Θ, then the mutual helicity [Fig. 8.21(b)] is proportionally reduced to

$$H = \frac{\Theta}{\pi}F_1F_2. \tag{8.65}$$

Finally, suppose we have two thin coronal flux tubes of fluxes F_A and F_B and having positive polarity footpoints A_+, B_+ and negative footpoints A_-, B_-. Then in general the mutual helicity of the two coronal loops is

$$H = (\angle B_+A_-B_- - \angle B_+A_+B_-)\frac{F_AF_B}{\pi}$$

in terms of the angles $B_+A_-B_-$ at footpoint A_- and $B_+A_+B_-$ at footpoint A_+. In particular, for the configurations in Fig. 8.22 the helicities are

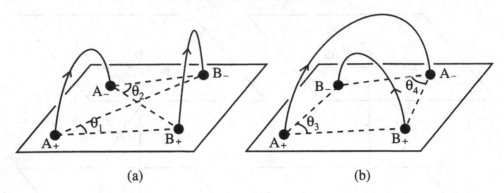

Fig. 8.22. Mutual helicity of (a) $(\theta_1 - \theta_2)F_A F_B/\pi$ for nearby flux tubes and (b) $(\theta_3 - \theta_4)F_A F_B/\pi$ for crossing flux tubes (after Berger, 1998).

$(\theta_1 - \theta_2)F_A F_B/\pi$ and $(\theta_3 - \theta_4)F_A F_B/\pi$ as shown, where $\theta_1, \theta_2, \theta_3, \theta_4$ are measured clockwise.

8.5.4 Magnetic Helicity and Three-Dimensional Reconnection

We have seen that the total magnetic helicity is conserved to a high degree of approximation during reconnection. However, it can be changed from mutual helicity to self-helicity or vice versa. Furthermore, reconnection can redistribute the magnetic helicity within a configuration and can often make the distribution more uniform so that it approaches a minimum-energy state having the same total magnetic helicity, but with the helicity uniformly distributed (i.e., a linear force-free field).

However, Schindler et al. (1988), in their study of general magnetic reconnection (§8.1.1), have discovered a more subtle and intimate connection between magnetic helicity and a particular type of reconnection, namely undriven finite-B reconnection that is global and spontaneous (Fig. 8.4), for which there is no energy supply from outside (i.e., the asymptotic electric field at large distances vanishes). Consider the relative helicity defined by

$$\overline{H} = \int_V (\mathbf{A} + \mathbf{A}_0)(\mathbf{B} - \mathbf{B}_0)\, dV, \tag{8.66}$$

where $\mathbf{B}_0$ is the magnetic field at some reference time (t_0) and $\mathbf{A}_0$ is the corresponding vector potential. Then, after using $\mathbf{E} = -\nabla\Phi - \partial\mathbf{A}/\partial t$, we find

$$\frac{d\overline{H}}{dt} = -2\int_V \mathbf{E}\cdot\mathbf{B}\, dV - \int_S \mathbf{n}\cdot[(\mathbf{B} - \mathbf{B}_0)\Phi + \mathbf{E}\times(\mathbf{A} - \mathbf{A}_0)]\, dS.$$

Assuming that $\mathbf{E} = \mathbf{0}$ and $\mathbf{B} = \mathbf{B}_0$ on S and remembering that $\mathbf{E}\cdot\mathbf{B} = 0$

outside the diffusion region D_R, we find this reduces to

$$\frac{d\overline{H}}{dt} = -2 \int_{D_R} \mathbf{E} \cdot \mathbf{B} \, dV.$$

Using as coordinates the distance (s) along the magnetic field and the Euler potentials (α, β), for which $\mathbf{B} = \nabla\alpha \times \nabla\beta$ by Eq. (8.2) and $E_\parallel = -\partial\Psi/\partial s$ by Eq. (8.4), we find this reduces to

$$\frac{d\overline{H}}{dt} = 2 \int \nabla\Psi \cdot \nabla\alpha \times \nabla\beta \, dV = 2 \int \frac{\partial\Psi}{\partial s} \, ds \, d\alpha \, d\beta.$$

Thus, if the magnetic helicity $\overline{H}$ changes with time, $\int E_\parallel \, ds$ must be non-vanishing on a measurable set of field lines and so by Eq. (8.12) global finite-B reconnection occurs.

In addition, from the above equation, if the magnetic helicity changes by $\delta\overline{H}$ in a time δt, the resulting parallel electric field is roughly

$$E_\parallel \simeq \frac{\delta\overline{H}}{2V_R B_0 \, \delta t},$$

where V_R is the volume of the diffusion region (D_R) and B_0 is the size of the magnetic field.

8.6 Reconnection at a Three-Dimensional Null Point

Building on earlier work by Lau and Finn (1990), the kinematics of steady reconnection at three-dimensional null points have been studied by Priest and Titov (1996), who proposed three distinct types of reconnection, namely spine reconnection, fan reconnection, and separator reconnection. (In addition, it is possible to develop a three-dimensional version of the almost-uniform family (§5.1) of two-dimensional models; see Jardine, 1994.) Priest and Titov solved the kinematic equations

$$\nabla \times \mathbf{E} = 0 \tag{8.67}$$

and

$$\mathbf{E} + \mathbf{v} \times \mathbf{B} = 0 \tag{8.68}$$

in three dimensions, where it is natural to consider the motions of flux surfaces rather than field lines. Equation (8.67) implies that the electric field can be written as $\mathbf{E} = -\nabla\Phi$ in terms of a potential (Φ) and then from Eq. (8.68) we can make two deductions. The first is that $(\mathbf{B} \cdot \nabla)\Phi = 0$ or

$$\Phi = \Phi(C, K), \tag{8.69}$$

where C and K are constants describing a field line. The second deduction from Eq. (8.68) is that the velocity normal to the magnetic field is

$$\mathbf{v}_\perp = \frac{\mathbf{E} \times \mathbf{B}}{B^2}. \qquad (8.70)$$

Thus, the boundary conditions determine $\Phi(C, K)$ and then the values of $\mathbf{E}$ and $\mathbf{v}_\perp$ throughout the volume may be calculated. At locations where $\mathbf{E}$ becomes singular, the inference is that Eq. (8.68) fails and diffusive effects must come into play to resolve the singularities. This approach has been applied to a single null and a pair of nulls by Priest and Titov (1996), as follows. (It is also used for fields without nulls in Section 8.7.)

8.6.1 Spine Reconnection

First of all, let us examine a three-dimensional null point and ask: What is the nature of reconnection near such a point? How do magnetic flux surfaces reconnect? If you impose continuous footpoint motions on any cylindrical surface surrounding the spine and crossing the fan, as shown in Fig. 8.23(a), then a singular motion is driven along the spine axis. Such a process is therefore referred to as spine reconnection. For example, consider field lines in any vertical plane and suppose the footpoints move down on the right side of the cylinder and up on the left. Then the field lines in Fig. 8.23(a) approach the null, break, and reconnect, while the other ends move across the top and bottom of the cylinder through the spine. Suppose the same process occurs in all other vertical planes but is modulated in θ: then what are the implications for flux surfaces formed from footpoints that lie initially on the top edge and move downwards? The other ends of the field lines (on the top or bottom surface) move in and through the spine. The flux surface therefore moves in and touches the null, forming a fold along the spine [Fig. 8.23(c)]. It then reconnects with an oppositely placed flux surface and unfurls from the spine like a cylindrical bubble [Fig. 8.23(d)]. A simple solution for the equations of the flux surfaces in cylindrical coordinates is

$$r^2 z = \pm 1 - t \sin \theta, \qquad (8.71)$$

and a simple solution for the field-line velocity is

$$(v_{\perp r}, v_{\perp \theta}, v_{\perp z}) = \frac{E_0(\phi)}{r(r^2 + 4z^2)}(2z, 0, r). \qquad (8.72)$$

It can be seen that a singularity exists all along the spine ($r = 0$) and so a key question for the future is whether it can be resolved by diffusion (see §8.6.4). We know that in at least one case it can, because the exact solutions

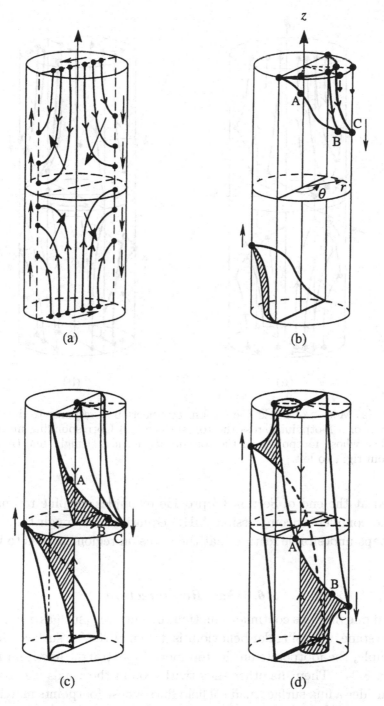

Fig. 8.23. Motions of (a) field lines in a plane θ = constant and of (b)–(d) flux surfaces during spine reconnection in response to the downwards motion of a set of footpoints (A, B, C) on the curved cylindrical surface.

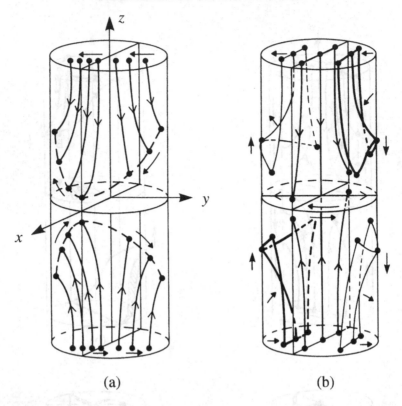

(a)   (b)

Fig. 8.24. (a) Motion of a field line for fan reconnection in response to the straight-line motion of a footpoint across the top surface. (b) Corresponding motion of a flux surface whose footpoints on the top surface form a straight line that moves across from right to left.

described at the end of Section 3.5 provide examples of spine reconnection using the complete incompressible MHD equations, although they possess the perhaps undesirable feature that the diffusion region extends to infinity.

8.6.2 Fan Reconnection

If instead one imposes continuous motions on surfaces (such as $z = \pm 1$) that cross the spine, then singular behaviour is driven at the fan surface. Suppose, for example, the footpoints on the top move in a straight line from right to left (Fig. 8.24). Then the other ends twirl around the z-axis like a swirling kilt. Consider a flux surface made of field lines whose footpoints march across the top in a straight line: the flux surface distorts and becomes a vertical surface plus a semicircle. It then breaks and reconnects with a similar flux surface on the opposite side of the null point. During this process, magnetic field lines rotate rapidly in one direction above the fan and in the opposite direction below it. A simple solution has field-line velocity

$$(v_{\perp x}, v_{\perp y}, v_{\perp z}) = \frac{1}{(x^2 + y^2 + z^2)(4 + y^2 z)^{3/2}}$$

$$\times \left[\frac{2xy(z^3 - 1)}{z^{1/2}}, \frac{2(x^2 + 4z^2 + y^2 z^3)}{z^{1/2}}, (4 + y^2 z + x^2 z)yz^{1/2} \right], \quad (8.73)$$

from which we note that there is a singularity at the fan ($z = 0$), and so again the question is: Can it be resolved by diffusion? As for spine reconnection, the exact three-dimensional solution of Craig et al. (1995) shows that there is at least one case where it can.

8.6.3 Separator Reconnection

A third class of reconnection can often exist when two or more null points are present. Any fan surface consists of field lines linking to a null point, and so, when there are two nulls, their two fans will in the generic (structurally stable) case intersect in a special curve called a separator. This is a field line that joins one null to the other, as shown in Fig. 8.25.

In separator reconnection, a current sheet forms along the separator, so that in planes across the separator the flow and field resemble those of classical two-dimensional reconnection. A simple example is to consider

$$(B_x, B_y, B_z) = [x(z - 3), y(z + 3), 1 - z^2], \quad (8.74)$$

which possesses null points at $z = \pm 1$ on the z-axis. Now, suppose, for example, we surround this configuration by a cube with a side whose length is two units long, centred at the origin, and impose a horizontal flow on the sides $y = \pm 1$ with $v_z = 0$. This drives fan reconnection at the upper null and spine reconnection at the lower null. The field lines are of the form

$$x(z - 1)^{-1}(z + 1)^2 = C, \quad (8.75)$$

$$y(z - 1)^2(z + 1)^{-1} = K, \quad (8.76)$$

and so a field line through the footpoint ($x_0, 1, z_0$) has

$$C = \frac{x_0(z_0 + 1)^2}{z_0 - 1}, \quad K = \frac{(z_0 - 1)^2}{z_0 + 1}. \quad (8.77)$$

The condition that $v_z = 0$ on $y = 1$ then implies that

$$E_x B_y = E_y B_x, \quad (8.78)$$

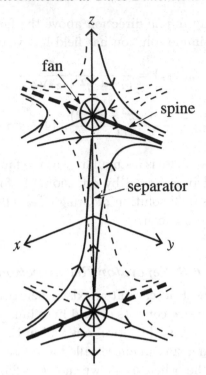

Fig. 8.25. A magnetic configuration with two nulls including a separator that joins one null to the other [see also Fig. 8.8(b)].

and this in turn implies that the potential has the form

$$\Phi = f\left[\tfrac{1}{2}x_0^2 + \frac{(z_0+1)^{1/2}(z_0-3)^{9/2}}{(z_0-1)^4}\right], \qquad (8.79)$$

where $x_0(C,K), z_0(C,K)$ are given by Eq. (8.77), while $C(x,z), K(x,z)$ follow from Eqs. (8.75) and (8.76).

8.6.4 Anti-Reconnection Theorem in Three Dimensions

For those who care to try and build models for three-dimensional reconnection, the anti-reconnection theorem in two dimensions of Section 5.3.1 may be extended to three dimensions, to provide some constraints on what is allowable (Priest et al., 1994c; Priest and Titov, 1996). In general we can state it as

> *Steady MHD reconnection, with convective plasma flow across a separatrix, spine or fan, is impossible in an inviscid plasma with highly sub-Alfvénic flow and a uniform resistivity.*

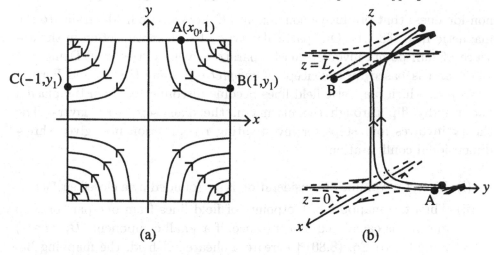

Fig. 8.26. The mapping of footpoints for (a) a two-dimensional X-field from the top or bottom boundary of a square to the side boundary and (b) a three-dimensional sheared X-field from the plane $z = 0$ to the plane $z = L$.

The theorem implies that for linear reconnection, when the flow speed is everywhere much smaller than the Alfvén speed, the above singularity that we have found at the spine during spine reconnection, and at the fan during fan reconnection, cannot be resolved by magnetic diffusion alone. One way around the theorem is to allow Alfvénic flow, as in the classical two-dimensional models, but this nonlinearity severely complicates the theory.

8.7 Quasi-Separatrix Layer Reconnection: Magnetic Flipping

When there is a null point, two-dimensional reconnection is associated with the fact that the mapping of field lines from one footpoint to another is discontinuous. For example, with the simple X-point field

$$B_x = x, \quad B_y = -y \tag{8.80}$$

the point (x_0, y_0) on one part of a boundary will map to (x_1, y_1) on another part in such a way that, when (x_0, y_0) crosses a separatrix, the point (x_1, y_1) suddenly jumps in location (Fig. 8.26). In their classic paper Schindler et al. (1988) realised that, when no nulls or bald patches are present, the mapping is continuous, so that separatrices do not exist and the two-dimensional concept of reconnection based on flux transfer across separatrices is no longer applicable. As we mentioned in Section 8.1.1, they proposed instead a concept of general magnetic reconnection to include all effects of local non-idealness that produce a component $(E_\parallel)$ of electric field parallel to the magnetic field (§8.1.1). One particular type of such reconnection in the absence of null points involves quasi-separatrix layers (Priest and Demoulin, 1995) and is based on a concept of *magnetic flipping* (Priest and Forbes, 1992a), in which magnetic field lines become disconnected from the plasma and rapidly flip through the plasma in the quasi-separatrix layers. The theory involves four steps for investigating reconnection in a given three-dimensional configuration:

(i) First, the volume of consideration is surrounded by a closed surface S.

(ii) Then the mapping of footpoints of field lines from one part of S to another is calculated. For instance, if a small component $(B_z = l \le 1)$ is added to Eq. (8.80) to create a sheared X-field, the mapping becomes continuous, so that, as the point (x_0, y_0) crosses the y-axis in the plane $z = 0$, the other end (x_1, y_1) in the plane $z = 1$ moves continuously [Fig. 8.26(b)].

(iii) Next, so-called quasi-separatrix layers are identified where the gradients of the mapping are very large.

(iv) Finally, reconnection occurs when there is a breakdown of ideal MHD and a change of connectivity of plasma elements: this takes place in the quasi-separatrix layers where the field-line velocity greatly exceeds the plasma velocity – it may be driven by slow regular boundary motions of footpoints (across a quasi-separatrix layer).

It may be noted that these definitions of a quasi-separatrix layer and magnetic reconnection involve a mapping to a boundary and therefore refer to global properties of a configuration. The concept of a quasi-separatrix layer may be defined formally as follows. Split the surface into parts S_0 and S_1 where the field lines enter and leave the volume, respectively, and set up orthogonal coordinates (u, v) in S and w normal to S. Then field lines map (u_0, v_0) in S_0 to (u_1, v_1) in S_1. Next, form the displacement gradient tensor

$$\mathcal{F} = \begin{pmatrix} s_1 \partial u_1 / \partial u_0 & s_2 \partial u_1 / \partial v_0 \\ s_3 \partial v_1 / \partial u_0 & s_4 \partial v_1 / \partial v_0 \end{pmatrix} \tag{8.81}$$

from the gradients of the mapping functions $u_1(u_0, v_0)$ and $v_1(u_0, v_0)$ and the scaling factors s_i and evaluate the norm

$$N = \sqrt{\left(s_1 \frac{\partial u_1}{\partial u_0}\right)^2 + \left(s_2 \frac{\partial u_1}{\partial v_0}\right)^2 + \left(s_3 \frac{\partial v_1}{\partial u_0}\right)^2 + \left(s_4 \frac{\partial v_1}{\partial v_0}\right)^2}. \tag{8.82}$$

Finally, define a quasi-separatrix layer as the region where

$$N \gg 1.$$

The properties of F are as follows. First of all, a difference (δu_0 and δv_0) in footpoint positions maps to

$$\begin{pmatrix} \delta u_1 \\ \delta v_1 \end{pmatrix} = F \begin{pmatrix} \delta u_0 \\ \delta v_0 \end{pmatrix}.$$

Secondly, a surface element dS_0 transforms to

$$dS_1 = \mathcal{J} \, dS_0,$$

where $\mathcal{J} = s_1 s_4 \left(\partial u_1 / \partial u_0 \right) \left(\partial v_1 / \partial v_0 \right) - s_2 s_3 \left(\partial u_1 / \partial v_0 \right) \left(\partial v_1 / \partial u_0 \right)$ is the Jacobian. Thus, flux conservation ($B_1 dS_1 = B_0 dS_0$) implies that

$$B_1 = \frac{B_0}{\mathcal{J}},$$

where $\mathcal{J}$ is finite and nonzero if the field has no nulls or singularities. Thirdly, the displacement gradient tensor may be written as the product

$$F = F_R F_0$$

of one matrix (F_R) representing a rotation through an angle and another (F_0) representing a stretching by λ_+ (the largest eigenvalue) along $\mathbf{e}_+$ (the corresponding eigenvector) together with a compression by λ_- (the other eigenvalue) along $\mathbf{e}_-$ (the other eigenvector). Thus, a quasi-separatrix layer (where $N \gg 1$) is associated with a large expansion along one direction and a large compression along the other, such that N is approximately equal to the largest eigenvalue

$$N \approx \lambda_{\max}.$$

Consider, as an example, the sheared X-field

$$(B_x, B_y, B_z) = (x, -y, l) \tag{8.83}$$

inside a cube with $l \ll 1$. The mapping from the base (S_0) to the top and sides (S_1) is given by

$$x_1 = x_0 \, e^{z_1/l}, \quad y_1 = y_0 \, e^{-z_1/l}. \tag{8.84}$$

Thus, when the point A($x_0, y_0, 0$) on S_0 is so close to the y-axis that $2x_0 < \epsilon$, A maps to a point B on the top ($z_1 = 1$) and

$$F = \begin{pmatrix} \epsilon^{-1} & 0 \\ 0 & \epsilon \end{pmatrix},$$

while

$$N \approx \frac{1}{\epsilon},$$

where

$$\epsilon = e^{-1/l} \ll 1.$$

In contrast, when $\epsilon < 2x_0 < 1$, A maps to C on the side ($x_1 = \frac{1}{2}$), while the elements of $\mathcal{F}$ and the value of N are of order unity. The resulting variations of x_1, y_1, z_1, N with x_0 are shown in Fig. 8.27, which reveals the quasi-separatrix layer as a very narrow region of width ϵ where $N \gg 1$. When $l = 0.1$ the value of N in the quasi-separatrix layer is 10^4, and even when l is as large as 0.3, N is about 28 in the quasi-separatrix layer. If the cube is replaced by a hemisphere or sphere, similar forms are produced but the functions become continuous and differentiable.

Having located a quasi-separatrix layer, we may now consider kinematic reconnection satisfying Eqs. (8.67) and (8.68) and producing a potential [Eq. (8.69)] and field-line velocity [Eq. (8.70)]. Thus, suppose we impose the field-line velocity components $v_{\perp 1x}$ and $v_{\perp 1y}$ on the top side ($z = 1$) of the cube and deduce the function $\Phi(x_1, y_1)$ together with $\mathbf{E}$ and $\mathbf{v}_\perp$ throughout the cube. The resulting electric field on the base ($z = 0$) of the cube has components

$$E_{x0} = -\frac{\partial \Phi}{\partial x_1}\frac{\partial x_1}{\partial x_0} - \frac{\partial \Phi}{\partial y_1}\frac{\partial y_1}{\partial x_0},$$

$$E_{y0} = -\frac{\partial \Phi}{\partial x_1}\frac{\partial x_1}{\partial y_0} - \frac{\partial \Phi}{\partial y_1}\frac{\partial y_1}{\partial y_0},$$

which depend partly on the electric field components on the top ($E_{x1} = -\partial\Phi/\partial x_1$, $E_{y1} = -\partial\Phi/\partial y_1$) and partly on the gradients of the mapping functions $x_1(x_0, y_0)$ and $y_1(x_0, y_0)$. Thus $\mathbf{E}_0$ is large where the gradients of the mapping are large, namely in the quasi-separatrix layer. For example, if on the top ($z = 1$) and side ($x = \frac{1}{2}$) of the cube we impose

$$v_{\perp 1x} = 0, \quad v_{\perp 1y} = v_0\, x_1$$

and

$$v_{\perp 1x} = 0, \quad v_{\perp 1y} = \tfrac{1}{2}v_0,$$

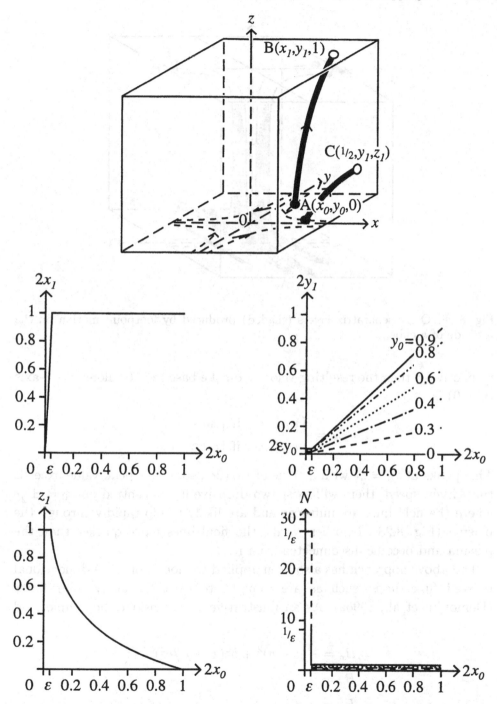

Fig. 8.27. Sheared X-field in a cube together with the variations of the end-point coordinates (x_1, y_1, z_1) on the top and side boundaries and of the norm (N) with the initial footpoint coordinates $(x_0$ and $y_0)$ on the bottom boundary.

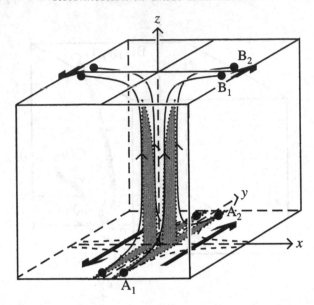

Fig. 8.28. Quasi-separatrix layers (shaded) produced by footpoint motion on the top side of a cube.

respectively, then the resulting velocity on the base ($z = 0$) along the x-axis ($y = 0$) is

$$v_{\perp y0} = \begin{cases} v_0 x_0/\epsilon^2 & \text{if } |x_0| < \tfrac{1}{2}\epsilon, \\ v_0/(4x_0) & \text{if } |x_0| > \tfrac{1}{2}\epsilon. \end{cases}$$

This peaks at $x_0 = \tfrac{1}{2}\epsilon$ with a value of $v_0/(2\epsilon)$, so, if this peak value exceeds the Alfvén speed, there will exist two diffusive layers centred on $x_0 = \pm\tfrac{1}{2}\epsilon$ where the field lines are unfrozen and are likely to flip rapidly through the plasma (Fig. 8.28). In other words, the field lines move quicker than the plasma and become disconnected from it.

The above approach has also been applied to models of three-dimensional twisted flux tubes, such as are thought to exist for many solar flares (Démoulin et al., 1996a). As an illustrative case consider the simple field

$$B_x = -(z - a)^2 + b^2(1 - y^2/c^2),$$
$$B_y = d,$$
$$B_z = x,$$

where a, b, c, and d are constants. The configuration is shown diagrammatically in Fig. 8.29 with three types of field line, namely those below the

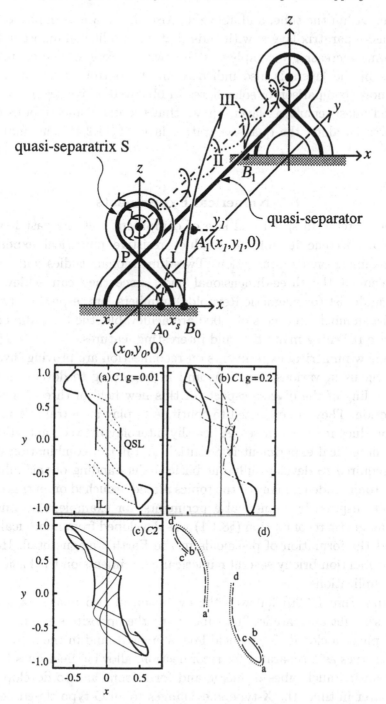

Fig. 8.29. (i) Twisted flux tube showing three types of field line I, II, and III. (ii) Quasi-separatrix layers viewed from above together with sample field lines of types (a) I (dotted curve), (b) III (dashed–dotted curve) and (c) II (solid curve) and (d) the connectivity of points on the quasi-separatrix layers.

flux tube, within the tube, and above it. Also shown are examples of calculated quasi-separatrix layers, with ends that curl up like an umbrella handle and become increasingly complex as the twist increases. The connectivity of points in the layers is also indicated in the bottom right-hand panel. Furthermore, comparison of solar flares in observed active regions with the calculated quasi-separatrix layers shows that the knots and ribbons of confined flares lie along the quasi-separatrix layers (§11.2.1; Démoulin et al., 1996b).

8.8 Numerical Experiments

With the increase in speed and size of computers over the past few years it has now become feasible to conduct resistive numerical experiments on three-dimensional reconnection. Two-dimensional studies will continue to complement the three-dimensional ones, since they can achieve much higher Lundquist (or magnetic Reynolds) numbers, but, especially now that Lundquist numbers in excess of 1,000 are achievable, the three-dimensional studies are revealing many new and interesting features.

These new numerical experiments on reconnection are proving invaluable in pointing us in various directions and in beginning to develop a partial understanding of the diverse aspects of this new field of three-dimensional reconnection. They are expected to continue to play a central role in future and to produce many surprises. A healthy interplay between analytical theory and numerical experiment is essential, and it is a combination of both that is required to develop fully our basic understanding of this inherently complex topic. Indeed, many of the topics already touched on in this chapter have been inspired by numerical experiments: for example, the concept of general magnetic reconnection (§8.1.1) was developed from numerical experiments of the formation of plasmoids in the Earth's magnetotail. Here, we shall just mention briefly several numerical experiments on the basic theory and its applications.

Lau and Finn (1996) followed the evolution in two phases of a pair of initially straight antiparallel flux tubes with the opposite sense of twist. In the first phase a closed X-type field line is created, and in the second phase this line serves as a separator for reconnection, allowing field lines from the initially antiparallel tubes to merge and form loops and to develop spatial chaos. Later in time, the X-type line changes to an O-type closed field line, surrounded by a ring of toroidal flux surfaces. Reconnection continues until there emerges a final steady state (Fig. 8.30) having two reconnected loops and a toroidal ring of flux surfaces with zero current between them.

Fig. 8.30. Representative magnetic field lines in the final steady state of two initially straight antiparallel flux tubes of opposite twist. The final state consists of one closed flux tube ($a\,b$) attached to the bottom boundary and another ($a'b'$) to the top boundary, together with a central torus. All the other field lines shown originate from region a' and pass through the torus (from Lau and Finn, 1996).

Galsgaard and Nordlund (1997b) considered an initial force-free configuration with eight magnetic null points [Fig. 8.31(a)]. They stressed two of the side boundaries, imposing periodic conditions on the remaining boundaries, and followed the evolution to see what kind of reconnection would occur. The null points tend to collapse into sheets and current sheets develop in weak-field tunnels along the separators joining nearby null points [Fig. 8.31(b)]. In other words, separator reconnection takes place, with jets of plasma being ejected out of the sides of the current sheets at Alfvénic speeds by Lorentz and pressure gradient forces. Eventually, most of the magnetic connectivity between the two boundaries is lost and long-lived arcade-like structures are set up, in which small-scale current sheets continually appear and disappear as they dissipate the energy supplied by the boundary driving.

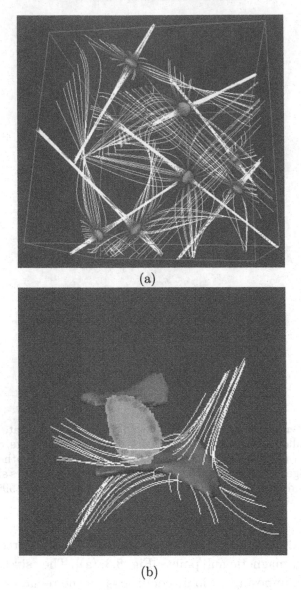

(a)

(b)

Fig. 8.31. (a) An initial force-free field with eight null points, showing the field lines near their spines and fans. (b) The resulting evolution in response to boundary motions, showing separator reconnection at the separator joining two of the nulls (darkly shaded), one in front of the other (from Galsgaard and Nordlund, 1997b).

Dahlburg et al. (1997) have investigated the collision and interaction of two twisted force-free flux tubes that are initially perpendicular and are being forced towards one another by a stagnation-point flow. For low twist the tubes tend to bounce elastically off one another without reconnecting. For

$t = 7.8$ $t = 21.1$ $t = 30.6$

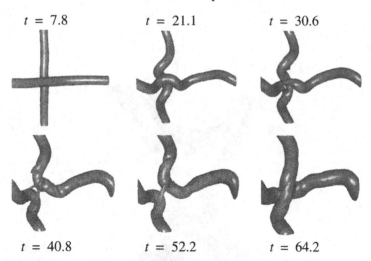

$t = 40.8$ $t = 52.2$ $t = 64.2$

Fig. 8.32. Isosurfaces of magnetic field strength (B) at several times during the tunnelling of one magnetic flux tube through another. The isosurfaces are chosen with B equal to half the maximum field strength at each time (Dahlburg et al., 1997b).

high twist and low Lundquist number ($L_u = 576$), the tubes instead reconnect in the standard way to give two bent right-angled tubes moving away from each other. However, when both the twist and the Lundquist number ($L_u = 2880$) are high, the tubes tunnel by a double-reconnection mechanism that allows the two straight tubes to pass right through one another, as if by magic (Fig. 8.32). Other studies of the basic theory include the development of fast reconnection with slow shocks in response to a local resistivity enhancement (Ugai and Wang, 1998); and collapse of null points in three-dimensional MHD turbulence (Politano et al., 1995; Matthaeus et al., 1996).

Several applications of three-dimensional reconnection have been studied numerically, both in laboratory (Chapter 9), magnetospheric (Chapter 10), solar (Chapter 11), and astrophysical (Chapter 12) contexts. For example, reconnection in tokamaks, reversed-field pinches, and spheromaks has been modelled in great detail by many authors, including Sato (1985), Schnack et al. (1979, 1985), Baty et al. (1992), and Finn and Sovinec (1998). The Earth's magnetosphere too has inspired many numerical experiments, including global models of the magnetosphere (e.g., Ogino et al., 1989), flux transfer events (see §10.3), and magnetospheric substorms (e.g., Birn and Hones, 1981; Sato et al., 1984; Birn and Hesse, 1990; Birn et al., 1996, 1997; see §10.5). For instance, Otto (1995) has modelled the creation of interlinked flux tubes at the Earth's magnetopause due to reconnection at two patches of enhanced resistivity of two inclined sets of magnetic field lines (Fig. 8.33).

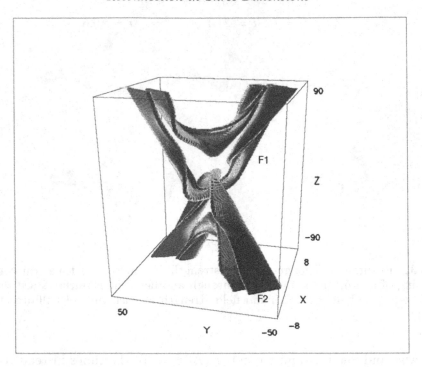

Fig. 8.33. The creation of interlinked flux tubes at a model of the Earth's magnetopause (Otto, 1995).

One important result concerning three-dimensional reconnection that has come out of laboratory reconnection experiments (Chapter 9) is the demonstration that the two-dimensional tearing mode can lead to three-dimensional turbulence. From their experience with three-dimensional MHD simulations, Dahlburg et al. (1992) have proposed that three-dimensional turbulence develops as a result of a secondary MHD instability that appears when the two-dimensional tearing mode saturates. Gekelman and Pfister (1988) observed something similar to this in their LCD experiment (§9.2). Their experimental results show that the extended X-lines of the two-dimensional mode are broken up into short segments, as predicted by the three-dimensional tearing-mode theory developed by Kan (1988) to explain patchy reconnection at the dayside magnetopause.

Solar applications (see Chapter 11) include: three-dimensional dynamo action and magnetoconvection (e.g., Proctor et al., 1993; Matthews et al., 1995); reconnection in braided or sheared solar coronal fields (e.g., Dahlburg et al., 1991; Strauss, 1993; Galsgaard and Nordlund, 1996; Longcope and Strauss, 1994; Kusano et al., 1995; Suzuki et al., 1997; Karpen et al., 1998;

Lionello et al., 1998); and kink instability of twisted flux tubes (e.g., Galsgaard and Nordlund, 1997a; Einaudi et al., 1997).

From the above examples, we can see that three-dimensional systems do in some circumstances exhibit features found in their two-dimensional counterparts. However, in general, three-dimensional configurations contain a much richer variety of behaviour than exists in two dimensions. Discovering this behaviour and determining which aspects are important for phenomena in laboratory, space, and astrophysical plasmas are tasks which will probably occupy researchers for many years to come.

9

Laboratory Applications

Reconnection is not difficult to achieve in a laboratory environment. When two simple dipole magnets are held near each other in air, two null points will generally be present, and when the magnets are moved relative to each other, their field lines easily reconnect. It is only when a conducting plasma is present in the vicinity of a null point that reconnection starts to become difficult and therefore interesting.

The principal application of reconnection theory in the laboratory has been in the development of magnetic containment devices for controlled thermonuclear fusion, but plasma experiments have also been designed specifically to study reconnection dynamics. Containment devices try to confine a sufficiently hot plasma inside a magnetic bottle for a period long enough to achieve a sustained nuclear reaction. Reconnection can both hinder and help in this regard. For example, in one device (known as the *tokamak*, §9.1.2) reconnection is involved in several different instabilities which degrade the confinement, but in another device (known as the *spheromak*, §9.1.3) reconnection is necessary to create the field configuration which actually confines the plasma.

Laboratory experiments specifically designed to study reconnection dynamics are motivated by a desire to understand reconnection as a general physical process, in the hope that this knowledge can be applied to both fusion, space, and astrophysical applications. However, as with numerical simulations, laboratory experiments cannot easily replicate the conditions that occur outside the Earth, primarily because of the problem of scale. Laboratory devices typically have dimensions of a metre or less, which is many orders of magnitude smaller than occurs in cosmical applications. Consequently, parameters which depend on the scale-size of the plasma, such as the magnetic Reynolds number, may be many orders of magnitude different from elsewhere in the universe.

Table 9.1. *Parameters for Some Plasma Containment Devices*

Parameter[1]	Tokamak (JET)	Reversed-Field Pinch (ZT-40M)	Spheromak (SPHEX)	Field-Reversed Configuration (HBQM)
L_e	1	0.2	0.4	0.1
n	10^{20}	3×10^{19}	5×10^{19}	7×10^{20}
T	10^8	2×10^6	2×10^5	10^5
T_e/T_i	0.1	≈ 1	0.7	≈ 1
B	2	0.2	0.04	0.1
$\tilde{\mu}$	2 (deuterium)	1 (hydrogen)	1 (hydrogen)	2 (deuterium)
R_{gi}/L_e	7×10^{-3}	0.04	0.03	0.04
λ_i/L_e	0.03	0.2	0.08	0.1
$\lambda_{\mathrm{mfp}}/L_e$	80	50	0.07	0.01
E_A/E_D	2×10^5	4×10^3	5	0.3
E_{SP}/E_D	20	6	0.09	0.03
β	0.09	0.05	0.02	0.2
$R_{me}(L_u)$	9×10^7	4×10^5	8×10^5	200

[1] MKS units are used, with lengths in metres, densities in inverse cubic metres, temperatures in degrees Kelvin, and magnetic field strengths in tesla. The length L_e is the smallest dimension of the chamber, R_{gi} is the ion gyro-radius, λ_i is the ion-inertial length, λ_{mfp} is the electron–ion mean-free path, E_A is the Alfvén convective electric field, E_D is the Dreicer field, E_{SP} is the Sweet–Parker electric field, β is the plasma beta, and R_{me} is the magnetic Reynolds number.

The use of collisional theory in laboratory applications is sometimes easier to justify than in space and astrophysical applications because the Coulomb mean-free path can be smaller than the global scale-size of the plasma. However, there are some laboratory devices such as JET (Joint European Torus) in which the collisional mean-free path is almost two orders of magnitude greater than the scale-size (Table 9.1). Even when the collisional mean-free path is small, non-collisional effects are usually present because it is not so small as to make all kinetic effects negligible everywhere throughout the plasma. This becomes evident when we consider the operating parameters shown in Tables 9.1 and 9.2 for some typical laboratory devices. The first six rows of entries in these tables show the characteristic plasma properties in each device. Here L_e is the smallest overall dimension of the plasma chamber (identified with the external scale of reconnection theory, §4.2.4), while $n, T, T_e/T_i, B$ and $\tilde{\mu}$ are the average density, temperature, electron–ion temperature ratio, magnetic field strength, and mean atomic weight, respectively. These operating parameters are used to calculate the remaining parameters in the table by using the formulae in Section 1.7.

Consider, for example, the DIPD (double inverse-pinch device) in Table 9.2. The mean-free path [λ_{mfp}, Eq. (1.62)] for electron–ion collisions is

Table 9.2. *Parameters for Some Reconnection Experiments*

Parameter[1]	Multiple X-Line Reconnection (Yamada)	LCD (Stenzel)	Double Inverse-Pinch Device (Bratenahl)	TS-D3 (Frank)	UN-Fenix (Altyntsev)
L_e	0.1	0.3	0.1	0.1	0.2
n	10^{20}	10^{18}	10^{20}	5×10^{20}	10^{18}
T	2×10^5	10^5	2×10^4	10^5	3×10^6
T_e/T_i	≈ 1	5	≈ 1	≈ 1	≈ 1
B	0.05	2×10^{-3}	0.7	0.5	0.04
$\tilde{\mu}$	1 (hydrogen)	40 (argon)	40 (argon)	4 (helium)	1 (hydrogen)
R_{gi}/L_e	0.08	2	0.01	0.01	0.2
λ_i/L_e	0.2	3	1	0.2	1
$\lambda_{\mathrm{mpf}}/L_e$	0.3	3	3×10^{-3}	0.02	3×10^3
E_A/E_D	3	2	10	5	4×10^4
E_{SP}/E_D	0.1	0.3	1	0.2	40
β	0.3	0.9	10^{-4}	7×10^{-3}	0.07
$R_{me}(L_u)$	900	60	70	700	10^6

[1] The same notation is used as for Table 9.1.

3×10^{-4} m and so is about 300 times smaller than the plasma scale-length of 0.1 m, so clearly Coulomb collisions are important here. Yet, the ion-inertial length λ_i [or ion-skin depth, Eq. (1.57)] in this device is also 0.1 m, which implies that Hall currents are likely to be significant. In the generalized Ohm's law of kinetic theory [see Eq. (1.53)] the ratio of the Hall electric field [$\mathbf{j} \times \mathbf{B}/(ne)$] to the convective electric field ($\mathbf{v} \times \mathbf{B}$) is of order $j/(nev)$. If one assumes that j is of order $B/(\mu L_e)$ and the flow speed (v) is on the order of the Alfvén speed, then this ratio is λ_i/L_e.

Another indication of the importance of kinetic effects is obtained by comparing the convective electric field with the Dreicer electric field [E_D; Eq. (1.66)], that is, the field required to produce runaway electrons (§13.1.2). In many devices the convective flow is close to the Alfvén speed after instabilities arise, so the convective electric field can, in principle, be as large as $v_A B$, and this field almost always exceeds the Dreicer field. Even for a slow Sweet–Parker type of reconnection, where the convection is of order $v_A B/\sqrt{R_{me}}$, the electric field may still exceed the Dreicer field (e.g., in the JET tokamak).

9.1 Controlled Thermonuclear Fusion

For many years, there has been a sustained effort to develop a controlled fusion reaction in order to generate electrical power. The most practical reaction for this purpose is

$$D + T \rightarrow {}^4He + n + 17.6\,MeV,$$

corresponding to the fusion of the two heavy isotopes of hydrogen, deuterium (D) and tritium (T), into an alpha particle (^{4}He) and a neutron (n). Tritium, though rare in nature, can easily be bred by neutron interactions in lithium. Both lithium and deuterium are so abundant in the terrestrial environment that the DT reaction could, in principle, satisfy global energy consumption at current rates for millions of years. However, achieving the conditions needed to sustain the reaction is not easy.

To achieve fusion, a thermal plasma must have a temperature in excess of about 10^8 K and sustain a sufficient number of reactions to recover the energy required to heat it. Approximately 80% of the energy released in the DT reaction is deposited in the neutrons, while the remaining 20% is deposited in the helium nuclei. Because the neutrons have an extremely low collisional cross section, they readily escape from the plasma, and their energy can only be extracted in a thick blanket of material outside the plasma volume. The energy deposited in the helium nuclei heats the plasma and plays a major role in sustaining high temperatures within the plasma.

The conditions required to ignite a self-sustaining reaction are shown in Fig. 9.1, and from this figure we can see that a necessary condition for a net energy yield for the DT reaction is

$$n\tau \gtrsim 1.5 \times 10^{20}\,m^{-3}s,$$

where n is the plasma density and τ is the energy confinement time of the plasma (Pease, 1993). This condition, which is known as *Lawson's criterion*, states that the plasma must be held at a temperature above 10^8 K at a density n and for a time τ in order for enough energy to be released to recover the energy required to heat the plasma in the first place. One way to do this is to contain plasma in a magnetic "bottle", which prevents the plasma from coming into contact with its surroundings. The magnetic field must be strong enough to prevent significant diffusion of plasma across the field to the vessel walls, and the field must be stable enough to maintain confinement for the time required by Lawson's criterion.

MHD is one of the principal tools for understanding the stability of the magnetic field within containment devices. Several proposed configurations have had to be abandoned because they suffer from ideal MHD instabilities (such as kinking and ballooning modes) or from resistive MHD instabilities (such as the tearing mode; §6.2). Since the magnetic Reynolds numbers for the containment devices lie in the range from 10^2 to 10^8, it may seem

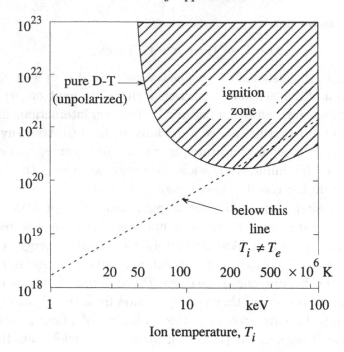

Fig. 9.1. Ignition conditions for a deuterium–tritium (DT) plasma, showing the Lawson criterion parameter $(n\tau)$ as a function of ion temperature (T_i) (after Pease, 1993).

surprising that resistive instabilities or reconnection could be important. In most devices the resistive time-scale based on classical Spitzer conductivity is an order of magnitude greater than the time required for containment. However, the rapid formation of thin current sheets and the onset of turbulence can cause the time-scale for resistive instabilities to be much closer to the ideal ones.

Some of the key questions about reconnection in containment devices are how fast does it occur, when does it occur, and what is the final configuration after it has taken place? An important development regarding the latter question (from Taylor, 1974, 1986) is as follows.

9.1.1 *Taylor Relaxation*

Taylor argued that the primary effect of reconnection in a turbulent plasma is to evolve the magnetic configuration towards a minimum-energy state prescribed by the linear force-free equation

$$\mathbf{j} \times \mathbf{B} = 0,$$

where

$$\mu \mathbf{j} = \nabla \times \mathbf{B} = \alpha_0 \mathbf{B} \qquad (9.1)$$

and α_0 is a constant (Taylor, 1986). According to Taylor, rapid reconnection does not lead to the potential state ($\mathbf{j} = \mathbf{0}$), but to a linear force-free one. His argument has two parts (Heyvaerts and Priest, 1984). The first is the assertion that during the reconnection process the global magnetic helicity (see §8.5),

$$H = \int_V \mathbf{A} \cdot \mathbf{B} \, dV, \qquad (9.2)$$

is conserved, where $\mathbf{A}$ is the vector potential of the magnetic field ($\mathbf{B}$) and V indicates integration over the total volume of the plasma. This expression assumes that field lines do not extend outside the closed volume (V) and that the volume is simply connected, but, as we discussed in Section 8.5, the definition can be generalized to include such fields as well. The main point of Taylor's original argument is that the infinite number of invariants of ideal MHD, namely the integrals of $\mathbf{A} \cdot \mathbf{B}$ along individual field lines, are broken in a resistive plasma, and only one, namely the total helicity, remains.

The conservation of global magnetic helicity is a consequence of the difference between the decay rates of magnetic energy and helicity in a turbulent plasma. Magnetic energy (W_m) decays at the rate

$$\frac{dW_m}{dt} = -\int_V \frac{j^2}{\sigma} \, dV, \qquad (9.3)$$

while for magnetic helicity

$$\frac{dH}{dt} = \int_V \left(\frac{\partial \mathbf{A}}{\partial t} \cdot \nabla \times \mathbf{A} + \mathbf{A} \cdot \nabla \times \frac{\partial \mathbf{A}}{\partial t} \right) dV. \qquad (9.4)$$

After application of the vector identity for the divergence of the cross product ($\mathbf{A} \times \partial \mathbf{A} / \partial t$) and the divergence theorem, this reduces to

$$\frac{dH}{dt} = 2 \int_V \frac{\partial \mathbf{A}}{\partial t} \cdot \nabla \times \mathbf{A} \, dV - \int_S \mathbf{A} \times \frac{\partial \mathbf{A}}{\partial t} \cdot \hat{\mathbf{n}} \, dS, \qquad (9.5)$$

where the second integral on the right-hand side is taken over the surface S of the volume V (see Section 8.5.4 for the equivalent expression using a generalized definition of helicity). If field lines do not penetrate the containment wall (or if the field lines are frozen into the wall), then the surface integral in Eq. (9.5) is zero. Using Eqs. (1.10a) and (1.18) to eliminate $\partial \mathbf{A} / \partial t$ then reduces Eq. (9.5) to

$$\frac{dH}{dt} = -2 \int_V \frac{\mathbf{j} \cdot \mathbf{B}}{\sigma} \, dV, \qquad (9.6)$$

since $\mathbf{B} = \nabla \times \mathbf{A}$.

For a non-turbulent plasma both energy and magnetic helicity decay at a rate which is proportional to σ^{-1}, but when turbulence is present they decay at different rates. Turbulent fluctuations produce numerous small current sheets with thicknesses of order $\sigma^{-1/2}$ and current densities proportional to $B\sigma^{1/2}$. Therefore, the energy decay rate,

$$\frac{dW_m}{dt} \propto \int_V B^2 \, dV,$$

is independent of σ. By contrast, global magnetic helicity decays as

$$\frac{dH}{dt} \propto \frac{2}{\sigma^{1/2}} \int_V B^2 \, dV,$$

so that as the resistivity σ^{-1} tends to zero the helicity dissipation becomes negligible. A more thorough discussion of why helicity does not decay in a turbulent, low-resistivity plasma is given by Montgomery et al. (1978).

The second part of Taylor's argument is a proof from Woltjer (1958) that minimisation of the magnetic energy under the constraint of magnetic helicity conservation leads to a linear force-free field. The proof uses the method of Lagrange multipliers, which states that, at a constrained minimum, the variation of the magnetic energy is equal to a constant (the Lagrange multiplier) times the variation in the helicity (e.g., Kaplan, 1952), that is,

$$\delta W_m = \frac{\alpha_0}{2\mu} \delta H, \tag{9.7}$$

where $\alpha_0/(2\mu)$ is the Lagrange multiplier. Substituting

$$W_m = \int_V \frac{B^2}{2\mu} \, dV$$

for the magnetic energy and Eq. (9.2) for H in (9.7) yields

$$\int_V \delta[(\nabla \times \mathbf{A}) \cdot (\nabla \times \mathbf{A})] \, dV = \alpha_0 \int_V \delta[\mathbf{A} \cdot \nabla \times \mathbf{A}] \, dV$$

or

$$\int_V [2(\nabla \times \mathbf{A}) \cdot (\nabla \times \delta\mathbf{A}) - \alpha_0 (\delta\mathbf{A} \cdot \nabla \times \mathbf{A} + \mathbf{A} \cdot \nabla \times \delta\mathbf{A})] \, dV = 0. \tag{9.8}$$

Applying the vector identities

$$\nabla \cdot [(\nabla \times \mathbf{A}) \times \delta\mathbf{A}] = \delta\mathbf{A} \cdot \nabla \times (\nabla \times \mathbf{A}) - (\nabla \times \mathbf{A}) \cdot (\nabla \times \delta\mathbf{A})$$

and

$$\nabla \cdot (\delta\mathbf{A} \times \mathbf{A}) = \mathbf{A} \cdot \nabla \times \delta\mathbf{A} - \delta\mathbf{A} \cdot \nabla \times \mathbf{A}$$

in Eq. (9.8) and using the divergence theorem produces

$$\int_V [2(\nabla \times \nabla \times \mathbf{A}) \cdot \delta\mathbf{A} - 2\alpha_0\, \delta\mathbf{A} \cdot \nabla \times \mathbf{A}]\, dV = 0, \qquad (9.9)$$

where, as before, the surface integrals vanish. Equation (9.9) can be rewritten as

$$\int_V [\nabla \times (\nabla \times \mathbf{A}) - \alpha_0 \nabla \times \mathbf{A}] \cdot \delta\mathbf{A}\, dV = 0$$

or

$$\int_V (\nabla \times \mathbf{B} - \alpha_0\, \mathbf{B}) \cdot \delta\mathbf{A}\, dV = 0. \qquad (9.10)$$

Since $\delta\mathbf{A}$ is arbitrary, the integrand of Eq. (9.10) must be identically zero, and so we finally deduce the linear force-free condition

$$\nabla \times \mathbf{B} = \alpha_0\mathbf{B},$$

as required.

A strictly force-free configuration, linear or otherwise, cannot contain a fusion plasma, since it exerts no counter-force to the gas pressure. Therefore, Taylor's theory, as originally formulated, does not provide the information needed to determine how much containment a particular field provides. Efforts to extend Taylor's theory to include a non-force-free magnetic field have been carried out by Bhattacharjee et al. (1983), Edenstrasser and Schuurman (1983), Turner and Christiansen (1981), Bhattacharjee and Dewar (1982), Hameiri and Hammer (1982), and Finn and Antonsen (1983), but none of these extensions appear to be as robust as Taylor's original formulation (Taylor, 1986). To some people this lack of success in generalising Taylor's analysis casts doubt on the original theory, which applies to an isolated system that is marginally stable and has small dissipation. By comparison, the reversed-field pinch (§9.1.3), to which the theory has been applied with at least partial success, is a system that is driven hard and has relatively large dissipation and large fluctuations. (Typical loop voltages are 40 V compared with 1 V for a tokamak.)

Taylor relaxation is fundamentally a turbulent process. As we shall see in the next section, the tokamak has a magnetic field structure which tends to inhibit the development of strong turbulence, so it is not normally susceptible to Taylor relaxation.

9.1.2 Tokamaks

In the late 1960s, Russian scientists discovered a toroidal configuration which has remarkable stability properties (see Wesson, 1997, for a review). The

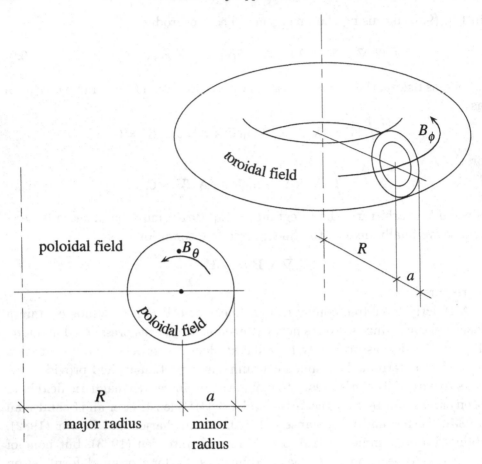

Fig. 9.2. The magnetic field geometry of a tokamak. Here R is the major radius, a is the minor radius, and B_θ and B_ϕ are the poloidal and toroidal field components (from O'Brien and Robinson, 1993).

geometry of the field of their device, which they referred to as a tokamak (a Russian acronym for "toroidal magnetic chamber"), is shown in Fig. 9.2. In a tokamak the poloidal field is generated by inducing a toroidal current by a transformer action rather than by external windings as in the *stellarator*, which is a similar, but more complex, toroidal device. Because of the induced nature of the current, the tokamak is inherently time-dependent, and its confinement-time cannot exceed the cycle-time of the transformer action. Despite this limitation, energy confinement-times in excess of 1 s have been achieved in large tokamaks such as JET – more than an order of magnitude greater than in any other device at the present time.

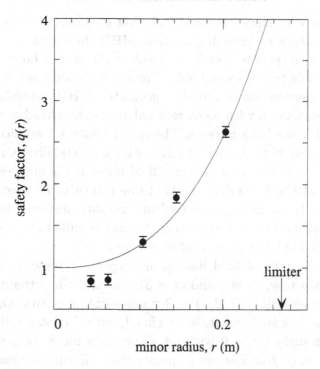

Fig. 9.3. Safety factor (q) as a function of radial distance (r) in a tokamak. The circles with error bars show values inferred from electron temperature in the DITE tokamak (from Pease, 1993).

The occurrence of MHD instabilities in tokamaks is closely related to the safety factor q (§6.4), defined as

$$q(r) \equiv \frac{B_\phi}{B_\theta} \frac{r}{R},$$
(9.11)

where B_ϕ and B_θ are the toroidal and poloidal field components, and R is the major radius of the torus as shown in Fig. 9.2. As we mentioned in §6.4, the safety factor is the number of times a field line on a flux surface goes around toroidally for one poloidal circuit, and it is also called the winding number. At surfaces where $q = m/n$, perturbations going like

$$e^{i(m\theta + n\phi)}$$

are resonant; that is, they have $k_{\|} = 0$, which means that the perturbation is constant along the field line, so that the field-line bending term vanishes at this surface. Usually $q(r)$ increases monotonically from the magnetic axis to the edge of the plasma as shown in Fig. 9.3, but it is also possible to create configurations with shear reversals such that $q(r)$ decreases for small r and increases for larger r. According to ideal MHD theory, the configuration is stable for $q > 1$ (if there is a conducting wall at the plasma boundary) but it is marginally stable (up to second order) for $q < 1$, depending on the details of the pressure distribution and toroidal geometry. MHD instabilities are most likely to be associated with mode-rational surfaces, defined as the locations where m and n are both integers. The $m = 1$ and $n = 1$ surface is the most important in this regard, since it has $q = 1$ where stability is marginal. On surfaces where q is irrational, a single field line fills the entire surface, since in no circuit of the torus does the field line return to its starting location. If the plasma becomes highly turbulent, the flux surfaces themselves can disappear and lead to the formation of what is called an *ergodic* field, in which a single field line fills an entire volume.

Ergodic behaviour of field lines is an important topic, in which neighbouring islands (say, $m = 2$ and $m = 3$) overlap. When the magnetic field is helically symmetric $[B(r, \theta, z) = B(r, m\theta + kz)]$, a constant of the motion exists and the flux surfaces are well-defined, with the intersections of a field line with the surface $z = 0$ lying on a curve. In more complex fields such intersections may give a series of points that fill out an area as the field line goes round the torus many times. Such a stochastic behaviour tends to start near the neutral points of the field and the breakdown of the magnetic surfaces there suddenly enhances diffusion of heat and particles across the field (Rosenbluth et al., 1966; Finn, 1975; Rechester and Rosenbluth, 1978; Rechester and Stix, 1976; Laval and Gresillon, 1979; Finn et al., 1992). For field lines lying on mode-rational surfaces, the general definition of field-line velocity [Eq. (1.42) or (8.25)] may produce multiple values at any radial location r. However, it is still possible to define a unique velocity for the magnetic flux surfaces as long as such surfaces exist. Therefore, reconnection is still a useful concept for describing the interaction of these surfaces.

A standard tokamak is a torus with $q \sim 1$, a small aspect ratio $(a/R \ll 1)$ and a plasma beta $(\beta) \sim (a/R)^2$, so that $B_\theta \ll B_\phi$ and $p \sim B_\theta^2/(2\mu)$. The values of q at the axis and surface are denoted by q_0 and q_a, respectively (§6.4). The aim is to confine plasma at typical values of $T \sim 15$ keV$(10^8$ K), $n \sim 10^{20}$ m^{-3}, $B \sim 2 - 10$ tesla $(20 - 100$ kG), and $\beta \sim 5\%$ for an energy confinement time $\tau \sim 1$ s. The principal interest in reconnection in tokamaks is the role it plays in various types of resistive instability that can destroy or degrade the confinement of the plasma. There are three types of resistive MHD instability in which it is involved (e.g., Wesson, 1981, 1997). One type is Mirnov oscillations, another is the sawtooth, and the third is the major disruption, as follows:

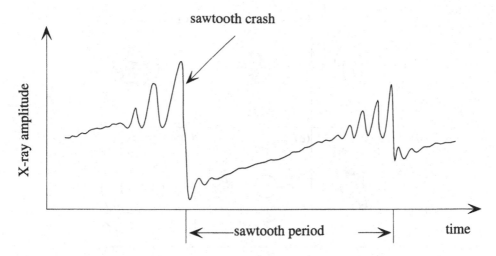

Fig. 9.4. A typical X-ray emission vs. time curve for sawtooth oscillations in a tokamak. The sawtooth period can range from milliseconds in small machines to seconds in JET (from O'Brien and Robinson, 1993).

(a) *Mirnov oscillations* show up with magnetic pickups as a small regular vibration at several values of $m > 1$. They are due to resistive modes near the $q = m$ surfaces that have saturated nonlinearly to a steady state containing islands. Finite-Larmor-radius and electron diamagnetic effects create a finite propagation frequency of linear tearing modes and can stabilise them. Nonlinearly, $\mathbf{E} \times \mathbf{B}$ rotation causes the steady helical structure to rotate around the torus.

(b) *Sawtooth Oscillations.* The sawtooth instability manifests itself primarily as a slow increase in temperature followed by a rapid drop or "crash," as shown in Fig. 9.4. The amplitude of the oscillations is typically 15%, and the oscillations are associated with the $q = 1$ surface. There is often a secondary, sinusoidal oscillation, which is most pronounced just before the crash, imposed on top of the sawtooth pattern.

According to the theory of Kadomtsev (1975), sawtooth oscillations occur because ohmic heating, which is greatest at $r = 0$, leads to a decrease in the electrical resistivity more or less as predicted by the classical Spitzer formula [Eq. (1.13)]. As the electrical resistivity decreases, the current rises so as to maintain the imposed electric field, which can be considered constant over the lifetime of a single sawtooth cycle. The increasing current enhances the poloidal field and causes the safety factor to drop below unity in the centre of the plasma. This in turn triggers an internal kink mode which expels plasma out of the region where $q < 1$, as shown in Fig. 9.5. The kink mode is an internal mode because it leaves the external region (where $q > 1$)

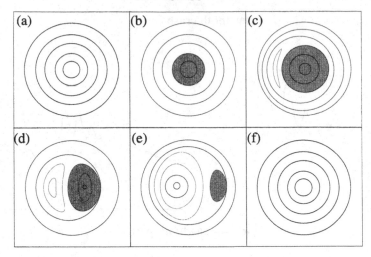

Fig. 9.5. Evolution of the magnetic flux surfaces during a sawtooth crash according to the theory of Kadomtsev (1975).

undisturbed, except near the $q=1$ surface. The secondary oscillations are explained by rotation of the asymmetrical helical structure associated with the developing kink.

Kadomtsev's theory predicts that after the reconnection is complete the value of q on the toroidal axis returns to a stable value and the cycle starts again. In fact, measurements show that q remains below unity throughout the sawtooth cycle (cf. Fig. 9.3), ranging from 0.7 just before the crash to 0.8 afterwards (Yamada et al., 1994). Also, the reconnection time-scale in Kadomtsev's model is not fast enough to explain the observed crash times, which are on the order of 10 to 100 μs, depending on the size of the toka-mak (Biskamp, 1994). Resistive MHD simulations (e.g., Paré, 1984; Aydemir et al., 1989) based on Kadomtsev's model suggest that the reconnection is of the slow Sweet–Parker type, which gives a crash time approximately an order of magnitude slower than observed. Prior to the crash, the plasma resis-tivity roughly matches the predictions of classical or neo-classical transport theory, depending on whether the plasma is in a collisional or semi-collisional regime (Pease, 1993). However, during the crash, it is likely that non-MHD (i.e., kinetic) effects are important (Wesson, 1990).

The discrepancies between the observed and predicted values of the safety factor and the crash time have led to alternative models. For example, the partial reconnection model of Goedheer and Westerhof (1988) invokes the onset of turbulence during the crash, and it appears to explain both the crash

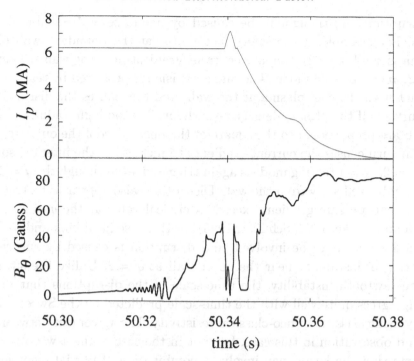

Fig. 9.6. Current (I_p) and poloidal magnetic field (B_θ) evolution during a high-density disruption in JET. MHD activity, shown here by the fluctuations in B_θ near the wall, rises strongly before the disruption of the current (from O'Brien and Robinson, 1993).

time-scale and the observed q-profile (Yamada et al., 1994). However, it does not explain the drop in temperature in the unreconnected core (Wesson, 1989). Instead, Wesson (1990) suggested that the reconnection rate may be determined by electron inertia and so may lead to a faster sawtooth collapse than Kadomtsev reconnection.

(c) *Major Disruptions.* The other instability in tokamaks which involves reconnection is the major disruption. It is potentially more serious than the sawtooth, because it leads to a complete loss of containment and unleashes forces which can cause structural damage in large machines (O'Brien and Robinson, 1993). During the beginning of the disruption the current flow is steady, but there are large-scale ($m = 2$) fluctuations in the poloidal field that grow rapidly over a few milliseconds, usually in the outer regions near the wall of the vessel. These fluctuations are followed by the emission of hard X-rays, a fall in soft X-ray emission, and a sudden and catastrophic drop in the current within a period of approximately 10–20 ms, as shown in Fig. 9.6.

Disruptions are thought to be caused by one of several possible mechanisms. For example, if the safety factor $q(a)$ at the boundary wall of the plasma is too low (<2), the large current gradient near the wall triggers an $m = 2$, $n = 1$ tearing mode. The magnetic islands produced by tearing then interact with the cool plasma at the wall, and this causes the disruption of the current. If the plasma density near the wall is too high, radiation losses (which are proportional to the square of the density) cool the outer regions. This in turn causes the current gradients to increase in the interior, so that the $m = 2$, $n = 1$ tearing mode is again triggered, even though the $q = 2$ surface may be well away from the wall. Disruptions also appear to be triggered if the local pressure gradient exceeds a critical value at the $q = 1$ surface (Kleva and Drake, 1991; Schüller, 1995). In this case ideal kink and ballooning modes appear to be involved, and disruption is caused by subsequent triggering of instability near the outer wall at $q = 2$. Unlike the situation for the sawtooth instability, the time-scale of the disruptions that involve tearing agrees quite well with the time-scale predicted by the Sweet–Parker theory using classical or neo-classical resistivity. The agreement between theory and observation in this case, but not in the case of the sawtooth crash, suggests that the latter may involve a combination of partial reconnection and ideal MHD instability.

Alternatively, a disruption may be evidence of a magnetic catastrophe. Usually, the tearing mode is self-stabilising with a saturated state that reduces the current gradient (dj_ϕ/dr), but if the current is too large Wesson (1986) has demonstrated that the saturated equilibrium no longer exists. The current profile $j_\phi(r)$ is modified by the $m=1$ and $m=2$ modes, whose islands flatten it near the resonant surfaces at $q(r)=1$ and $q(r)=2$. As the current grows, so the island widths grow until, at some critical width, the equation $\Delta'(w)=0$ no longer has a solution. Once the magnetic catastrophe is reached it presumably leads to mode coupling. Another explanation for tokamak disruptions was given by Biskamp and Welter (1989), who suggested that the high-m Alfvén-like modes ($m > 6$) create small-scale magnetic turbulence with a negative resistivity that drives the $m=2$ mode unstable.

More recently, it has been realised that there are at least three different types of disruption (density limit, current limit, and beta limit). The density-limit ones have Mirnov oscillations as precursors and appear to involve tearing modes, but the others involve ideal modes (Fredrickson et al., 1993; Finn and Sovinec, 1998). Field-line chaos is probably involved in all kinds, but in some cases it may only come in after the plasma is doomed anyway.

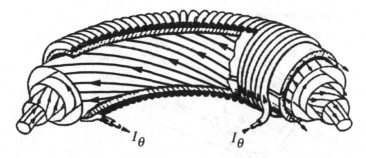

Fig. 9.7. Schematic diagram of the toroidal field windings in an RFP. The direction of the toroidal field reverses as one travels from the inner to the outer flux surfaces (from Baker, 1984).

Lessons that astrophysicists may learn from tokamak studies include the following: The details of the magnetic structure $[B_p(r)$ and $j_\phi(r)]$ are important for determining the relevant instability. The resistive modes occur at lower thresholds than the ideal ones, but they are not necessarily destructive since they may instead produce gentle oscillations (Mirnov or sawtooth). Also, it is crucial to study the nonlinear development of any instability to see whether it saturates (like the sawtooth) or grows explosively (as in a disruption). Nevertheless, laboratory plasma physicists should beware of applying their theories uncritically to astrophysics, since often the parameter ranges are quite different and there may be new effects that dominate (such as photospheric line tying in the case of the solar flare).

9.1.3 Other Containment Devices

In a tokamak, reconnection is something to be avoided, since it is always associated with instabilities that are deleterious to the confinement of the plasma. However, there are several devices in which reconnection is essential for the formation or maintenance of a stable configuration. One of these, the reversed-field pinch (RFP), has the same toroidal shape as the tokamak, but, unlike the tokamak, the toroidal field is not much stronger than the poloidal field (Caramana et al., 1983; Baker, 1984; Ejiri and Miyamoto, 1995). Instead, the toroidal and poloidal fields are of the same order, and gross stabilization of the field relies on a conducting chamber wall (Fig. 9.7). Figure 9.8 compares the field configuration of the RFP with a tokamak. The name "reversed-field pinch" arises from the fact that the toroidal field actually reverses direction.

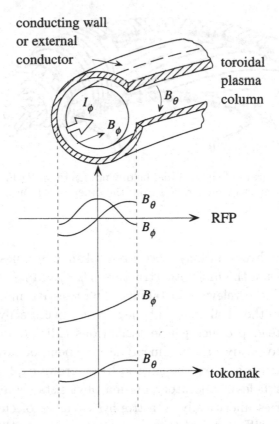

Fig. 9.8. Comparison of the profiles of the poloidal (B_θ) and toroidal (B_ϕ) field components in an RFP with those in a tokamak (from Baker, 1984).

The formation of the reversal is associated with helical instabilities that cause a partial relaxation to the linear force-free state

$$B_r = 0, \tag{9.12}$$

$$B_\theta = B_0 \, J_1(\alpha r), \tag{9.13}$$

$$B_z = B_0 \, J_0(\alpha r), \tag{9.14}$$

for large aspect ratios ($R/a \gg 1$) and for $\alpha < 3.11/a$, where J_0 and J_1 are Bessel functions. The parameter α is determined by

$$\frac{H}{F^2} = \frac{\alpha R \left[J_0^2(\alpha a) + J_1^2(\alpha a) \right] - J_0(\alpha a) J_1(\alpha a)}{J_1^2(\alpha a)}, \tag{9.15}$$

where F is the total toroidal flux, H is the magnetic helicity (see §8.5), and the quantities F and H can be independently controlled by the injection of toroidal and poloidal fields produced in external circuits. Some care must

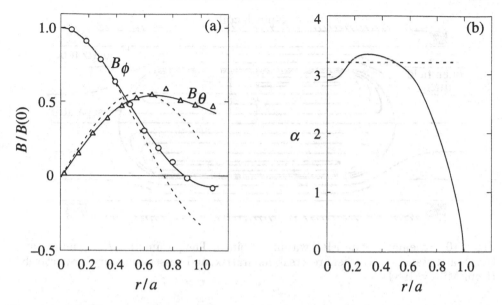

Fig. 9.9. (a) Magnetic field profile and (b) force-free parameter (α) for the RFP device ETA-BETA II (after Ortolani, 1987).

be taken to use a definition of H which is gauge-invariant because a torus is a multiply connected domain (Taylor, 1986). According to Taylor's theory, a field reversal appears when α exceeds $2.4/a$, which is the first zero of the Bessel function J_0. For values of H/F^2 that correspond to $\alpha \geq 3.11/a$, the symmetric solution of Eqs. (9.12)–(9.15) no longer applies, because the toroidal current column starts to kink. In this regime there is an alternative, non-axisymmetric solution with lower energy whose helical deformation increases as H/F^2 increases (Taylor, 1986).

However, Taylor's theory is only partially successful in explaining the behaviour of the RFP because the magnetic field profile in practice differs substantially from a linear force-free state, as Fig. 9.9 shows. Although α is roughly constant for $r < 0.75a$, it decreases near the wall (Ortolani, 1987), and the helical state predicted to occur when $\alpha > 3.11/a$ has never been observed (Finn et al., 1992). The principal problems with Taylor's theory are that the turbulent reconnection process it requires is suppressed near the wall and the theory does not properly take into account the fact that the RFP is continually driven by external forcing. Alternative theories, such as the minimum dissipation principle (Montgomery and Phillips, 1988), have been proposed which work better in some respects.

An interesting feature of the RFP is that in actual experiments the toroidal magnetic field does not decay as would be predicted by a simple ohmic

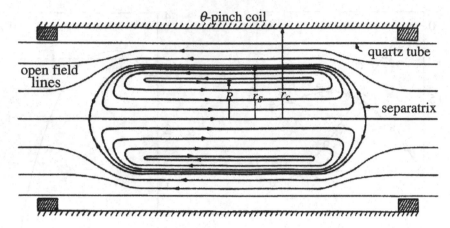

Fig. 9.10. Schematic of an FRC within a θ-pinch. The parameters R, r_s, and r_c are the axial distances to the field reversal, separatrix, and vessel boundary, respectively (from Milroy, 1984).

decay model. Within the RFP the externally induced currents maintain the poloidal field over the operating time, but they do not maintain the toroidal field. Sustainment of the toroidal field against decay is predicted by Taylor's theory because magnetic helicity would not be conserved if the toroidal field decayed while the poloidal field did not (Baker, 1984). During operation, turbulence within the plasma converts poloidal flux into toroidal flux so as to preserve the magnetic helicity, and this process has been referred to as a magnetic dynamo.

The partial success of Taylor's theory in describing the behaviour of the reversed-field pinch has two important implications for the rate of three-dimensional reconnection. First of all, the theory implies that any three-dimensional reconnection process which occurs in a turbulent plasma and which conserves helicity will occur at a rate that is independent of the electrical resistivity of the plasma. That is, the reconnection will be very fast, occurring on the Alfvén time-scale. The second implication is that the reconnection will not be so fast if it requires violation of global helicity conservation.

Reconnection and relaxation are also important for non-toroidal containment devices such as the field-reversed configuration (FRC; Milroy, 1984) and the closely rated spheromak. These devices, whose field configurations are shown schematically in Figs. 9.10 and 9.11, use magnetic reconnection to create a three-dimensional plasmoid in either spherical (spheromak) or cylindrical (FRC) geometry. All the field lines within the plasmoid are self-contained, with the exception of the separator (§8.4.1) running through the

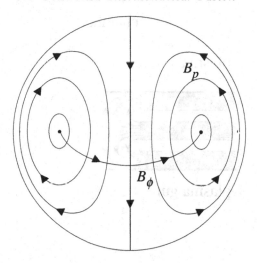

Fig. 9.11. A diagram illustrating the structure of the toroidal (B_ϕ) and poloidal ($\mathbf{B}_p$) fields in a spheromak (from Taylor, 1986).

two null points at the plasmoid's surface. One of the more common ways to form the spheromak field is by flux injection from a coaxial plasma gun, as shown in Fig. 9.12.

The Taylor state of a spherical spheromak is

$$B_r = 2B_0(\alpha r)^{-3/2} J_{3/2}(\alpha r) \cos \theta, \tag{9.16}$$

$$B_\theta = -\frac{B_0}{\alpha r} \frac{d}{dr} \left[r(\alpha r)^{-1/2} J_{3/2}(\alpha r) \right] \sin \theta, \tag{9.17}$$

$$B_\phi = B_0(\alpha r)^{-1/2} J_{3/2}(\alpha r) \sin \theta, \tag{9.18}$$

in spherical coordinates, where $\alpha = 4.49/a$ (Rosenbluth and Bussac, 1979). As in the RFP, a dynamo action is present because the poloidal field of the relaxed state in the spheromak is higher, typically by a factor of 5, than the injected poloidal field (Browning et al., 1992). The amplification of the poloidal field is impossible in an axisymmetric formation process, such as Fig. 9.12 implies, and the actual injection process involves a highly asymmetric configuration like the one shown in Fig. 9.13 (Hammer, 1984). Finn and Guzdar (1991) have developed a theory which demonstrates how reconnection, by means of the tearing mode, can amplify the poloidal field by the amount required to maintain constant magnetic helicity.

In a real spheromak, some field lines may remain attached to an external boundary, such as the plasma gun shown in Fig. 9.12. The connection of the field to the boundary makes it possible to sustain the spheromak field

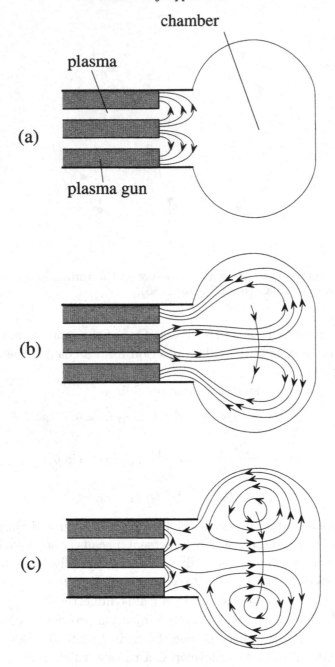

Fig. 9.12. Steps in the formation of a spheromak field by injection from a plasma gun. In (a) a ring of plasma containing a toroidal component is accelerated into the chamber by $\mathbf{j} \times \mathbf{B}$ forces. In (b) the plasma carries the field into the receiving vessel (or flux conserver), while finally in (c) reconnection leads to the formation of closed toroidal field lines and amplification of the poloidal flux (after Browning, 1992).

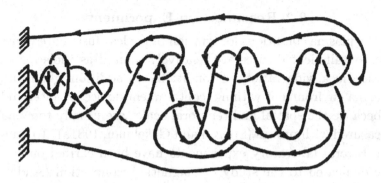

Fig. 9.13. A proposed mechanism for flux amplification during formation of a spheromak by plasma-gun injection (after Hammer, 1984).

indefinitely against resistive decay (Turner, 1984; Dixon et al., 1988). For a gun, the rate of helicity injection with $\nabla \times \mathbf{B} = \alpha\mathbf{B}$ is prescribed by

$$\frac{dH}{dt} = 2V_{\text{gun}}F_{\text{gun}} - 2\int_V \frac{\mathbf{j}\cdot\mathbf{B}}{\sigma}\,dV,$$

where V_{gun} is the gun voltage, F_{gun} is the injected flux, and the last term of the right-hand side is the ohmic decay from Eq. (9.6). Since the injected flux (F_{gun}) is entirely toroidal, the poloidal flux, which actually confines the plasma, is maintained only by conversion of the toroidal field to poloidal field by magnetic reconnection. Thus, magnetic reconnection is essential, not only for the formation of the spheromak field, but also for its maintenance.

Direct measurement of the reconnection rate during relaxation has been attempted by Sevillano and Ribe (1984), using an FRC device. The plasma properties of their machine, known as the High-Beta Q-Machine (HBQM) are shown in Table 9.1. Applying Ohm's Law to the diffusion-region current sheet, they estimate that the electrical resistivity (σ^{-1}) varies from approximately 5.9×10^{-6} ohm-m during the early phase of the reconnection to 3.3×10^{-5} ohm-m during the late phase. By comparison, the classical value calculated with Spitzer's formula is 1.8×10^{-6} ohm-m. Thus, the measured resistivity during the early phase is within a factor of 2 of the classical value, but during the late phase it is enhanced by an order of magnitude. The rough agreement with classical theory during the early phase is consistent with the fact that the plasma regime in this particular machine is collisional, as the parameters in Table 9.1 indicate. During the late phase, the regime becomes less collisional, and this transition may be why the resistivity then increases significantly above the classical value (Brunel et al., 1982; Hassam, 1984).

9.2 Reconnection Experiments

Because the plasma conditions for ignition of nuclear fusion are very specific, magnetic containment devices are relatively limited as tools for exploring reconnection in a wide variety of conditions. In addition, good diagnostics of the plasma environment within a containment device are often difficult to achieve because the spatial and temporal scales are usually too small to be easily measured with probes (Stenzel and Gekelman, 1981a). To remedy this situation, several laboratory experiments have been carried out which are expressly dedicated to the study of magnetic reconnection (Andersen and Kunkel, 1969; Ohyabu et al., 1974; Bratenahl and Baum, 1976; Overskei and Politzer, 1976; Zukakishvili et al., 1978; Stenzel and Gekelman, 1979; Irby et al., 1979; Altyntsev et al., 1986; Bulanov and Frank, 1992; Yamada et al., 1997a). Some examples illustrating the range of conditions which have been explored are listed in Table 9.2. The reconnection experiments cover a wider range of parameters than the containment devices shown in Table 9.1. For example, the plasma β in the experiments ranges from 10^{-4} to nearly one, compared with 0.02–0.2 in the containment devices.

Of the five experiments listed in Table 9.2, the multiple X-line reconnection (MRX) is the closest to a collisional regime. That is, its ion gyro-radius, ion-inertial length, and collision mean-free path are all smaller than the smallest overall dimension (L_e) of the plasma, although not by very much. The other experiments have one or more of these parameters greater than, or equal to, L_e. In the LCD experiment, all three parameters are greater than L_e, while in the UN-Fenix only the mean-free path is larger than L_e. Although the DIPD has a gyro-radius and a mean-free path which are both about a 100 times smaller than L_e, its ion-inertial length is of the order of L_e, which indicates that Hall currents are likely to be important (see §1.7). The large ion gyro-radius (R_{gi}), ion-inertial length (λ_i), and electron–ion mean-free path (λ_{mfp}) in the LCD are a deliberate feature of its experimental design. These large scale-lengths make it much easier to use probes to measure parameters within current sheets and boundary layers that develop within the plasma. An important feature of the LCD, as well as other discharge experiments, is that the current flow through the reconnection layer is primarily controlled by the interaction between the external circuit and the space-charge separation layer that develops near the cathode (Stenzel and Gekelman, 1986).

9.2.1 Anomalous Resistivity

In laboratory reconnection experiments the resistivity ($\eta_e = \sigma^{-1}$) of the plasma in the diffusion region is normally computed from the ratio of the observed electric field to the current density along the X-line; that is,

$$\eta_e = \frac{E}{j},$$

(9.19)

where E and j are the magnitudes of the components lying along the X-line in the z-direction. Using Eq. (1.18), we can rewrite Eq. (9.19) as

$$\eta_e = -\frac{1}{j}\left(\frac{\partial A}{\partial t} + \frac{\partial \Phi}{\partial z}\right).$$

(9.20)

If the electrostatic field ($\partial \Phi / \partial z$) associated with charge separation is negligible, then the resistivity is simply

$$\eta_e = -\frac{1}{j}\left(\frac{\partial A}{\partial t}\right),$$

(9.21)

which can be determined by measuring the temporal and spatial variation of the magnetic field. Unfortunately, in many experiments the charge-separation fields are not negligible, and therefore Eq. (9.21) is generally not a good way to estimate the electrical resistivity (Stenzel et al., 1982a). However, it does provide an upper limit to the resistivity because the charge separation field is oppositely directed to the induced electric field. In some previous experiments (e.g., Syrovatsky et al., 1973; Baum and Bratenahl, 1980) resistivities have been determined with Eq. (9.21) which are as much as a factor of 100 times larger than the Spitzer value, and this has led to claims of very high anomalous resistivities. Yet, in the same experiments, no corresponding enhancement in ohmic heating was found. This discrepancy is now thought to result from the neglect in Eq. (9.21) of the electrostatic field produced by charge separation (Stenzel et al., 1982a).

Stenzel and Gekelman (1981b) and Wild et al. (1981) have used probes to make direct measurements of the electric field. They found enhancements by a factor of 25 over the Spitzer value, but the resistivity is very non-uniform. The thickness of the diffusion-region current sheet in their experiment is on the order of the electron-inertial length (i.e., the collisionless skin-depth, c/ω_{pe}), which is in contrast to most other experiments where the thickness is on the order of an ion gyro-radius (Altyntsev et al., 1989; Yamada et al., 1991). This result can be understood as a consequence of the fact that the ion gyro-radius in the LCD experiment of Stenzel and Gekelman is larger than the scale-size of the plasma, while in the other reconnection experiments it is usually much smaller.

The design of some experiments is such that Eq. (9.21) can, in fact, give a fairly accurate estimate of the resistivity. The MRX, shown in Fig. 9.14, is an example of such an experiment. Because of the axial symmetry of the fields, the gradient of the electrostatic potential (Φ) along the circular separator

flux cores

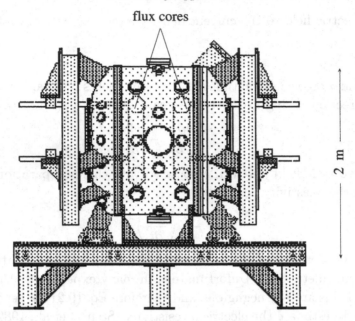

Fig. 9.14. Schematic diagram of the MRX reconnection experiment (courtesy of M. Yamada and the Princeton Plasma Physics Laboratory).

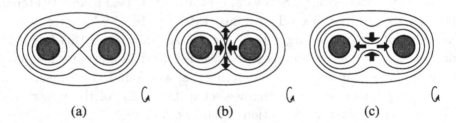

(a)   (b)   (c)

Fig. 9.15. Evolution of poloidal flux surfaces in the MRX experiment over a complete cycle. During the first half of the cycle the magnetic fields are pushed towards the X-line, while during the last half of the cycle the fields are pulled apart (after Yamada et al., 1997a).

line is effectively zero. Figure 9.15 diagrams the stages in the evolution of the flux surfaces of the poloidal magnetic field during a cycle of the MRX. In the first part of the cycle an increasing current in the two conducting rings drives magnetic flux towards the X-line, while in the second part of the cycle a decreasing current pulls the flux away from the X-line. Figure 9.16 shows the actual field in the experiment as determined from probe measurements during the pull part of the cycle.

The MRX provides insight into anomalous resistivity (§13.1.3) in plasmas that are intermediate between being completely collisional and completely

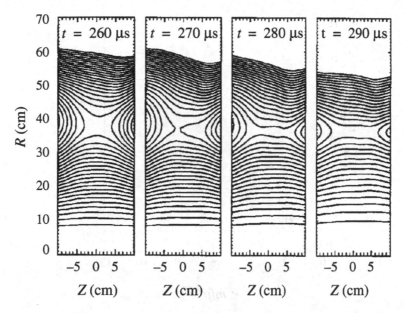

Fig. 9.16. Observed temporal behaviour of poloidal flux surfaces in the MRX during the "pull" phase of the cycle for the counter-helicity case (after Yamada et al., 1997b).

collisionless. In the MRX the mean-free path (λ_{mfp}) for electron–ion collisions is typically five times smaller than the smallest dimension of the plasma, but it is possible to vary the operating parameters so that the mean-free path is of the same order as the smallest dimension. Thus, significant deviations from classical, collisional transport can occur, as shown in Fig. 9.17, which compares the experimentally observed resistivity with the collisional resistivity predicted by Spitzer (1962) for different values of λ_{mfp}. For the smallest value, the observed resistivity is enhanced by a factor of about 1.5, but for the largest value, it is enhanced by a factor of 9. These values are consistent with the observed amount of heating produced by ohmic dissipation.

Another indication of collisionless effects in the MRX is that the thickness of the current sheet at the diffusion region is on the order of the ion gyro-radius (R_{gi}), which is greater than the thickness predicted by collisional theory. According to the Sweet–Parker relations (§4.2.1), the thickness (l) of the current sheet is

$$l = \frac{L}{R_{mi}^{1/2}},$$

where L is the current sheet length and R_{mi} is the magnetic Reynolds number based on L and Spitzer conductivity. Now, in the MRX, L is

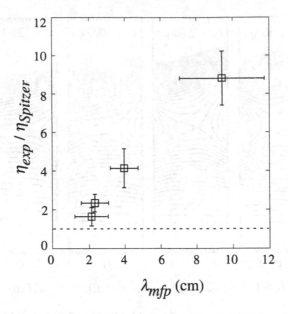

Fig. 9.17. Anomalous resistivity measured in the MRX as a function of the mean-free path (λ_{mfp}) for electron–ion collisions (M. Yamada, private communication).

~ 0.1 m, while $R_{mi} = R_{me}L/L_e$ is about 10^3 (Table 9.2). Consequently, the current sheet thickness (l) predicted by collisional theory is approximately 3×10^{-3} m, but the observed thickness is 8×10^{-3} m, namely, the value of the ion-gyro radius (Yamada et al., 1990; Yamada et al., 1991). A theoretical explanation of why the current sheet thickness should be on the order of the ion-gyro radius has been put forward by Drake et al. (1997).

Some of the reconnection experiments show evidence for energetic particle acceleration (Chapter 13). Stenzel and Gekelman (1984) observed electron distribution functions with nonthermal tails in their experiment, which is perhaps not surprising given the large mean-free path for electron–ion collisions there. However, these tails do not appear to be caused by runaway electrons, since the electric field in their experiments remains well below the Dreicer field. Both Gekelman et al. (1982) and Altyntsev et al. (1990) have reported ion acceleration in their experiments as well. In both cases the acceleration appears to result from turbulent fields within the current sheet, but the nature of the turbulence has not been fully explored.

In many of the experiments the current flow is terminated by some kind of disruption. The exact cause of the disruption is often not easy to identify, and it is probably not the same in all devices. In the LCD the disruption is associated with the formation of electrostatic double layers (Stenzel et al., 1982b; 1983), but for other devices alternative mechanisms have been proposed (Bratenahl and Yeates, 1970; Altyntsev et al., 1989; Velikanova et al., 1993).

9.2.2 Types of Reconnection

The idea that different boundary and initial conditions lead to different types of reconnection has been a central theme of the previous chapters, so it is natural to ask what types of reconnection occur in laboratory experiments. Generally, the operating time of the experiments is anywhere from 10 to 100 times longer than the Alfvén travel-time, so the closed nature of the vessels in which the experiments are carried out has a strong effect on the reconnection process. The exception is the LCD, where the Alfvén and operating times are both are on the order of 10^{-4} s. Thus, except for the LCD, high-speed outflow from reconnection tends to be blocked by the vessel walls. Nevertheless, rapid reconnection still occurs, probably because of the turbulence (Altyntsev et al., 1986, 1989) or kinetic turbulence (Gekelman and Stenzel, 1984) within the current sheet.

Even though the LCD operates in the collisionless regime, many aspects of its behaviour can be understood in terms of MHD theory or electron MHD theory. The LCD has external coils which produce a solenoidal magnetic field, but within the plasma chamber transverse, reconnecting fields are produced by pulsing current along the two plates, as indicated in Fig. 9.18. When the LCD is configured so as to create a current sheet which is thin enough to tear (§6.2), tearing does indeed occur, as shown in Fig. 9.19. Since the ions in the LCD are not frozen to the field lines, the process corresponds to an electron tearing mode. Despite the fact that Hall currents and whistler-wave turbulence are present, the growth-rate of the tearing, as shown in Fig. 9.20, is nearly the same as that predicted by the simple formula for the electron tearing mode (e.g., Laval et al., 1966; Cross and Van Hoven, 1976).

Claims to have detected slow-mode shocks and Petschek-type reconnection in the laboratory were published by the DIPD experimental team in the 1970s (Bratenahl and Yeates, 1970; Bratenahl and Baum, 1976; Baum and Bratenahl, 1976, 1977). However, these claims have not generally been accepted because of the difficulties involved in this experiment with obtaining accurate diagnostics of the magnetic field. Figure 9.21 shows a schematic of the DIPD along with the magnetic field geometry expected to be produced by the currents driven through the two rods that connect the upper and lower

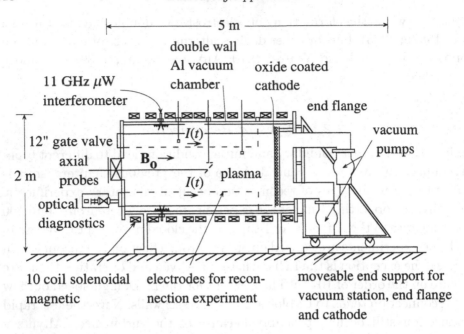

Fig. 9.18. Schematic diagram of the LCD reconnection experiment (from Stenzel and Gekelman, 1981a).

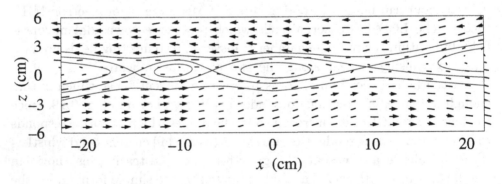

Fig. 9.19. Magnetic field vectors (arrows) and inferred flux surfaces (contours) in the LCD experiment, showing the magnetic structures produced by the electron tearing mode (from Stenzel, 1992).

walls of the cylindrical chamber. Although direct measurements of the magnetic field in the device were made at various locations within the chamber, the number of locations was relatively small (135 points in a 9 by 15 array), and the locations were all in a single quadrant of the mid-plane between the end-walls of the device. These rather sparse field measurements were then used to calculate a current-density distribution under the assumption that

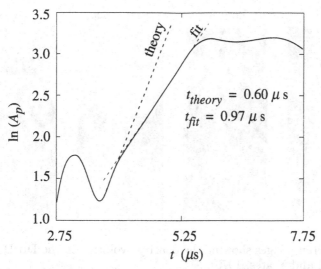

Fig. 9.20. Growth of the wave amplitude (A_p) as a function of time in the LCD experiment. The slope of the dashed line labelled "theory" is for the two-dimensional electron tearing mode (after Gekelman and Pfister, 1988).

(a)   (b)

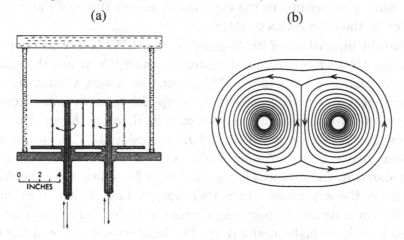

Fig. 9.21. The DIPD reconnection experiment of Baum and Bratenahl (1976): (a) schematic diagram of the apparatus (from Bratenahl and Yeates, 1970) and (b) corresponding magnetic flux surfaces calculated by Green (1965) for a similar current configuration.

the field is essentially two-dimensional and symmetrical. However, from the LCD results, we now know that such assumptions are unlikely to be valid. Furthermore, the relatively large ion inertial length of the DIPD means that Hall currents are significant, and additional complications arise from the fact that the magnetic wave-fronts propagating outward from the current rods also ionize the gas through which they pass. Thus, the claim that the

4.5 m sec   5.6 m sec   6.6 m sec   7.6 m sec

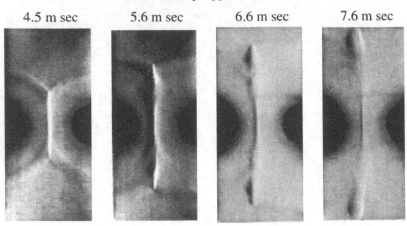

Fig. 9.22. Schlieren images showing the density evolution in the DIPD experiment (after Bratenahl and Yeates, 1970).

current-density structures in the experiment match those of the Petschek reconnection theory remains doubtful.

Despite the limitations of its diagnostics, the DIPD is still an interesting experiment. Of all the experiments listed in Table 9.2, it has the smallest collisional mean-free path $(3 \times 10^{-3} L_e)$ and the lowest plasma $\beta (10^{-4})$. The small mean-free path implies that, except for possible Hall effects, the plasma will tend to behave like a collisional fluid, while the small value of β implies that gas pressure will be negligible outside the thin current sheet that forms between the colliding fields. Indirect evidence of this current sheet is shown in the schlieren images in Fig. 9.22, which show the density variations in the experiment integrated along a line of sight from above. Where the two outward-propagating, circular wave-fronts meet one another, there is a prominent high-density ridge. The location and extent of this ridge are roughly similar to those predicted by Green (1965) for the quasi-static collision of two circular flux systems [Fig. 9.21(b)]. However, as might be expected from an experiment which is far from being quasi-static, there are features which are not consistent with Green's solution. Firstly, the ridge at later times extends beyond the length predicted by Green's solution, and secondly, the ends of the density ridge are thicker than its centre. It may be, on the one hand, that the ridge is in fact a thin current sheet that is bifurcated at its ends as in the Petschek solution. On other hand, it could also be that these features are simply a consequence of the collision of the two circular wavefronts, especially since the density within the ridge is higher than expected for slow-mode shocks. Bratenahl and Yeates (1970) estimated the density in the regions at the ends of the ridge to be approximately a factor of 10 higher than in the upstream region, and this value is several times higher than the maximum density jump of 2.5 predicted by compressible Petschek theory (Soward and Priest, 1982; Forbes, 1986).

10

Magnetospheric Applications

Although the existence of the Earth's magnetic field has been known since ancient times, the fact that it does not extend indefinitely into space, but is confined to a cavity, is a discovery of the twentieth century. This cavity is created by the solar wind, which shapes the Earth's field into a comet-like structure as shown in Fig. 10.1. The region in which the Earth's field is confined is called the *magnetosphere*, a term coined by T. Gold in 1959 before the true shape of this region had become known (Gold, 1959). Upstream of the magnetosphere there is a detached *bow shock*, which occurs because the solar wind flows past the Earth at a speed approximately eight times greater than either the Alfvén or sound speeds. The bow shock is located about 15 R_E sunward of the Earth, where one Earth-radius (1 R_E) is about 6,370 km. In front of the bow shock is a region called the *fore shock*, which contains energetic particles and waves. In MHD terms, the bow shock and fore shock together constitute a fast-mode MHD shock, while the bow shock itself may be thought of as a subshock (Kennel, 1988), where the solar wind is compressed. The region downstream of the bow shock is called the *magnetosheath*, and it is here that the shocked solar wind is deflected around the magnetospheric cavity.

The concept of the magnetosphere was first introduced by S. Chapman and V. Ferraro (1931), who thought the Earth's magnetic field would be confined to a cavity whenever the Earth is impacted by a plasma cloud ejected from the Sun. They argued that plasma clouds would be produced after large solar flares and would occasionally hit the Earth and lead to a temporary confinement of the terrestrial magnetic field. The idea that the confinement might be permanent did not arise until L. Biermann (1951) and E. Parker (1958) began to argue for the existence of a permanent solar wind. Indisputable evidence that the solar wind and magnetosphere are indeed permanent features did not become available until Mariner II crossed the

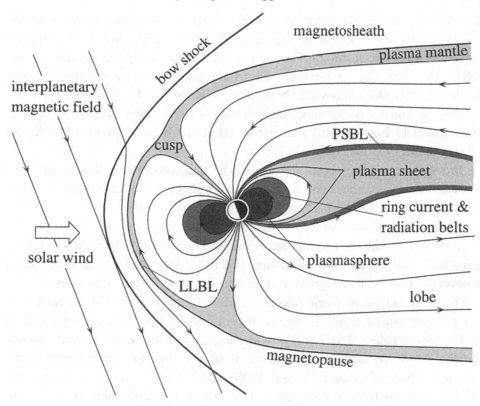

Fig. 10.1. Schematic cross-section of the terrestrial magnetosphere in the noon–midnight plane. LLBL and PSBL are acronyms for low-latitude boundary layer and plasma-sheet boundary layer, respectively (after Parks, 1991).

boundary of the magnetosphere and entered the solar wind on its way to Venus in 1962 (Neugebauer and Snyder, 1962).

The model proposed by Chapman and Ferraro in 1931 for the occasional confinement of the Earth's magnetic field contains two magnetic null points located in the *cusp regions* (Fig. 10.1). These null points occur on the magnetopause, the boundary of the cavity, and they are not the standard X-type points that we typically associate with reconnection. They are produced by an unmagnetized plasma flowing around the magnetosphere rather than by the interconnection between the terrestrial and interplanetary magnetic fields.

The shape and distance of the dayside magnetopause is determined primarily by a balance between the ram pressure of the solar wind and the magnetic pressure of the terrestrial field. Typically, the dayside magnetopause is located at about 10 R_E away from the Earth. In the early days of the space age, it was assumed that the nightside magnetopause would be determined by a balance between the magnetic pressure of the terrestrial field and the thermal pressure of the solar wind. This assumption led to predictions that the magnetosphere would be egg-shaped, with a nightside magnetopause about $50\,R_E$ away from the Earth (e.g., Johnson, 1960; Axford and Hines, 1961). However, the force which shapes the nightside magnetosphere is not predominantly the solar-wind thermal pressure but rather the Maxwell (and possibly viscous) stress acting on the flanks of the magnetosphere. This force pulls the field lines out into an extended tail, which is about $1{,}000\,R_E$ in length (Dungey, 1965; Hughes, 1995).

One of the most important aspects of the magnetospheric plasma environment is that it is almost completely collisionless (§11.7), except at its inner boundary, the ionosphere. This collisionless state occurs because of the extremely low density of the solar wind, which makes up the bulk of the plasma in the outer reaches of the magnetosphere. The rate of binary particle collisions is so low in the solar wind that the mean-free path for such collisions is several hundred times greater than the size of the magnetosphere.

Thus, the magnetospheric plasma exhibits many kinetic effects which cannot be understood solely in terms of a single-fluid formulation of resistive MHD. Nevertheless, MHD still provides a general framework for understanding the gross transport of momentum, magnetic flux, and energy within the magnetosphere (§1.7 and Parker, 1996). Furthermore, by generalizing the MHD equations to treat electron and ion fluids separately and to include an anisotropic pressure tensor, the range of phenomena that can be understood is greatly extended. The main limitation of the MHD formalism is that it cannot be used to calculate transport coefficients such as the electrical and thermal conductivities. A separate kinetic theory is required to calculate such quantities, whether the plasma is collisional or collisionless. A difficulty in using resistive MHD in a collisionless plasma arises from the fact that the kinetic processes are no longer localized by collisions but may be coupled to the global behaviour of the plasma in a nontrivial way. However, the validity of the MHD equations themselves does not depend on whether the kinetic processes are local or global.

The standard unit for measuring magnetic fields in MKS units is the tesla (T) and in CGS units is the gauss (G), where $1\,\mathrm{G} = 10^{-4}\,\mathrm{tesla}$, but in the solar wind typical field strengths are a few nanotesla (nT), where $1\,\mathrm{nT} = 10^{-9}\,\mathrm{tesla} = 10^{-5}\,\mathrm{G}$, and so this is the unit we shall often adopt in this chapter.

10.1 Dungey's Model of the Open Magnetosphere

Once it was realized that the solar wind is magnetised, Dungey (1961) proposed a new topological model for the magnetosphere which is quite different

(a) (b)

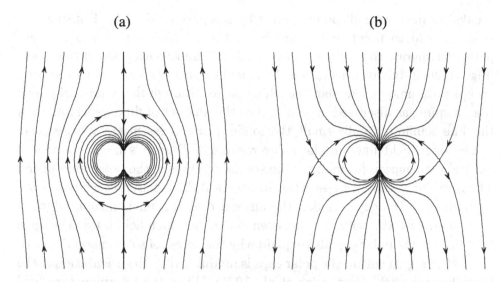

Fig. 10.2. Schematic closed and open magnetospheric topologies, obtained by combining a uniform magnetic field with a dipole field. In (a) the uniform field is parallel to the dipole field, while in (b) the uniform field is antiparallel.

from that of Chapman and Ferraro. Dungey's model also has two magnetic null points, but, unlike Chapman and Ferraro's model, these null points are formed by the interconnection between the terrestrial field and the field of the solar wind. The crucial point about Dungey's model is that it contains magnetic field lines which map into interplanetary space and thus back to the Sun. By contrast, all of the field in the Chapman and Ferraro model maps from one hemisphere to the other. Thus, their model is referred to as closed, while Dungey's model is referred to as open.

In Dungey's model, one neutral point occurs on the dayside magnetopause and the other occurs on the nightside. The amount of reconnection depends strongly on the North–South orientation of the interplanetary magnetic field of the solar wind. The field at the dayside nose of the magnetosphere is oriented northward, so when the interplanetary field is southward the field geometry is favourable for the type of reconnection that generates new open flux. When the interplanetary field is northward, some reconnection is possible, but it only reconfigures flux that is already open.

The effect of the orientation of the interplanetary field is illustrated in Fig. 10.2, which shows the magnetic field configurations that occur when a purely uniform field is combined with a dipole field. When the equatorial field of the dipole's North–South axis is exactly opposite to the uniform field, there is a magnetic neutral line encircling the dipole's equator [Fig. 10.2(b)]. As soon as the uniform field is tilted in any direction, this neutral line breaks up into two null points joined by a separator (§8.4.1). Rotating the uniform field to point in the purely northward direction causes the null points to migrate to positions above the North and South poles of the dipole [Fig. 10.2(a)]. In this orientation the magnetosphere is confined to a sphere, except for a unique, singular line which passes through the polar null points (i.e., a spine, as discussed in §8.2). Since the amount of flux associated with this line is infinitesimally small, the configuration is essentially a closed one.

The open field lines in the polar regions of Dungey's model provide a natural explanation for many of the special features of the polar ionosphere. The most important of these is the auroral oval that encircles each polar cap. According to Dungey's model, the auroral oval is a phenomenon which is associated with the boundary between closed and open field lines. The open field lines in the polar cap also explain why the access of solar energetic particles to the region within the polar caps is modulated by the orientation of the interplanetary field (Van Allen et al., 1971). When the interplanetary field is southward, particles can penetrate into the polar regions, but prolonged periods of northward interplanetary field cut off the entry of these particles.

Because of the direct connection between the polar-cap field and the interplanetary field, the orientation of the interplanetary field has a direct influence on the orientation of the surface fields in the polar caps (Svalgaard, 1973; Mansurov, 1969). This influence, which is called the Svalgaard–Mansurov effect, is a rather subtle one, because the polar-cap fields (6.4×10^{-5} tesla $= 0.64$ G) are about four orders of magnitude greater than the interplanetary field (typically $\lesssim 10$ nT $= 10^{-4}$ G). Nevertheless, the effect is strong enough to allow a rough estimate of the orientation of the interplanetary field in space. This has made it possible to use ground-based observations to infer the behaviour of the interplanetary field over time-periods stretching as far back as 150 years ago.

10.1.1 Length of the Geomagnetic Tail

The magnetic connection between the terrestrial and interplanetary magnetic field has important dynamical consequences for the magnetosphere. The polar-cap field lines which are connected to the solar wind are stretched into an extended tail. How far back they extend depends on the speed of the solar wind and the rate at which the field lines reconnect at the dayside and nightside null points. The solar-wind drag on the field lines also sets up a reverse flow near the Earth, which returns field lines to the dayside magnetopause. This flow in turn creates field-aligned currents mapping from the magnetosphere to the polar caps.

Dungey's open model provides a simple explanation for the existence of a long geomagnetic tail (Dungey, 1965). The tail is created by the Maxwell stresses caused by the open magnetic field lines that cross the magnetopause. The geomagnetic tail length (L_{tail}) can be estimated by relating the position of a field line in the polar cap to its corresponding position in the solar wind at the bow shock (Hughes, 1995). For a steady-state situation, the time that a field line spends in the polar cap is just the length (L_{pc}) of the polar cap divided by the velocity (V_{pc}) at which field lines are convected across the polar cap from the dayside to the nightside. This time must be approximately the same as the time it takes for the other end of the field line in the solar wind to move through the distance L_{tail}. Hence

$$\frac{L_{\text{pc}}}{V_{\text{pc}}} \approx \frac{L_{\text{tail}}}{V_{\text{sw}}}, \tag{10.1}$$

where V_{sw} is the velocity of the solar wind. The velocity of convection in the polar cap is just the polar-cap electric field divided by the average polar-cap magnetic field (B_{pc}). In terms of the polar-cap potential (Φ_{pc}) this velocity is

$$V_{\text{pc}} = \frac{\Phi_{\text{pc}}}{L_{\text{pc}} B_{\text{pc}}}. \tag{10.2}$$

The polar-cap potential is related to the electrical potential applied by the solar wind across the width of the magnetosphere. This width is essentially the width of the tail (W_{tail}). If all of the solar-wind field lines that intercept the magnetopause became reconnected, then the applied potential would be $V_{\text{sw}} W_{\text{tail}} B_{\text{sw}}$, where B_{sw} is the magnetic field in the solar wind. However, most field lines brought up to the nose of the magnetosphere bend and slip to the side rather than reconnect. Thus, only a fraction (f) of the intercepted field lines become connected to the polar cap. Consequently, the cross-cap potential is related to the solar-wind potential (Φ_{sw}) by

$$\Phi_{\text{pc}} = f \, \Phi_{\text{sw}} = f \, V_{\text{sw}} W_{\text{tail}} B_{\text{sw}}. \tag{10.3}$$

Combining Eqs. (10.1)–(10.3) leads to

$$L_{\text{tail}} = \frac{B_{\text{pc}} L_{\text{pc}}^2}{f \, W_{\text{tail}} B_{\text{sw}}} \tag{10.4}$$

for the length of the geomagnetic tail.

The fraction (f) depends on the rate of reconnection, and, as we have already mentioned, this cannot be determined without considering a full analysis of the plasma dynamics at the reconnection site. However, the polar-cap velocity (V_{pc}) can be directly measured by satellites flying over the polar cap, and from this measurement one can deduce that f is about 0.1. That is, only about 10% of the electric potential in the solar wind is transferred into the polar cap. Setting $B_{sw} = 5\,\text{nT}\,(= 5 \times 10^{-5}\,\text{G})$, $L_{pc} = 0.52\,R_E = 3{,}300\,\text{km}$, $f = 0.1$, $W_{tail} = 40\,R_E$, and $B_{pc} = 0.64\,\text{G}$ gives

$$L_{tail} = 860\,R_E$$

for the length of the magnetotail, where R_E is the radius of the Earth. The above value of L_{tail} corresponds to a polar-cap convection speed of $230\ \text{m s}^{-1}$ and a cross-cap potential of $51\,\text{kV}$, both of which are consistent with the observed values. Dungey's formula is really a lower limit on the tail length because it does not take into consideration the time required for disconnected field lines in the tail to catch up with the solar wind.

10.1.2 Effect of the Interplanetary Magnetic Field

One of the most important aspects of Dungey's model is that it predicts that the whole structure of the magnetosphere is strongly modulated by the North–South direction of the interplanetary magnetic field. When the interplanetary field turns southward, the kinetic energy of the solar wind is more efficiently tapped, and the convection and currents within the magnetosphere increase. Confirmation of this prediction by Fairfield and Cahill (1977), Schatten and Wilcox (1967), Arnoldy (1971), and Foster et al. (1971) was instrumental in the general acceptance of Dungey's model.

The strong correlation between the North–South direction of the interplanetary field and the currents flowing in the auroral ionosphere is shown in Fig. 10.3, which presents a superposed epoch analysis of 86 isolated substorms (see §10.5). AE stands for the "auroral electrojet" index, which is a measure of the ionospheric currents flowing in the auroral zone that encircles the northern polar cap (McPherron, 1995). As Dungey's model predicts, the auroral ionospheric currents increase when the interplanetary magnetic field points southward and decrease when it points northward. There is approximately a 30-min delay between the southward turning of the field and the increase in the currents, and this delay is due to the time it takes the magnetosphere to respond globally to the enhanced energy input caused by the reconnection. The overall shape of the curves in Fig. 10.3 will be discussed in Section 10.5 when we consider magnetospheric substorms.

10.2 Dayside Reconnection

Even though the reconnection process in Dungey's open model of the magnetosphere starts on the dayside magnetopause, most early work focused on

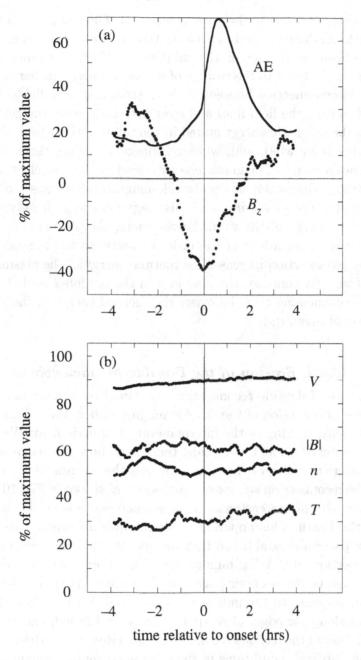

Fig. 10.3. Comparison of the AE index, measuring the current flowing in the auroral ionosphere, with (a) the North–South component (B_z) of the solar-wind magnetic field and (b) the solar-wind speed (V), field magnitude ($|B|$), density (n), and temperature (T). The data show the time-signatures obtained by averaging 86 isolated substorms that are synchronized with respect to their onset time. Only B_z shows a correlation with AE (from Foster et al., 1971).

the reconnection process in the geomagnetic tail. The main reason for this is that particle acceleration and plasma heating at the dayside magnetopause are not as obvious as they are in the tail (Cowley, 1980). Reconnection at the dayside primarily leads to the storage of magnetic energy rather than its release. The interconnection between the terrestrial and solar fields allows the solar wind to drag the field lines and stretch them to form the geomagnetic tail. Thus, the magnetic energy stored in the tail is derived from the kinetic energy of the solar wind, and, when reconnection occurs there, it releases the enormous amount of magnetic energy stored in the stretched field lines.

In contrast, reconnection at the dayside magnetopause taps the magnetic energy stored in the solar wind, and this magnetic energy is only about 1% of the kinetic energy of the wind. Furthermore, the plasma β of the solar wind and magnetosheath is of the order of unity, so the magnetic energy released by reconnection increases the thermal energy of the plasma only by a factor of two. By contrast, the plasma β in the tail lobes is of the order of 10^{-3}, and reconnection there increases the thermal energy of the plasma by three orders of magnitude.

10.2.1 Erosion of the Dayside Magnetosphere

Early evidence of dayside reconnection was found by Aubrey et al. (1970), who observed the erosion of the dayside magnetosphere immediately following a southward turning of the interplanetary magnetic field. When reconnection is switched on at the dayside, there is no flow of plasma within the magnetosphere to transport field lines towards the reconnection site. Consequently, the reconnection site moves earthward, as shown in Fig. 10.4. As the reconnection site moves earthward, the magnetopause is eroded and moves closer to the Earth. This apparent motion stops only when a steady-state convection pattern is established that transports field lines towards the dayside reconnection site. A few minutes after the onset of reconnection at the magnetopause, the ionospheric plasma in the polar-cap region starts to move tailward in response to the pull of the solar wind. Within about 30 min, a return flow along the edges of the polar cap is established, and this flow returns field lines to the dayside reconnection site (Holzer and Reid, 1975; Reid and Holzer, 1975). If conditions in the solar wind remain roughly constant, then a steady state may be established and the erosion process stops.

The time required to establish a steady state depends on the inertial time-scale of the ionosphere and the wave travel-time between the ionosphere and the magnetopause. Most of the delay in establishing a steady state is due to the wave propagation time across the system. Since this time is large

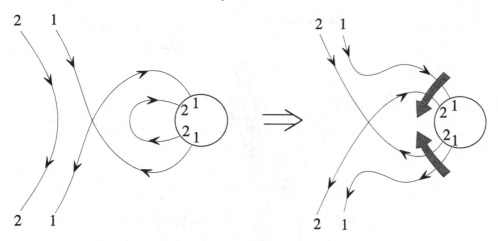

Fig. 10.4. Schematic diagram of the erosion of the dayside magnetosphere caused by the onset of magnetic reconnection at the nose of the magnetosphere (Forbes, 1988).

compared with the time-scale of fluctuations in the solar wind, a true steady state is rarely established. According to Coroniti and Kennel (1973), the inertial time-scale of the ionosphere is only on the order of 5–10 min. Unlike the solar photosphere (see Chapter 11), the ionosphere is not sufficiently massive to offer more than a momentary resistance to the forces set up by reconnection, and thus field lines there are not as inertially line-tied as they are at the surface of the Sun.

10.2.2 Quasi-Steady Reconnection

Additional indirect evidence of reconnection at the dayside magnetosphere was found by Sonnerup (1971a) shortly after Aubrey's discovery of dayside erosion. Using a minimum-variance analysis of magnetometer data from Explorer 12, Sonnerup found that there is a magnetic field component normal to the magnetopause. In a closed magnetosphere the normal component at the magnetopause is zero, but in an open magnetosphere the normal component cannot be zero, except at the location of a null point.

The presence of a magnetic field component which penetrates the magnetopause means that the magnetopause is not simply a tangential discontinuity as it would be in a closed model. Instead, it is a rotational discontinuity combined with a narrow slow-mode expansion downstream (Levy et al., 1964; Sonnerup, 1984). This configuration differs from the familiar one for symmetric reconnection (i.e., Petschek, 1964), because the plasma regimes on the two sides of the magnetopause are quite different. On the magnetosheath

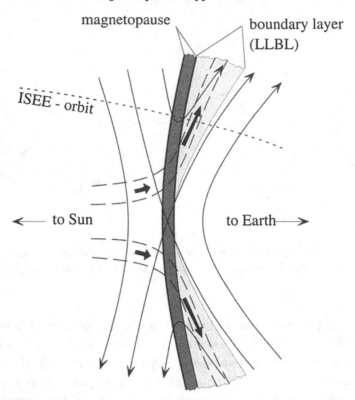

Fig. 10.5. Structure of a dayside reconnection site in the meridional plane as inferred from eleven crossings of the magnetopause by the ISEE-1 and ISEE-2 satellites (after Sonnerup et al., 1981).

side, the magnetic field is weak and the plasma pressure is strong, while
on the magnetospheric side the opposite is generally true (Fig. 10.5). To
accomplish the transition from one side to the other requires the combination of a rotational discontinuity and a slow-mode expansion fan (Yang
and Sonnerup, 1977). Across the rotational discontinuity the magnetic field
and plasma pressure are constant, but these quantities change within the
expansion region so as to match the external conditions. The flow through
the rotational discontinuity is at the Alfvén speed based on the component
of the magnetic field normal to the magnetopause.

A boundary layer is in fact observed on the interior side of the magnetopause, but it is more complex than the simple expansion fan predicted
by the Levy et al. (1964) model. As one approaches the dayside cusps, the
boundary layer becomes quite thick and actually extends downward like a
funnel to the dayside ionosphere. Outside the cusp regions, the boundary
layer slowly expands as the plasma in it flows ever tailward. By a distance

of about 400 R_E on the nightside of the Earth, the boundary layer expands to the point where it completely fills the tail and thus ceases to be a layer (see Fig. 10.10 below).

The first direct evidence of the large-scale global reconnection at the dayside that Dungey had predicted was found by Paschmann et al. (1979). This team constructed a pair of high-time-resolution plasma detectors on the ISEE-1 and ISEE-2 satellites that were able, for the first time, to make detailed measurements of the plasma flow and magnetic field during a magnetopause crossing. Such crossings are nearly always very rapid, because they typically occur when the magnetopause is responding to changes in the magnetic field direction or ram pressure of the solar wind. By analyzing the observations within the framework of MHD reconnection, Sonnerup et al. (1981) were able to show that global reconnection does indeed exist at the magnetopause. These observations put to rest claims by some researchers that the required signatures of reconnection at the dayside magnetopause are not present, and that Dungey's model must therefore be wrong (Heikkila, 1978; Haerendel et al., 1978).

Figure 10.5 shows the structure of dayside reconnection site that was inferred by Sonnerup et al. (1981). The accelerated outflow from the reconnection site lies entirely within the boundary layer just inside the magnetopause. A typical separator field line (§8.4.1) inferred by Sonnerup et al. (1981) is shown in Fig. 10.6, which is a view looking from the Sun towards the nose of the magnetopause. At the times of the ISEE crossings, the interplanetary field is usually rotated southward from the ecliptic plane, and thus the separator line is rotated relative to the ecliptic plane. Along the separator the reconnecting fields are not generally antiparallel. Sonnerup et al. (1981) assumed that it is only the component of the field perpendicular to the separator that actually reconnects. Consequently, this type of reconnection is sometimes referred to as *component merging* (Sonnerup et al., 1981; Soward, 1982). Simple quantitative models for predicting the location of the separator line have been developed by Voight (1978), Alekseyev and Belen'kaya (1983), and Crooker et al. (1990).

As an alternative to the concept of a global separator line with component merging, Crooker (1979) proposed a model with a broken separator line as shown in Fig. 10.7. This broken configuration results if reconnection occurs only at locations where the interplanetary field is antiparallel to the magnetospheric field. An attractive consequence of this *antiparallel merging* hypothesis is that the energy transfer into the magnetosphere drops precipitously when the southward component of the interplanetary field disappears. Statistical studies (Luhmann et al., 1984; McPherron, 1995) show that

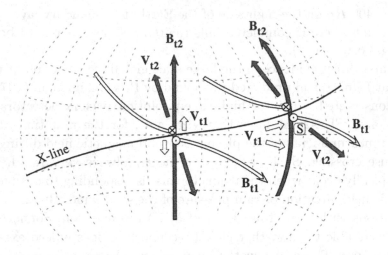

Fig. 10.6. Front view of the reconnection site at the dayside magnetopause when there is a substantial positive y-component in the solar-wind magnetic field. Letters with a subscript 1 refer to parameters in the magnetosheath, while those with a subscript 2 refer to parameters inside the magnetosphere. The subscript t indicates a tangential component, and the letter S shows the approximate location of the two ISEE satellites, which are typically separated by less than 2,000 km (after Sonnerup et al., 1981).

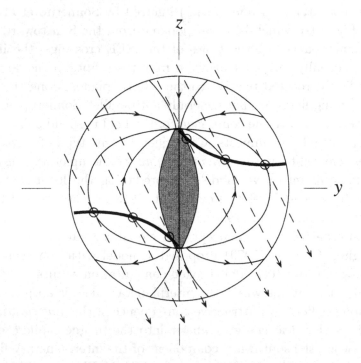

Fig. 10.7. Schematic view from the Sun of the dayside magnetopause. Field lines attached to the Earth (thin solid lines) form two cusp-like configurations. The thick solid lines are the loci of points where the interplanetary magnetic field lines (dashed lines) are antiparallel to the terrestrial field lines (from Crooker, 1979).

the energy transfer to the magnetosphere behaves like a half-wave rectifier in response to the North–South orientation of the interplanetary magnetic field. When the field is northward, the energy input is small and relatively constant, but when the field is southward, the energy input increases in proportion to the strength of the field. The antiparallel merging hypothesis reproduces this behaviour, because the reconnection sites move onto field lines mapping into the tail when the interplanetary field becomes northward. For a northward interplanetary field the antiparallel field in the magnetosphere lies tailward of the dayside cusps where field lines map into the tail. Reconnection still occurs in this case, but the energy input is greatly reduced. By contrast, the component merging hypothesis produces a fall-off in energy input which decreases smoothly as $\sin(\theta/2)$, where θ is the rotation angle from due North ($\theta = 0$) of the North–South component of the interplanetary magnetic field.

Simulations provide a way to test which, if either, of the above hypotheses is correct. Using three-dimensional MHD simulations, Fedder et al. (1991), Crooker et al. (1998) and White et al. (1998) have found that steady dayside reconnection exhibits some aspects predicted by both hypotheses, but that neither really provides a satisfactory description. This is because the reconnection of field lines occurs over the entire expanse of the dayside magnetopause and cannot be adequately characterized by considering what happens only in the vicinity of the separator line. Component-type merging does occur, but it is strongly modulated by how anti-parallel the fields are at any given location and does not depend simply on the situation at the separator line.

The antiparallel merging hypothesis appears to work best when the interplanetary field is northward, and it provides some understanding of why the energy transfer behaves like a half-wave rectifier in both the simulations and the real magnetosphere. However, the component merging hypothesis works better and provides a more accurate description of the flows when the field is southward.

10.3 Flux Transfer Events

Although Dungey's dayside reconnection site was confirmed by the observations of Paschmann et al. (1979), the number of magnetopause crossings which showed global reconnection were much smaller than expected and the reconnection was rarely steady. In an independent search for reconnection at the dayside magnetopause, Russell and Elphic (1978) noticed that there appear to be localized, transient reconnection events which are quite different from Dungey's steady-state picture. These events last only about a minute and have a magnetic field signature which suggests that there is some

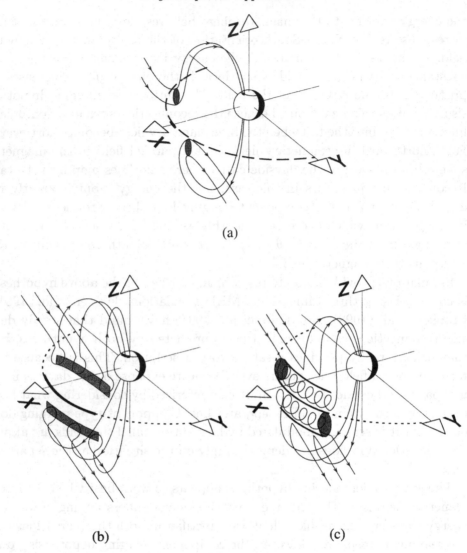

(a)

(b)   (c)

Fig. 10.8. Three models for FTEs: (a) the original flux-tube model of Russell and Elphic (1978), (b) the impulsive two-dimensional model of Scholer (1988a) and Southwood et al. (1988), and (c) the multiple X-line reconnection model of Lee and Fu (1985). The grey-shaded areas indicate the cross-sections of the reconnected field lines at the surface of the magnetopause (after Lockwood et al., 1990).

kind of bubble or tube-like disturbance moving along the magnetopause at a speed higher than the flow in the magnetosheath. Russell and Elphic (1978) interpreted these events as isolated tubes of magnetic flux interconnecting the interplanetary and magnetospheric fields, as shown in Fig. 10.8(a). They called them *flux-transfer events*, or FTEs for short.

Several observational studies have revealed various properties of FTEs. Saunders et al. (1984a) determined that the tube radius of a typical flux transfer event is on the order of one Earth radius ($1\,R_E$) at the point where the tube penetrates the magnetopause. The occurrence of FTEs is strongly correlated with the direction of the North–South component of the interplanetary field, with virtually no FTEs present when the field is purely northward (Rijnbeek et al., 1982, 1984; Southwood et al., 1988; Lockwood, 1991). Within the tube of the FTE the magnetic field has a strong axial component, but there is often an azimuthal component as well (Paschmann et al., 1982). Thus, the interior field is typically helical, as in a flux rope. The direction of twist depends on the North–South locations and the sign of B_y (the lateral component, as indicated in Fig. 10.8). Saunders et al. (1984b) showed that in some events the twist appears to propagate as an Alfvén wave. Studies of the velocity field within and around FTEs have revealed that the tubes are moving faster than the flow in the magnetosheath (see, e.g., Papamastorakis et al., 1989). The observed speeds in the rest frame of the magnetosheath are typically about 80% of the local Alfvén speed, which suggests that the outflow from the reconnection site is fast, as in Petschek's model (see §4.3), rather than slow, as in the Sweet–Parker model.

The origin of the internal twist was a key question in the 1980s and was the main motivation for the Lee and Fu model. Wright (1987) showed that the Russell and Elphic model would naturally produce a half-twist that propagates away as an Alfvén wave. Southwood et al. (1988) realised how an extended reconnection line can produce a structure with apparently more twist. Magnetic helicity calculations (Wright and Berger, 1989; Song and Lysak, 1989) put quantitative limits on the direction and amount of twist from all the reconnection models.

Three types of reconnection model have been proposed for FTEs. The first type, proposed by Russell and Elphic (1978), is that FTEs are reconnected flux tubes penetrating the magnetopause, as illustrated in Fig. 10.8(a). The second type, proposed by Scholer (1988a) and Southwood et al. (1988), produces FTEs by episodic two-dimensional reconnection, as shown in Fig. 10.8(b). Finally, the third type, proposed by Lee and Fu (1985), assumes that an FTE is a two-dimensional magnetic island that lies along the surface of the magnetopause as shown in Fig. 10.8(c). In this model the magnetic island is not isolated but is part of an interconnected chain of two-dimensional X-points and O-points. Thus, it is often referred to as multiple X-line reconnection, or MXR for short. The main limitation of the MXR model is that its two-dimensional geometry does not really fit the observations of typical FTEs very well. One of the most attractive aspects of the MXR model is that it is much easier to analyse theoretically. This tractability has made it possible to show that the shear flow at the magnetopause can produce a Kelvin–Helmholtz instability that interacts via tearing to produce patchy reconnection (La Belle-Hamer et al., 1987; Truemann and Baumjohann, 1997).

Wright (1999) has proposed that the two single X-line models [Figs. 10.8(a) and 10.8(b)] may be viewed as extremes of the same process. If reconnection occurs at a suitably long X-line for a short time, the quasi-two-dimensional model of Fig. 10.8(b) is appropriate, but if the reconnection occurs at a short X-line, then a configuration like that of Fig. 10.8 is the result.

At the present time, there is very little quantitative theory for a time-dependent reconnection model with an X-line of finite length. Such a model is inherently three-dimensional, which makes it quite difficult to analyse or simulate numerically. The basic question of why isolated flux tubes should form in the first place has never been fully answered, but some effort has been made to model the dynamics of their propagation within the magnetopause. Sonnerup (1987) showed how an internal twist can be produced and he analysed the forces acting on a kinked flux tube penetrating the magnetopause. Because the tubes are created by reconnection at the magnetopause, they have a sharp kink in them at that location. As in any reconnecting current sheet, this kink produces a strong $\mathbf{j} \times \mathbf{B}$ force which tends to accelerate the plasma. Very roughly, the kink force ($\mathbf{F}_{kink}$) is

$$\mathbf{F}_{kink} = \frac{\mathbf{B}_2 - \mathbf{B}_1}{\mu} B_n \pi a^2, \tag{10.5}$$

where $\mathbf{B}_2$ is the magnetospheric field, $\mathbf{B}_1$ is the magnetosheath field, B_n is the component of the field within the flux tube that is normal to the magnetopause, and a is the radius of the flux tube. As the kink force accelerates the flux tube, the magnetic field ($\mathbf{B}_{mp}$) lying within the plane of the magnetopause is compressed at the forward edge of the accelerating tube, as shown in Fig. 10.9(a). This compression exerts a counterforce ($\mathbf{F}_{comp}$), acting in a direction opposite to $\mathbf{F}_{kink}$ of

$$\mathbf{F}_{comp} = l_{mp} \, \mathbf{I}_{hole} \times \mathbf{B}_{mp}, \tag{10.6}$$

where l_{mp} is the thickness of the magnetopause and $\mathbf{I}_{hole}$ is the current associated with the compression of the field at the location of the hole.

Within a few seconds after the formation of a reconnected flux tube, the kink and compression forces tend to balance each other, so that

$$\mathbf{F}_{kink} + \mathbf{F}_{comp} = \mathbf{0}. \tag{10.7}$$

However, this balance cannot be achieved or maintained if the kink force is too strong. As $\mathbf{F}_{kink}$ increases, the surface current ($\mathbf{I}_{hole}$) must increase as

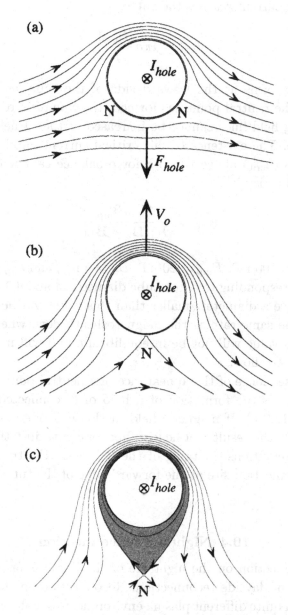

Fig. 10.9. Three alternatives for the deformation of magnetopause field lines by an FTE flux tube with a current I_{hole}. The circle is the cross-section of the tube in the plane of the magnetopause [see Fig. 10.8(a)]. In (a) the force caused by the kink in the flux tube at the magnetopause is counterbalanced by the compression of the magnetopause field lines, and the FTE convects with the ambient magnetosheath plasma. In (b) I_{hole} equals the critical value at which the neutral points N in (a) meet, while in (c) I_{hole} exceeds the critical value and creates a detached neutral point (after Sonnerup, 1987).

well. When this current reaches the value

$$I_{\text{crit}} = 4\pi a \frac{|B_{\text{mp}}|}{\mu},$$ (10.8)

a neutral point forms at the leeward side of the tube, as shown in Fig. 10.9(b). Once the neutral point has formed, the counter-stress produced by the surrounding field lines cannot be increased further if field lines rapidly reconnect there. The existence of this critical current implies that there is a critical radius (a_{crit}) above which a force-balance cannot be maintained. The critical radius is

$$a_{\text{crit}} = \frac{4\,l_{\text{mp}}B_{\text{mp}}^2}{|B_n(\mathbf{B}_2 - \mathbf{B}_1)|}.$$ (10.9)

For $|\mathbf{B}_2 - \mathbf{B}_1| = 100$ nT, $B_{\text{mp}} = 20$ nT, $B_n = 5$ nT, and $l_{\text{mp}} = 10^3$ km, a_{crit} is 3,200 km, corresponding to a flux-tube diameter of about 1 R_E. Thus, flux tubes which have a diameter smaller than this value will be in equilibrium and move at the same rate as the magnetosheath flow, whereas flux tubes larger than this value will not be in equilibrium and will move faster than the magnetosheath flow.

Another consequence of the appearance of a neutral point at the leeward side of the tube is the formation of a halo of disconnected field lines, as shown in Fig. 10.9(c). If magnetic field in this halo merges with the field in the flux tube, the result is a helical flux rope with just the right sign to explain the observations. Merging of the halo with the tube requires that there be a reconnection site at the forward edge of the tube.

10.4 Nightside Reconnection

Magnetic reconnection on the nightside of the Earth is far from being the mirror process of dayside reconnection. Reconnection on the nightside not only occurs in a quite different plasma environment, but the whole energetics of the process is different as well. The important aspect of dayside reconnection is that it converts the kinetic energy of the solar wind into magnetic energy stored in the stretched field lines of the tail. At the dayside, the small amount of solar-wind magnetic energy which is converted to heat and kinetic energy is relatively unimportant. In contrast, for nightside reconnection, it is exactly this process, namely the conversion of stored magnetic energy into thermal and kinetic energy, that is important.

10.4.1 Morphology

The different role of reconnection on the nightside of the magnetosphere is reflected in the structure of the geomagnetic tail, as shown in Figs. 10.1 and 10.10. The northern and southern polar-cap fields are stretched out by the solar wind into two long flux tubes that are separated by a current sheet flowing across the mid-plane of the tail. This current sheet is imbedded in a region of relatively hot (1 keV) plasma, called the tail *plasma sheet*. In the reconnection model of the tail, the plasma sheet corresponds to the outflow region from a distant, cross-tail X-line lying at a distance of approximately 100–150 R_E down the tail where the plasma has been heated by the reconnection process. The plasma sheet extends earthward to about 6 R_E (i.e., synchronous orbit) and has a thickness in the near-Earth region of about 8 R_E.

Above and below the plasma sheet lie the *tail lobes,* which contain a cold plasma (50 eV) with a plasma beta (β) on the order of 10^{-3} compared with β of the order of unity in the plasma sheet. As one moves down the tail, the lobes gradually become filled with a tailward-streaming plasma that comes from the leakage of the magnetosheath plasma along the open field lines, as shown in Figs. 10.10 and 10.11. In the near-Earth region this leakage of plasma constitutes a boundary layer called the *plasma mantle*, but as one moves down the tail the thickness of the layer increases until it fills the tail. In MHD models (e.g., Coroniti and Kennel, 1979; Siscoe and Sanchez, 1987) the plasma mantle is an extension of the dayside expansion fan (§10.2.2), which continually increases in thickness with distance away from the dayside X-line. Figure 10.11 shows the relation of the plasma mantle to the magnetopause and the open field lines crossing it. Along the region where the field is parallel to the boundary, the magnetopause forms a tangential discontinuity instead of a rotational one. The MHD models of the mantle are useful, but to understand the observed particle distributions it is necessary to consider time-of-flight effects (Cowley and Southwood, 1980; Liu and Hill, 1990).

At somewhere near 400–500 R_E down the tail, there is a critical location where the speed of the tailward flow exceeds the ambient Alfvén speed (Cowley, 1984). This critical location represents an upper limit to how far down the tail the X-line may be located. Because the outflow from the X-line in a frame moving with the X-line cannot exceed the ambient Alfvén speed, field lines cannot return to the dayside magnetopause, as they must, unless the X-line is earthward of the critical location.

On either side of the plasma sheet proper lies the *plasma-sheet boundary layer,* or PSBL for short. Within this layer, energetic ions and electrons

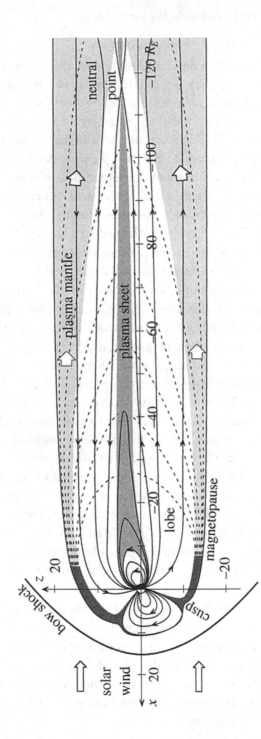

Fig. 10.10. Noon–midnight cross section of the magnetosphere and geomagnetic tail drawn to scale. The orbit of the moon is at 60 R_E, while the location of the tail neutral point during quiet conditions is at 115 R_E. Dashed lines indicate the drift paths of particles of different energies as they convect towards the plasma sheet (after Pilipp and Morfill, 1978).

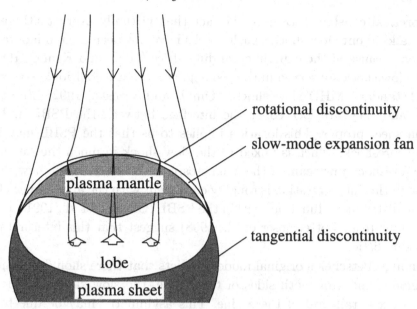

Fig. 10.11. Cross-sectional structure of the geomagnetic tail as predicted by the MHD model of Siscoe and Sanchez (1987). The high-latitude boundary along the plasma mantle consists of a rotational discontinuity and a variable-thickness slow-mode expansion fan, which touch at the pole and at their juncture with the tangential discontinuity that comprises the low-latitude boundary layer.

stream outwards, either towards or away from the Earth, depending on whether one is earthward or tailward of the X-line. On the earthward side of the X-line, there is a counter-streaming flow of particles created by the mirroring of particles at the Earth (Cowley, 1980; Forbes et al., 1981). The energetic particles in the PSBL move through the much colder tail-lobe population without interacting with it (Onsager et al., 1991), and it is not until one reaches the plasma sheet proper that the lobe population and energetic particles are replaced by a thermalised distribution convecting at the electric-field drift-speed.

10.4.2 Slow-Mode Shocks

One of the most momentous discoveries concerning reconnection in the tail was made about 10 years ago when Feldman et al. (1984) and Smith et al. (1984) found clear evidence of slow-mode shocks, whose configuration is essentially the one predicted by Petschek (1964). Although there has been some evidence that slow shocks may exist in the inner heliosphere (Richter et al., 1985), the tail observations provide unambiguous evidence that slow-mode MHD shocks do actually occur in nature and are not simply a theoretical construct. Despite the fact that virtually every textbook on MHD talks about slow shocks, such shocks have not been seen in laboratory plasmas because of the experimental difficulties (Hada and Kennel, 1985).

The slow shocks observed in the geomagnetic tail are collisionless versions of the standard MHD slow shocks (Omidi and Winske, 1992; Lee et al., 1989), and they are located at the interface between the PSBL and the plasma sheet proper. This location implies to us that the PSBL may be a kind of foreshock, which is linked to the slow shock in much the same way as the foreshock upstream of the magnetosphere is linked to the bow shock. Although this interpretation is roughly consistent with the appearance of the particle distribution functions within the PSBL (Onsager et al., 1991), recent simulations (e.g., Lottermoser et al., 1998) suggest that this interpretation is too simplistic.

Although Petschek's original model predicts that there should be pairs of slow-mode shocks on both sides of the X-line, the slow shocks in the tail tend to occur tailward of the X-line. This asymmetry may be due to the obstacle posed by the inner magnetosphere to the earthward-directed flow from the X-line. As we have discussed in Section 5.4 and will discuss in Section 11.1.2, an obstacle to the reconnection outflow in an MHD reconnection model causes the outflow to be diverted along the separatrices. If the outflow is supermagnetosonic with respect to the fast-mode wave speed, then the diversion occurs downstream of a termination shock at the tip of the jet (see Fig. 11.14). In either case, the Petschek-type shocks are no longer symmetric around the X-line. In the submagnetosonic case, strong slow-mode shocks occur only along the unobstructed outflow jet. In the supermagnetosonic case, the shocks do occur on both sides, but on the obstructed side, they extend only as far as the termination shock. In essence, the Petschek shocks have a one-to-one correspondence with a high-speed outflow from the X-line. If this outflow is blocked, then the Petschek-type slow-mode shocks which flank it are not present.

Observations by Onsager and Mukai (1995) support the view that earthward-directed outflow from the distant-tail X-line is in fact obstructed by the inner magnetosphere. Using a time-of-flight analysis, they infer that the earthward outflow from the X-line appears to be a jet which extends only a distance of about 5 R_E. Within the jet the flow speed is about 1,000 km s^{-1}, but at 5 R_E the earthward flow slows to less than 100 km s^{-1}.

10.5 Magnetospheric Substorms

To those who are not familiar with the distinction between a magnetic storm and a magnetospheric substorm, it always seems strange that there is so much more interest in the latter than the former. The prefix "sub" in the word "substorm" may seem to imply that a substorm is somehow less important than a storm, but this is not the case. Despite its name, the magnetospheric substorm can occur even where there is no magnetic storm. (For a discussion of how these names came about, see Stern, 1989.) A substorm is a disturbance that results from the injection of energy into the magnetosphere caused by a southward turning of the interplanetary magnetic field. It typically lasts approximately an hour, and the resulting magnetic field disturbances at the Earth's surface are concentrated in and around the auroral zones. For these reasons, substorms have also been termed "auroral or polar" substorms. Substorms occur at a rate of about one a day, but they occur more frequently (one every few hours) during a magnetic storm.

A magnetic storm is a much longer-lived disturbance, lasting on the order of several days. Storms are typically produced by the impact on the magnetosphere of plasma ejected from the Sun as a coronal mass ejection (§11.1). The impacting material compresses the magnetosphere and causes an abrupt enhancement of the field at the Earth's surface known as a *sudden commencement*. After approximately an hour into a magnetic storm, the surface magnetic field begins a prolonged decrease which is greatest in the equatorial region. The maximum disturbance of the field occurs about a day after the start of the storm, and it usually takes several days for the Earth's field to return to its normal state. The prolonged decrease in the surface field is caused by the development of a ring current due to the injection of energetic particles from tail reconnection. The ring current circles around the Earth at an altitude of about 4.5 R_E, just above the magnetic equator (Hargreaves, 1993).

The ring current is produced by repeated injection of energetic particles from the magnetotail by substorms occurring more frequently and with more vigour than normal. The enhanced frequency of substorms can be caused by an increase in solar-wind speed, or a sustained southward turning of the interplanetary field, or both. If there is no increase in the solar-wind speed, then the storm begins without a sudden commencement. In this latter case, the storm is no more than a series of substorms which occur so rapidly that the magnetosphere cannot fully recover between each event. The overlapping of individual events allows the ring current to grow to the point where it dominates the overall disturbance of the surface field. Thus, the substorm is really the fundamental dynamical process.

10.5.1 Substorm Phases

A substorm can be subdivided into three distinct phases, which are called the *growth, expansion,* and *recovery* phases, respectively. The growth phase starts with the beginning of reconnection at the dayside magnetopause due to a southward turning of the interplanetary field. This leads to the erosion of the dayside magnetopause and an enhancement of the current flowing in the tail current sheet. During the growth phase, magnetic field lines are dragged from the dayside into the tail and are stretched, so the magnetic field in the tail lobe increases as shown in Fig. 10.12(a). At some point in time, the current in the tail reaches a critical threshold which causes a dynamic release of the magnetic energy stored in the tail. This point is the *substorm onset*, which marks the beginning of the expansion phase. In the original substorm model proposed by Hones (1973), Russell and McPherron (1973), and Nishida and Nagayama (1973), substorm onset is due to the formation of a near-Earth X-line at about 15 R_E [Fig. 10.12(b)], although more recent surveys show it generally forms between 20 and 30 R_E downtail (Nagai et al., 1998). Why an X-line should develop at this particular location is not well understood, but it may be related to the overall convective flow pattern in the tail. Erickson and Wolf (1980), for example, have proposed that the convection during the growth phase leads to a pinching of the sheet at approximately this location in the tail.

At the start of the expansion phase, bright aurorae are generated at local midnight at the lower boundary of the auroral zones at each pole ($\sim$67° magnetic latitude on the nightside). With time, the aurorae expand poleward and westward until they cover a large area, and within this area they are very dynamic (Akasofu, 1968). New arcs continually form and develop folds, while brightly coloured patches of emission appear and pulsate. The propagation of the dynamic region westward and poleward occurs in a step-like fashion (McPherron, 1995).

During the expansion phase, a strong current appears in the ionosphere above the region of dynamic auroral activity. This current is the auroral electrojet, and it produces a large perturbation in the horizontal magnetic field at the surface of the Earth. The auroral electrojet occurs at the time of the peak value of the auroral electrojet (AE) index shown in Fig. 10.3. The AE index is based on the horizontal magnetic field component from a dozen or so stations around the North Pole between latitudes 65° and 75°. The increase in AE prior to the onset of the expansion phase is due to ionospheric currents generated during the growth phase, but, once the expansion phase starts, the dominant contribution comes from the electrojet. The current flowing in the electrojet comes from field lines mapping to the outer edges of an arc lying along the inner edge of the plasma sheet, as shown in Fig. 10.13. The region where the diversion of the cross-tail current to the ionosphere occurs is known as the *substorm current wedge*.

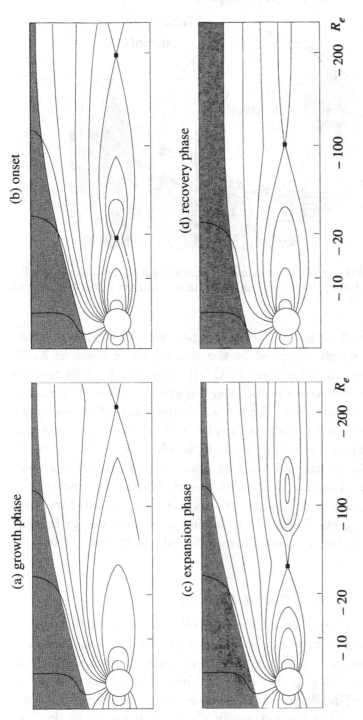

(a) growth phase

(b) onset

(c) expansion phase

(d) recovery phase

Fig. 10.12. Structure of the geomagnetic tail during the different phases of a geomagnetic storm according to the ideas of Hones (1973) and Russell and McPherron (1973). (a) The growth phase starts with the beginning of extensive reconnection at the dayside magnetopause when the solar-wind field turns southward. (b) This leads to a build-up of magnetic flux in the tail, which continues until the sudden formation of a near-Earth neutral line at substorm onset. (c) Rapid reconnection at the near-Earth neutral line occurs during the expansion phase, when a plasmoid is released down the tail. (d) Finally, during the recovery phase the neutral line moves to its usual quiet-time location in the distant tail, and the field relaxes back to the structure it had before the start of the growth phase.

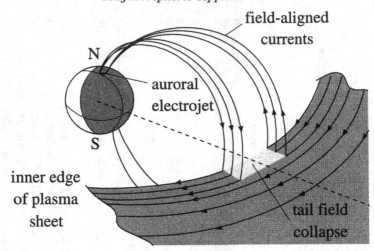

Fig. 10.13. Diagram of the three-dimensional current system that is responsible for the auroral electrojet during the expansion phase of a substorm (after Clauer and McPherron, 1974).

Sometime shortly after the start of the expansion phase, the magnetic island, or plasmoid, which is formed by the appearance of the near-Earth X-line, starts to move tailward [Fig. 10.12(c)]. The field lines in the tail relax and become more dipolar (a process called *dipolarization*), but auroral activity and the electrojet persist for a few minutes more. Eventually, the activity and electrojet start to fade as the plasmoid reaches the distant tail. This declining phase is the substorm recovery, which eventually restores the magnetosphere to the configuration it had prior to the start of the growth phase [Fig. 10.12(d)]. The exact conditions which trigger the recovery are still quite uncertain, but in some cases the recovery appears to start when the interplanetary magnetic field switches northward and the energy transfer from the solar wind to the magnetosphere is halted (McPherron, 1995).

10.5.2 Tail Plasmoid

Most of the magnetic energy released by the collapse of the tail field during the expansion phase goes into kinetic energy of the tailward-moving plasmoid. In two dimensions, it is rather easy to visualize the formation of the plasmoid as illustrated in Fig. 10.12, but in three dimensions it is not so easy. As we mentioned in Chapter 8, the simple process of adding a uniform field out of the plane to the fields shown in Fig. 10.12 has a dramatic effect on the field-line topology of the plasmoid. Field lines within the plasmoid become helical and are connected either to the Earth or to the solar wind or both, and so true magnetic islands are no longer present (Birn, 1991; Wright and Berger, 1989; Hughes and Sibeck, 1987).

Interestingly, the field-line structure within a three-dimensional plasmoid can exhibit a highly complex behaviour. Birn et al. (1989) have shown this explicitly for the following simple model of a magnetotail containing a plasmoid:

$$B_x = B_0 z / L_0, \tag{10.10}$$

$$B_y = B_0 k, \tag{10.11}$$

$$B_z = B_0 \lambda_1 / L_0^3 [x - x_1(y)][x - x_2(y)][x - x_3(y)], \tag{10.12}$$

$$x_1 = x_n - \bar{a} \left\{ 1 - [1 - (y/\bar{b})^2]^{1/2} \right\}, \tag{10.13}$$

$$x_2 = x_n - \bar{a} \left\{ 1 + [1 - (y/\bar{b})^2]^{1/2} \right\}, \tag{10.14}$$

$$x_3 = \begin{cases} x_f - \lambda_2 [1 - (y/\bar{c})^2]^{1/2}, & |y| \le \bar{c}, \ \bar{d}/4 \le \bar{a} \\ x_f, & \text{otherwise}, \end{cases} \tag{10.15}$$

where

$$\bar{c} = \bar{b} \{ 1 - [(\bar{d} - \bar{a})/(3\bar{a})]^2 \}^{1/2},$$

$$\bar{d} = x_n - x_f \text{ and } \lambda_2 = 2\bar{a} - \bar{d}/2.$$

In Equations (10.10)–(10.15), x_1, x_2, and x_3 are the locations of the near-Earth X-line, the plasmoid O-line, and the distant-tail X-line, respectively; B_0 is a typical magnetic field strength and L_0 a length-scale, while k, λ_1, and λ_2 are parameters. The distances x_n and x_f correspond to the intersection points of the near-Earth and distant-tail X-lines with the noon-midnight meridian plane. The parameter $\bar{a}$ is the distance between the O-line and the near-Earth X-line in the noon-midnight plane. Field lines mapping to the $x = 0$ plane are considered as mapping to the Earth.

Within the plasmoid described by the relatively simple equations (10.10)–(10.15), field lines with both ends connected to the Earth are mixed together with fields that have only one end connected and field lines which have neither end connected. As B_y decreases, the topological mixture increases in complexity until in the limit $B_y = 0$, the mapping becomes fractal in nature. Figure 10.14 shows a three-dimensional picture of the plasmoid and its mapping from a numerical simulation which starts with a field similar to that of Eqs. (10.10)–(10.15) (*cf.* Fig. 8.3).

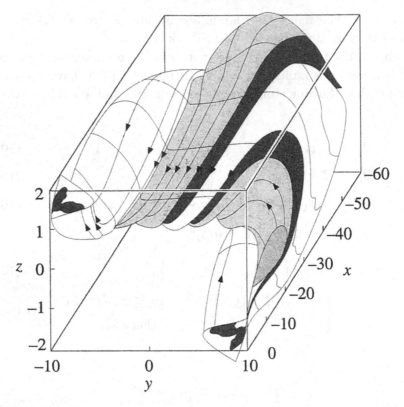

Fig. 10.14. Perspective view of a plasmoid in a three-dimensional model of the geomagnetic tail during the expansion phase of a substorm (Birn and Hesse, 1990). Unshaded regions have closed field lines with both ends connected to the earthward boundary at $x = 0$. Field lines in the shaded regions are open field lines with one end connected to the earthward boundary but the other end connected to either the distant boundary at $x = -60$ (dark shading) or the flank boundaries at $|y| = 10$ (light shading).

10.5.3 Determining the Reconnection Rate

Direct determination of the reconnection rate in the magnetotail by measuring the electric field in situ at the X-line is very difficult. Accurate electric field measurements by spacecraft are not easy to obtain, but even more limiting is the fact that the probability of a spacecraft's intercepting the X-line in the tail is extremely small. However, there does exist an indirect method using ground-based observations which can provide some idea of the reconnection rate in the tail as a function of time (de la Beaujardière et al., 1991 and Blanchard et al., 1996, 1997). This method relies upon the fact that, according to Dungey's model, the magnetic flux within the polar caps is related to the rates of reconnection at both the dayside and nightside

X-lines as follows. Let the open magnetic flux at the polar cap be defined as

$$F_B = \int\int_S B_n \, dS,$$

where B_n is the normal magnetic field component at the ground and the area of integration is over the region (S) of open field lines in the polar cap. Since temporal variations in B_n during a substorm are very small $(\Delta B_n/B_n \approx 10^{-3})$, the time rate of change of F_B is to good approximation

$$\dot{F}_B = \oint_C B_n(s) \, V_\perp(s) \, ds, \tag{10.16}$$

where s is measured along the boundary (C) between the open and closed field lines and $V_\perp(s)$ is the velocity of the boundary in a direction perpendicular to it. Now, if we assume that there is a single separator line connecting a dayside neutral point with a nightside neutral point, the total potential drop (Φ_E) along this line is, according to Faraday's equation (1.7),

$$\Phi_E = \oint_{C_S} \mathbf{E} \cdot \mathbf{ds} \equiv \dot{F}_B,$$

where C_S is the path along the separator (Vasyliunas, 1984). Substituting this expression into Eq. (10.16) we obtain

$$\Phi_E = \oint_C B_n(s) \, V_\perp(s) \, ds \tag{10.17}$$

for the total potential drop along the separator.

We can obtain a similar expression for the potential drop along just part of the separator by choosing a path which encloses only a portion of the open flux. For example, let us assume that the open-field region is circular and consider the open flux in a pie-shaped region with a longitudinal angular width $\Delta\phi_c$ and a latitudinal extension from $\theta = 0$ (the pole) to the open–closed boundary at θ_c. The magnetic flux in this region is

$$F_B(t) = \Delta\phi_c \, R_E^2 \int_0^{\theta_c(t)} B_r(\theta) \sin\theta \, d\theta$$

and the corresponding rate of change is roughly

$$\dot{F}_B = \Delta\phi_c \, R_E^2 \, B_r(\theta_c) \, \dot{\theta}_c \sin\theta_c. \tag{10.18}$$

The net potential associated with this rate of change is

$$\Phi_E = \int_{X\text{-line}} E_{rec} \, ds + \int_C E_c \, ds = \dot{F}_B, \tag{10.19}$$

where the first term is integrated along the X-line (i.e., the separator) and

the second along the open–closed boundary in the polar-cap ionosphere
(Blanchard et al., 1996, 1997). This expression assumes that the net electric field along the field lines connecting the X-line and the polar cap is
zero; ideal MHD applies everywhere except in the immediate vicinity of the
X-line; and a well-defined separator line exists (see Vasyliunas, 1984 for more
discussion).

Expressing the ionospheric electric field as $E_c = V_c(t)B_r(\theta_c)$ and combining Eqs. (10.18) and (10.19), we obtain

$$\Phi_E \approx l_c\, B_r(\theta_c)\,(U_s - V_c), \tag{10.20}$$

where $l_c = \Delta\phi_c\, R_E \sin\theta_c$ is the length of the ionospheric boundary mapping
to the X-line, U_s is the apparent velocity of the open-closed boundary, and V_c
is the true plasma velocity at this boundary. When the dayside and nightside
reconnection rates are equal, U_s is zero and a steady-state occurs. By further
assuming that Φ_E is approximately equal to $l_s E_{\text{rec}}$, we have

$$E_{\text{rec}} \approx \frac{l_c}{l_s}\, B_r(\theta_c)\,(U_s - V_c), \tag{10.21}$$

where l_s is the length of the portion of the separator mapping to l_c. In
other words, l_c/l_s is the ratio of the field-line spacing along the open-closed
boundary to that along the separator.

The velocity (U_s) of the boundary between open and closed field lines
can be determined by using incoherent radar to locate the poleward limit
of auroral electrons. Although the main auroral arcs during a substorm are
located on field lines mapping to the inner edge of the plasma sheet, there
are precipitating electrons located at very high latitudes that originate on
field lines mapping to the vicinity of the X-line in the tail (Winningham
et al., 1975; de la Beaujardière et al., 1991). These electrons create a sharp
boundary in the amount of radar scatter that occurs in the E-region of
the ionosphere, and thus U_s can be determined by tracking this boundary.
The polar-cap convection velocity (V_c) can be measured directly by determining the Doppler shift of the reflected signal. Using these methods de la
Beaujardière et al. (1991) and Blanchard et al. (1996, 1997) have found that
the net potential drop (Φ_E) across the width of the magnetosphere is about
50 kV on average. This value is only approximately a tenth of the maximum
possible potential drop of 510 kV ($400\ \text{km s}^{-1} \times 5\ \text{nT} \times 40\, R_E$) that would occur if every solar-wind field line that intercepted the magnetopause became
reconnected. Thus, these results confirm the statement in Section 10.1.1 that

only about 10% of the incoming field lines are reconnected on average, while the remaining 90% simply pass around the flanks of the magnetopause.

In order to deduce the rate of reconnection (E_{rec}) in the magnetotail from ground-based measurements, some assumption must be made about the value of the ratio l_c/l_s. Various models (e.g., Birn, 1991; Walker et al., 1993) suggest that during quiet times l_c/l_s is roughly of the order of 0.1, which implies $E_{rec} \approx 10^{-4}\,\mathrm{V\,m^{-1}}$ on average. During the expansion phase of the substorm, the ground-based observations and the models imply that E_{rec} may be as high as $10^{-3}\,\mathrm{V\,m^{-1}}$, since $(U_s - V_c)$ is typically six times larger than its average value, and l_c/l_s is probably a factor of 2 or 3 smaller than its average value.

We can determine whether these rates are slow or fast by comparing E_{rec} with the Alfvén electric field defined by $E_A = V_A B = B^2/\sqrt{\mu\rho}$. For steady-state theories such as that by Sweet and Parker or Petschek, E_{rec}/E_A is the same as the Alfvén Mach number (M_A) often used to characterize the reconnection rate. Over a sufficiently long time-scale, the rate in the tail is equal to the rate at the dayside magnetopause, since over the long term the total amount of flux in the magnetosphere is constant. Thus, the average reconnection rate observed by Blanchard et al. (1996) should reflect the reconnection rate at the nose of the magnetosphere – in other words, the rate at which the field lines in the magnetosheath are connected to terrestrial field. In the magnetosheath, B is typically about 20 nT and n about $2 \times 10^7\,\mathrm{m^{-3}}$, so E_A there is approximately $2 \times 10^{-3}\,\mathrm{V\,m^{-1}}$. Therefore, the average reconnection rate for the magnetosphere corresponds to $M_A \approx 0.05$ (i.e., $M_A = 10^{-4}\,\mathrm{V\,m^{-1}}/2 \times 10^{-3}\,\mathrm{V\,m^{-1}} \approx 0.05$). For the peak reconnection in the tail during the substorm expansion phase, B is again about 20 nT, but the density (n) in the upstream region of the tail lobes is approximately a thousand times smaller, so, for the peak reconnection rate in the tail, the Alfvén Mach number is about 0.02 (i.e., $M_A \approx 10^{-3}\,\mathrm{V\,m^{-1}}/6 \times 10^{-2}\,\mathrm{V\,m^{-1}} \approx 0.02$).

It is difficult to compare these values of M_A directly with the MHD theories because such theories depend on the magnetic Reynolds number (R_m), whose value in the collisionless environment of the magnetosphere is difficult to calculate (§1.7). However, we can invert the problem and ask what value of R_m is implied by the value of M_A inferred from observations? On the one hand, if the reconnection rate is of the Sweet–Parker type, then $R_m = M_A^{-2}$ and the effective R_m is approximately 400 for the average reconnection rate and 2,500 for the peak rate in the tail during the expansion phase. On the other hand, if the reconnection is of the Petschek type, then the effective magnetic Reynolds numbers $\{R_m \approx \exp[\pi/(4\,M_A)]\}$ are very much larger, ranging from 10^7 for the average dayside value to 10^{17} for the peak value in the tail.

10.5.4 Initiation of the Expansion Phase

One of the important questions remaining in substorm research is, What is the physical process which triggers the onset of the expansion phase? For many years proponents of the reconnection model of substorms identified onset with the formation of a near-Earth X-line. According to one of the more popular scenarios, the X-line appears when the current sheet in the near-Earth region thins to the point that its thickness is on the order of an ion-gyro radius (Schindler, 1974). Such thinning is in fact observed (Mitchell et al., 1990), and it is thought to be due to enhanced convection during the growth phase (Erickson and Wolf, 1980; Walker et al., 1993). Once the current sheet becomes so thin, it may be unstable to the ion-tearing mode (Lakhina and Schindler, 1988).

Recently, some researchers have argued that substorm onset precedes the formation of the near-Earth X-line by a few minutes, and that the first sign of onset appears to occur at the inner edge of the plasma sheet at about $7\,R_E$, rather than at the $15\,R_E$ where the X-line forms (Takahashi et al., 1987; Lopez et al., 1989). The inner edge of the plasma sheet maps to the equatorial edge of the auroral oval, where the first brightening occurs that signals the start of the expansion phase. Lui et al. (1992) have suggested that onset is caused by a current instability at the inner edge of the tail current sheet, and that the formation of the near-Earth X-line is a by-product of this instability. However, the energy released by the current diversion is far from sufficient to account for the energy produced during the expansion phase, and the only source which contains enough energy to do this is the energy released by reconnection at the near-Earth X-line (Baker et al., 1997).

As an alternative to the current-instability model, Schindler and Birn (1993) and Wiegelman and Schindler (1995) have suggested that the event which triggers onset is the sudden formation of a thin current sheet near the inner edge of the plasma sheet. Using a simple tail model, they show that a slow increase in the lobe field can lead to the dynamic collapse of a broad current layer into a thin layer whose thickness is on the order of the dissipation scale-length. Thus, onset corresponds to the beginning of this dynamic thinning process rather than the formation of a near-Earth X-line. As the collapse nears completion, the near-Earth X-line appears and sustains the energy release begun by the collapse. A three-dimensional resistive MHD simulation of a substorm together with particle-orbit computations

(Birn et al., 1996, 1997) have been highly successful in explaining many of the observed properties of substorms, including particle acceleration (§13.4.4). Observations by Maynard et al. (1996) reinforce the idea that the inter-actions between the ionosphere, the inner edge of the plasma sheet, and the near-Earth X-line are indeed strongly coupled and that their behaviour cannot be understood independently of one another.

10.6 Magnetospheres of Other Planets and of Comets

Mercury and the gas giants (Jupiter, Saturn, Uranus, and Neptune) all have an intrinsic magnetic field and are thus surrounded by magnetospheres which are similar in many respects to that of the Earth. There are, however, im-portant differences. Because Mercury has the weakest field of this group (4×10^{-4} of the Earth's) and is closest to the Sun (0.39 AU), where the solar wind is strongest, its magnetosphere is quite small. The distance from its magnetopause to the surface of the planet is only about half a planetary radius compared with 10 radii for the Earth and 80 radii for Jupiter. It is surprising that Mercury has a magnetosphere at all. According to present understanding of dynamo theory, it is too small and rotating too slowly to have a magnetic field (the Mercury rotation period is 59 Earth days). Nevertheless, during its fly-by of Mercury in 1974, Mariner 10 found a well-developed magnetosphere, complete with a bow shock and a tail. There is even some evidence that a magnetospheric substorm might have been in progress at the time of the fly-by (Siscoe et al., 1975). The unique feature of Mercury's magnetosphere is that it has no ionosphere.

Jupiter is at the opposite extreme of Mercury in that it has a magneto-sphere that is much larger than the Earth's. The Jovian magnetic dipole moment is approximately 2×10^4 times that of Earth's. The strength of this source combined with the rapid rotation of the planet (one rotation every 10 h) creates a magnetosphere whose plasma corotates as a unit all the way out to the dayside magnetopause (Hill, 1983). Only in small regions above the magnetic poles and in the distant tail does the plasma convect in re-sponse to the drag induced by the solar wind. By contrast, in the Earth's magnetosphere the corotating region is a rather small, donut-shaped re-gion, which extends outwards from the equator to approximately only $4\,R_E$ (Fig. 10.1). This region is known as the *plasmasphere*, and it does not play a significant role in the reconnection dynamics. However, in Jupiter's mag-netosphere the corotation is so strong that it creates a current sheet in the shape of a flat disk encircling the planet. This sheet is susceptible to mag-netic tearing and thus may give rise to a dynamic reconnection process which has no equivalent in the Earth's magnetosphere (Zimbardo, 1989). Another important feature of Jupiter's magnetosphere, which is absent in the Earth's, is the interaction with its large satellites. Much of the plasma in Jupiter's magnetosphere comes from the satellites, particularly Io, whose active volcanoes continually inject material into the corotating region (Kennel and Coroniti, 1975). Also, the Jovian system contains the only satellite (namely Ganymede) known so far to have its own magnetosphere, which the Galileo spacecraft penetrated in 1996 (Kivelson et al., 1996).

Substorm-like events are likely to exist in the magnetospheres of all the gas giants (Zimbardo, 1993). They all have extended tails formed by their interaction with the solar wind, and, at a sufficiently large distance down the tail, the field geometry tends to look much the same as the Earth's. Evidence of dayside reconnection has been found at both Uranus (Richardson et al., 1988) and Neptune (Desch et al., 1991), which implies that field lines are dragged into the tail in the same way as at the Earth.

Although comets do not have an intrinsic magnetic field, they nevertheless can entrap solar-wind magnetic field in the comas which surround their nuclei. The portion of a field line which intercepts the coma diffuses slowly through it, while the rest of the field line is swept into a cometary plasma tail, as shown in Fig. 10.15 (Niedner, 1984). From time to time, the plasma tails of comets are seen to disconnect suddenly from the coma and fly off into space with the flow of the solar wind. As the old tail moves away from the coma, a new one starts to grow, and within a few days the plasma tail is completely restored. This phenomenon is called a *disconnection event*, and its existence is one of the main reasons for believing that reconnection occurs in comets.

It is possible, at least in principle, to create a disconnection event without reconnection by abruptly enhancing the diffusivity of the field lines through the coma. However, the occurrence of disconnection events is correlated almost exclusively with the passage of a comet through the interplanetary current sheet (Delva et al., 1991; Yi et al., 1994). This current sheet is created by the outward flow of the solar wind, and it lies very roughly in the equatorial plane of the Sun. Whenever a spacecraft or comet passes through it, the interplanetary field reverses direction by about 180°. In the jargon of heliospheric observations, the reversal is called a *sector boundary*.

To explain the correlation of disconnection events with the crossing of a sector boundary, Niedner and Brandt (1978) proposed that reconnection occurs at the sector boundary whenever it encounters the coma. The process is shown in Fig. 10.15 and is referred to as a dayside reconnection model because the reconnection starts on that side of the comet. Verigin et al.

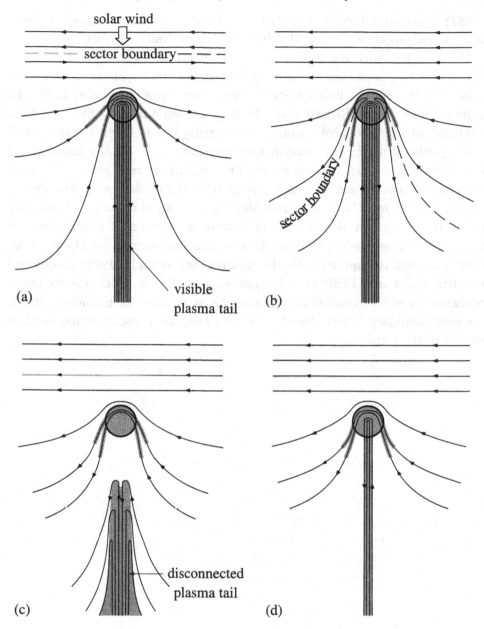

Fig. 10.15. Different stages in the disconnection of a comet tail according to the scenario of Niedner and Brandt (1978). (a) Prior to a collision with the sector boundary the comet's tail is in a steady state. (b) As the sector boundary enters the coma, reconnection causes the tail to disconnect. (c) The disconnected tail is expelled away from the Sun at close to the Alfvén speed. (d) It is eventually replaced by a new tail.

(1987) reported evidence of particle acceleration, possibly caused by such dayside reconnection, during the fly-by of comet Halley in 1986 by Vega-1.

One problem with the Niedner and Brandt model seems to be that there appear on rare occasions to be comet disconnection events even when the comet is far from the heliospheric current sheet (Ip and Mendis, 1978). To explain these events, Russell et al. (1986) proposed a model which is a close analogue of the reconnection process occurring in the Earth's magnetotail during substorms. In this model, reconnection occurs on the nightside of the comet in the tail current sheet. The conditions required to trigger a disconnection event are not really specified by the model, but Malara et al. (1989) have shown that the current sheet in the tail of a comet is especially prone to the Kelvin–Helmholtz and tearing instabilities if it becomes too long. Interestingly enough, three-dimensional simulations by Ogino et al. (1986) appear to support both the Niedner and Brandt (1978) model and the Russell et al. (1986) model. Spontaneous disconnection events occur occasionally in the simulation even when there is no sector boundary, but, if a sector boundary is introduced, it always triggers a disconnection as soon as it hits the coma.

11

Solar Applications

Much of the impetus for the development of reconnection theory began with R. G. Giovanelli's efforts to develop an electromagnetic theory of solar flares. His first paper on this subject had the title "A Theory of Chromospheric Flares" (Giovanelli, 1946), which is somewhat outdated because flares are now thought of as primarily a coronal phenomenon rather than a chromospheric one. The prodigious advances in X-ray and radio astronomy since Giovanelli's time have made it increasingly clear that what is seen in the chromosphere by ground-based optical telescopes is a response to coronal activity. During a flare, the heated plasma in the corona becomes so hot ($T > 10^7$ K) that the emitted radiation lies outside the visible portion of the spectrum. Thus, the coronal flare is imperceptible to telescopes observing in the visible spectrum.

Despite the title of his paper, Giovanelli proposed what is essentially a coronal theory of flares. He argued that the chromospheric emissions seen during a flare are produced by a bombardment of energetic electrons accelerated along field lines by electric fields in the vicinity of coronal magnetic null points. This concept still prevails today. However, Giovanelli thought that the electric fields could be produced by induction as the magnetic fields in the photosphere shifted their position slowly over time. Nowadays, such a process is considered to be far too slow to produce a sufficiently strong and impulsive electric field.

After Dungey's (1953) pioneering work on reconnection, Sweet (1958a) and Parker (1957) reconsidered Giovanelli's neutral-point model as a reconnection process which taps the energy stored in the magnetic field associated with coronal currents. The question which they focused on was whether reconnection can convert magnetic energy into kinetic and thermal energy at a rate fast enough to account for the observed energy release in solar flares. This question is still being vigorously debated.

If one assumes that reconnection has the same time-scale as a simple diffusion process, then it is not possible to explain the rapid energy release in flares if the resistivity of the plasma is the classical one given by Spitzer (1962). To see this explicitly, let us consider a large flare which releases 10^{25} J (10^{32} ergs) in an active region, where the average free magnetic field is on the order of 10^{-2} tesla (100 G). In order to generate 10^{25} J (10^{32} ergs), the size (L_e) of the coronal volume involved in the flare must therefore be 6×10^4 km. In the corona, the Spitzer magnetic diffusivity [Eq. (1.13)] is approximately

$$\eta \approx 10^9 \; T^{-3/2} \; m^2 \; s^{-1},$$

which, for a temperature (T) of 2×10^6 K, gives $\eta = 0.35$ m^2 s^{-1}. Now the time-scale (τ_d) for a simple diffusion process is just [Eq. (1.16)]

$$\tau_d = \frac{L_e^2}{\eta}. \tag{11.1}$$

Substituting $L_e = 6 \times 10^4$ km and $\eta = 0.35$ m^2 s^{-1}, we then find $\tau_d = 3 \times 10^8$ yr! This incredibly long time is the main reason why initially there was strong theoretical opposition to the idea that any diffusive process, such as reconnection, could be involved in flares.

However, Eq. (11.1) does point the way to the two routes for fast reconnection that we discussed in Chapter 1, namely small scale-lengths and anomalous resistivity. Knowing that the time-scale for the impulsive phase of the flare is on the order of 100 s, we can deduce that the diffusive scale-length must be about 10^7 times smaller than the global (or external) scale-length (L_e), in other words, about 6 m. This value is more than two orders of magnitude smaller than the electron–ion mean-free path in the corona, and it implies that the dissipation processes involved are collisionless ones (§1.7). Thus, Spitzer's collisional formula is not really appropriate, and both anomalous resistivity and a small scale-length are likely to be important for reconnection in the corona.

Reconnection is crucial for the generation of the Sun's magnetic field by a dynamo, for which the reader is referred to the comprehensive books by Moffatt (1978) and Proctor et al. (1993). In addition, the solar corona has been revealed by spacecraft observations [especially Yohkoh, the Solar and Heliospheric Observatory (SOHO) and the Transition Region and Coronal Explorer (TRACE)] to be an MHD world, where reconfigurations of coronal structures by magnetic reconnection are common. Indeed, reconnection appears to be the key to a wide variety of dynamic coronal phenomena and even to be responsible for the very existence of a hot corona. In Section 11.1

we describe eruptive solar flares and other related eruptive phenomena. Section 11.2 describes confined flares and similar smaller-scale dynamic energy-release processes. Then in Section 11.3 we discuss the role of reconnection in coronal heating, and finally in Section 11.4 we move outwards to the outer corona. As with other applications of reconnection, it would be easy to devote a whole book summarizing the many examples of reconnection on the Sun which have been published. Since our own research interests are on solar coronal applications, we have taken the authors' prerogative to focus on those applications that we know best.

11.1 Large-Scale Eruptive Phenomena

Traditionally, there are three different types of large-scale eruptive phenomenon occurring in the solar atmosphere, namely *coronal mass ejections* (CMEs), *prominence eruptions*, and large *two-ribbon flares*. It has become increasingly clear with time that these phenomena are closely related and may, in fact, be different manifestations of a single physical process.

CMEs are defined as large-scale ejections of mass and magnetic flux from the lower corona into interplanetary space. Measurements from coronagraphs and spacecraft show that a typical CME injects roughly 10^{15} Wb (10^{23} Mx) of magnetic flux and 10^{13} kg of plasma into space (Hundhausen, 1988; Gosling, 1990; Webb et al., 1994). During the quiet phase of the solar cycle there are approximately two CMEs per week, but during the active phase the rate can exceed one per day. The fact that the continual ejection of magnetic flux by CMEs does not lead to an indefinite build-up of magnetic flux in interplanetary space is incontrovertible evidence that magnetic reconnection is occurring somewhere in the corona (Gosling, 1975; McComas et al., 1992). Reconnection must be occurring because it is, by definition, the only process which allows magnetic field lines attached to a CME to disconnect and thereby reduce an otherwise indefinite magnetic energy build-up in the solar wind.

Most CMEs are not associated with large flares, but, if a large flare does occur, it is almost always associated with a CME and the eruption of an active-region prominence. The opening of the field lines in such an active region by a CME leads to the formation of flare ribbons and loops. Flare ribbons are chromospheric patches of Hα emission which lie at the feet of the flare loops, and both ribbons and loops are formed by the process of magnetic energy release as open field lines reconnect to become closed loops again. The apparent motion of the ribbons and loops constitutes some of the best evidence for reconnection in the solar atmosphere.

Doppler-shift measurements show conclusively that the motions of flare loops and ribbons are not due to mass motions of the solar plasma, but rather to the upward propagation of an energy source in the corona (e.g., Schmieder et al., 1987). As we shall show in Section 11.1.2, such motions are exactly what one expects for a reconnection model of the flare loops.

More than half of all CMEs are associated with the eruption of large quiescent prominences. As observations have improved, it has become increasingly clear that erupting prominences have many features typical of large flares. Like large flares, erupting prominences produce loops and ribbons which move apart in time, but, unlike large flares, the ribbons are usually too faint to be seen in $H\alpha$. However, the ribbons can often be seen in the He 10,830-Å line, which is a more sensitive indicator of chromospheric excitation (Harvey and Recely, 1984). The eruption of quiescent prominences does not usually produce hard X-ray or γ-ray emissions, probably because of the relatively weak fields involved (<10 G).

High-resolution images obtained by the soft X-ray telescope on Yohkoh have made it clear that all CMEs and prominence eruptions create faint X-ray loops [Fig. 11.21 below]. Although these loops are much larger than typical flare loops, they rise with time in much the same manner to create a helmet streamer with a cusp at the summit. If there is an active region underneath the CME, then bright flare loops will also appear in the active region.

Priest (1992) and Švestka and Cliver (1992) have proposed that all such solar phenomena which expel material into interplanetary space should be called *eruptive flares*. Some researchers prefer to confine the term "flare" to those phenomena which produce bright chromospheric emissions that are at least a factor of two greater than the background emission. However, the latter definition is not in keeping with our current thinking that all flares are basically coronal, rather than chromospheric, phenomena (see also Harrison, 1995). Therefore, we prefer to define a flare as a rapid increase in the coronal X-ray emission rather than chromospheric $H\alpha$ emission. With this newer definition all CMEs are necessarily eruptive flares.

Initially, some people thought that a large solar flare or an erupting prominence drives a coronal mass ejection, but this is clearly not the case since careful study of the timing shows that a CME can start typically half an hour before the impulsive phase of a flare (Harrison, 1995), namely, just at the time that the prominence begins to be agitated and start to rise slowly prior to its rapid eruption. Also, there may be more energy in a CME than an accompanying flare. In our view, we are dealing here with a complex large-scale MHD configuration with rapid communication by MHD waves between the different parts of the system, and it is the eruption of the whole structure that is of importance. The prominence, the CME and sometimes the flare represent different manifestations of the "eruptive flare".

Figure 11.1 shows time-profiles at various wavelengths of a very large eruptive event which occurred on 28 August 1966. This event had intense $H\alpha$, X-ray, and radio emissions, and it produced a high-speed coronal mass ejection preceded by an intense shock wave which manifested itself in the chromosphere as a *Morton wave*. The Morton wave is the chromospheric footprint of a fast-mode MHD shock in the corona, and it is caused by a slight disturbance of the chromosphere at the location where the fast-mode wave intersects the solar surface (Dodson and Hedeman, 1968; Zirin and Lackner, 1969). Energetic protons ($\gtrsim 10\,\mathrm{MeV}$), produced by the shock wave when it was still quite close to the Sun, were seen at the Earth within 30 min after the flare onset at 15:21 UT (Arnoldy et al., 1968).

The $H\alpha$ emission comes from one of the two flare ribbons whose appearance signals the onset of the flare. The $H\alpha$ emission becomes quite intense within 5 min after onset, but it takes a long time to decay. Even after 6 h, it still exceeds the pre-flare emission by almost a factor of two. During the rapid rise-phase of the $H\alpha$ emission, the flare ribbons move apart at a rate of more than $100\ \mathrm{km\ s^{-1}}$, but as soon as the peak is reached they quickly slow to a speed of the order of $4\ \mathrm{km\ s^{-1}}$.

The soft X-rays, which are thermal in origin, are produced by the hot ($\gtrsim 10^7\,\mathrm{K}$) flare loops whose footpoints map to the $H\alpha$ ribbons. Like the $H\alpha$ emission, the soft X-rays persist for many hours, and in fact soft X-rays can still be produced as long as two days after a large event (Švestka, 1976).

Hard X-rays (>20 keV) appear only during the impulsive phase when $H\alpha$ and the soft X-ray emissions are rapidly increasing in intensity. The hard X-rays are generally thought to be produced by nonthermal electrons, and they are accompanied by radio emissions which support this interpretation (Švestka and Simon, 1969). During the impulsive phase, γ-rays and neutrons also appear which indicate the presence of high-energy protons with energies in excess of 100 MeV.

11.1.1 Eruptive Mechanisms – the Flux-Rope Model

When CMEs were first directly observed by Skylab in 1972, many researchers assumed that they were caused by the outward expansion of hot plasma produced by large flares. We now know that this is not the case, for several reasons. Firstly, less than 20% of all CMEs are associated with large flares (Gosling, 1993). Secondly, CMEs that are associated with flares often

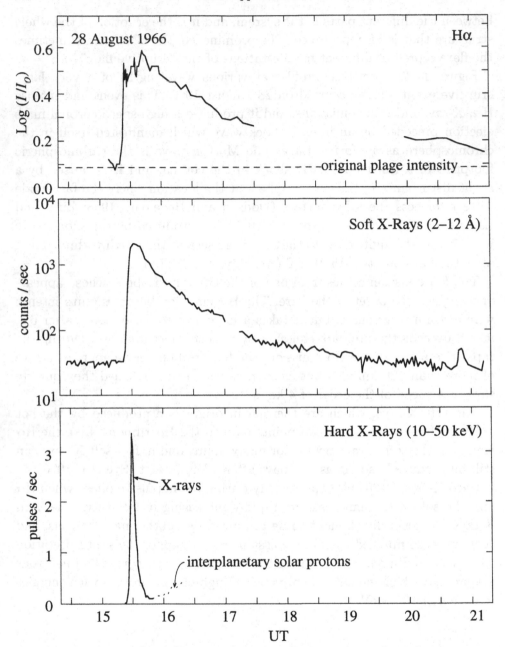

Fig. 11.1. (a) Hα, (b) soft X-ray, and (c) hard X-ray emissions for the large, two-ribbon flare of 28 August 1966. The Hα data are from the McMath–Hulbert Observatory and show the logarithm of the intensity of one of the Hα ribbons in units of the undisturbed Hα background intensity (after Dodson and Hedeman, 1968). The soft X-ray data are from Explorer 33 (from Van Allen and Krimigis, as published in Zirin and Lackner, 1969). The hard X-rays were measured by an ion chamber which is sensitive to both X-rays and MeV protons.

appear to start before the onset of the flare (Harrison et al., 1985; Wagner et al., 1981). Finally, the thermal energy produced by the flare is not sufficient to propel the CMEs at the high speeds (500–1200 km s^{-1}) that are sometimes observed (Canfield et al., 1980; Linker et al., 1990; Webb et al., 1980). At the present time, the most generally accepted explanation for the cause of CMEs is that they are produced by a loss of stability or equilibrium of the coronal magnetic field. The continual emergence of new flux from the convection zone and the shuffling of the footpoints of closed coronal field lines causes stresses to build up in the coronal field. Eventually, these stresses exceed a threshold beyond which a stable equilibrium cannot be maintained, and the field erupts. The eruption releases the magnetic energy stored in the fields associated with coronal currents, so models based on this mechanism are sometimes referred to as *storage models*.

During a CME, magnetic field lines mapping from the ejected plasma to the photosphere are stretched outwards to form an extended, open-field structure. This opening of the field creates an apparent paradox, since the stretching of the field lines implies that the magnetic energy of the system is increasing, whereas storage models require it to decrease (Sturrock et al., 1984). Barnes and Sturrock (1972) argued that this paradox can be resolved because the relaxation of the stressed field (which exists prior to the eruption) releases more magnetic energy than is consumed in stretching the field lines. In other words, the magnetic energy required to open the field should be less than the free magnetic energy stored in the corona. Following this line of thought Kopp and Pneuman (1976) proposed the scenario shown in Fig. 11.2(a) for a three-stage model of an eruptive flare. Prior to the eruption, energy is stored in a force-free arcade or flux rope. Eventually, the field erupts outwards to form a fully opened magnetic field configuration. Finally, the opened configuration reconnects to form a closed, nearly current-free field. According to Barnes and Sturrock (1972) the evolution from the first stage to the second would be an ideal MHD process, while the evolution from the second stage to the third would be a resistive MHD process. Thus, the middle stage would constitute a metastable state at an intermediate magnetic energy level.

In 1984, J. J. Aly conjectured that the above scenario is energetically impossible (Aly, 1984). Using quite general arguments, he argued that the fully opened field configuration shown in Fig. 11.2(a) must always have a higher magnetic energy than a corresponding force-free magnetic field, as long as the field is simply connected. Theoretical proofs of this conjecture have now been developed by both Aly (1991) and Sturrock (1991). Aly's result caused consternation among MHD theorists because it seems at first to imply that

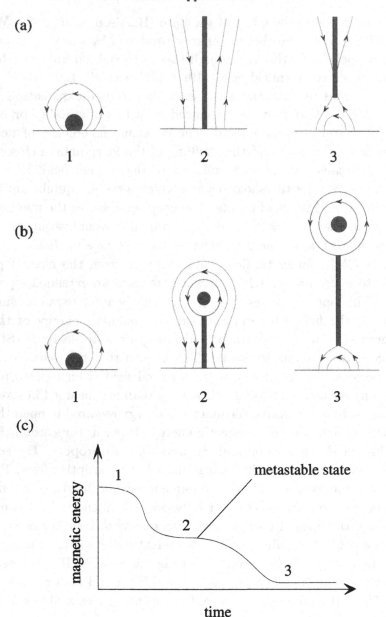

Fig. 11.2. Two scenarios which have been proposed for opening the magnetic field. (a) In the first scenario an ideal MHD process changes the closed-field configuration (1) into an open configuration (2) during the impulsive phase, and reconnection recloses the field (3) during the gradual phase. (b) In the second scenario an ideal MHD process creates a relatively short current sheet without opening the field, but magnetic flux can still escape into space if rapid reconnection occurs in this sheet. If there is no input of magnetic energy during the eruption, then the magnetic energy continually decreases during both the impulsive and gradual phases of the flare, as shown in (c).

eruptive flares are energetically impossible. However, there are several ways to avoid this predicament (Priest and Forbes, 1990). Firstly, the magnetic fields may not be simply connected but contain X- and O-points. Secondly, an ideal MHD eruption can still extend field lines as long as it does not open them all the way to infinity [Fig. 11.2(b)]. Finally, an ideal MHD eruption may be possible if it only opens a portion of the closed field lines (Wolfson and Low, 1992; Low and Smith, 1993).

One type of storage model that has received much attention tries to create an eruption by shearing the footpoints of an arcade of loops (Mikic et al., 1988; Martinell, 1990; Steinolfson, 1991; Inhester et al., 1992; Aly, 1994; Kusano et al., 1995; Amari et al., 1996). In two-dimensional force-free configurations with translational symmetry (sometimes referred to as $2\frac{1}{2}$-D configurations), shearing causes the arcade to expand smoothly outwards towards a fully opened state without ever producing an eruption. However, it is not yet known whether this is still true for three-dimensional configurations with axial symmetry. Wolfson and Low (1992) have analyzed an axially symmetric arcade of loops located along the equator of a sphere which is sheared by turning the northern and southern hemispheres in opposite directions. Their analysis, although not conclusive, suggests that an initially closed arcade of loops may erupt to form a partly opened configuration containing a current sheet.

Even if shearing an arcade does not produce an ideal loss of equilibrium or stability, it is still possible to create a rapid eruption by invoking a resistive instability. Shearing an arcade leads to the formation of a current sheet which can undergo reconnection. If the reconnection occurs rapidly, at a rate which is on the order of, say, a few Alfvén time-scales (τ_A), then a rapid eruption occurs (Aly, 1994). Figure 11.3 shows a simulation of this process by Mikić and Linker (1994). From $t = 0$ to $540\ \tau_A$, the arcade is sheared with the resistivity as near to zero as possible. After $540\ \tau_A$ the shearing is stopped and the resistivity is instantaneously increased to a value which gives an effective magnetic Reynolds number of about 10^4. This increase leads to reconnection and the formation of a flux rope which is expelled outwards, away from the Sun.

In order for the above mechanism to work, the reconnection rate must undergo a sudden transition. Prior to the eruption it must be much slower than the time-scale of the photospheric motions, so that energy can be stored in the coronal currents. After the eruption it must be fast, so that energy can be released rapidly. Thus, a complete model of the eruption process must explain why the reconnection rate suddenly changes at the time of the eruption. There are several possible mechanisms which could

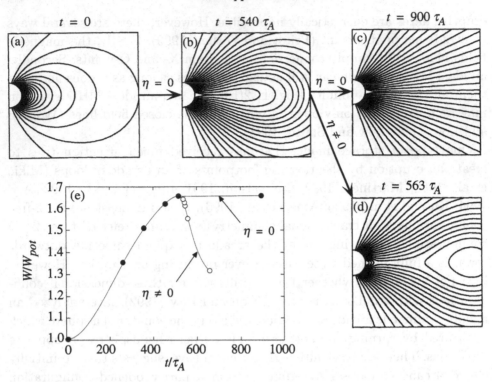

Fig. 11.3. Quasi-static evolution of an axially symmetric arcade in response to shearing of its footpoints. (a) The initial field is a Sun-centred dipole which (b) evolves into a force-free field when its footpoints in the upper and lower hemispheres are rotated in opposite directions. (c) After a rotation of 126°, the field becomes fully opened as long as the diffusivity (η) remains zero. (d) However, an eruption of a plasmoid occurs, if η is suddenly increased. (e) A plot of the corresponding time evolution of the total energy divided by the potential energy (from Mikić and Linker, 1994).

do this. For example, if the current sheet is subject to the tearing mode instability (§6.2), then reconnection will not occur until the length of the current sheet becomes longer than about 2π times its width (Furth et al., 1963). Alternatively, as the current sheet builds up, its current density may exceed the threshold of a micro-instability which creates an anomalous resistivity (Heyvaerts and Priest, 1976). The anomalous resistivity subsequently triggers rapid reconnection and the ejection of a flux rope, as in Fig. 11.3.

Sheared arcades can also be formed by the emergence of a flux rope from below the photosphere as long as the central axis of the rope lies on or below the photosphere. However, if the flux-rope axis lies above the photosphere, then the arcade will contain a flux rope whose field lines are anchored only at its ends.

A two-dimensional Flux-Rope Model for eruptive behaviour has been proposed by Priest and Forbes (1990) and developed by Forbes and Priest (1995). This model has the limitation that, because it is two-dimensional, the flux rope is not anchored at its ends, as we would expect in a realistic three-dimensional configuration. Nevertheless, it does demonstrate the basic concepts involved for a flare model based on a loss of ideal MHD equilibrium. In the model the flux rope loses equilibrium when the photospheric sources of the field approach one another, as shown in Fig. 11.4. The configuration is obtained by solving the Grad–Shafranov equation

$$\nabla^2 A + \frac{1}{2}\frac{dB_z^2}{dA} = 0 \tag{11.2}$$

in the semi-infinite xy-plane with $y \geq 0$, where B_z is the field perpendicular to the xy-plane and $A(x,y)$ is the flux function defined by $(B_x, B_y, B_z) = [\partial A/\partial y, -\partial A/\partial x, B_z(A)]$. The surface at $y=0$ corresponds to the photosphere. Equation (11.2) is used to construct an evolutionary sequence of force-free equilibria by assuming that changes in the photospheric boundary conditions occur more slowly than the Alfvén time-scale in the corona.

The photospheric boundary condition is

$$A(x,0) = A_0\,\mathcal{H}(\lambda - |x|), \tag{11.3}$$

where $\mathcal{H}$ is the Heavyside step-function and A_0 is the net flux through the photosphere in the region $x \geq 0$ (or, equivalently, the value of A at the origin). This boundary condition corresponds to two sources of opposite polarity located at $x = \pm\lambda$. If these sources are slowly moved together, then the flux rope eventually erupts upwards to create a vertical current sheet, as shown in Fig. 11.4. The field configuration with both the flux rope and current sheet present has the flux function

$$A(x,y) = \text{Re}\left[\frac{\mu I}{2\pi}\ln\left(\frac{\sqrt{\zeta^2+b^2}+i\sqrt{h^2-b^2}}{\sqrt{\zeta^2+b^2}-i\sqrt{h^2-b^2}}\right)\right.$$

$$\left. + i\frac{A_0}{\pi}\ln\left(\frac{\sqrt{\zeta^2+b^2}+\sqrt{b^2+\lambda^2}}{\sqrt{\zeta^2+b^2}-\sqrt{b^2+\lambda^2}}\right)\right] \tag{11.4}$$

for $|\zeta - ih| > a$, where $\zeta = x + iy$ is the complex coordinate, h is the flux-rope height, a is the flux-rope radius, I is the flux-rope current, and b is the height of the current sheet that lies below the flux rope. For $|\zeta - ih| \leq a$, the flux function A has the form

$$A(x,y) = f(r,I), \tag{11.5}$$

where

$$r = \sqrt{x^2 + (y-h)^2}$$

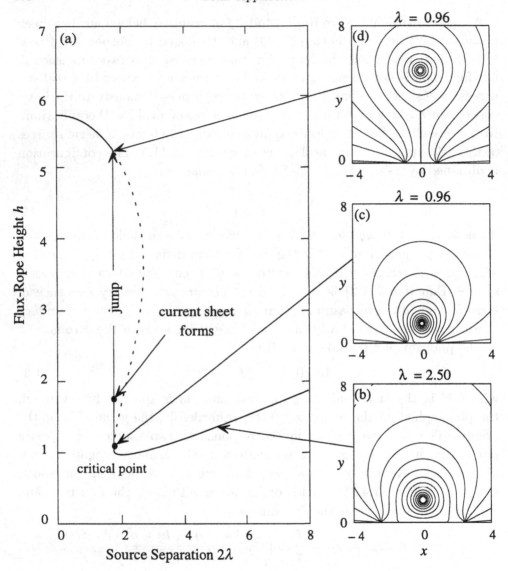

Fig. 11.4. A Flux-Rope Model. (a) Ideal MHD evolution of a two-dimensional arcade containing an unshielded flux rope of height h as the source separation (2λ)
decreases. (b), (c) The flux rope and arcade move upwards when the two photospheric field sources are pushed too close to one another. (d) In the absence of
reconnection the eruption leads to a new equilibrium containing a current sheet.

is the radial coordinate of the flux rope and $f(r, I)$ is the solution for the
equilibrium of an isolated flux rope in the absence of other current sources
as given by Parker (1974).

The axial field (B_z) inside the flux rope ($r < a$) is determined by substituting Eq. (11.5) into the Grad–Shafranov equation (11.2) and integrating.

For $r > a$ the flux function [Eq. (11.4)] is potential, and so everywhere outside the flux rope $B_z = 0$.

The equilibrium value for h is given parametrically as a function of λ by

$$\lambda = \frac{1 + \xi^2}{4\xi^2} a_0 \exp\left[\left(\frac{\pi}{4} + \ln\frac{2\lambda_0}{a_0} - \tan^{-1}\frac{1}{\xi}\right)\left(\frac{1 + \xi^2}{2\xi}\right)\right] \tag{11.6}$$

for $\xi \leq \sqrt{3}$ and

$$\lambda = \frac{(3\xi^2 - 1)\sqrt{2(\xi^2 - 1)}}{4\xi^2(\xi^2 + 1)} a_0 \exp\left\{\left[\frac{\pi}{4} + \ln\frac{2\lambda_0}{a_0} - \tan^{-1}\sqrt{\frac{\xi^2 - 1}{2\xi^2}}\right]\right.$$
$$\left. \times \left[\frac{\sqrt{2(\xi^2 - 1)}}{\xi}\right]\right\} \tag{11.7}$$

for $\xi \geq \sqrt{3}$, where

$$h = \lambda\xi \tag{11.8}$$

and ξ is the parametric variable. The constants a_0 and λ_0 are the values of a and λ at the point on the equilibrium curve where the flux-rope current has its maximum value of

$$I_0 \equiv \frac{4A_0}{\mu}$$

at $\xi = 1$.

The corresponding flux-rope current is

$$I = I_0\frac{2\xi}{\xi^2 + 1} \tag{11.9}$$

for $\xi \leq \sqrt{3}$, and

$$I = I_0\frac{\xi}{\sqrt{2(\xi^2 - 1)}}$$

for $\xi \geq \sqrt{3}$, while the radius is just

$$a = a_0\frac{I_0}{I}$$

for all ξ. The height (b) of the current sheet is

$$b = \xi\lambda\sqrt{\frac{\xi^2 - 3}{3\xi^2 - 1}}. \tag{11.10}$$

Equations (11.6)–(11.10) assume that the flux-rope radius (a) is much smaller than the flux-rope height (h). Making this assumption allows one to take advantage of the fact that the magnetic field due to all sources external to the flux rope vanishes at the equilibrium location of the flux rope (Forbes and Isenberg, 1991). For this flux-rope model there are three external sources. The first is the photospheric source current, which creates the background field, the second is the current sheet, and the third is the surface current, which arises from line tying. In the absence of the flux rope, the combined field of these three sources creates an X-line at $x = 0, y = h$. Thus, if $a \ll h$, the equilibrium inside the flux rope is approximately the same as it is when there are no external sources, and it can be solved independently of the global equilibrium. As shown in Forbes and Priest (1995), the error is of order $(a/h)^2$.

Figure 11.4(a) shows the equilibrium flux-rope height as a function of the source separation (2λ) for $a_0 = 0.1\lambda_0$. The S-shaped curve is characteristic of cusp-type catastrophes, where the highest and lowest branches are stable, but the middle branch is unstable (Poston and Stewart, 1978). If one starts with a configuration corresponding to a point on the lower branch, as shown in Fig. 11.4(b), and then moves the source regions towards each other, a catastrophe will occur when λ reaches the point where the lower and middle branches of the equilibrium curve meet. At this point the configuration [Fig. 11.4(c)] loses magnetic equilibrium, and the flux rope is thrown upwards by the imbalance which develops between the magnetic compression and tension forces acting on it (Van Tend and Kuperus, 1978; Yeh, 1983; Martens and Kuin, 1989; van Ballegooijen and Martens, 1989).

What happens after the loss of equilibrium depends on the dynamics. On the one hand, if one assumes that there is no reconnection and all the kinetic energy released by the loss of equilibrium is dissipated, then the flux rope restabilizes at the upper equilibrium shown in Fig. 11.4(d). On the other hand, if this energy is not dissipated, then the flux rope oscillates up and down between the lower equilibrium at the catastrophe point and a height somewhere above the height of the upper equilibrium. Finally, if reconnection does occur, then the flux rope continues to move upwards indefinitely, although its upward motion may be slowed once its altitude exceeds the upper equilibrium height.

The model shown in Fig. 11.4 is somewhat idealized because all of the shear in the system is contained in a flux rope that does not intercept the photosphere. However, an eruption still occurs even if the radius of the flux rope is much larger (Forbes and Priest, 1995). Shearing the arcade overlying the flux rope also enhances the energy released in the ideal MHD jump (Démoulin et al., 1991).

Insight into how reconnection is driven by an eruption (caused by a loss of ideal MHD equilibrium or stability) can be obtained by analyzing the dynamical evolution of the flux rope under the assumption that the reconnection process is not inhibited by the conductivity of the plasma. This assumption is the opposite of ideal MHD, because the field lines are not frozen to the plasma but instead move as they would through a vacuum. By assuming such rapid reconnection we can obtain an idea of how the reconnection is driven in time by the magnetic force which propels the flux rope upwards.

In the absence of a current sheet, the field outside the flux rope is determined by Eq. (11.4) with b set to zero. Although we no longer assume that field lines are frozen to the plasma at the X-line, we do assume that they are still frozen in the flux rope. Setting $A(0, h-a)$ constant provides an expression for the flux-rope current (I) as a function of its height (h), namely

$$\left(\frac{I}{I_0}\right) \ln\left(\frac{2h}{a}\right) + \tan^{-1}\left(\frac{\lambda}{h}\right) = \ln\left(\frac{2\lambda_0}{a_0}\right) + \frac{\pi}{4}, \qquad (11.11)$$

where $a = a_0 I/I_0$ as before. The magnetic energy per unit length during the eruption is (see Forbes and Priest, 1995)

$$W_m(h) = \frac{4A_0^2}{\pi\mu}\left[\left(\frac{I}{I_0}\right)^2 \ln\left(\frac{2h}{a}\right) + \frac{1}{4} + \frac{I^2}{2I_0^2}\right].$$

Ignoring gravity and assuming that all of the energy released goes into kinetic energy, we find the velocity of the flux rope is

$$\dot{h} = \sqrt{2[W_s - W_m(h)]/m}, \qquad (11.12)$$

where W_s is the stored magnetic energy (i.e., W_m at the nose-point) and m is the effective mass-per-unit length. Expression (11.12) assumes that the mass is concentrated in the flux rope and neglects the fluid motions and waves arising from a distributed mass.

Figure 11.5 shows the trajectories of the flux rope and X-line obtained by numerically integrating Eq. (11.12). The flux rope rises slowly at first because it starts from equilibrium. Its speed then increases rapidly for a while, but eventually it levels off at the velocity obtained by setting the kinetic energy equal to the total free magnetic energy stored in the configuration. By contrast, the speed of the X-line, which appears at approximately $t = 7$, is extremely fast at first but it slows rapidly with time. In this particular example, the initial speed of the X-line is infinite because the photospheric field is represented by two point sources. For non-point sources

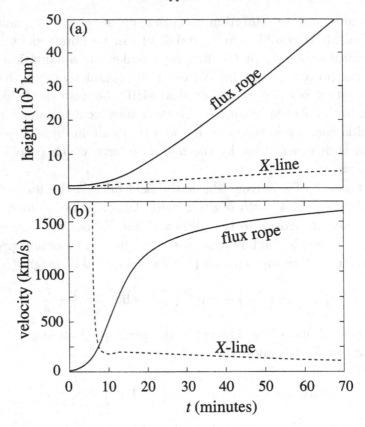

Fig. 11.5. (a) Heights and (b) velocities of the flux rope and X-line as functions of time if there is no limit to the reconnection rate.

the initial speed is finite, but in either case, the speed decreases with time except for a very slight increase during the period of rapid flux-rope acceleration.

To obtain the particular values shown in Fig. 11.5, the following scalings have been used: $\lambda_0 = 5 \times 10^4$ km, $L = 10^5$ km, $a_0 = 0.4\,\lambda_0$, $n = 5 \times 10^{16}$ m^{-3}, and $B_0 \equiv A_0/\lambda_0 = 10^{-2}$ tesla (100 G), where λ_0 and a_0 are as previously defined, L is the length-scale in the z-direction (out of the plane of Fig. 11.4), and n is the effective total density. For these values, the magnetic field strength at $x = 0$, $y = 0$ on the photosphere is $2B_0/\pi$ or 6.4×10^{-3} tesla (64 G), and the strength of the photospheric potential field at the initial height of the flux rope is B_0/π or approximately 3.2×10^{-3} tesla. The free magnetic energy stored in the configuration is 6.1×10^{25} J (6.1×10^{32} ergs), and the energy released by $t = 70$ min, the last time shown in Fig. 11.5, is 2.2×10^{25} J. The total mass of the ejecta (including the plasma in the medium surrounding the flux rope) is roughly $n\overline{m}L\lambda_0^2 = 2.1 \times 10^{13}$ kg, and the velocity of the flux rope at $t = 70$ min is about 1.5 times the Alfvén speed $v_{A0} \equiv B_0/\sqrt{\mu n \overline{m}}$, where $\overline{m}$ is the mean particle mass. The corresponding Alfvén scale-time ($\tau_{A0} = \lambda_0/v_{A0}$) is 51 s. The flux-rope speed at $t = 70$ min is ~60% of the terminal velocity of 2,400 km s^{-1}.

The speeds shown in Fig. 11.5(b) represent upper limits to the actual speeds, since a large fraction of the energy (perhaps as much as a half) will be lost to dissipative processes that heat the flare plasma and to processes such as shock-wave formation. In addition, the flux rope will have to do work against gravity, which will further lower the velocity.

Figure 11.6(a) shows the rate of magnetic energy release during the eruption for the same parameters used in Fig. 11.5. There is a low-level output prior to the formation of the X-line at $t = 7$ min, at which time the power output rises steeply. After the peak at about $t = 13$ min, the power output slowly declines to a low level, which persists for quite a long time. Although the power-output curve appears to imply that the energy release is virtually over by $t = 70$, only about 40% of the total energy has been released by this time. Thus, even when the reconnection rate is unrestrained by the conductivity of the plasma, magnetic energy is released over an extended period of time.

To understand the rapid rise and slow decay of the energy output, it is helpful to consider the mathematically simpler equations which result when the flux-rope radius is very much smaller than the scale-length λ_0. As $a \to 0$, the nose-point in the equilibrium curve occurs at $h = \lambda_0$, $\lambda = \lambda_0$. We find that an expansion about $\ln(2\lambda_0/a_0)$ reduces (11.12) to

$$\dot{h} \approx \sqrt{\frac{8}{\pi}} v_{A0} \left[\ln\left(\frac{h}{\lambda_0}\right) + \frac{\pi}{2} - 2 \tan^{-1}\left(\frac{h}{\lambda_0}\right) \right]^{1/2} + \dot{h}_0, \qquad (11.13)$$

where $\dot{h}_0$ is an initial perturbation velocity (set to 1% of the Alfvén speed at the base when calculating the curves shown in Figs. 11.5 and 11.6). The corresponding power output is

$$\dot{W}_m = -\frac{4A_0^2}{\pi\mu} \frac{\dot{h}(h - \lambda_0)^2}{h\,(h^2 + \lambda_0^2)}.$$

At early times, when $h/\lambda_0 \ll 2$, these expressions simplify even further to

$$\dot{h} \approx \sqrt{\frac{4}{3\pi}} v_{A0} \left(\frac{h}{\lambda_0} - 1\right)^{3/2} + \dot{h}_0$$

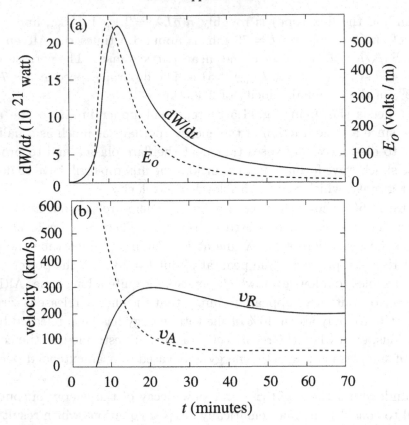

Fig. 11.6. Behaviour of the Flux-Rope Model when there is no limit to the reconnection rate. (a) The rate of change (dW/dt) of the magnetic energy (solid curve)
and the electric field (E_0) at the X-line (dashed curve) as functions of time ($1\,\mathrm{W} =
10^7\,\mathrm{ergs\ s^{-1}}$). (b) The reconnection speed (v_R, solid curve) and ambient Alfvén
speed (v_A, dashed curve) at the height of the X-line as functions of time.

and

$$\dot{W}_m \approx -\frac{2A_0^2}{\pi\mu}\left(\frac{h}{\lambda_0} - 1\right)^2 \frac{\dot{h}}{\lambda_0}. \tag{11.14}$$

Solving for h as a function of time and assuming $t \ll \lambda_0/v_{A0}$ gives

$$h \simeq \lambda_0 + \dot{h}_0 t + \frac{4v_{A0}}{5\sqrt{3\pi}}\left(\frac{\dot{h}_0}{\lambda_0}\right)^{3/2} t^{5/2} \tag{11.15}$$

$$\dot{h} \simeq \dot{h}_0 + \frac{2v_{A0}}{\sqrt{3\pi}}\left(\frac{\dot{h}_0}{\lambda_0}\right)^{3/2} t^{3/2}.$$

These expressions show that the initial power output grows algebraically in
time with a time-scale determined by the size of the initial perturbation.

As h/λ_0 tends to 2, $\dot{h}$ tends to $2v_{A0}/\sqrt{3\pi} + \dot{h}_0$, i.e., roughly the ambient Alfvén speed if $\dot{h}_0 \ll v_{A0}$. Thus, the subsequent evolution is rapid, occurring on the Alfvén time-scale (τ_{A0}).

At late times, when $h/\lambda_0 \gg 1$, but $|\ln h|$ is still much smaller than $|\ln a|$, Eq. (11.13) reduces to

$$\dot{h} \approx \sqrt{\frac{8}{\pi}} v_{A0} \left[\ln\left(\frac{h}{\lambda_0}\right) - \frac{\pi}{2} \right]^{1/2}, \tag{11.16}$$

which has the solution

$$\frac{2h}{\lambda_0} \, \mathrm{daw} \left[\sqrt{\ln\left(\frac{h}{\lambda_0}\right) - \frac{\pi}{2}} \right] \approx \sqrt{\frac{8}{\pi}} \frac{t}{\tau_{A0}} + \text{constant}, \tag{11.17}$$

where daw is the Dawson integral function (see Spanier and Oldham, 1987), and the constant is a constant of integration which can only be determined from the exact solution. If we further assume that $\ln(h/\lambda_0) \gg 1$, then Eq. (11.16) implies

$$h \approx \lambda_0 \sqrt{\frac{8}{\pi}} \left[\ln\left(\frac{h}{\lambda_0}\right) \right]^{1/2} \frac{t}{\tau_{A0}},$$

and the power output is

$$\dot{W}_m \approx \frac{4A_0^2}{\pi\mu} \frac{1}{t}. \tag{11.18}$$

Therefore, the power decays inversely with time until the flux rope reaches a height on the order of $h \approx \lambda_0^2/a_0$. For $h \gg \lambda_0^2/a_0$, the flux rope's velocity is essentially constant, having nearly reached its terminal value of

$$\dot{h}_{\text{term}} \approx \sqrt{\frac{8}{\pi}} v_{A0} \left[\ln\left(\frac{2\lambda_0}{a_0}\right) \right]^{1/2}.$$

In this two-dimensional model, the $1/t$ decay results from the fact that the compressive driving force decreases as $1/h$. We might expect, therefore, that in a three-dimensional model, where this force would decrease as $1/h^3$ at large distances, the fall-off might be more rapid. However, in a three-dimensional model there is an additional driving force due to the curvature of the flux rope (Shafranov, 1966; Garren and Chen, 1994), and this force decreases as $h^{-1} \ln(h/a_0)$, which is slightly slower than $1/h$. Thus, a three-dimensional model of an eruption would not necessarily have a faster decay.

The rate at which the power output decays is much slower than the rate at which the power increases during the growth-phase. The time required for the power to increase by a factor of 100 during the growth-phase is about 1.5 Alfvén time-scales, while the time required for the power to decrease by the same factor during the decay-phase is about 100 Alfvén time-scales. It has sometimes been assumed that the difference between the rates of rise and decay in flare intensity reflects the time-scales of different physical processes. For example, some flare models assume that the rise-time indicates the rate of reconnection, while the decay-time indicates the rate of radiative or conductive cooling. In our example above, both the rise-time and decay-time result from the energy injection process alone. The rise-rate is determined by the Alfvén scale-time at the base of the corona, since the eruption is driven by the loss of MHD equilibrium that occurs in this region. The decay-time, in contrast, is determined by the fall-off of the magnetic driving force with height as $1/h$.

The variation of the electric field at the X-line is shown by the dashed curve in Fig. 11.6. This electric field is a direct measure of the reconnection rate in two-dimensional systems, since the electric field is the rate at which magnetic flux passes though the X-line. The electric field does not rise above zero until $t = 7$ min, because the eruption starts in this particular model before any reconnection occurs.

For the flux-rope model of the previous section, the X-line location is

$$y_0 = \left[\frac{\lambda h(hI_0 - 2I\lambda)}{\lambda I_0 + 2Ih} \right]^{1/2} \tag{11.19}$$

and the electric field (E_0) there is

$$E_0 = -\frac{\partial A(0, y_0)}{\partial t} = -\frac{\partial A}{\partial y_0}\frac{\partial y_0}{\partial t} - \frac{\partial A}{\partial h}\frac{\partial h}{\partial t} = -\frac{\partial A}{\partial h}\dot{h},$$

since $\partial A/\partial y_0 = B_x(y_0) = 0$ at the X-line. Substituting

$$A(0, y_0) = \frac{A_0}{\pi}\left[2\frac{I}{I_0}\ln\left(\frac{y_0 + h}{h - y_0}\right) + 2\tan^{-1}\left(\frac{\lambda}{y_0}\right) \right], \tag{11.20}$$

which is obtained by setting $b = 0$ and $y = y_0$ in Eq. (11.4), leads to

$$E_0 = -\frac{2}{\pi}A_0\dot{h}\left[\frac{2y_0(I/I_0)}{y_0^2 - h^2} + \frac{\partial(I/I_0)}{\partial h}\ln\left(\frac{h + y_0}{h - y_0}\right) \right], \tag{11.21}$$

where y_0 is given by Eq. (11.19) and I/I_0 by Eq. (11.11).

Let us now assess the effect that plasma conductivity would have on reconnection in the above model by considering the velocity at which field lines are moving towards the X-line. At the X-line itself, the field-line velocity is infinite (§1.4), but in the same spirit as for the steady-state analysis of Chapter 4, we can define a reconnection velocity (v_R) as the X-line electric field divided by a characteristic field strength. The simplest choice is to use the background photospheric field at the height of the neutral line. To within a factor of two, or better, the strength of this field is the same as the maximum value of the total field along the line $y = y_0$. Therefore, the photospheric contribution to the field at the X-line is representative of the ambient field at the height of the X-line. With this definition we have

$$v_R = \frac{E_0}{B_p(0, y_0)},$$

which is shown in Fig. 11.6(b) as a function of time. We now compare this velocity to a simple model of the ambient Alfvén speed. Assuming the gravitational acceleration to be constant with height and the corona to be isothermal gives an Alfvén speed which varies as

$$v_A = \frac{B_p(0, y_0)}{\sqrt{\mu \, \overline{m} \, n_0}} \, e^{y_0/(2H_g)},$$

where $H_g = T_0/g$ is the gravitational scale-height and n_0 is the density at the base ($y = 0$). Here we assume the temperature (T_0) is such that $H_g = 10^5$ km.

For the example shown in Fig. 11.6, v_R exceeds v_A at approximately $t = 15$ min, near the peak in the power output. Since v_R cannot be expected to exceed v_A in a realistic plasma, we can deduce that at $t = 15$ a current sheet must form at the X-line.

The inability of the plasma to reconnect at a rate which exceeds v_A will reduce the power output near the peak but prolong the time-scale of the decay. Recently, Lin and Forbes (1999) have reworked the above model without making the assumption that the reconnection rate is unlimited. They find that limiting the reconnection rate to more realistic values (i.e., $v_R < v_A$) leads to the formation of an extended current sheet and an increase by a factor of 10, or more, in the decay time-scale.

A numerical simulation of an eruption caused by a loss of magnetic equilibrium is shown in Figs. 11.7 and 11.8 (Forbes, 1991). The simulation uses a flux-corrected transport code to solve the resistive MHD equations for a magnetic Reynolds (Lundquist) number of about 200. In this simulation the initial magnetic field is the sum of a quadrupole background at depth d, a simple current filament centred at $y = h$, and a corresponding image filament at $y = -h$. The initial field configuration is the unstable equilibrium at the estimated location of the nose point, and the maximum field strength along the photospheric boundary is 0.01 tesla (100 G). The panel labelled $t = 0$ in Fig. 11.7 shows the initial magnetic field configuration when the filament is positioned at the catastrophe point at $h = d$. Here a, the radius of the filament, is set at 0.05 times d, and initially the plasma is at rest and has uniform entropy. The boundary conditions at the photosphere ($y = 0$) are

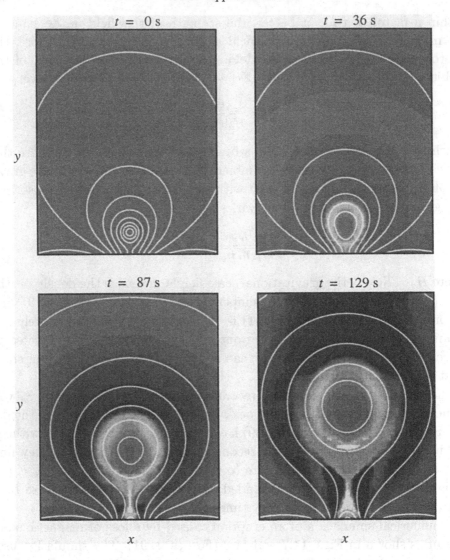

Fig. 11.7. Resistive MHD simulation based on the two-dimensional flux-rope model of Fig. 11.4. White curves are magnetic field lines, while the grey scale corresponds to temperature variations. White regions have the highest temperature ($>10^8$ K in the absence of cooling processes), while black regions have the lowest. The magnetic Reynolds (Lundquist) number is about 200, many orders of magnitude smaller than expected for the Sun.

$j = 0$, $v_x(x, 0, t) = 0$, and $v_y(x, 0, t) = 0$. The line-tying condition ($j = 0$) follows from the requirement that field lines be anchored to the base (see Forbes, 1990, for further details on the boundary conditions).

Figure 11.8 shows the shock, flux-rope, and X-line trajectories as functions of time. Here the numerical results have been scaled using a length-

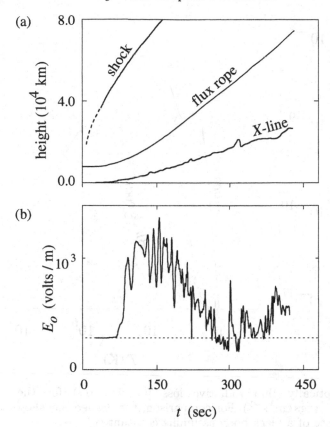

Fig. 11.8. (a) Trajectories of the shock, flux rope, and X-line for the simulation shown in Fig. 11.7. (b) The electric field at the X-line as a function of time.

scale of 10^4 km, a maximum photospheric field strength of 0.01 tesla (100 G), a coronal density of 10^{15} m^{-3}, and an ambient coronal temperature of 10^6 K. The flux rope is initially in equilibrium, but the equilibrium is unstable. Consequently, the rope rises very slowly at first. As it gains speed, it moves higher, and rapid reconnection starts to occur at the X-line which forms below it. By $t \approx 400$ s the flux rope is ejected from the top of the numerical box, and the calculation is stopped. Similar results have been achieved in a numerical simulation by Wu et al. (1995), which also incorporates a helmet streamer lying above the flux rope.

The principal limitation of the flux-rope model as discussed above is that it is two-dimensional. This makes it easier to create an eruption because the ends of the flux rope are not anchored to the photosphere as they would be in a three-dimensional model. It seems unlikely that the additional anchorage could completely prevent an eruption because the poloidal flux driving the eruption can be made arbitrarily large relative to the flux passing out

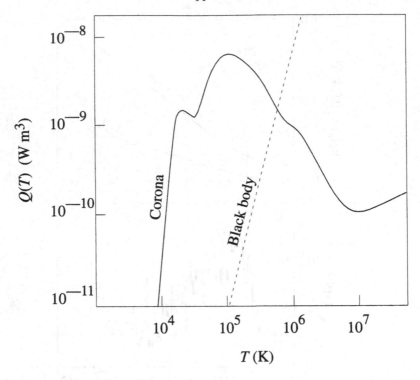

Fig. 11.9. Optically thin radiative loss function (Q) for the solar corona ($1\,\mathrm{W\,m^3} = 10^{13}\,\mathrm{ergs\,cm^3\,s^{-1}}$). For comparison, the dashed line shows the temperature dependence of a black body assuming constant volume.

of the ends of the rope. Furthermore, there is an additional outward driving force caused by the curvature of the flux rope, and this force can compensate for the effect of the anchoring at the ends (Steele et al., 1989). Even more important, perhaps, is that in three dimensions the flux rope can be kink unstable, and this can also facilitate an eruption.

11.1.2 Flare Loops

Flare loops range from temperatures of 10^4 K to 3×10^7 K, with the cooler loops nested below the hotter ones. They have traditionally been called "post"-flare loops, but this is a misnomer which we prefer to avoid, since it is now clear that the energy release continues through most of their lifetime. Remarkably, the loops at the low temperature are two orders of magnitude cooler than the surrounding coronal plasma. These super-cool loops are formed from the hot loops by a radiative thermal instability, which is possible in the solar atmosphere because of the non–black-body behaviour of the corona (Parker, 1953). As shown in Fig. 11.9, the coronal radiative

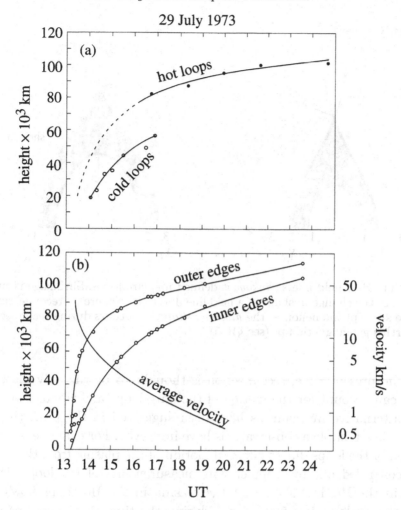

Fig. 11.10. Evolution of flare loops and ribbons for an eruptive flare on 29 July 1973 that was well observed by Skylab. (a) Loop heights and (b) ribbon separation distance and velocity as functions of time (after Moore et al., 1980).

loss function (§1.2.1) decreases with temperature in the range from 10^5 K to 10^7 K, and in this range a temperature perturbation can lead to run-away cooling (Cox, 1972).

As we mentioned in Section 11.1, the footpoints of the flare loops map to $H\alpha$ emission features in the chromosphere that are known as flare ribbons. The outermost edge of the hot X-ray loops maps to the outer edge of the ribbons (Schmieder et al., 1996), while the inner edge of the cool $H\alpha$ loops maps to the inner edge of the ribbons (Rust and Bar, 1973). During the course of a flare, the separation between the ribbons increases, and the loops grow larger with time. Figure 11.10 shows an example of such an evolving loop system.

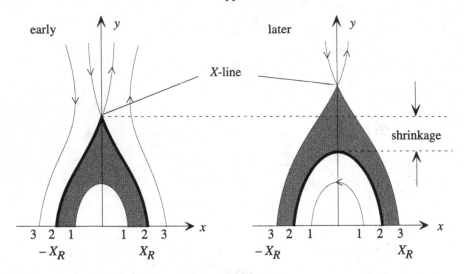

Fig. 11.11. How field line *shrinkage* is defined for flare loops. Shrinkage is simply a measure of the change in shape of a field line due to its closure by reconnection, and it is the same phenomenon as the *dipolarization* that occurs during reconnection in the Earth's geomagnetic tail (see §10.5).

At the present time, reconnection is thought to be the only mechanism which can account for the motion of the flare loops and ribbons. Although some alternative mechanisms have been suggested in the past, they have become less tenable as observations have improved. For example, simple expansion of the loops due to outward motion of the plasma from the flare site has been ruled out by Doppler-shift measurements of Hα loops. Doppler shifts in the Hα line show that the plasma in the Hα loops flows downward at speeds of 100–500 km s^{-1} during the time that the loops appear to be expanding (e.g., Schmieder et al., 1987). Thus, the loop motions are not due to mass motions of the plasma, but rather to the continual propagation of an energy source onto new field lines. Furthermore, the objection to reconnection that Alfvénic outflows are not observed is flawed on two accounts, namely, that absence of evidence is not evidence of absence and also that enhanced outflow pressure (especially below a downflowing jet) can make the outflowing jet move much slower than the Alfvén speed (§4.2.2).

In a two-dimensional configuration, there is a simple relation between the rate of reconnection and the motion of the chromospheric footpoints of the separatrices (Forbes and Priest, 1984b). Consider the configuration shown in Fig. 11.11. Using Faraday's equation, we can express the electric field (E_0)

at the X-line as

$$E_0 = -\frac{\partial A_0}{\partial t} - \nabla\Phi = -\frac{\partial A_0}{\partial X_0}\frac{\partial X_0}{\partial t} - \nabla\Phi$$

or

$$E_0 = B_y(X_0, 0)\,\dot{X}_0, \tag{11.22}$$

where A_0 is the magnitude of the flux function at the separatrix, X_0 is the location of the separatrix footpoint, $B_y(X_0, 0)$ is the vertical magnetic field located at X_0, and $\dot{X}_0$ is the apparent velocity of the separatrix footpoint. The $\nabla\Phi$ term is absent from Eq. (11.22) because the configuration is two-dimensional. Consequently, $\nabla\Phi = \partial\Phi/\partial z\,\hat{\mathbf{z}}$ must be a constant in order for it not to depend on z. Since the massive inertia of the photosphere combined with its high conductivity means that the convective electric field there vanishes, this constant must be zero. Relation (11.22) is model-independent except for the assumptions that an X-line exists, that the configuration is two-dimensional, and that the field lines are tied to a stationary photosphere.

If one assumes that the outer edges of the flare ribbons lie at the separatrix footpoints, then it is possible to use Eq. (11.22) to determine the electric field in the corona as a function of time (Forbes and Priest, 1982a). This exercise has been carried out by Poletto and Kopp (1986), who obtained the results shown in Fig. 11.12 for the 29 July 1973 flare observed by Skylab. A few minutes after flare onset the inferred electric field for this event reaches peak values of nearly $200\,\mathrm{V\,m^{-1}}$, but then it declines until it is an order of magnitude weaker a few hours later.

The electric fields deduced from the ribbon motion are exceedingly strong. If electrons or protons were to travel unobstructed along the entire length of a typical arcade (10^5 km), they would gain an energy of 10 GeV (see §13.4.2). The electric fields are also several orders of magnitude larger than the Dreicer electric field (Sturrock, 1994), so it is not surprising that strong particle acceleration occurs during a flare.

Although loop and ribbon motions are strong evidence for reconnection (Kopp and Pneuman, 1976), the evidence they provide is circumstantial. In principle, X-ray telescopes of sufficient spatial resolution and sensitivity should be able to provide direct evidence by imaging the reconnection site itself as it moves upwards in the corona. However, the ability to determine whether or not there is a reconnection site in the corona depends very much on theoretical expectations of what such a site should look like. During the past few years, high-resolution images obtained from the Hard X-ray Telescope (HXT) and the Soft X-ray Telescope (SXT) on

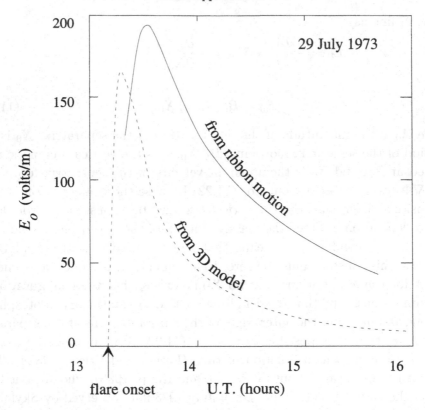

Fig. 11.12. Coronal electric field inferred by Poletto and Kopp (1986) for the flare on 29 July 1973, using magnetograms and Hα observations. The solid curve is obtained from the simple two-dimensional formula (11.22), while the dashed curve is based on a three-dimensional potential extrapolation of the surface field.

the Japanese–US–UK Yohkoh satellite show several features that are highly suggestive of a reconnection site in the corona. These features include: a hard X-ray source located above the soft X-ray loops (Sakao et al., 1992; Masuda et al., 1994; Bentley et al., 1994); cusp structures suggestive of either an X-type or a Y-type neutral line (Acton et al., 1992; Tsuneta, 1993; Doschek et al., 1995); bright features at the tops of the soft X-ray loops (Feldman et al., 1984; Tsuneta et al., 1992; McTiernan et al., 1993); and high-temperature plasma along the field lines mapping to the tip of the cusp (Tsuneta, 1996).

Some of the cusp-shaped loops observed by Yohkoh have a linear, trunk-like feature which extends from the top of the cusp all the way down to the inner arch of the flare loop system. An example of such a trunk feature is shown in Fig. 11.13 for a flare that occurred on 21 February 1992. The hottest regions in the loop system do not lie in the trunk feature but along

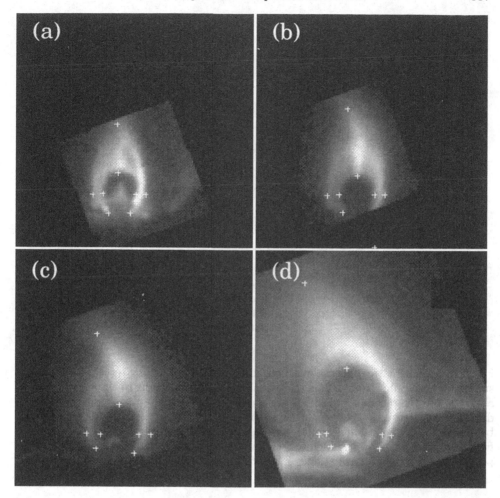

Fig. 11.13. Images of the flare system of 21 February 1992 at four different times: (a) 3:12 UT, (b) 4:11 UT, (c) 5:02 UT, and (d) 12:19 UT. The intensity levels in each frame have been renormalized to maintain high contrast. The crosses show the locations of the points used to measure the widths and heights of the inner and outer edges of the loop system as well as the locations of the apparent footpoints.

the edges of the cusp formed by the outermost loop (Tsuneta, 1996). Along these edges the temperature is 1.6×10^7 K, and the density inferred from the emission measure assuming a line of sight of 5×10^4 km is 4×10^{15} m^{-3}. The temperatures in the trunk feature itself range from 9×10^6 k at the top to 6×10^6 K at the bottom. For the same line of sight as above, the observed emission measure leads to a density of 5×10^{15} m^{-3} at the top of the trunk to one of 2.5×10^{16} m^{-3} at the bottom. Thus, the trunk feature is both cooler and denser than the plasma surrounding it.

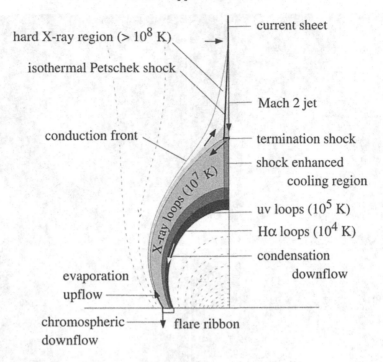

hard X-ray region (> 10^8 K)

current sheet

isothermal Petschek shock

Mach 2 jet

conduction front

termination shock

shock enhanced
cooling region

X-ray loops (10^7 K)

uv loops (10^5 K)

Hα loops (10^4 K)

condensation
downflow

evaporation
upflow

chromospheric
downflow

flare ribbon

Fig. 11.14. Schematic diagram of a flare loop system formed by reconnection in the supermagnetosonic regime. This regime is most likely to occur in the early phase of a flare when the reconnecting fields are strong. It has both upward- and downward-directed jets, but only the region below the downward jet has high-density plasma, because in two-dimensional models chromospheric evaporation occurs on just those field lines that lie below the X-line. Solid curves indicate boundaries between various plasma regions, while dashed ones indicate magnetic field lines.

Figures 11.14 and 11.15 show a theoretical model which explains the loop structures in terms of the processes of reconnection and evaporation. These figures incorporate the early ideas of Carmichael (1964), Sturrock (1968), Hirayama (1974), and Kopp and Pneuman (1976), as well as Cargill and Priest (1982), who were the first to suggest reconnection slow-mode shocks as the main source of heating. However, most of the details of the figures come from the results of various simulations of reconnection (Forbes and Malherbe, 1991), evaporation (Nagai, 1980; Somov et al., 1982; Cheng, 1983; Doschek et al., 1983; Pallavicini et al., 1983; Fisher et al., 1985) and condensation (Antiochos and Sturrock, 1982). According to this model, flare loops are created by chromospheric evaporation on field lines mapping to slow-mode shocks in the vicinity of the neutral line (Forbes and Malherbe, 1986). These slow shocks are similar to those proposed originally by Petschek

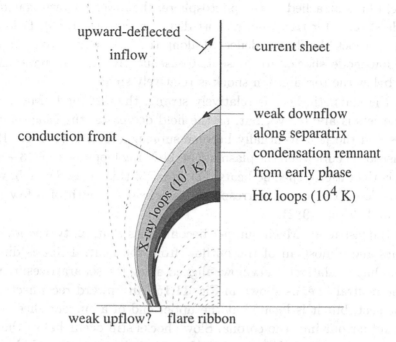

Fig. 11.15. Schematic diagram of a flare loop system formed by reconnection in the submagnetosonic regime. This regime is most likely to occur when the reconnecting fields are weak. Here the downward jet of Fig. 11.14 is replaced by a weak bifurcated flow along the field lines mapping from the tip of the current sheet to the chromosphere. Because of the weaker fields, the evaporation process is greatly reduced and the plasma density in the loops becomes too low to trigger a thermal condensation. However, condensation remnants may remain lower down as a result of an earlier supermagnetosonic phase.

(1964), except that the conduction of heat along the field lines causes them to dissociate into isothermal shocks and conduction fronts, as shown in Fig. 11.14. The shocks annihilate the magnetic field in the plasma flowing through them, and the liberated thermal energy is conducted along the field to the chromosphere. This in turn drives an upward flow of dense, heated plasma back towards the shocks and compresses the lower regions of the chromosphere downward.

In order for strong slow shocks to form on the field lines below the neutral line, the outflow from it must be supermagnetosonic with respect to the fast-mode wave speed. If the magnetic fields are sufficiently strong, the outflow from the neutral line produces two supermagnetosonic jets, one directed upward and the other downward. Because of the obstacle presented by the closed field-lines attached to the photosphere, the lower jet terminates at a fast-mode shock after travelling a short distance (Forbes, 1986). Below the termination shock the flow is deflected along the field, and only weak field-aligned slow-mode shocks are present. Consequently, the magnetic energy released below the termination shock is relatively small.

When the magnetic field is relatively strong, the fast-mode Mach number of the jets is about two, but, as the field decreases, the Mach number decreases and the jets eventually become submagnetosonic (Forbes, 1986). The transition occurs when the plasma β in the X-ray loops exceeds $(3 - \gamma)/\gamma$, where γ is the ratio of specific heats. For $\gamma = 5/3$ this gives $\beta = 4/5$, which for typical loop parameters corresponds to a field strength of a few gauss (Soward and Priest, 1982).

When the fast-mode Mach number becomes less than unity, the lower jet disappears and almost all of the outflow from the neutral line is directed upwards. Only a relatively slow flow, aligned along the separatrices, remains below the neutral line, as shown in Fig. 11.15. An upward reconnection jet is still present, but it is located at the upper end of a current sheet which may extend far out into the corona. Slow shocks still occur below the neutral line at the lower tip of the current sheet, but they are aligned along the separatrices and are much weaker than the Petschek-type shocks. Because of the weaker shocks, the evaporation process is greatly reduced and the plasma density in the loops becomes too low to trigger a thermal condensation. For a while, condensation remains on the loops at lower altitude, but, eventually, as the older loops continue to cool and disappear, the trunk feature vanishes. This process accounts for the early disappearance of the trunk feature in Fig. 11.13.

Švestka et al. (1982) argued convincingly that the ribbon width (W, say) is approximately determined by

$$W = V_R \tau_{\text{cr}}, \qquad (11.23)$$

where V_R is the ribbon velocity and τ_{cr} is the cooling time due to conduction and radiation acting together. Švestka et al. (1982) calculated τ_{cr} from the observed ribbon width and ribbon velocity and showed that this value is consistent with the cooling times inferred from the emission measure and the thermal temperature. For the 29 July 1973 flare observed by Skylab, they found $\tau_{\text{cr}} \approx 15$ min ($n \approx 10^{17}\,\text{m}^{-3}$) shortly after flare onset, and $\tau_{\text{cr}} \approx 240$ min ($n \approx 10^{16}\,\text{m}^{-3}$) 10 h after onset. Cooling by thermal conduction is important for the hottest loops, but radiative cooling becomes significant as the loops cool (Cargill et al., 1995).

The thickness (W_j) of the region of the ribbon which maps to the jet is roughly

$$W_j = \left(\frac{V_R}{V_j}\right) L_j, \qquad (11.24)$$

where V_j is the speed of the plasma flow in the jet and L_j is its length. According to the reconnection-evaporation model, V_j is approximately the Alfvén speed in the dense loops, which is about a factor of 10 smaller than the Alfvén speed in the corona (Forbes et al., 1989). Assuming $V_j = 500$ km s^{-1}, $V_R = 10$ km s^{-1}, and $L_j = 5 \times 10^4$ km gives $W_j = 10^3$ km, which is a little more than 1 arc sec. Because the value of W_j is of the same order as the thin (≈ 2 arc sec) region of red-shifted chromospheric plasma observed by Švestka et al. (1980), it is likely that this red-shifted region maps to the jet.

The mapping of Fig. 11.14 implies that the trunk feature seen in the SXT images cannot be a supermagnetosonic reconnection jet along its entire length. If a jet were present, it would have to be confined to the uppermost part of the feature. A more likely explanation, at least for the lower part of the trunk feature in Fig. 11.13, is that it is created by the onset of the thermal instability which leads to the cool Hα loops lower down. A very similar feature has been found in numerical simulations that include radiation but not conduction (Forbes and Malherbe, 1991). In these simulations (see Figs. 11.16 and 11.17) the termination shock triggers a thermal instability in the downstream region, and, as the reconnection site travels upward, the condensing region forms a trunk-like feature on the static loops.

More recently, Yokoyama and Shibata (1997) have carried out simulations which include field-aligned conduction and a chromosphere but do not yet include radiation. These simulations provide convincing support that the slow-mode shocks seen in Fig. 11.17 do indeed dissociate into isothermal slow-mode shocks and conduction fronts as indicated in Fig. 11.14. As in this figure, dense loops are created in the simulation by the ablation (i.e., evaporation) of the chromosphere by the conduction of heat along field lines mapping from the shocks to the chromosphere.

Just as for reconnection in the geomagnetic tail, reconnection in flare loops leads to a relaxation of stretched field lines. In the geomagnetic tail, this relaxation is sometimes referred to as *dipolarization* (see §10.5.1), but in flare loops it is referred to as *shrinkage* – a term coined by Švestka et al. (1987). These authors noticed that the altitude of the cool Hα loops never reaches the altitude of the hot X-ray loops, and they inferred that this was because the field lines on which the cool loops lie have shrunk.

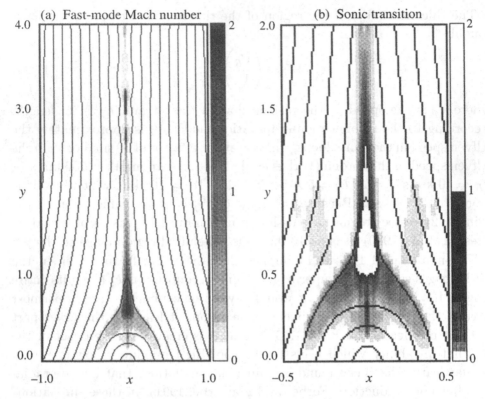

Fig. 11.16. Magnetic field lines and fast-mode Mach number (grey scale) in a numerical experiment of reconnection with radiative cooling (Forbes et al., 1989). The field lines are tied to the base but are free to move at the other boundaries; (a) shows the fast-mode Mach number in the full computational domain, while (b) shows a close-up of the downward reconnection jet. In (b) the grey scale has been reset to show the region of supermagnetosonic flow as a white triangle embedded within the lower jet. The long sides of this triangle are slow-mode shocks, while the short side is a fast-mode shock.

What causes the shrinkage is the upward retreat of the neutral line. When a field line is connected to the neutral line, it is cusp-shaped, but, as the neutral line moves further up, the field line becomes rounded. The expected shrinkage from this effect is approximately 20–30% for typical flare loops, and this matches the observed values fairly closely (Forbes and Acton, 1996). The downward motion of individual field lines has been inferred from plasma motions by Hiei and Hundhausen (1996). They have found that there are small downward-moving features in the apexes of the large X-ray loops formed after a CME. These features are relatively cool and move downwards very slowly at a rate of only 1–2 km s^{-1} – just the speed expected

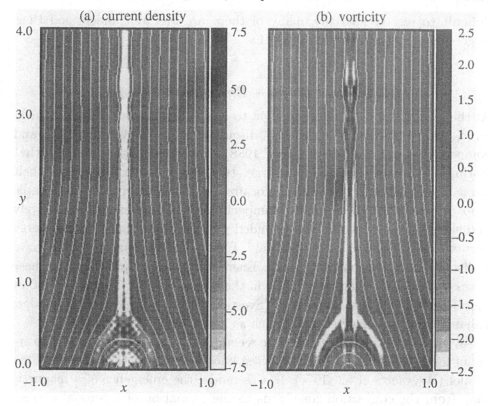

Fig. 11.17. (a) Current density and (b) vorticity for the same simulation as shown in Fig. 11.16. Petschek-like slow-mode shocks are apparent in (a) as a split in the current sheet below the lowest X-line. The small angle ($\approx 7°$) between the shocks agrees with the angle predicted by Petschek theory, even though a steady state has not been achieved. The simulation uses a resistivity that is uniform in space and constant in time ($R_m \approx 10^3$), and the current sheet is well resolved numerically by using a nonuniform grid.

for shrinkage. The features may be nascent condensations, which are frozen to the field and therefore track the true motion of the field lines rather than the apparent motion of the loop system.

11.2 Impulsive, Compact Phenomena

There is a host of small-scale impulsive phenomena in the corona and chromosphere that are thought to involve reconnection. These include small flares (§11.2.1), X-ray bright points (§11.3.1), transition-region explosive events (§11.2.3), surges, and sprays (§11.2.4). Because these phenomena are difficult to resolve spatially, many of them are less well understood than larger-scale phenomena such as CMEs.

11.2.1 Small Flares – Emerging Flux Model

At the present time, there are thought to be two basic types of flare, namely large, two-ribbon flares, such as we discussed in the previous section, and *compact flares* (Priest, 1982; Zirin, 1988). There are several features that distinguish compact flares from large, two-ribbon flares. Firstly, as their name implies, these flares are more localized, often appearing to form simple loops (Švestka, 1976). Secondly, compact flares release energy impulsively and show little evidence of the extended gradual phase that occurs in large events (Jakimiec et al., 1987; Simnett, 1992). Thirdly, compact flares have a distinct compositional signature. Energetic particles produced in these flares typically show an enrichment in the isotope He^3 by a factor of 100 or more over the coronal abundance (Švestka, 1976), as well as more modest enhancements of heavy elements such as iron (Meyer, 1993).

Several models involving magnetic reconnection have been proposed to account for compact flares. One of the best known of these is the Emerging Flux Model (Heyvaerts et al., 1977). In this model the emergence of a magnetic loop from the convection zone leads to the formation of a coronal current sheet as it pushes up against the pre-existing fields as shown in Fig. 11.18. The model assumes that the plasma resistivity in the current sheet remains very low until the current sheet is forced to an altitude where the ratio of the current to mass density exceeds the threshold for a micro-instability. When this happens the resistivity becomes anomalous, and a rapid energy release ensues. Unlike the flux-rope model discussed in Section 11.1, the trigger mechanism in the emerging flux model is not an ideal MHD process.

If the emerging flux comes up into a region of closed magnetic loops, as in Fig. 11.18, then the footpoints of all four separatrices lie in the photosphere (Syrovatsky, 1982). This means that reconnection may lead to four ribbons rather than two, and such multiple-ribbon events have in fact been observed (Machado et al., 1988). The Emerging Flux Model has also proposed a scenario for large two-ribbon flares, which it suggests may either occur spontaneously when a threshold for eruption is exceeded (§11.1.1) or may be triggered by emerging flux.

Numerical simulations of the flux emergence process have been carried out by Forbes and Priest (1984a) and Shibata et al. (1990). These simulations

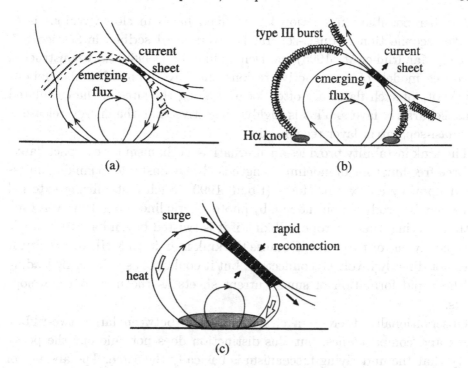

Fig. 11.18. The Emerging Flux Model of Heyvaerts et al. (1977) for a small flare. (a) The preflare phase when the emerging flux slowly reconnects with the overlying field. Slow-mode shocks (dashed curves) radiate from a small current sheet and heat the plasma that passes through them (striped region). (b) The impulsive phase caused by the onset of turbulence and anomalous resistivity in the current sheet when it reaches a critical height. The electric field generated by the sudden enhancement in the reconnection rate accelerates the particles, which produce hard X-rays and type III radio bursts. In the main phase (c), quasi-steady reconnection leads to extensive heating.

show that, during the active phase of emergence, the reconnection is of flux pile-up type, but it becomes Petschek-like once the emergence has stopped. If the plasma β is sufficiently low, supermagnetosonic reconnection jets are produced with a fast-mode Mach number of about 2. The reconnection in these simulations is generally quite fast, so that the current sheet created by the emergence quickly dissipates. Although this rapid dissipation can be attributed to the relatively low magnetic Reynolds numbers used in the simulations ($R_{me} < 10^3$), the fact that the reconnection is of flux pile-up and Petschek types implies that it will remain fast as the magnetic Reynolds number increases. This in turn suggests that it is difficult to maintain a large-scale current sheet in the corona and so store energy in it.

Another possible explanation for confined flares in closed regions is to invoke reconnection in quasi-separatrix layers, as described in Section 8.7. Indeed, Mandrini et al. (1995) and Démoulin et al. (1996b) have constructed force-free models of many active regions and have found that the ribbons and knots of such flares are often located along the sites of the calculated quasi-separatrix layers. This is highly suggestive that the energy release is by quasi-separatrix layer reconnection.

The kink instability provides an alternative mechanism for compact flares. A force-free flux rope of indefinite length is always unstable to kinking unless acted upon by an external force (Hood, 1990). Such a stabilizing external force may be produced on the Sun by photospheric line-tying. However, even with line-tying, the flux rope will kink if it is twisted beyond a critical angle of typically one or two turns. Because the kink is an ideal MHD instability, it does not directly involve reconnection, but it could act as a driver by leading to the rapid formation of small current sheets as the flux rope develops knots.

Observationally, there is a clear distinction between large, two-ribbon flares and compact ones, but this distinction does not rule out the possibility that the underlying mechanism is basically the same. The absence of a sustained gradual phase in small flares might simply be a consequence of the fact that small flares do not release enough energy to open the field structure. Compositional differences in the energetic particle spectrum might be due to the fact that compact flares are confined to the lower corona and are less likely to produce an interplanetary shock wave, which can serve as a site for diffusive particle acceleration (Lee and Ryan, 1986).

Support for a single underlying mechanism comes from the fact that the frequency distribution of both small and large flares follows a power-law function of the form

$$\frac{dN}{dW} \sim W^{-a},$$

where dN is the number of events with energy between W and $W + dW$ and $a \approx 1.8$. Lu et al. (1993) in their avalanche model argued that this continuous power-law distribution occurs because large flares are simply the result of a cascade of small, elemental reconnection events. Different explanations have been proposed by Rosner and Vaiana (1978), Litvinenko (1994), and Vlahos et al. (1995) which suggest that the power law is a result of a stochastic trigger mechanism.

Extremely small flares, known as *nanoflares*, have been proposed by Parker (1988) as a mechanism for heating the corona. Continual shuffling of the footpoints of coronal field lines by photospheric convection leads to the spontaneous formation of numerous current sheets, which dissipate the energy by means of reconnection. In order for sufficient energy to be stored, the reconnection process must not be too fast. Otherwise, the work done by photospheric convection is not sufficient to heat the corona. The reconnection rate required is related to the heating flux (F_H) by

$$v_R = \frac{v_0^2 B^2}{\mu F_H} \frac{d}{L},$$

where v_0 is the characteristic convection speed, B is the coronal magnetic field strength, F_H is the power required to maintain a hot corona, d is the diameter of a twisted flux tube, and L is its length. For $v_0 = 0.4$ km s^{-1}, $B = 0.01$ tesla $(100$ G$)$, $F_H = 10^4$ W m^{-2} (i.e., 10^7 ergs m^{-2} s^{-1}), $d = 10^4$ km, and $L = 10^5$ km, we find $v_R = 0.13$ km s^{-1}. This speed is just a third of the convective velocity, which means that the reconnection rate only has to be a factor of about 3 smaller than the rate at which the fields are driven together by convection. Since the reconnection speed (v_R) is 4–5 orders of magnitude smaller than the Alfvén speed in the corona, the reconnection rate is considerably slower than the maximum Petschek rate. Nonetheless, it is still faster than the Sweet–Parker speed of $R_{me}^{-1/2} v_A \approx 10^{-3}$ km s^{-1}.

11.2.2 Triggering of Turbulence in a Solar Flare Current Sheet

The importance of plasma turbulence in current sheets has been stressed for the *emerging flux model* of solar flares (§11.2.1, 1977) and has been investigated further by Heyvaerts and Kuperus (1978). The threshold for micro-instabilities is about 80 A m^{-2} in the solar corona and is only likely to be exceeded in intense current concentrations such as sheets. Heyvaerts and Kuperus delineated the different regimes of current-sheet behaviour, depending on its length (L) and width (l). They defined a short hydrodynamic time-scale $(\tau_l = l/c_s = 10^{-2} l/T^{1/2}$ s$)$, a long hydrodynamic time $(\tau_L = L/c_s = 10^{-2} L/T^{1/2}$ s$)$, a diffusion or Joule heating time $(\tau_d = l^2/\eta = 10^{-9} l^2 T^{3/2}$ s$)$, a radiative time $[\tau_r = p/(n^2 Q) = 5 \times 10^{-52} T^2 B_\infty^{-2} Q^{-1}$ s$]$ and a conduction time $[\tau_\kappa = pL^2/(T\kappa)]$, where total pressure balance across the sheet has been assumed. The current density is

$$j = \frac{B_\infty}{\mu l} = 7.96 \times 10^5 \frac{B_\infty}{l} \text{ A m}^{-2},$$

where B_∞ is measured in tesla (1 tesla $= 10^4$ G), l is in metres and the critical current density for the onset of micro-instabilities (§13.1.3) is roughly

$$j^* = n e v_{Te} = 6.24 \times 10^{-16} n T^{1/2} \text{ A m}^{-2},$$

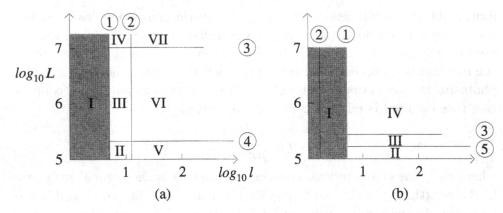

Fig. 11.19. The different regimes of behaviour of a current sheet of width l (in m), length L (in m), external magnetic field $B_\infty = 0.01$ tesla (100 G), and temperature $T = 10^6$ K. (a) Slow external changes ($\tau_{\text{ext}} = 100$ s). Region I is turbulent, II is 2D and dissipative with $C = J$, III is 2D and dissipative with $R = J$, IV is 1D dissipative with $R = J$, V is 2D frozen with C, VI is 2D frozen with R, and VII is 1D frozen with R (C, J, R refer to conduction, Joule heating, and radiation). (b) Fast external changes ($\tau_{\text{ext}} = 2$ s). Region I is turbulent, II is 2D frozen with C, III is 2D frozen with adiabatic changes, and IV is 1D frozen and adiabatic (after Heyvaerts and Kuperus, 1978).

where n is in m^{-3}, T is in degrees Kelvin and v_{Te} is the electron thermal speed (§13.1.1). Figure 11.19 shows how the sheet behaviour varies with l and L for external changes on a time-scale (τ_{ext}) that is (a) long or (b) short. Line 1 denotes $j = j^*$, so that the shaded region represents a turbulent sheet. Line 2 denotes $\tau_d = \tau_{\text{ext}}$, to the left of which the sheet is dissipative and to the right of which it responds to the external changes with the magnetic flux frozen to the plasma. Line 3 denotes $\tau_L = \tau_{\text{ext}}$, above which the sheet behaves one-dimensionally and below which the behaviour is two-dimensional. Line 4 denotes $\tau_r = \tau_\kappa$, such that above it radiation dominates conduction. Line 5 denotes $\tau_\kappa = \tau_{\text{ext}}$, such that below it conduction dominates and above it thermal instability occurs. For slow external changes, diffusive processes are important as turbulence is approached (II–IV in Fig. 11.19). When a neutral sheet is compressed rapidly (Fig. 11.19), turbulence can be approached without dissipation being important, and so Heyvaerts and Kuperus studied this regime by assuming adiabatic, frozen-in behaviour.

In practice, they found it is very difficult to trigger microturbulence during the one-dimensional development of a sheet that is initially far from criticality. They also considered the two-dimensional, isothermal, frozen-in flow induced by compression and found that the region of the sheet where pinching is strongest is rapidly evacuated and conditions for microturbulence are easily fulfilled.

B. Somov, V. Titov, and co-workers (see Somov, 1992, Chapter 3) have modelled the energy release site in a solar flare as a Sweet–Parker current sheet with integral or order-of-magnitude relations between input and output parameters, as sketched in Section 4.6.3. Their advance is to consider several important extra effects, including an energy equation and the possibility of anomalous diffusion and heating due to current-induced micro-instabilities (§13.1.3).

First of all, they studied a neutral current sheet and found that a hot sheet with coronal temperatures is not possible with a classical coulomb diffusivity. Furthermore, a high enough energy release rate for a solar flare cannot be produced by ion-acoustic turbulence, although anomalous Bohm diffusion due to gradient instabilities with a lower threshold is adequate. Next, they included the effect of a weak transverse magnetic field in the y-direction of Fig. 4.3: it leads to a higher outflow of mass and energy which can account for the flare energy; at the same time they incorporated a two-fluid model and calculated anomalous resistivity due to ion-cylotron or ion-acoustic turbulence.

Then they proceeded to consider under what circumstances the sheet is unstable to the tearing mode instability, including the strong stabilizing influence of the transverse component (see also Forbes and Priest, 1987). Finally, the effect of a longitudinal magnetic field component (B_z) in the sheet was incorporated, since reconnection can take place at separators as well as null points in three dimensions (see Chapter 8). This is especially important in pre-flare current sheets. The flux of the longitudinal component is conserved as it is carried through the sheet, but there is a strong compression of this component and a corresponding current (j_y); the result is an additional ohmic dissipation from the annihilation of the main (B_x) component.

11.2.3 Transition-Region Explosive Events

Spectral lines which form in the transition region between the corona and the chromosphere often show impulsive Doppler shifts indicating velocities in excess of 100 km s^{-1} (Brueckner and Bartoe, 1983). Because the duration of these velocities is less than 100 s, the phenomena that produce them have been termed *explosive events*. Explosive events are also found to be associated with the cancellation of magnetic flux in the photosphere (Dere et al., 1991) and they have a duration which is almost 100–1,000 times shorter than X-ray bright points (XBPs – see §11.3.1). This difference in time-scale means that transition-region explosive events must involve some process which operates on a faster time-scale than the cancellation process itself. One plausible suggestion (Dere et al., 1991) is that the reconnection process involved in cancellation is inherently impulsive and bursty (Priest, 1986). The short time-scale of the explosive events would then be due to individual bursts of reconnection during the cancellation process. Another possibility is that the reconnection process may from time to time involve a stressed magnetic flux tube which becomes unstable as the large-scale cancellation proceeds.

The explosive events in the transition region are more numerous by a factor of 10 or more than XBPs. Habbal and Grace (1991) have suggested that this difference reflects the difference in height between explosive events and XBPs. Explosive events are seen in the CIV line at a temperature of 10^5 K, which forms at a height of approximately 2×10^3 km. In contrast, XBPs form in the corona at heights well above 10^4 km, where the plasma has an ambient temperature in excess of 10^6 K. Habbal and Grace (1991) found that there is nearly always an impulsive feature in the extreme ultraviolet which is associated with an XBP, but that the reverse is not true. It may be that there are many more reconnection events at lower altitudes than at higher ones, so that, as one looks at emissions from higher altitudes, fewer events are found.

11.2.4 Surges and Sprays

Observations in Hα show that many flares and prominence eruptions are accompanied by the ejection of cool material at less than 10^4 K. If the ejected material returns to the Sun, it is called a *surge*, but if it escapes it is called a *spray* (Švestka, 1976). A surge is well-collimated and can usually be seen to fall back to the Sun. Most surges have upward velocities of less than 200 km s^{-1}. By contrast, most sprays have velocities in excess of the escape velocity (670 km s^{-1}), and the ejected material spreads out in a wide cone. Most of them are associated with eruptions of active-region prominences.

Because reconnection heats and accelerates plasma in nearly equal amounts, it has long been thought that surges and sprays, which are relatively cool, are not produced by reconnection. However, in the past few years surges have been discovered to occur in association with nearby X-ray features (Schmieder et al., 1988; Švestka et al., 1990). The amount of X-ray emission is quite small, so that, even if it is included, the kinetic energy of the surge is still a factor of 10 greater than the thermal energy (Schmieder et al., 1994). Švestka et al. (1990) have argued that the relatively small X-ray emission

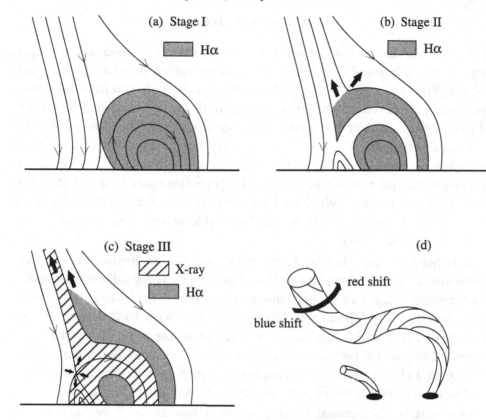

Fig. 11.20. The reconnection scenario proposed by Canfield et al. (1996) for the generation of Hα surges. (a) Prior to onset, cool Hα material is trapped in small loops jutting out from the chromosphere. (b) At the onset of reconnection, cool material is ejected upward by the disconnection of the field, but it is not heated because it does not pass through the reconnection site. (c) Plasma that is heated by the reconnection process appears as thermal X-rays lying along field lines adjacent to the Hα region. (d) The observations of red and blue Doppler shifts can be accounted for by incorporating three-dimensional untwisting of the reconnected flux tubes.

may be an observational artifact caused by the relatively low emission measure of hot material injected into large-scale loops. Their interpretation is supported by the fact that surges in small-scale loops have a stronger X-ray emission (Martin and Švestka, 1988).

Canfield et al. (1996) have also observed that surges are closely associated with the appearance of X-ray jets. Furthermore, they found that some surges produce both upward and downward flows which are suggestive of counter-directed reconnection jets. They propose that the cool Hα plasma in a surge is accelerated by a whiplash effect, which occurs when field lines are severed by the reconnection as shown in Fig. 11.20.

11.3 Coronal Heating

The solar corona consists of several types of structure. Magnetically open regions called *coronal holes* show up as dark regions in white light or soft X-rays [Figs. 11.21 and 11.2(a)]. Magnetically closed *coronal loops* have both footpoints anchored in the solar surface and are of many different types (Priest, 1981): in eclipse images large-scale loops show up as the lower parts of helmet streamers [Fig. 11.21 (top panel)], but in soft X-rays many more types are apparent, including active-region loops, flaring loops, interconnecting loops, and quiet-region loops [Figs. 11.21 (bottom panel) and 11.29(a)]. A third type of structure, which we have referred to earlier, is the *X-ray bright point*, a few hundred of which may be visible at one time in soft X-rays (Figs. 11.21 and 11.24).

It is possible that these different structures are all heated by the same mechanism, but it is also possible that they are heated by different processes. Two general classes of model have been proposed for heating the corona (e.g., Priest, 1993). The first is Alfvén waves (Hollweg, 1983; Roberts, 1984; Goossens, 1991), which may dissipate either by phase mixing (Heyvaerts and Priest, 1983) or by resonant absorption (Tataronis and Grossman, 1973). The second class is magnetic reconnection, either at null points (§8.6) or in the absence of null points (§8.7), especially by braiding (Parker, 1972).

One part of the coronal heating problem appears to have been solved, since it has been shown convincingly that XBPs are probably heated, according to the Converging Flux Model, by magnetic reconnection driven in the corona by footpoint motions (§11.3.1). However, the heating mechanisms of the other structures are at present unknown. The most likely mechanism for heating coronal holes is probably magnetic waves: a particularly attractive option is by high-frequency waves between $1\,\mathrm{Hz}$ and $1\,\mathrm{kHz}$, generated by rapid, tiny reconnection events in supergranule boundaries (Axford and McKenzie, 1996; McKenzie et al., 1995, 1997, 1998), especially since the resulting ion-cyclotron waves may also be driving the fast solar wind and explaining the huge line-broadening that is seen with the UVCS instrument on SOHO (Kohl et al., 1997).

In the next section we discuss XBPs, which, although they do not provide all the energy for the corona, represent elementary and isolated heating events by reconnection. Furthermore, the fact that they are now known to be small coronal loop systems is suggestive that perhaps larger loops are also heated by reconnection. We describe briefly three theoretical ideas that could be involved in coronal heating, namely driven reconnection (§11.3.1), current-sheet production by braiding of field lines (§11.3.2), and MHD

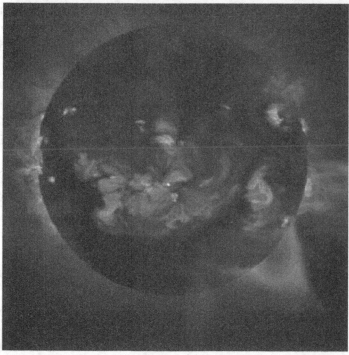

Fig. 11.21. The solar corona (above) during the 26 February 1998 eclipse, showing magnetically closed coronal loops and magnetically open coronal holes (courtesy of the High Altitude Observatory) and (below) in soft X-rays from Yohkoh, showing a loop (bottom right) that is reforming by after reconnection an eruption (courtesy of S. Tsuneta).

turbulence (§11.3.3). Then we discuss the large-scale diffuse corona (§11.3.4) and summarise recent observations from Yohkoh and SOHO (§11.3.5) that provide highly suggestive evidence in favour of reconnection as the prime cause of coronal heating.

11.3.1 X-Ray Bright Points: the Converging Flux Model

All over the surface of the Sun one finds very small regions ($\approx$3 arc sec) which are bright in X-rays and have a duration of a few hours or less. Because these regions appeared point-like in early X-ray telescopes, they are called X-ray bright points or XBPs for short. Some XBPs are located above emerging flux and are explained by the Emerging Flux Model (§11.2.1), but most are situated in the corona above pairs of opposite-polarity magnetic fragments that are approaching one another. As they collide, these fragments appear to annihilate one another in a process which observers refer to as *cancellation* (Martin et al., 1985). Although it has been suggested that cancellation might simply be the result of the submergence of a simple loop, this is thought to be rather unlikely (Priest, 1987), because there is a sharp edge which forms between the opposite-polarity fragments as they disappear. This sharp edge suggests that a current sheet forms as the fragments push into one another, and their disappearance is a result of reconnection in the current sheet. Furthermore, there is no obvious way that submergence would produce an overlying coronal brightening. In the mid-1980s, K. Harvey (Harvey, 1985) discovered that there is a close correlation between the appearance of XBPs and cancelling polarities.

The Converging Flux Model shown in Figs. 11.22 and 11.23 explains how cancellation can lead to the appearance of an XBP (Priest et al., 1994a). Because of the overlying field in the cancelling region, a null-point does not form until the opposite polarities are sufficiently close. The null-point first appears at the surface and then moves upwards as the polarities approach. However, continued motion eventually causes the null-point to reverse direction and sink back into the photosphere. In most cases magnetic flux emerges in a supergranule cell and then moves to the boundary, where one polarity tends to accumulate while the other reconnects with opposite-polarity network and forms a bright point (Fig. 11.23). The model predicts that high-resolution X-ray images of XBPs should show considerable structure, and such structure has in fact been found by Golub et al. (1994). Figure 11.24 shows how this structure matches the predictions made by a three-dimensional version of the model (Parnell et al., 1994).

Falconer et al. (1999) hunted for X-ray bright points with the EIT (Extreme ultra-violet Imaging Telescope) instrument on SOHO in a large

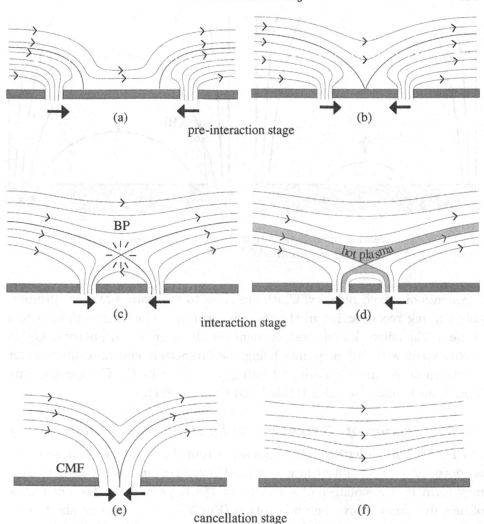

Fig. 11.22. The Converging Flux Model for an X-ray bright point, showing the stages in the approach and interaction of two equal and opposite magnetic fragments. (a) In the pre-interaction stage an X-point does not form because the fragments are too far apart. (b) Once they are close enough, an X-point appears at the base. (c) As the fragments move yet closer, the X-point initially rises upward into the corona to create an X-ray bright point (BP) with filamentary extensions, but then (d) the X-point starts to move downwards again. (e) Finally, when the fragments meet in the photosphere as a cancelling magnetic feature (CMF), the coronal X-point disappears (f) and so eventually do the fragments.

square region of side 0.6 $R_\odot$. They applied a filter to remove the background haze, and this showed up many smaller bright points than normal, which they called micro-bright points. Comparison with a Kitt Peak magnetogram (Fig. 11.25) showed that the normal bright points lie over large magnetic

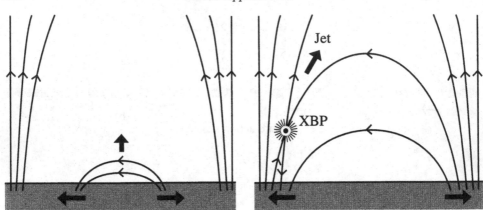

Fig. 11.23. The creation of an X-ray bright point at the edge of a supergranule cell, according to the Converging Flux Model.

fragments of mixed polarity, which are close to each other and are presumably driving reconnection in the overlying corona. The micro-bright points all lie in the network and most of them also lie over mixed polarity, so this is consistent with bright points being the large-scale end of a much larger spectrum of reconnection events heating the corona by the Converging Flux Model mechanism (see also Habbal and Grace, 1991).

11.3.2 Creation of Current Sheets by Braiding and Other Means

In Chapter 2 we described in detail how current sheets may form in response to footpoint motion at a surface to which they are connected. A current sheet may form by the collapse of a null point (§2.1, §2.5) or by the emergence of new flux from below the photosphere (Fig. 2.7). Such current sheets may form in otherwise potential fields (§2.2) or force-free fields (§2.3) or by magnetic relaxation (§2.4). They may be created along separatrices or separators by shearing (§2.6), both in fields without null points and with null points (cf. Longcope and Silva, 1998). Finally, they may be formed according to Parker (1972) by the braiding of magnetic fields around one another (§2.7). A three-dimensional resistive MHD numerical experiment by Galsgaard and Nordlund (1996) has shown that the resulting current sheets are highly complex, that the braiding is roughly by one turn before reconnection sets in, and that the resulting heating along a loop is rather uniform (Galsgaard et al., 1999), as shown in Fig. 11.26. Furthermore, Démoulin and Priest (1997) have shown how coronal heating may occur at the quasi-separatrix layer boundaries (§8.7) between the twisted coronal fields that emanate from many intense flux tubes in the photosphere.

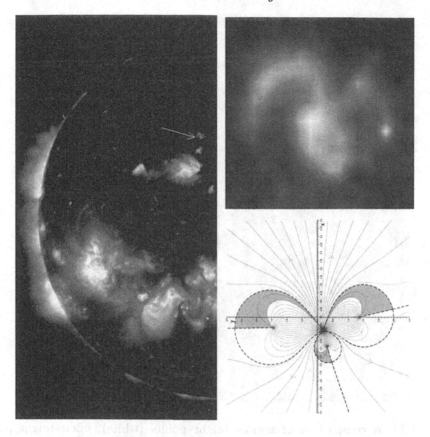

Fig. 11.24. Left: X-ray image of the Sun's disk. Most of the bright emission comes from active regions close to sunspots. However, the arrow shows a faint transitory X-ray bright point which has just switched on. Top right: close-up of this bright point showing filamentary fine structure. Bottom right: location of the lines of force extrapolated from the observation of the surface magnetic field. There are three null points at the locations where the separatrices (dashed curves) intersect. The shaded regions show the regions where plasma heated by the reconnection process should be dense enough to be visible in X-rays. (From Parnell et al., 1994).

11.3.3 Self-Consistent Model for Coronal Heating by MHD Turbulence

MHD turbulence theory has been applied in great detail to the solar dynamo problem (Moffatt, 1978) and has just begun to be applied to the solar wind (Matthaeus et al., 1984; Velli et al., 1990), but not yet in any detail to the coronal heating problem.

Important physical quantities are the global ideal invariants, which are conserved in the absence of dissipation. In 2D MHD the global invariants

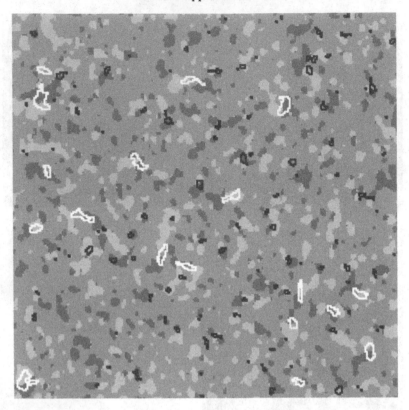

Fig. 11.25. A comparison of normal bright points (white), micro-bright points (black), and magnetic fragments of positive (light grey) and negative (dark grey) polarity over a square of the quiet Sun of side $0.6\,R_\odot$ (Falconer et al., 1999).

(Montgomery, 1983; Frisch et al., 1975) are the energy, correlation (or mean-square vector potential), and cross helicity

$$W_2 = \int \tfrac{1}{2}\rho v^2 + \frac{B^2}{2\mu}dS, \quad a = \int \tfrac{1}{2}A^2 dS, \quad H_2 = \int \mathbf{v}\cdot\mathbf{B}\,dS,$$

while in 3D we have the energy, magnetic helicity, and cross helicity

$$W_3 = \int \tfrac{1}{2}\rho v^2 + \frac{B^2}{2\mu}dV, \quad H = \int \mathbf{A}\cdot\mathbf{B}\,dV, \quad H_3 = \int \mathbf{v}\cdot\mathbf{B}\,dV,$$

where $\mathbf{B} = \nabla \times \mathbf{A}$.

In Fourier space these global invariants undergo *cascades*, which are direct if the transfer is from large to small wavelengths and indirect (or inverse) if it is in the other direction. The energy has either a Kolmogorov spectrum

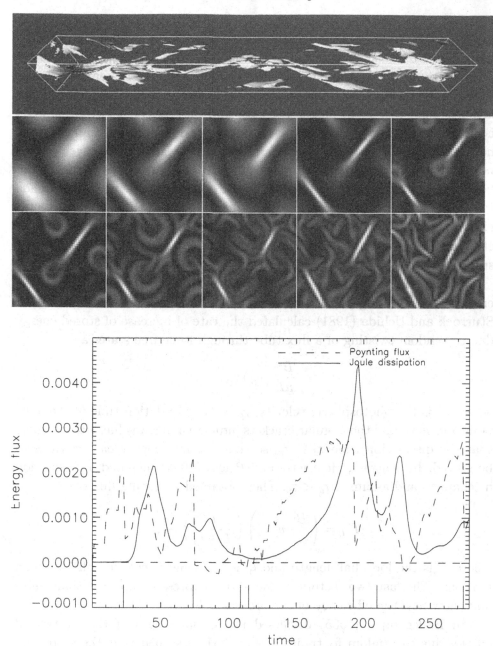

Fig. 11.26. The current filamentation due to braiding shown (top) in a 3D perspective plot at one time and (middle) in a section at several times. The figure at the bottom shows the impulsive release of energy as a function of time (Galsgaard and Nordlund, 1996).

$(\sim k^{-5/3})$ or a Kraichnan spectrum $(\sim k^{-3/2})$ and has a direct cascade towards small wavelengths, whereas the correlation a $(\sim k^{-1/3})$ in 2D and magnetic helicity $H(\sim k^{-2})$ have indirect cascades towards large wavelengths.

Ting et al. (1986) have conducted a series of experiments on two-dimensional MHD in which they find *selective decay* when the initial kinetic energy is much smaller than the magnetic energy and the cross helicity (normalised) is less than the energy. Here the magnetic energy decays faster than the magnetic helicity and so has a direct cascade towards small wavelengths, since nonlinear interactions tend to replenish it. At the same time the magnetic helicity has an indirect (or inverse) cascade, and the magnetic field tends towards a force-free state. In contrast, when the kinetic and magnetic energies are similar and the cross-helicity is of a similar size to the energy, Ting et al. find a process of dynamic alignment with a tendency towards Alfvénic states having $\mathbf{v} = \pm\mathbf{B}/\sqrt{\mu\rho}$ (see also Biskamp, 1994).

Applications of the theory to coronal heating have been made by several authors. Hollweg (1983) reviewed applications to wave-heating theories. Sturrock and Uchida (1981) calculated the rate of increase of stored energy due to random twisting of a flux tube through force-free states as

$$\frac{B^2}{\mu L} \left\langle v_p^2 \right\rangle \tau_p,$$

where v_p is the photospheric velocity, τ_p is the correlation time of the footpoint motions, and the angular brackets indicate a mean value. They assume that the dissipation time $\tau_d > \tau_p$, so that the free energy can continue to be stored. In contrast, Heyvaerts and Priest (1984) included a dissipation mechanism and assumed $\tau_d < \tau_p$. They obtained a heating flux of

$$F_H = \left(\frac{B^2}{\mu L} v_p^2 \tau_p \right) \left(\frac{L}{L + l_p} \right)^2 \frac{\tau_d}{\tau_p},$$

where L is the loop half-length and l_p is the length-scale for photospheric motions. The last two factors in the above expression are less than unity and show how the efficiency of the process is limited.

Van Ballegooijen (1985) discussed the initial stages of the cascade of energy due to random footpoint motions. He assumed that the states are force-free and the length-scale (l_p) for photospheric motions is much smaller than the loop length. He found that the mean-square current density $\langle j^2 \rangle$ increases exponentially $[\sim\exp(t/\tau_p)]$. Gomez and Fero Fontan (1988) have applied two-dimensional MHD turbulence theory to twisted coronal loops and have suggested the injection of energy at a specific wave-number (k_p), followed by a cascade of energy like $k^{-5/3}$ to a dissipation wave-number k_d, together with an inverse cascade of mean-square potential like $k^{-1/3}$. Also, Gomez et al. (1993) have measured fine-scale structure in NIXT (Normal Incidence X-ray Telescope) images and find the intensity has a k^{-3} spectrum, although only over one order of magnitude.

Many coronal heating mechanisms, such as braiding and current-sheet formation or resistive instabilities or waves, all lead to a state of MHD turbulence, so how can we analyse such a state? Heyvaerts and Priest (1984) made a start by adapting Taylor's relaxation theory (§9.1.1) to the coronal environment, in which the field lines thread the boundary rather than being parallel to it.

In Taylor's model the global magnetic helicity is conserved, but in the Heyvaerts–Priest model, footpoint motions make the coronal field evolve through a series of linear force-free fields, satisfying $\nabla \times \mathbf{B} = \alpha_0 \mathbf{B}$. The footpoint connections are not preserved, but instead the constant α_0 is determined from the evolution of the magnetic helicity (§8.5) due to the injection by boundary motions according to Eq. (8.52). The resulting heating flux is of the form

$$F_H = \frac{B^2 v}{\mu} \frac{\tau_d}{\tau_0},$$

where, as before, τ_d is the dissipation time and τ_0 the time-scale for footpoint motions. This is the same as Parker's result is Section 11.2.1 when τ_d is replaced by the reconnection time (d/v_R) and τ_0 is replaced by the convection time (L/v_0).

The basic analysis of Heyvaerts and Priest (1984) has been applied by Browning et al. (1986) to a set of closely packed flux tubes and by Dixon et al. (1988) to an axisymmetric flux tube. Also, Vekstein et al. (1990) have modelled the simultaneous stressing and relaxing of an arcade.

Although many mechanisms produce a turbulent state, they are incomplete in the sense that there is a free parameter present, such as τ_d in the above equation, or a correlation or a relaxation time. In other words, the mechanisms do not determine the heating flux (F_H) in terms of photospheric motions alone. Heyvaerts and Priest (1992) therefore began a new approach in which they assume photospheric motions inject energy into the corona and maintain it in a turbulent state with a turbulent magnetic diffusivity (η^*) and viscosity (ν^*). There are two parts to their theory. First of all, they calculate the global MHD state driven by boundary motions, which gives F_H in terms of ν^*. Secondly, they invoke cascade theories of MHD turbulence to determine the ν^* and η^* that result from F_H. In other words, the circle is completed and F_H is determined independently of ν^* and η^*. They applied their general philosophy to a simple example of one-dimensional random photospheric motions producing a two-dimensional coronal magnetic field.

Suppose the dimensionless boundary motions are $\pm V(x)\hat{\mathbf{y}}$ (with Fourier coefficients V_n) at $z = \pm L$ and produce motions $v(x, z)\hat{\mathbf{y}}$ and field $B_0\hat{\mathbf{z}} + B_y(x, z)\hat{\mathbf{y}}$ within the volume between $z = -L$ and $z = L$. Then the steady MHD equations of motion and induction reduce simply to

$$0 = \frac{B_0}{\mu}\frac{\partial B_y}{\partial z} + \rho\nu^*\nabla^2 v_y, \quad 0 = B_0\frac{\partial v_y}{\partial z} + \eta^*\nabla^2 B_y.$$

The solutions may easily be found and the resulting Poynting energy flux through the boundary is

$$F_H = \frac{B_0^2 v_{A0}}{\mu} \sum_0^\infty \frac{V_n^2 H^*}{\eta/(l v_{A0})}\left(1 + 2\lambda_n^2/\sqrt{1 + 4\lambda_n^2}\right)$$

$$\times \frac{\sinh\left(\sqrt{1 + 4\lambda_n^2/H^*}\right) + \sinh(1/H^*)}{\cosh\left(\sqrt{1 + 4\lambda_n^2/H^*}\right) - \cosh(1/H^*)},$$

where $H^* = \sqrt{\eta^*\nu^*}/(Lv_{A0})$ is the inverse Hartmann number. For the second step, invoking Pouquet theory gives $\nu^* = \eta^*$ and, if a is the half-width of the loop,

$$F_H = \frac{27\,[\nu^{*2}/(Lv_{A0})]\,\pi^3}{2\,a^3/L^3}\,\frac{B_0^2\,v_{A0}}{\mu},$$

so that equating the two above expressions for F_H gives a single equation for ν^*. They found typically for a quiet-region loop that a density of 2×10^{16} m^{-3} and a magnetic field of 3×10^{-3}–5×10^{-3} tesla (30–50 G) produces a heating (F_H) of 2.4–5.5 $\times$ 10^2 W m^{-2} and a turbulent velocity of 24–33 km s^{-1}, whereas values of 5×10^{16} m^{-3} and 10^{-2} tesla (100 G) for an active-region loop give 2×10^3 W m^{-2} for the heating and 40 km s^{-1} for the turbulent velocities. Given the limitations of the model, these reasonable values are very encouraging. Inverarity and Priest (1995a,b) then went on to apply the theory to a twisted flux tube and to turbulent heating by waves due to more rapid footpoint motions.

11.3.4 Large-Scale Corona

The large-scale corona seen in soft X-ray images from the Japanese satellite Yohkoh [e.g., Figs. 11.21(b) and 11.29(a) below] consists of large magnetic

Fig. 11.27. The so-called "magnetic carpet" showing observed photospheric magnetic fragments and calculated overlying coronal magnetic field lines (courtesy of MDI consortium).

loop systems that dominate the corona at solar minimum. They are also present outside active regions and coronal holes at solar maximum when the global X-ray intensity is an order of magnitude higher (Acton, 1996). How is this large-scale corona heated?

A clue has come from SOHO (Solar and Heliospheric Observatory), which has shown the surface of the quiet Sun to be covered with a *magnetic carpet* (Fig. 11.27), consisting of positive and negative magnetic fragments that show up in MDI (Michelson Doppler Imager) magnetograms and lie along the boundaries of supergranule cells. These fragments continually emerge and travel to the cell boundaries, where they either cancel with flux of opposite polarity or merge or fragment. The surprising result is that the magnetic flux at the surface is replaced every 40 h (Schrijver et al., 1997). This replacement, according to the Converging Flux Model (§11.3.1), would probably be by *reconnection submergence* (Priest, 1987), in which reconnection takes place above the photosphere and the resulting loop below the reconnection site is then ejected or pulled down through the photosphere.

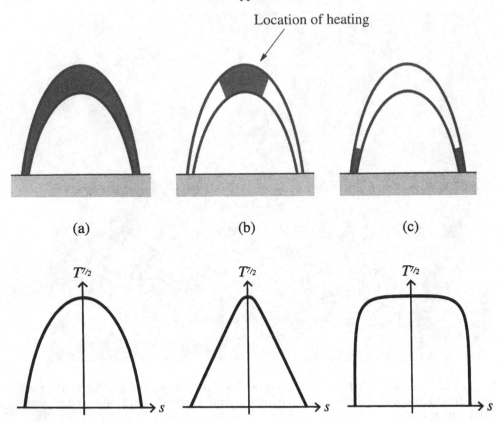

Fig. 11.28. Three types of heating in a coronal loop: (a) heating located throughout the loop, (b) heating near the summit, or (c) heating near the feet. Each produces a different temperature profile, as indicated.

   This process which replenishes the magnetic carpet is also likely to be driving coronal heating, but how? Part of the energy will go directly into X-ray bright points and micro-bright points, and perhaps these represent the large-scale tail of much more numerous and smaller-scale nano-bright points or nano-flares that are heating the corona by the same mechanism (Fig. 11.23). However, another possibility is that the replenishing mechanism drives an in-situ heating process throughout the corona. An understanding as to which process dominates has come from a recent comparison with Yohkoh observations as follows.

   If, on the one hand, the large-scale coronal loops were heated by turbulent reconnection in many small current sheets due to Parker-braiding, the heat would tend to be deposited fairly uniformly through the loop [Fig. 11.28(a)]. If, on the other hand, the heating were by long-wavelength Alfvén waves standing in the loop and possessing a maximum amplitude at the summit, it would tend to be dumped near the summit [Fig. 11.28(b)]. Furthermore, if the heating were by XBPs or other reconnection processes near the solar surface, then the heat would be liberated mainly near the loop feet [Fig. 11.28(c)]. A steady-state thermal balance between such heating and thermal conduction would then produce a temperature profile along the loop from one coronal footpoint to another that has the variable $T^{7/2}$ being a quadratic function in the first case, or a pointed function in the second case, or having a steep footpoint rise and a flat summit profile in the third case.

Realizing that the temperature profile is highly sensitive to the nature of the heating mechanism, Priest et al. (1998) used the Yohkoh Soft X-ray Telescope to compare the temperature along a large loop (Fig. 11.29) with a series of models in order to deduce the likely form of heating. The most likely values of the parameters of the models were found by minimising χ^2.

The observed temperatures rise from about 1.4 MK near the feet to about 2.2 MK at the summit, and it can be seen in Fig. 11.30(a) that the model with heat concentrated near the feet gives a very poor fit, whereas heat focused at the summit [Fig. 11.30(b)] produces a better fit. However, uniform heating [Fig. 11.30(c)] fits best of all and therefore provides strong evidence that the heating mechanism deposits the energy fairly uniformly along the length of the loop, at least for this example.

Of the existing models, the one which can most easily explain the uniform heating is Parker's mechanism of turbulent reconnection by braiding. Another more speculative but very interesting mechanism is the Axford–McKenzie model of high-frequency ion-cyclotron waves (Axford and McKenzie, 1996; McKenzie et al., 1997). However, to generate such waves requires smaller reconnection sites in the network than are observed at present. For example, with an Alfvén speed of 10^4 km s^{-1}, reconnection of a flux tube 100 km wide would tend to launch a wave of period 0.01 s.

11.3.5 General Evidence from SOHO in Favour of Reconnection

Yohkoh and SOHO have given important clues about the nature of coronal heating. When large-scale fields close down after eruptions, Yohkoh discovered that they do so in characteristic cusp-shaped structures [Figs. 11.13 and 11.21 (bottom panel)]. In addition, observations of the temperature in active regions show that all the hottest loops are either cusps (see §11.1.2) or pairs of apparently interacting structures, which is highly suggestive of

3 OCTOBER 1992
Exposure : 15.1 secs
Filter : Al 1

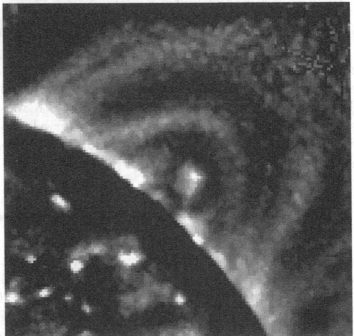

Fig. 11.29. (top) Global image of the Sun in soft X-rays from the Japanese satellite Yohkoh. (bottom) Close-up of the loops analysed in the top right corner of the global image.

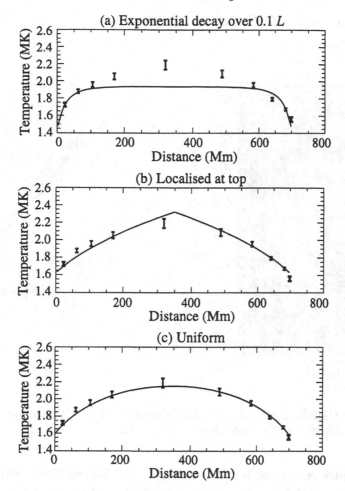

Fig. 11.30. A comparison of the observed coronal loop temperature profile with three models, in which the heat is (a) dumped near the feet, (b) localised at the summit, and (c) deposited uniformly through the loop.

reconnection (Yoshida and Tsuneta, 1996). In addition, many X-ray jets have been discovered by Shibata et al. (1992, 1994), some of them associated with X-ray bright points, and have been beautifully modelled (Shibata et al., 1996) – another clear signature of reconnection at work. These jets can extend for more than half a solar radius with a flow speed in excess of 200 km s^{-1}.

SOHO has three instruments observing the low corona (EIT, SUMER, and CDS) and two observing the outer corona (UVCS and LASCO), all of which have spotted the results of reconnection. Indeed, SOHO has demonstrated that reconnection gives an elegant explanation for many diverse phenomena,

Evolution of a jet in Si IV 1393 A

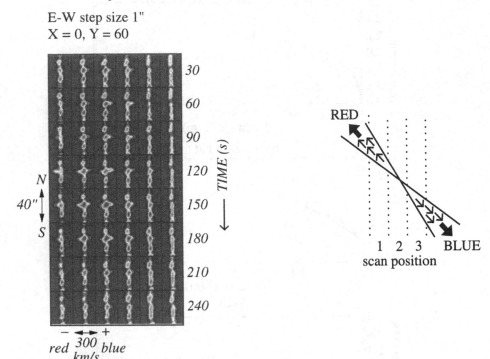

Fig. 11.31. SUMER spectra of an explosive event in Si IV with a step-size from left to right in each row of 1 arc sec (Innes et al., 1997).

such as explosive events, blinkers, possibly tornadoes, X-ray bright points, the variation of the magnetic carpet, and the existence of the large-scale corona.

You can, of course, never prove a theory with observations – you can only disprove it, which is rather sad for theorists! But now there has been a real paradigm shift. Whereas previously reconnection was a fascinating concept which exercised the imagination of theorists like ourselves, now so many SOHO and Yohkoh observations fit beautifully into place with the eyes of reconnection that it has become the natural explanation for many coronal heating phenomena. Let us then say a little about each of the above manifestations of energy release by magnetic reconnection.

Many explosive events (see §11.2.3) have been observed with SUMER by Innes et al. (1997). They have presented an example in Si IV with a step-size of 1 arc sec, which reveals bidirectional jets that have been interpreted as jets accelerated in opposite directions from a reconnection site (Fig. 11.31).

Furthermore, Chae et al. (1998) have compared an explosive event with Big Bear magnetograms and found that it is located over a site where magnetic fragments are approaching one another and, presumably, driving reconnection in the overlying atmosphere.

With the CDS instrument Harrison (1997) has discovered brightenings in the transition region which he has christened *blinkers*. They are located in the network and last typically 10 min. Berghmans et al. (1998) has considered brightenings with EIT in He II at 80,000 K. They are similar to blinkers but have a wide variety of sizes and time-scales. From observations of 10,000 such events they have determined their statistical properties and found clear correlations of duration, intensity, and area with energy. Furthermore, Krucker et al. (1998) have observed microflares with EIT. They have an energy of $10^{18} - 10^{19}$ J ($10^{25} - 10^{26}$ ergs), an energy spectrum of $W^{-2.6}$, where W is the energy of an event, and they contribute about 20% of the quiet-Sun heating. Even smaller-scale events have been studied with the Transition Region and Coronal Explorer (TRACE) data by Parnell and Jupp (1999), who find that the spectrum continues down to at least 10^{24} ergs at a slope between W^{-2} and $W^{-2.6}$. If it continued further to 4×10^{22} ergs at a slope of $W^{-2.6}$, nanoflaring events of such an energy could provide all the heating of the quiet corona.

Pike and Mason (1998) have observed macrospicules in coronal holes with CDS. The macrospicules are found to be rotating at typically 20 km s^{-1} but occasionally at 150 km s^{-1}, and they christened such structures *tornadoes*. Are these a consequence of 3D reconnection, since twisted structures are naturally produced by the conservation of global magnetic helicity? Or are they torsional waves driven by photospheric twisting motions? Furthermore, do ordinary spicules also possess such twist – and is the same true for even finer-scale structures that could be driving ion-cyclotron waves and accelerating the fast solar wind?

11.4 The Outer Corona

With spacecraft it is possible, at least in principle, to make in situ measurements of reconnection processes ranging from distances as close to the Sun as four solar radii (the proposed Solar Probe) to as far away as 125 AU (where Voyager II is expected to lose contact with Earth in the year 2020). The structures which are accessible to spacecraft include CMEs, the heliospheric current sheet, planetary magnetopauses (§10.6), and the heliopause separating the interstellar and solar magnetic fields.

11.4.1 Coronal Mass Ejections

As mentioned at the start of this chapter, the injection of magnetic flux into the outer heliosphere would cause the magnetic field there to increase indefinitely were it not for reconnection. Without reconnection the field at 1 AU would double every few weeks, so there can be little doubt that reconnection is occurring somewhere (McComas et al., 1992).

However, there are two possible ways in which the reconnection could occur. It could occur directly behind a CME shortly after its appearance. Alternatively, it might occur at a much later time at the boundary layer separating the field lines opened by the CME from the closed field lines of nearby active regions (Hammond et al., 1996). These two cases can be distinguished observationally from one another by following the evolution of the field lines opened by the CME. Reconnection behind a CME requires the opened field lines to reclose in a manner that starts in the centre of the opened region, while reconnection with adjacent structures requires the field to reclose at the edges of the opened region. He 10,830 images of the evolution of the coronal hole formed by the opened field lines almost always show that the majority of field lines reclose starting at the centre rather than the edges (Kahler, 1992). Thus, the predominant mode of reconnection appears to be reconnection directly beneath a rising CME. This result is further confirmed by the quick reformation of helmet streamers disrupted by CMEs and the appearance of large-scale X-ray loops (Hiei et al., 1993) and He 10,830 ribbons (Harvey and Recely, 1984).

The same argument that we used in Section 10.4 to determine a maximum distance for the location of the X-line in the Earth's magnetotail can also be used to determine a maximum altitude at which reconnection can occur below a CME. If we assume, as before, that the maximum speed of outflow from the X-line is the Alfvén speed of the ambient field, then newly reconnected field lines will return to the Sun only if the speed of the X-line itself is less than the ambient Alfvén speed. The slowest possible speed for the X-line is just the speed of the local solar wind. Once the X-line reaches the Alfvén critical point, where the solar-wind speed exceeds the ambient Alfvén speed, it will be impossible for newly reconnected field lines to return. For normal coronal conditions, the Alfvén critical point lies somewhere between 10 and 20 solar radii from the Sun (Weber and Leverett, 1967), so any reconnection which occurs beyond this distance seems unlikely to return flux to the Sun. Enhanced outflow behind a CME might further reduce the distance at which the speed of the X-line exceeds the ambient Alfvén speed.

At least 10% of coronagraph observations of the wake behind CMEs show circular plasma structures or arcs which appear (at least in the two-dimensional plane) to lie on field lines that have become disconnected from the photosphere by reconnection (Webb and Cliver, 1995). The structures are reminiscent of the lower part of the magnetic island or plasmoid shown in Fig. 11.7, but their relative rarity has led some observers to conclude that reconnection behind coronal mass ejections is an uncommon event (Illing and Hundhausen, 1983; McComas et al., 1991). Observations from the LASCO instrument on SOHO of disconnection events have, however, shown many clear signatures of reconnection (Simnett et al., 1997).

McComas et al. (1989) and Gosling et al. (1995) have reported possible spacecraft signatures of the reconnection behind CMEs. Despite the fact that most of the reconnection is likely to occur within a few solar radii of the Sun, it is still possible to distinguish the global field topology by using spacecraft that are located at 1 AU (220 solar radii). Suprathermal electron fluxes are only seen at these distances as long as the field lines on which they travel remain connected to the Sun. If field lines become disconnected at any point between the base of the corona and the spacecraft, the disconnection shows up as an absence of the suprathermal electrons. Furthermore, it is possible to distinguish whether both ends of a field line, or only one, are connected to the Sun. When both ends are connected, the suprathermal electron distribution is bimodal, but if only one end is connected then the distribution is peaked along the field in the direction away from the Sun.

Using observations from the ISEE spacecraft and Ulysses, Gosling et al. (1995) have deduced the three-dimensional topological sequence shown in Fig. 11.32 for reconnection in CMEs. Initially, reconnection creates a rising flux rope which is still connected at both its ends to the Sun [Fig. 11.32(a)]. During this stage, large-scale X-ray loops are generated by the reconnection, but the field lines still remain connected at both ends to the Sun. Eventually, reconnection with the open field lines of the normal solar wind disconnects field lines from the Sun, first at one end and then at the other [Figs. 11.32(b) and 11.32(c)]. Reconnection between the open solar wind field lines can also occur as shown in Fig. 11.32(d). However, eventually, all the current structure dissipates, and the CME disappears.

By counting the number of CMEs with different types of connectivity at different distances from the Sun, it is possible, at least in principle, to determine the rate at which reconnection occurs behind and around a CME. After 2–3 days, when most CMEs have reached 1 AU, about 80% of them

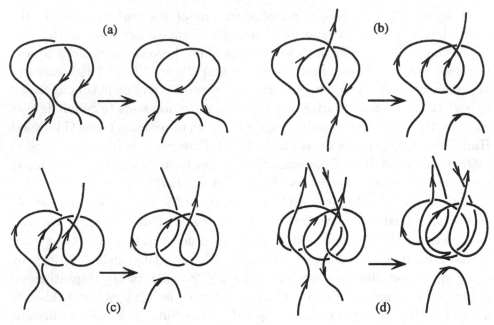

Fig. 11.32. Different stages in the development of reconnection in CMEs. (a) Two sheared loops reconnect to form a helical field line connected to the Sun at both ends. (b) Reconnection occurs between a helical line within the flux rope and an open field line of the normal solar wind. This produces a helical field line which is connected at only one end to the Sun, but (c), if the process is repeated, complete detachment can occur. (d) Finally, reconnection may occur between open field lines surrounding the flux rope (from Gosling et al., 1995).

show a signature which implies they are still connected at both ends to the Sun, as in Fig. 11.32(a). Most of the remaining 20% show a connection at one end, as in Fig. 11.32(b), and there are only a few, rare cases which show complete disconnection from the Sun, as in Fig. 11.32(c) (Gosling, 1996). Even these few cases may not correspond to completely disconnected events, since the CME flux ropes are likely to contain a mixture of both closed and open field lines, as occurs in the creation of a plasmoid in the geomagnetic tail during a substorm (see §10.5). After approximately 10–15 days (at 5 AU), the percentage of CMEs still showing connection at both ends is smaller, on the order of 50%, but cases showing complete disconnection are still rare (Gosling, private communication). After 30 days (beyond 10 AU) CMEs become increasingly difficult to identify, so it is likely by this time that all field lines within the CME have become part of the ambient field of the solar wind (McComas et al., 1995).

11.4.2 Helmet Streamers and the Heliospheric Current Sheet

At heights greater than a solar radius above the photosphere, the solar wind begins to stretch the solar magnetic field to such extent that the field becomes mostly radial. Above an active region this stretching creates a helmet streamer which is topped by a current sheet extending outward into interplanetary space (Fig. 11.21). By 10–20 solar radii the current sheets from individual helmet streamers merge to form a global heliospheric current sheet. Both the helmet-streamer current sheets and the heliospheric current sheet they form are potential locations for magnetic reconnection.

At one time it was thought that solar flares might be produced by reconnection in the current sheet at the top of a helmet streamer (Sturrock, 1966), but it is now recognized that the energy is released at a much lower altitude. Yet, the possibility remains that there might be a class of separate phenomena which arise from reconnection in a streamer current sheet. Reconnection near the base of a streamer current sheet would disconnect the sheet and cause it to move rapidly away from the Sun (Linker et al., 1992). Possible cases of such helmet-streamer disconnection events have been reported by McComas et al. (1991) and Cliver and Kahler (1991), but these events appear to be relatively rare compared to CMEs and are not associated with flares.

Reconnection is also thought to occur in the heliospheric current sheet. Spacecraft orbiting the Sun in the ecliptic plane routinely cross through this current sheet as well as other current sheets whose sources are not as well understood. Some of the additional current sheets may arise from CMEs and corotating interaction regions, where slow solar wind is overtaken by fast solar wind coming from a coronal hole. There are also numerous small current sheets (Neugebauer et al., 1986), which may be created by shuffling of the footpoints of magnetic field lines, as suggested by Parker (1983). Several analyses (Formisano and Amata, 1975; Pudovkin and Shukhova, 1988; Wu and Lin, 1989) of the spacecraft observations of these sheets indicate that some of them may be undergoing reconnection, but the evolution of this reconnection process is poorly understood.

11.4.3 The Heliopause

Somewhere near 100 AU the solar wind becomes subsonic at the termination shock, which is formed by the resistance of the interstellar medium to the outflow of the solar wind (Suess, 1990). The subsonic solar plasma downstream of the termination shock eventually collides with the interstellar plasma at a boundary at about 120 AU (Steinolfson and Gurnett, 1995). This boundary is known as the *heliopause*, and, just as in the case of the interaction between the Earth's magnetosphere and the solar wind, reconnection is likely to occur there (Macek, 1989, 1990).

As a result of the solar cycle, the polarity of the field above and below the heliospheric current sheet reverses every 11 years. Thus, if the orientation of the interstellar magnetic field is roughly constant in time, the region of the heliopause most favourable for reconnection will oscillate between the northern and southern hemispheres of the Sun (Nerney et al., 1993). If reconnection is important for the entry of cosmic rays into the heliosphere, as proposed by Suess and Nerney (1993), then part of the 22-year modulation cycle of cosmic rays could be a result of reconnection, although at the moment there is no evidence that this is the case.

12

Astrophysical Applications

The application of reconnection theory to astrophysical systems is a relatively recent development in comparison with applications to the terrestrial magnetosphere and the solar corona. The extreme remoteness of objects outside our solar system presents an enormous challenge for plasma physicists, because there are few spatially resolved observations on stellar scales with which to constrain theory. However, advances in Doppler imaging and the development of high-resolution instruments such as the Hubble Space Telescope are beginning to provide some help. Astrophysical magnetism is a huge field which we can only touch upon briefly here, but for an in-depth account the reader is referred to the new monograph by Mestel (1999).

The two astrophysical topics to which reconnection theory has been extensively applied are stellar flares (Mullan, 1986) and accretion disks (Verbunt, 1982). The analysis of stellar flares relies heavily upon the assumption that they are basically similar to solar flares except more energetic (e.g., Gershberg, 1983; Poletto et al., 1988). Flare stars can release 10^4 to 10^6 times the amount of energy seen in a large solar flare, but only modest increases in magnetic field strengths and scale-sizes are required to account for this extra amount. The use of reconnection theory in accretion disks has a dual purpose. One is to explain flare-like outbursts generated within disks, and the other is to account for the viscosity needed to allow material in the disks to fall inwards.

Reconnection theory also enters into models of protostellar collapse (Mestel and Strittmatter, 1967; Norman and Heyvaerts, 1985), extragalactic jets (Romanova and Lovelace, 1992), galactic magnetic fields (Kahn and Brett, 1993; Lesch and Reich, 1992), and even galactic clusters (Jafelice and Friaca, 1996). In fact, it seems that, anywhere magnetized plasmas are found, reconnection is likely to occur. So numerous are the applications that it is difficult to find a magnetized plasma where reconnection can safely be ignored.

12.1 Flare Stars

The discovery of a flare in a star other than the Sun occurred almost ninety years after the first solar flare was discovered by Carrington and Hodgson in 1859. On the night of 25 September 1948, Joy and Humason observed a flare on the red dwarf L726-8 (Haisch et al., 1991), and since that time more than a hundred similar flare stars have been discovered (Lang, 1992).

Many different types of star produce flares right across the Hertzsprung–Russell diagram (see Pettersen, 1989, for a review), but the classical flare stars are red dwarfs such as the one observed by Joy and Humason in 1948. Red dwarfs constitute about 65% of the total number of stars in our galaxy, and of these about 75% are flare stars (Rodonò, 1986), including our nearest stellar neighbour, Proxima Centauri. These flare stars are also known as UV Ceti stars (the classic prototype) and dMe or dKe stars (their spectral classification). As their name implies, red dwarfs are small stars, having masses of 0.08–0.8 $M_{\odot}$ and radii of 0.15–0.85 $R_{\odot}$, which means they are intrinsically faint (Lang, 1992).

The optical radiation produced by a dMe red dwarf during a flare is especially noticeable in the blue continuum, where the contrast with the star's red photospheric light is greatest. There are also Balmer-line enhancements which precede the rise in the blue continuum, but once the continuum appears it dominates the emission lines (Haisch, 1989). The emission lines remain enhanced after the continuum fades. As Fig. 12.1 shows, the time profile of the continuum is very impulsive, which suggests that the continuum radiation is the stellar counterpart of white-light emission in solar flares. Although solar flares rarely emit any white light at all, when they do, the white light appears only during the impulsive phase in small patches at the photosphere (Neidig, 1989). Similarly, the blue continuum in dMe flares is also associated with the impulsive phase and originates at the star's surface (van den Oord et al., 1996).

Since the mid-1970s, soft X-rays have been detected in numerous flare stars (Haisch, 1996). These X-ray emissions occur in the star's corona after the impulsive burst of the optical radiation and are cotemporal with long-lived emission lines such as Hα. The X-rays are thermal in origin and exhibit the same temporal profile as soft X-rays in solar flares, as shown in Fig. 12.1.

Red dwarfs that produce flares are typically fast rotators with deep convection zones. Periods as short at 10 h (compared with 26 days for the Sun)

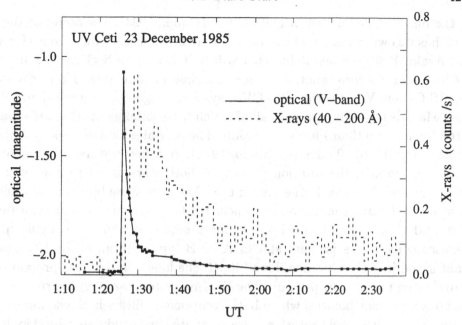

Fig. 12.1. Optical and X-ray light curves for a flare on the red dwarf UV Ceti. The optical curve is for emissions in the V band, which has an effective wavelength of 5,441 Å and a bandwidth of 708 Å (after de Jager et al., 1989).

are not unusual, and stars of less than $0.2\,M_\odot$ are thought to be convective throughout (Haisch and Schmitt, 1996). The fast rotation combined with the deep convection zone greatly enhance dynamo activity in these stars, although the nature of magnetic field generation in fully convective stars is still an open question. Direct measurements of the surface fields of red dwarfs, as well as other types of flare star, indicate that they have star spots with field strengths on the order of 1,000 G (Hartmann, 1987; Linsky, 1989; Saar, 1996), which is consistent with a rough balance between plasma and magnetic pressure at the stellar surface. These field strengths are of the same order as in the Sun, but the areas of the stellar spots are much larger, covering typically 20% of the stellar surface (Vogt and Penrod, 1983). In some cases the stellar spots appear to extend over as much as 50% of the surface (Bruning et al., 1987), whereas in the Sun the spot area never exceeds 2% (Zirin, 1988).

Flares also occur in early, pre–main-sequence stars known as T-Tauri stars. These stars have solar-like masses but are only 10^5 to 10^7 years old. Many are also rapid rotators (Montmerle and Casanova, 1996), but the features which make T-Tauri stars especially intriguing are their accretion disks and outflow jets. The jets emanate from the star's polar regions and are fed by the mass inflow from the accretion disk (Gahm, 1994). The accretion disk itself has its own corona, and this corona may be a source of disk flares (Levy and Araki, 1989) – a possibility that will be taken up in Section 12.2.3.

Close binaries form another important class of flare star. They include the RS Canum Venaticorum (RS CVn) systems, Algol-type binaries, and W Ursa Majoris (W Uma) systems, all of which have components that are separated by no more than a few stellar radii. The orbital period of these systems ranges from 0.5 to 50 days (Catalano, 1996), and, because the components are tidally locked, the rotation periods of both stars are the same as the orbital period. Thus, as in the case of the dMe stars, close binaries are rapid rotators with strong magnetic dynamos. RS CVn systems consist typically of K and G subgiants located on the main sequence, while the Algol-type binaries consist of a primary star of A or B type with an evolved K subgiant secondary that has overflowed its Roche lobe and is in the process of transferring mass to the primary. The W Uma systems are very short-period (<1 day), contact binaries where both components fill their Roche lobes.

Because of their rapid rotation, flare stars are ideal candidates for Doppler imaging (Cameron, 1995). Figure 12.2 shows Doppler images of the dKe flare star AB Doradus constructed from spectra of photospheric lines (Donati and Cameron, 1997). The large polar spot seen in the top panel may indicate the presence of a strong global dipole field aligned with the spin axis. Much effort has gone into verifying the reality of large polar spots, and, although in some cases they may be an observational artifact caused by the absence of rotation at the poles, their existence is now fairly well established. The evidence for them lies in the anomalously "flat-bottomed" profiles of rotationally broadened lines. This phenomenon is almost ubiquitous among those stars whose rotation rates are fast enough to permit Doppler imaging of their surfaces. The effect is stronger in stars with low-axis inclinations, whose polar regions are less foreshortened, than in high-inclination stars.

Table 12.1 compares observed quantities for coronal emission (primarily X-rays) from solar and stellar flares. Two separate solar classes are listed, one for compact flares and one for large flares of long duration. The latter are often referred to as eruptive flares (Priest, 1992a) and they are the long duration events (LDEs) which produce large-scale two-ribbon flares (§11.1). In Table 12.1, Lu is the peak luminosity, W_r is the total radiative energy from the flare, integrated over its lifetime, τ_{rise} and τ_{decay} are the rise and decay times of the luminosity curve, $T_{\max}$ is the maximum temperature, E_m is the volumetric emission measure, R_* is the stellar radius, h is the height of the flare loops (for solar flares only), and n_d is the plasma density deduced

Photospheric Occupancy

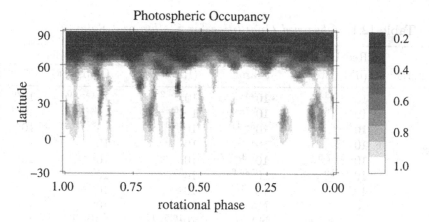

Radial Magnetic Field

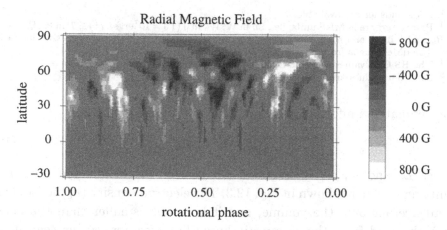

Fig. 12.2. Doppler images of AB Doradus obtained with the Anglo-Australian telescope on 7 December 1995. The top panel is the maximum entropy brightness, while the bottom panel is the radial magnetic field obtained by using the Zeeman effect. The apparent absence of a north polar magnetic field may be the result of suppression of the Zeeman signal, caused by the low surface brightness in the polar region (Donati and Cameron, 1997).

from the ratio of density diagnostic lines. The emission measure (E_m) is defined as

$$E_m = \int n_e^2 \, dV, \qquad (12.1)$$

where n_e is the electron plasma density and the integral is over the entire volume (V) of the observed region. If the electron density in the volume is

Table 12.1. *Observed Quantities for Stellar and Solar Flares*[1]

Parameter[2]	Red Dwarfs (dMe Stars)[3]	T-Tauri Stars	Binaries (RS CVn)[4]	Solar[5]	
				Compact	Eruptive
Lu	10^{21-23}	10^{23-26}	10^{23-24}	10^{18-20}	10^{20-21}
W_r	10^{24-26}	10^{27-30}	10^{26-29}	10^{22-24}	10^{24-25}
τ_{rise}	10^2	10^{3-4}	10^{2-3}	10^2	10^{2-3}
τ_{decay}	10^{2-3}	10^{4-5}	10^{3-4}	10^3	10^{4-5}
T_{max}	$10^{7.3-7.5}$	$10^{7.3-7.7}$	$10^{7.7-8}$	$10^{7-7.5}$	$10^{7-7.5}$
E_m	10^{56-59}	10^{58-63}	10^{59-60}	10^{53-55}	10^{55-56}
R_*	$10^{8.3}$	$10^{9.3}$	$10^{8.7}$	$10^{8.8}$	$10^{8.8}$
h	NA	NA	NA	10^{6-7}	10^8
n_d	10^{17-18}	NA	$10^{16.5-18}$	$10^{17-18.5}$	$10^{16.5-17}$

[1] NA stands for not available.
[2] Parameters are in MKS units, i.e., Lu in W, W_r in J ($1\,J = 10^7$ ergs), t in s, T in K, E_m in m^{-3}, R_* and h in m, and n_d in m^{-3}. The densities (n_d) are based on density-sensitive line ratios.
[3] The dMe value is from Schrijver et al. (1995).
[4] The RS CVn values are from Byrne (1995).
[5] Solar flare values are from Cook et al. (1995).

approximately uniform, then Eq. (12.1) can be rewritten as

$$E_m = n_e^2 \, L_0 \, A_0, \qquad (12.2)$$

where L_0 and A_0 are characteristic scales for the height and area of the emitting region, as shown in Fig. 12.3. The electron density (n_e) in Eq. (12.2) is an average over the volume, and it is usually smaller than the density (n_d) obtained from the diagnostic lines. The discrepancy between n_e and n_d can be expressed in terms of the filling factor ($\overline{f} = n_e/n_d$), which is necessarily less than, or equal to, unity. The flare luminosity (Lu) is directly proportional to E_m times a function which depends only on the temperature (T). Thus, if two flares have the same temperature, but different luminosities, the difference is due to the emission measure (E_m).

In addition to the quantities shown in Table 12.1, there are also observations of Doppler shifts indicating flows in the flaring plasma. As for solar flares, there are strong red shifts (velocities $>100\,\mathrm{km\,s^{-1}}$) in H$\alpha$ and other surface emission lines during the impulsive phase (Haisch, 1989; Linsky et al., 1989; Neff et al., 1989). These red shifts are the chromospheric signature of chromospheric downflows that occur in response to the onset of evaporation (see §11.1.1). Other types of Doppler shift are also observed, but these are more difficult to interpret (Houdebine et al., 1993).

For the most part, the stellar flares in Table 12.1 are considerably more energetic than their solar counterparts, although there is an overlap between

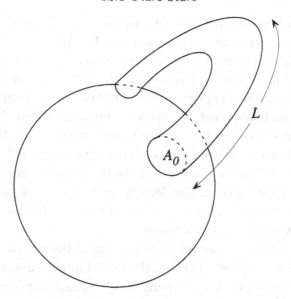

Fig. 12.3. Diagram showing a possible configuration for emitting plasma in a stellar flare. The area A_0 is the effective cross-sectional area of a loop of half-length L. If the loop has uniform thickness, then A_0 would be equal to the emission area at the stellar surface.

the smallest dMe flares and the largest solar flares. The flares occurring in T-Tauri stars and RS CVn binaries are the largest, with radiative energy outputs that are four to five orders of magnitude greater than the largest solar flares. Since the densities and temperatures of the stellar flares are comparable to those of solar flares, the larger energy output of the stellar flares must be due to the involvement of larger volumes.

12.1.1 Scaling Laws

One of the central problems in applying theory developed for solar flares to stellar flares is that the parameters n_e, L_0, and A_0 (which make up the emission measure) cannot be individually observed. Only their combination, as expressed by the emission measure [Eq. (12.2)], is known for sure. Thus, a great deal of effort has been made over the years to find theoretical relations which might allow a determination of n_e, L_0, and A_0 separately. Two physical principles have been used to obtain such relations. The first is that the density in the flaring loops is determined by chromospheric evaporation, and the second is that the decay time-scale of the thermal phase of a flare is determined by a cooling process. Both these principles work well in solar flares, except for large, eruptive events in which the decay time-scale is not determined by the radiative cooling time (see §11.1.1).

Chromospheric evaporation is an ablative process which occurs whenever energy is released in the corona (Katsova et al., 1997). Conduction along field lines mapping from the energy-release site in the corona heats the chromospheric plasma, and this creates a pressure imbalance that drives material back up into the corona. If the energy release is sustained for a sufficient time, an equilibrium is reached where the energy conducted into the chromosphere is balanced by energy lost to radiation (Hirayama, 1974). Because the radiation loss increases as the density increases, evaporation is choked off once the density reaches a point where the energy loss is comparable to the energy input into the chromosphere.

Since there is a time lag between the heating of the corona and the injection of heated chromospheric plasma, the emission measure, which depends on the density but not on the temperature, does not respond as quickly to the injection of energy as the temperature does. In both the Sun and flare stars, the peak emission measure occurs 10–20 s after the peak temperature (Pan et al., 1995). This delay is usually interpreted as the time required for evaporated plasma to fill a magnetic loop.

Numerous efforts have been made to develop simple algebraic formulae (known as scaling laws) to predict the density of the loop in response to steady heating (see the review by Jordan, 1992). These formulae are useful even when the heating is impulsive, because they provide an upper limit on the density produced by evaporation.

The scaling laws are based on the energy equation [Eq. (1.5)]:

$$\frac{1}{(\gamma - 1)} \frac{dp}{dt} = -\frac{\gamma}{\gamma - 1} p \nabla \cdot \mathbf{v} + \kappa_0 \nabla \cdot (T^{5/2} \nabla T) - n_e^2 Q(T) + E_h, \quad (12.3)$$

where $\kappa_0 T^{5/2}$ is the coefficient of thermal conduction, $Q(T)$ is the radiative loss function (§1.2.1), and E_h is a source term. The radiative loss function, which is shown in Fig. 11.9, is usually approximated by a piece-wise function of the form $Q(T) = \chi T^{\alpha^*}$, where χ and α^* are assigned different values for different temperature ranges (e.g., Cox, 1972). X-ray observations of flare stars typically lie in the temperature range from 3×10^7 to 10^8 K, and for this range $\alpha^* = 0.5$ and $\chi = 10^{-39.6}$ W m^3 K $^{-0.5}$ (Haisch, 1983). In this range the radiation losses are primarily due to X-ray bremsstrahlung, and Q is an increasing function of T rather than the decreasing one that is found in solar flares where the temperature range is lower.

A very simple scaling law was first obtained by Hirayama (1974) for evaporation in solar flares. He assumed that the flare energy is released at the tops of magnetic loops of half-length (L) and that a static equilibrium could be established between the energy conducted into the chromosphere and the radiative loss in the loop. With these assumptions Eq. (12.3) becomes

$$\kappa_0 \frac{\partial}{\partial x} \left(T^{5/2} \frac{\partial T}{\partial x} \right) = n_e^2 \chi T^{1/2}$$

or very roughly

$$\kappa_0 \frac{T^{7/2}}{L^2} \approx n_e^2 \chi T^{1/2}, \tag{12.4}$$

where $\chi = 10^{-39.6}$ W m^3 K$^{-0.5}$ and $\kappa_0 = 10^{-11}$ m^{-5} W K$^{-7/2}$. The source term (E_h) does not appear in this equation because the assumption that all of the energy input occurs at the top of the loop means that the energy input is treated as a boundary condition. Solving Eq. (12.4) for L and assuming $n_e = \bar{f} n_d$ (where $\bar{f}$ is the filling factor) gives the loop height as

$$L \approx \left(\frac{\kappa_0}{\chi} \right)^{1/2} \frac{T^{3/2}}{\bar{f} n_d} = 2.0 \times 10^{14} \frac{T^{3/2}}{\bar{f} n_d}, \tag{12.5}$$

which is equivalent to Hirayama's formula (1974) for evaporation in MKS units (for solar flares with a temperature range of 2×10^6 to 3×10^7 K, $\alpha^* \simeq -1$ instead of 0.5). A more general and precise formulation of this scaling law was developed by Rosner et al. (1978) which allows for a distributed heat source in the loop.

Equations (12.4) and (12.5) imply that conduction and radiation are of equal importance, but this will not necessarily be true if the heating is impulsive as in a flare. If the time-scale for material to flow up the field lines into the corona is longer than the time for which the heating is applied, then the radiation term will be smaller than the conduction term. In this case, Eq. (12.5) provides an upper limit for the loop scale-length (L) when the filling factor ($\bar{f}$) is set to one.

Having assumed that the density of the X-ray emitting plasma is due to evaporation, we can use the emission measure to obtain a rough estimate for the area of the flaring region. Combining Eq. (12.2) with Eq. (12.5), we find that the emitting area (A_0) is

$$A_0 = \frac{E_m}{(\bar{f} n_d)^2 L} \approx \left(\frac{\chi}{\kappa_0} \right)^{1/2} \frac{E_m}{\bar{f} n_d T^{3/2}} = 5.0 \times 10^{-11} \frac{E_m}{\bar{f} n_d T^{3/2}} \tag{12.6}$$

in MKS units. If the flaring region consists of only a single loop, then A_0 is the cross-sectional area of the loop, but, if there is more than one loop, A_0 is the net area of these loops (see Fig. 12.3).

Table 12.2. *Calculated Parameters for Large Flares, Using
Density-Diagnostic Lines and Assuming
Quasi-Steady Evaporation*

Parameter	Red Dwarfs (dMe Stars)	Binaries (RSCVn)	Solar Compact	LDE
n (m^{-3})	10^{18}	10^{18}	10^{18}	10^{17}
L (m)	$<10^{7.5}$	$<10^{8.3}$	$<10^{7.5}$	$<10^{8.5}$
A_0 (m^2)	$>10^{15.5}$	$>10^{15.8}$	$>10^{11.5}$	$>10^{13.5}$
L/R_*	<0.15	<0.4	<0.04	<0.4
A_0/A_*	$>10^{-2.2}$	$>10^{-2.7}$	$>10^{-7.3}$	$>10^{-5.3}$
B (G)	>300	>600	>300	>100
v_A (km s^{-1})	>650	>1400	>650	>650
v_{rise} (km s^{-1})	700	200	300	300
v_R (km s^{-1})	60	20	8	7
M_A	0.09	0.02	0.01	0.01

Table 12.2 shows the loop scale-length (L) and the flare area (A_0) for large flares in dMe dwarfs, RS CVn binaries, and the Sun obtained from Eqs. (12.5) and (12.6) with $\overline{f} = 1$. The magnetic fields are quoted in gauss (1G = 10^{-4} tesla). The solar flares (§§11.1, 11.2) are separated into compact flares (large, but non-eruptive) and eruptive flares (long-duration events), and the values in this table are based on the upper range of the values of T and E_m listed in Table 12.1. The values of L for solar flares are larger than observed (Table 12.1) by a factor of about three, implying that a quasi-steady evaporation has not been achieved and that Hirayama's relation overestimates the scale-length (L) even when the filling factor is assumed to be unity. Smale et al. (1986) and Cheng and Pallavicini (1991) have drawn a similar conclusion for dMe flares, and so the values of L in Table 12.2 are listed as upper limits and the corresponding values of A_0 as lower limits. Stellar-flare areas calculated from optical emission during the impulsive phase (see Kahler et al., 1982; de Jager et al., 1989) tend to give values smaller by an order of magnitude than shown in Table 12.2. Perhaps this is because, as in solar white-light flares, the region of continuum emission is smaller than the region occupied by X-ray loops, or maybe it is because of a rapid increase in the loop area with height.

Despite the fact that the energy released by large stellar flares is two to five orders of magnitude greater than the largest solar flare, the values of the loop length (L) in Table 12.2 are nearly the same. However, there are big differences in the flare area (A_0). The inferred areas of stellar flares are more than 10^2 times greater than the areas of solar LDE flares and 10^4 greater than the areas of compact solar flares. The much larger flare areas are consistent with the star spots having areas which are typically 10^3 to 10^4 times greater than sunspots (Rodonò and Lanza, 1996).

An alternative approach to using the line-ratio density (n_d) and assuming a filling factor of unity is to estimate the plasma density by assuming that the decay time-scale (τ_{decay}) is some kind of cooling time-scale. As we discussed in Section 11.1, this is not a good assumption for solar LDEs because the decay time-scale of these large, eruptive events is determined by the reconnection rate. Nevertheless, the assumption does appear to be valid for compact solar flares, and it could very well be that at least some stellar flares are scaled-up versions of this type of solar flare (van den Oord et al., 1996). The most commonly used approach for determining the density is to assume that the cooling is radiative and that the pressure within the loop is uniform. The assumption of uniform pressure is justified if the radiative time-scale is much larger than the dynamic time-scale required for plasma flows to eliminate pressure gradients along a magnetic field line. Such an approach is sometimes referred to as *quasi-static cooling*, but this term is somewhat misleading since "quasi-static" does not refer to the radiative time-scale but rather to the dynamic time-scale for flows. As has been pointed out by van den Oord and Mewe (1989), quasi-static cooling does not require an approximate balance between heating and cooling, and it is valid even when there is no heating at all.

Quasi-static cooling also assumes that there is no net heat flux into or out of a magnetic loop, so that, when Eq. (12.3) is integrated along the whole length of the loop, the heat conduction term vanishes. In other words, conduction serves only to redistribute heat within the loop and does not contribute to the net cooling. Thus, if both the plasma velocity ($\mathbf{v}$) and heating source (E_h) are set to zero, Eq. (12.3) reduces to

$$\frac{1}{(\lambda - 1)} \frac{dp}{dt} \approx -n_e^2 \chi T^{1/2}, \tag{12.7}$$

where n_e and T now represent average values. From the equation of state, the pressure for a hydrogen plasma is

$$p = 2 n_e k_B T,$$

where k_B is Boltzmann's constant. Replacing dp/dt by p/τ_r and solving for n_e leads to the estimate

$$n_e \approx \frac{3k_B}{\chi} \frac{T^{1/2}}{\tau_r} = 1.6 \times 10^{17} \frac{T^{1/2}}{\tau_r} \tag{12.8}$$

in MKS units, where τ_r is the radiative time-scale. If one assumes that the observed decay time-scale (τ_{decay}) is equal to the radiative time-scale (τ_r), then Eq. (12.8) provides an estimate for the average electron plasma density (n_e). The filling factor ($\overline{f}$) is in turn given by the ratio n_e/n_d.

The application of Eq. (12.8) poses several problems. First, as mentioned previously, the time-scale of the large eruptive events (LDEs) is not a measure of the radiative time-scale (τ_r). This problem can be side-stepped in solar flares by spatially resolving individual loops and measuring the rate at which they cool, but such a technique is not available in the stellar case. Another problem is that the flare heating is often too impulsive for the pressure to become uniform. Even in solar LDEs, where the total energy input lasts a long time, individual magnetic field lines move through the reconnection site so rapidly that there is insufficient time to equalize the pressure along them before they are disconnected from the reconnection site (see Fig. 11.14).

Sylwester et al. (1993) have found that only about 20% of all solar flares obey the quasi-static relation (12.8), and Haisch (1983) has found similar results for flares on dMe stars. An analysis by Smale et al. (1986) found that the quasi-static relation tends to predict too low a density in dMe flares, and thus to give unreasonably large scale-lengths.

Van den Oord and Mewe (1989) have attempted to extend the quasi-static method to allow for extended heating in the gradual phase. However, their analysis assumes that there is a sudden decrease in the heating rate at the end of the impulsive phase and that the decay time-scale is just the time required for the radiative losses to adapt to the reduced heating rate. Recently, Cargill et al. (1995) have developed a method which relaxes the assumption that the cooling is quasi-static, and this method works quite well for compact solar flares.

12.1.2 Role of Reconnection

At the present time it is not possible to measure the strength of the magnetic field in either solar or stellar coronae, but a lower limit can be estimated by assuming that the field is sufficient to contain the thermal pressure of the plasma, so that

$$B \geq \sqrt{2\mu n k_B T}. \tag{12.9}$$

The resulting values, which are shown in Table 12.2, are of the same order as the coronal field strength obtained by current-free extrapolation of the observed photospheric magnetic fields. The Alfvén speed implied by

Eq. (12.9) is

$$v_A = \frac{B}{\sqrt{\mu \, m_p \, n_e}} \geq c_s,$$

where $m_p = 1.67 \times 10^{-27}$ kg is the proton rest mass and $c_s = \sqrt{(2k_B/m_p)T}$ is the isothermal sound speed. If a flare is initiated by an ideal MHD process such as we discussed in Section 11.1.1, then the Alfvén speed (v_A) should be comparable with the speed (v_{rise}) obtained by dividing the loop scale-length (L) by the rise-time (τ_{rise}) of the luminosity curve; that is,

$$v_{\text{rise}} = \frac{L}{\tau_{\text{rise}}}.$$

The values of v_{rise} listed in Table 12.2 are of the same order as v_A except for the binary systems, where v_{rise} is about a factor of 10 smaller than v_A. This discrepancy may indicate that flares in binary systems are not simple analogs of solar flares.

We can use the peak luminosity (Lu) listed in Table 12.1 to estimate the reconnection velocity (v_R) of the plasma flowing into the reconnection site. The luminosity must be less than or equal to the rate at which magnetic energy is released, since not all of the released energy goes into radiation. If this energy is primarily released by a reconnection process, then

$$v_R \approx \frac{\mu \, \text{Lu}}{B^2 S_R}, \tag{12.10}$$

where S_R is the effective surface area of the plasma which is processed through the reconnection site. The total volume processed through reconnection is $S_R v_R \tau_R$, where τ_R is the reconnection time-scale. If we assume that the processed volume is of the order of the volume of the flare loops with $S_R \simeq L\sqrt{A_0}$, we still obtain the relatively modest values for v_R shown in Table 12.2. Dividing these numbers by the corresponding Alfvén velocities gives the Alfvén Mach number (M_A). In all cases M_A is much less than 0.1, and so, even though the volumes involved in flare stars are much larger than in the Sun, the reconnection rate required is essentially the same.

The similarity of the reconnection rates listed in Table 12.2 does not necessarily mean that the driving mechanism for reconnection in stellar and solar flares is the same. The similar rates reflect the similarity of the temperature, density, and magnetic field strength of the stellar and solar coronae, but there is no guarantee that the geometry of the field structure is the same. Stellar flares are much larger than solar ones, probably because they occur on stars with a more vigorous magnetic dynamo. The enhanced dynamo action is thought to result from the more rapid rotation and deeper

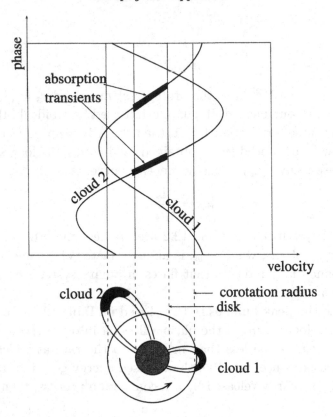

Fig. 12.4. Schematic illustration of the relationship between the drift rate of a prominence on AB Doradus and its distance from the surface (from Cameron, 1996).

convection zones of these stars, so their magnetic structure might be quite different from the Sun. What these differences are, is difficult to assess, but Doppler imaging provides some clues.

As mentioned previously, flare stars may have large polar spots (although this is far from certain) which have no counterpart on the Sun. Recent studies have also revealed the existence of unusually large prominences on some rapidly rotating flare stars (Cameron, 1996; Byrne et al., 1996). Figure 12.4 shows the geometry of these prominences inferred from Hα observations of the K-dwarf AB Doradus. The prominences lie between three and nine stellar radii from the rotation axis, beyond the star's co-rotation radius at 2.6 stellar radii. Tension in the magnetic loops connecting from the prominences to the stellar surface counterbalances the centrifugal pull of the star's rotation and prevents the prominences from escaping. In contrast to the Sun, the

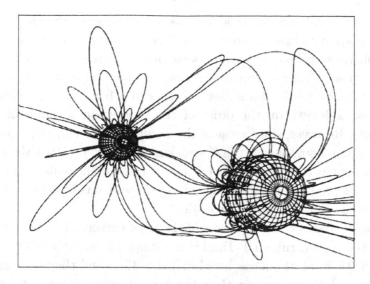

Fig. 12.5. Model magnetic field for an RS CVn binary system in which the two stars are tidally locked into rapid synchronous rotation (Uchida, 1986).

magnetic field helps to hold the prominences down against the rotation rather than holding them up against gravity. The prominences last for several stellar rotations, and the amount of mass contained in them approaches 10^{15} kg – a hundred times greater than the mass in a solar prominence. Jeffries (1993) has noted that the stress exerted by the rotating prominences on the field may be the predominant mechanism for storing magnetic energy in the star's corona. If so, then we would expect reconnection in the flare star to be driven by the rotational stresses, which is not the case for the Sun.

In the past there has been some debate as to whether the extremely large flares occurring in RS CVn binary systems are caused by a mass interchange between the components, but the spectroscopic and polarimetric analyses of the various emissions from these systems (see, e.g., Weiler et al., 1978) have eliminated this possibility. The flares are almost certainly magnetic in nature and involve the build-up of magnetic stresses, but where these stresses develop is not clear.

It has been proposed by Simon et al. (1980) and Uchida (1986) that the flares are produced by sporadic reconnection between the magnetic field of the two components, as shown in Fig. 12.5. Because the two components are tidally locked to each other, the connectivity of the field lines between the two stars changes only in response to the evolution of the surface fields. Thus, the stressing of the field in a binary system is similar to the solar case, in which stresses are built up by changes in the surface field.

There are several observational features which support the idea that flares in RS CVn systems involve both stars. Firstly, flare emissions occur in the chromospheres of both components; secondly, mapping of flare-related radio emissions shows that the emitting region includes both stars; and, finally, the long duration of the flares (several days in a few cases) implies a characteristic scale-length on the order of the separation distance of the two components. However, since most of the observed flare emissions come from the more active component (the K subgiant or dwarf), rather than the less active component (the G subgiant), it might be that the emissions occurring on the less active component occur only because it has field lines connected to the flare site on the more active star.

Perhaps the flares which are most likely to be driven differently from solar flares are those occurring in T-Tauri stars (Appenzeller, 1990). These are the most energetic of all the flares listed in Table 12.1, and they have an energy more than 10^5 times greater than the largest solar flares. The accretion disks surrounding these stars contain plasma and are therefore magnetically linked to their central star. Although the inner region of the accretion disk may corotate with the star, the outer regions (where the magnetic field is weaker) do not do so because the material in the disk follows Keplerian orbits (Mestel and Spruit, 1987). Consequently, the disk continually exerts a torque that tends to shear the magnetic field of the star.

12.2 Accretion Disks

It is likely that reconnection is an important process in *quasars*. According to current thinking, quasars are black holes surrounded by accreting material which releases gravitational energy at a rate of 10^{39} W as it falls into the black hole. However, most of the infalling material has sufficient angular momentum to go into orbit, so accretion does not occur unless there is some way for the angular momentum of the orbiting material to be dissipated. Many of the dissipation mechanisms which have been proposed invoke MHD turbulence, and, as we shall see in the next section, the efficacy of MHD turbulence as a dissipation mechanism depends upon magnetic reconnection.

Accretion disks also occur in other types of astrophysical system besides black holes. These include newly formed stars (protostars and T-Tauri stars) and stellar binaries with mass transfer (Algol-type systems and dwarf novae). Dissipation of the angular momentum in any accretion disk is thought to require some kind of effective viscosity, but the actual processes involved may be quite varied and include stellar and disk winds, magnetic torques between disk and star, field-aligned accretion flows and polar jets, and possible

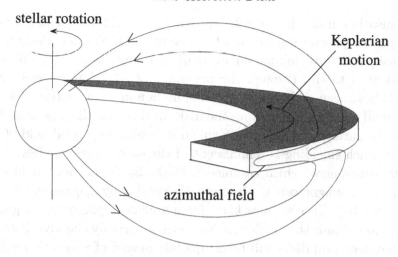

Fig. 12.6. Creation of an azimuthal field in an accretion disk by Keplerian motion. Here the velocity in the disk is everywhere slower than the local corotation velocity of the star, but this need not be the case in general.

momentum transfer by MHD waves (see Papaloizou and Lin, 1995, for a concise review). The kind of collisional viscosity that operates here on Earth is too minuscule to be effective (Verbunt, 1982), so there has been a great deal of effort to determine if an MHD process can provide the necessary viscosity (e.g., Eardley and Lightman, 1975; Galeev et al., 1979; Coroniti, 1981; Heyvaerts and Priest, 1989; Tout and Pringle, 1992; Brandenburg et al., 1996; Lesch, 1996).

One reason magnetic fields are likely to be involved is that, as material in the disk spirals in towards the central object, half of the gravitational energy released goes into kinetic energy while the other half goes into heating (Ostlie and Carroll, 1996). For T-Tauri stars, material in the disk is heated to a few 10^3 K, while in neutron stars and quasars temperatures as high as several 10^6 K occur. In the latter case there is sufficient ionization to create a highly conducting plasma, but in the former case the ionization may be fairly low in the centre of the disk (Shu et al., 1994). If there is a highly conducting plasma in the disk, then any magnetic field threading it will be frozen into the plasma, and the Keplerian motion in the disk will tend to shear and amplify the field. An example of this effect is shown in Fig. 12.6 for the case in which the field comes from a central star with a dipole field. Initially, the field from the star is perpendicular to the disk plane, but the differential rotation within the disk winds this field around to create a strong azimuthal field. The azimuthal field continues to grow until the material in the disk is brought into corotation or until the field structure is disrupted in some way.

If the ionization in the disk is sufficiently low, the field may be able to diffuse through the plasma fast enough to prevent the field from winding up. In low-density plasmas dominated by neutral atoms, *ambipolar diffusion* allows field lines to slip through the plasma (§12.2.2). Alternatively, Shu et al. (1994) have argued that the Keplerian motion in the accretion disk will act to expel all magnetic fields from the disk. In this case the disk would be surrounded by surface currents which shield it from the external field of the central star. Such shielding is reminiscent of the early closed models of the Earth's magnetosphere, which assumed that the Sun's magnetic field does not penetrate the magnetosphere because of shielding by magnetopause currents (Chapter 10). Since we now know from studies of planetary magnetospheres and comet tails that such shielding is only partially effective, it seems unlikely that accretion disks will be completely devoid of magnetic fields.

12.2.1 Relation Between Disk Viscosity and Magnetic Reconnection

In this section we review two different models which relate the rate of accretion to the rate of reconnection of magnetic fields within a disk. Both models use the shear flow within the disk to amplify the disk magnetic field and invoke MHD turbulence as the mechanism for the radial transport of material. The first model (Coroniti, 1981), based on ideas introduced earlier by Eardley and Lightman (1975), is the simpler but less realistic of the two. It is essentially two-dimensional and assumes that the only important magnetic field components are the ones lying in the plane of the disk. The second model (Tout and Pringle, 1992) is inherently three-dimensional and assumes that it is the reconnection of magnetic field lines vertical to the plane of the disk which is the important process. Furthermore, unlike the Coroniti model, the Tout and Pringle model incorporates a well-defined physical mechanism for the production of turbulence within the disk, namely the Balbus–Hawley instability.

To start the discussion let us assume that an accretion disk lies in the $r\phi$-plane located at $z = 0$ in cylindrical coordinates, and that the material in it orbits a central star in accordance with Kepler's third law

$$\Omega_a^2 = \frac{GM_*}{r^3},$$

where Ω_a is the orbital frequency in the accretion disk at radius r, G is the gravitational constant, and M_* is the mass of the central star. The rate of mass accretion $(\dot{M})$ at the radial location r in the disk can be expressed as

$$\dot{M} = 4\pi r \, h_d \, \rho \, v_r, \tag{12.11}$$

where h_d is the half-thickness of the disk, ρ is the density, and v_r is the infall velocity (Sakimoto and Coroniti, 1989).

The half-thickness (h_d) can be estimated by using the concept of a gravitational scale-height. For simplicity, let us assume that the gravitational force due to the mass of the disk itself is negligible, and that the primary forces acting in the z-direction, perpendicular to the plane of the disk, are just the z-components of the total pressure gradient (∇P) and the gravitational field ($\rho \mathbf{g}$) of the central star. Since $g = GM_*/r^2 = \Omega_a^2 r$, force balance in the z-direction requires

$$\frac{P}{z} \sim \rho g \frac{z}{r} \sim \rho \Omega_a^2 z.$$

After setting $z = h_d$, we obtain

$$h_d = \left(\frac{P}{\rho \Omega_a^2} \right)^{1/2}, \tag{12.12}$$

where the total pressure (P) is the sum of the gas (p), magnetic ($B^2/(2\mu)$), and radiation pressures.

Present-day calculations of accretion rates are mainly based on the pioneering work of Shakura and Sunyaev (1973), who set up a simple model for a thin disk. From the momentum equation [Eq. (1.2)] they found that for Keplerian flow, v_r is related to the $r\phi$-component of the stress tensor ($\mathbf{S}$) by

$$v_r = -\frac{S_{r\phi}}{\rho \Omega_a r}. \tag{12.13}$$

To construct an analytical model of the accretion process, they assumed that the stress component ($S_{r\phi}$) is related to the total pressure (or sometimes the gas pressure) by the simple, ad-hoc expression

$$S_{r\phi} = \alpha_S P, \tag{12.14}$$

where the disk parameter (α_S) is a constant. With this assumption, they could completely solve their model equations describing both the dynamical and radiative properties of the disk.

Although there has been an enormous amount written about whether the assumption (12.14) is truly justified, the α_S-formalism introduced by Shakura and Sunyaev has remained popular. Many studies have attempted to determine the parameter α_S from first principles based on various assumptions about the physical processes operating in an accretion disk. An important result from these studies is that the original accretion-disk model obtained by Shakura and Sunyaev (1973) is unstable to both thermal and secular perturbations. However, a stable model can be achieved either by assuming α_S to be a function of radius or by replacing the total pressure

(P) in Eq. (12.14) by the gas pressure (p). In the following analyses we shall replace P by p in (12.14).

The relation between the Shakura–Sunyaev α_S-parameter and the coefficient of kinematic viscosity (ν) in Eq. (1.4) is

$$\alpha_S = \frac{3\rho\,\Omega_a\,\nu}{2p}, \tag{12.15}$$

where ν is now the viscosity produced by MHD turbulence within the disk. Substitution of Eq. (12.15) into Eqs. (12.14) and (12.13) gives

$$S_{r\phi} = \frac{3}{2}\rho\,\nu\,\Omega_a \tag{12.16}$$

for the stress and

$$v_r = \frac{3}{2}\frac{\nu}{r} \tag{12.17}$$

for the inflow velocity. This expression for inflow has the form typical of a diffusion process.

Numerous analytical prescriptions have been proposed for calculating α_S (Verbunt, 1982), but at the present time there is no consensus that any of them is definitive. The complexity of the turbulent processes occurring in an accretion disk makes it likely that quantitatively accurate models can only be achieved by means of numerical simulations (e.g., Brandenburg et al., 1996; Hawley et al., 1996; Matsumoto et al., 1993). Nevertheless, from the point of view of understanding the role of reconnection in accretion, simple analytical models such as those developed by Coroniti (1981) and Tout and Pringle (1992) are highly instructive.

Coroniti's model is based on the concept introduced by Eardley and Lightman (1975) that the Keplerian motion in a disk eventually creates a magnetic field which lies in the plane of the disk (Fig. 12.6) and is organized in elliptical cells as shown in Fig. 12.7(a). These magnetic cells are continually created and destroyed in a turbulent process that leads to the radial diffusion of the plasma in the disk. From mixing-length theory, the effective viscosity (ν) associated with this turbulent diffusion is

$$\nu \approx \frac{L_0^2}{\tau_0}, \tag{12.18}$$

where L_0 and τ_0 are the length- and time-scales associated with the largest magnetic cells.

An example of a single cell is shown in Fig. 12.8. Initially, the magnetic field in the cell is assumed to be circular and too weak to have much effect on the Keplerian flow, so the flow rapidly distorts the field by stretching the

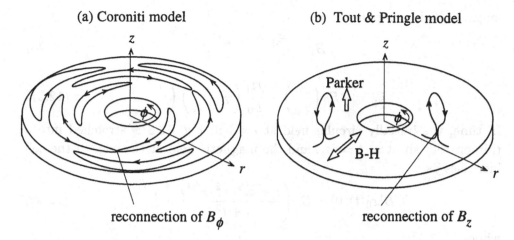

(a) Coroniti model   (b) Tout & Pringle model

reconnection of B_ϕ reconnection of B_z

Fig. 12.7. Magnetic reconnection geometry for the models of (a) Coroniti (1981) and (b) Tout and Pringle (1992). The labels Parker and B–H stand for the Parker buoyancy instability and the Balbus–Hawley rotational instability.

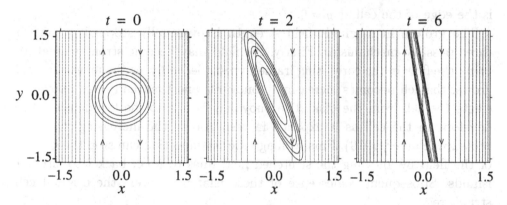

Fig. 12.8. The evolution of the magnetic field in Coroniti's cell model (1981) for Keplerian amplification with times (t) measured in units of the shear time-scale (τ_s).

field lines. For a field structure which is initially circular, the distorted field is prescribed by the vector potential

$$A_z = \frac{B_0}{2L_0}\left[\left(y + \frac{t\,x}{\tau_s}\right)^2 + x^2\right],\qquad (12.19)$$

where x and y define a local corotating coordinate system defined by $x = R - r$, $y = r(\phi - \Omega_a t)$. Also, $\tau_s = 2/(3\Omega_a)$ is the shear time-scale and $A_z \leq B_0 L_0/2$. The upper limit to this value of A_z corresponds to the edge of the cell, outside which the field is assumed to be zero. In the corotating coordinate system the local Keplerian velocity is $v_y = -x/\tau_s$. The corresponding magnetic field components are

$$B_x = \frac{B_0}{L_0}\left(y + \frac{xt}{\tau_s}\right), \tag{12.20}$$

$$B_y = -B_0\left(\frac{x}{L_0}\right) - \frac{B_0}{L_0}\left(y + \frac{xt}{\tau_s}\right)\left(\frac{t}{\tau_s}\right). \tag{12.21}$$

In time, the initially circular field at $t = 0$ in Fig. 12.8 is stretched into a thin current sheet at $t = 6\tau_s$, and the magnetic field at the edge of the cell increases as

$$B[x_0(t), 0] = B_0\left(\frac{1 + 3t^2/\tau_s^2 + t^4/\tau_s^4}{1 + t^2/\tau_s^2}\right)^{1/2}, \tag{12.22}$$

where

$$x_0 = \frac{L_0}{(t^2/\tau_s^2 + 1)^{1/2}}$$

is the edge of the cell at $y = 0$.

The growth in magnetic field strength is eventually stopped by one of three possible mechanisms. Firstly, the field may become strong enough to bring the flow in the corotating frame to rest. Secondly, as the magnetic field grows, the cell becomes increasingly buoyant, and eventually this buoyancy ejects the cell from the disk. The buoyancy is due to the low density that develops in the cell as it expands in response to the increasing magnetic pressure (Parker, 1979). Finally, magnetic reconnection can stop the growth of the field by allowing the stretched field lines to relax back into circular islands. Subsequent coalescence of these islands recovers the original cell structure.

Coroniti estimated the time (τ_{cr}) for the field to become strong enough for corotation to ensue, by balancing the deceleration of the flow against the Maxwell stress at the front of the cell. In other words,

$$\rho\frac{dv_y}{dt} = \frac{B_x}{\mu}\frac{\partial B_y}{\partial x},$$

evaluated at $x = L_0$, $y = L_0\left(1 - \tau_{cr}/\tau_s\right)$, which is roughly

$$\rho\frac{L_0}{\tau_{cr}\tau_s} \approx \frac{B_0^2}{\mu L_0}\left(1 + \frac{\tau_{cr}^2}{\tau_s^2}\right).$$

Solving for τ_{cr} gives

$$\frac{\tau_{cr}}{\tau_s}\left(1 + \frac{\tau_{cr}^2}{\tau_s^2}\right) \approx \left(\frac{L_0}{v_{A0}\tau_s}\right)^2, \tag{12.23}$$

where $v_{A0} = B_0/(\mu\rho)^{1/2}$ is the initial Alfvén speed in the disk. Since significant amplification of the field occurs only for $\tau_{cr} > \tau_s$, Eq. (12.23) requires that the cell dimension (L_0) be greater than $v_{A0}\tau_s$. Cells whose size is smaller than this value are brought into corotation too quickly to have any effect on the azimuthal flow.

As shown by Parker (1979), buoyancy becomes important when the plasma beta (β) in the cell becomes less than unity. The requirement that $\beta > 1$ for $t < \tau_{cr}$ leads to the condition that the cell length be less than c_s/λ, where $c_s = (p/\rho)^{1/2}$ is the isothermal sound speed. Combining the buoyancy condition with the corotation condition leads to

$$v_{A0} \leq \frac{L_0}{\tau_s} \leq c_s. \tag{12.24}$$

The isothermal sound speed is used because the thermal conductivity is large enough to produce a nearly isothermal plasma. The range of cell sizes prescribed by Eq. (12.24) corresponds to the cells which are large enough to avoid immediate corotation but small enough to prevent ejection from the disk by buoyancy. Condition (12.24) is equivalent to requiring the shear velocity at the edge of the cell to be greater than the initial Alfvén speed but less than the sound speed.

In any mixing-length formalism, the cells with the largest dimension are the ones which are most important for the radial diffusion of angular momentum. Thus, setting $L_0 = c_s\tau_s$, $v_{A0} = c_s\beta_0^{-1}$ and assuming $\tau_{cr} \gg \tau_s$, we find that the time required to bring the largest cells into corotation is

$$\tau_{cr} \approx \beta_0^{1/3}\tau_s, \tag{12.25}$$

where β_0 is the initial plasma beta in the disk. The weaker the initial magnetic field (i.e., the larger β_0), the longer it takes for a cell to be brought into corotation.

The time (τ_R) required for the field lines in the stretched cell to reconnect is estimated as

$$\tau_R = \frac{x_0(t)}{M_A v_A(x_0,0)} = \frac{x_0(t)}{M_A v_{A0}} \frac{B_0}{B(x_0,0)},$$

where x_0 is the half-width of the cell, v_{A0} is the Alfvén speed at x_0, and M_A is the inflow Alfvén Mach number at x_0. Coroniti assumes that M_A is constant in both space and time. Substituting for x_0 and $B(x_0,0)$, we find

$$\tau_R = \frac{\tau_{R0}}{(1 + 3t^2/\tau_s^2 + t^4/\tau_s^4)^{1/2}}, \tag{12.26}$$

where $\tau_{R0} = L_0/(M_A v_{A0})$ is the initial reconnection time-scale.

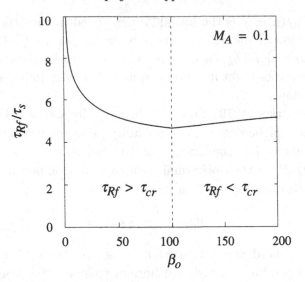

Fig. 12.9. The ratio of the reconnection time-scale (τ_{Rf}) to the shear time-scale (τ_s) versus the initial plasma beta (β_0) in Coroniti's cell model.

The final reconnection time-scale (τ_{Rf}) depends on whether the reconnection is completed before or after corotation occurs. If it is completed before corotation, that is if $\tau_R < \tau_{cr}$, then τ_{Rf} is estimated by setting $t = \tau_R = \tau_{Rf}$ in Eq. (12.26), so that

$$\tau_{Rf} \approx (\tau_{R0}/\tau_s)^{1/3}\tau_s \approx \frac{\beta_0^{1/6}}{M_A^{1/3}} \text{ for } \tau_R < \tau_{cr},$$

where $\beta_0 \equiv (c_s/v_{A0})^2$ and $t \gg \tau_s$ has been assumed as before. If corotation occurs first, that is if $\tau_R > \tau_{cr}$, reconnection still continues, but there is no further amplification of the field beyond $t = \tau_{cr}$. In this case, τ_{Rf} is estimated by setting $t = \tau_{cr}$ in Eq. (12.26). Upon substitution of Eq. (12.25) for τ_{cr}, this gives

$$\tau_{Rf} \approx \frac{\tau_s}{M_A \tau_{R0}^{1/3}} \approx \frac{\tau_s}{M_A \beta_0^{1/6}} \text{ for } \tau_R > \tau_{cr}.$$

The final reconnection time-scale (τ_{Rf}) is plotted in Fig. 12.9 as a function of β_0. Note that there is a minimum reconnection time-scale of

$$\tau_{Rf \min} \approx \frac{\tau_s}{M_A^{2/3}} \tag{12.27}$$

at $\beta_0 = M_A^{-2}$ when $\tau_R = \tau_{cr}$. Such a minimum exists because, when β_0 is too small, corotation is achieved quickly, before the field can undergo significant amplification. In contrast, when β_0 is too large, the magnetic field is so weak that reconnection remains slow despite amplification.

Finally, using the minimum reconnection time [Eq. (12.27)] for the scattering time (τ_0) in Eq. (12.18) and $L_0 = c_s \tau_s$ for the maximum scattering length, we obtain the maximum viscosity predicted by Coroniti's model, namely

$$\nu_{\max} \approx M_A^{2/3} c_s^2 \tau_s.$$

Substituting this into Eq. (12.14) and using $c_s^2 = p/\rho$ yields a viscous stress of

$$S_{r\phi} \approx M_A^{2/3} p. \tag{12.28}$$

This is the same form for the stress as used by Shakura and Sunyaev (1973), except that the total pressure (P) has been replaced by the gas pressure (p). The Shakura–Sunyaev viscosity parameter (α_S) is now expressed in terms of the reconnection parameter (M_A) as

$$\alpha_S \approx M_A^{2/3} \approx \frac{\tau_s}{\tau_R}, \tag{12.29}$$

which is also to say that α_S is just the ratio of the shear time-scale (τ_s) to the reconnection time-scale (τ_R). If the reconnection process is very slow then $\tau_R \to \infty$, $\alpha_S \to 0$, and the accretion process halts.

A key issue not explicitly addressed in Coroniti's model is the origin of the turbulence within the disk. One might think that the tearing and coalescence instabilities invoked by Coroniti would generate the required turbulence. However, Rädler (1986) argues that non-axisymmetric processes, such as Coroniti's, cannot sustain themselves in an accretion disk because they lead to growth of the axisymmetric magnetic component and decay of the non-axisymmetric component. Convection, such as occurs in the outer envelope of the Sun and other stars, is not considered a tenable process for accretion disks (Ryu and Goodman, 1992), and no other, purely hydrodynamic process has been found which is likely to generate turbulence (Mestel, 1999).

One of the important realizations of recent years is that there is an excellent MHD candidate for generating turbulence, namely the hydromagnetic shear instability discovered by Velhikhov (1959) and independently by Chandrasekhar (1961). In their studies of Couette flow (i.e., the flow between rotating cylinders) in MHD fluids, they realized that, in the presence of a magnetic field perpendicular to the plane of rotation, the flow would in certain circumstances be unstable. The importance of this instability for accretion disk flows has become evident as the result of several studies by Balbus and Hawley, and so the instability is now commonly referred to as the *Balbus–Hawley instability* when applied to accretion disks (see Hawley and Balbus, 1991, for references).

The Balbus and Hawley instability can occur when there is a magnetic field component perpendicular to the plane of the accretion disk and the disk rotation frequency (Ω_a) decreases with distance. However, the instability can be quenched even under these conditions if the magnetic field is sufficiently strong. When the instability is present, it causes radial displacements of the vertical field.

Figure 12.8(b) shows how the instability is incorporated in the model of Tout and Pringle (1992). Because the Balbus–Hawley instability requires the existence of a vertical magnetic field, Tout and Pringle make the vertical field an integral part of their model, rather than ignoring it as Coroniti does. To generate the vertical field, or to amplify the vertical field component of the stellar magnetic field threading the disk, they invoke the Parker buoyancy instability (Parker, 1979) to convert horizontal field lines into vertical ones. Loops of magnetic flux pop out of the surface of the disk [Fig. 12.7(b)] in a manner that is analogous to the formation of active regions at the surface of the Sun. The presence of the vertical fields then triggers the Balbus–Hawley instability within the disk, and reconnection occurs between the alternating up-and-down fields produced by the emerging loops.

Because the Balbus–Hawley instability is an ideal MHD process, it operates on the Alfvén time-scale. Thus, Tout and Pringle assume that it is the reconnection of the vertical field driven by strong radial and shear flows which is the important process, rather than the reconnection of the azimuthal field as Coroniti assumes [Fig. 12.7(a)]. Their approach also differs from Coroniti's in that they do not assume a specific field model but simply estimate the sources and losses of the various magnetic field components. The rate of change of the azimuthal (B_ϕ), radial (B_r), and vertical (B_z) components is described by

$$\frac{dB_\phi}{dt} \approx \frac{B_r}{\tau_s} - \frac{B_\phi}{\tau_P}, \tag{12.30}$$

$$\frac{dB_r}{dt} \approx \frac{B_z}{\tau_{\mathrm{BH}}} - \frac{B_r}{\tau_P}, \tag{12.31}$$

$$\frac{dB_z}{dt} \approx \frac{B_\phi}{\tau_P} - \frac{B_z}{\tau_R}, \tag{12.32}$$

where $\tau_s, \tau_P, \tau_{\mathrm{BH}}$, and τ_R are, respectively, the time-scales for shearing, Parker instability, Balbus–Hawley instability, and reconnection. The shear time-scale is just

$$\tau_s \approx \frac{2}{3\Omega_a},$$

(12.33)

as before, and the time-scale for the Parker instability is estimated as

$$\tau_P \approx 2.2 \frac{h_d}{v_{A\phi}},$$

(12.34)

where h_d is the gravitational scale-height given by Eq. (12.12) with the total pressure (P) replaced by the gas pressure (p), and $v_{A\phi}$ is the Alfvén speed based on the ϕ-component of the field. Since the isothermal sound speed (c_s) is just p/ρ, we can rewrite h_d as

$$h_d = \frac{c_s}{\Omega_a}.$$

(12.35)

The time-scale for the Balbus–Hawley instability is

$$\tau_{BH} = \frac{1}{\gamma_{BH}(v_{Az})\,\Omega_a},$$

(12.36)

where the function γ_{BH} of v_{Az} is

$$\gamma_{BH}(v_{Az}) \approx \gamma_{max} \left[1 - (1 - \sqrt{3})^{-2} \left(1 - \frac{\pi v_{Az}}{c_s \sqrt{2}} \right)^2 \right]^{1/2}.$$

Here $\gamma_{max} \approx 0.74$ and v_{Az} is the Alfvén speed based on the z-component of the field. This expression is valid for $\sqrt{2}/\pi < (v_{Az}/c_s) < \sqrt{6}/\pi$, but if $v_{Az}/c_s < \sqrt{2}/\pi$ then $\gamma_{BH} = \gamma_{max}$, and if $v_{Az}/c_s > \sqrt{6}/\pi$ then $\gamma_{BH} = 0$. The latter condition corresponds to the field strength required to quench the instability (i.e., $B_z = c_s \sqrt{6\mu\rho}/\pi$).

The first term on the right-hand side of Eq. (12.30) is the growth produced by the conversion of radial magnetic field into azimuthal field by the Keplerian shear, while the second term is the decay associated with the Parker instability. The first term on the right-hand side of Eq. (12.31) is the growth produced by the conversion of vertical magnetic field into radial field by the Balbus–Hawley instability, while the second term is again the decay associated with the Parker instability. Finally, the first term on the right-hand side of Eq. (12.32) is the growth of the vertical field produced by the conversion of the azimuthal field into vertical field by the Parker instability, while the second term is the decay caused by magnetic reconnection. The conversion of radial field into vertical field by the Parker instability is assumed to be negligible, so it does not appear in Eq. (12.32).

Equations (12.30)–(12.32) also show the key difference between the models of Coroniti (1981) and Tout and Pringle (1992). If the same equations were written for Coroniti's model, the equation for the change in azimuthal field would include a reconnection loss term, while the equation for the radial field would include a reconnection source term. In the Tout and Pringle model, reconnection of the azimuthal field is considered to be negligible compared with reconnection of the vertical field.

If the disk is in equilibrium, then the right-hand sides of Eqs. (12.30)–(12.32) must be zero, which is true if

$$v_{Ar} = v_{A\phi} \frac{\tau_s}{\tau_P}, \tag{12.37}$$

$$v_{Az} = v_{Ar} \frac{\tau_{BH}}{\tau_P}, \tag{12.38}$$

$$v_{A\phi} = v_{Az} \frac{\tau_P}{\tau_R}, \tag{12.39}$$

where v_{Ar} is the Alfvén speed based on the radial component of the field. Ignoring the functional dependence of τ_P and τ_R and solving for v_{Az} gives

$$v_{Az} = \frac{\sqrt{6}}{\pi} c_s \left(1 - \frac{\tau_s^4}{\gamma_{max}^2 \tau_P^2 \tau_R^2} \right)^{1/2} \approx \frac{\sqrt{6}}{\pi} c_s, \tag{12.40}$$

since the shear time-scale (τ_s) is almost certain to be smaller than either the Parker or reconnection time-scale.

For the Tout and Pringle model, the Shakura–Sunyaev disk parameter (α_S) can be expressed in terms of the Maxwell stress ($B_r B_\phi$) as

$$\alpha_S = \frac{v_{Ar} v_{A\phi}}{c_s^2},$$

which gives $S_{r\phi} = B_r B_\phi = \alpha_S P$ and, upon substitution of Eqs. (12.37)–(12.40), it yields

$$\alpha_S = \frac{6}{\pi^2} \frac{\tau_P \tau_s}{\tau_R^2}. \tag{12.41}$$

This is essentially Coroniti's expression (12.29) for α_S multiplied by additional factors of $(6/\pi^2)$ and (τ_P/τ_R). The factor $(6/\pi^2)$ is the effect of the Balbus–Hawley instability (which is an ideal MHD process and therefore operates on the Alfvén time), while the factor (τ_P/τ_R) is the effect of the Parker instability.

To evaluate the three time-scales in Eq. (12.41) we use Eqs. (12.37)–(12.40) to obtain

$$\tau_P = \frac{v_{A\phi}}{v_{Az}} \tau_R,$$

$$\tau_s = \frac{\sqrt{2}}{3c_s} h_d,$$

and write τ_R in terms of the reconnection scale-length (L_R) as

$$\tau_R = \frac{L_R}{M_A \, v_{Az}}.$$

Using arguments we shall not repeat here, Tout and Pringle estimate L_R as

$$L_R \approx 0.46 \frac{v_{A\phi}}{c_s} h_d,$$

which leads to

$$\tau_R = 0.46 \frac{v_{A\phi}}{v_{Az}} \frac{h_d}{c_s} \frac{1}{M_A} = 0.46 \frac{\tau_P}{\tau_R} \frac{h_d}{c_s} \frac{1}{M_A}.$$

Rewriting this as

$$\frac{\tau_R^2}{\tau_P} = 0.46 \frac{h_d}{c_s} \frac{1}{M_A}$$

and substituting into Eq. (12.41) yields

$$\alpha_S \approx 0.6 \, M_A. \tag{12.42}$$

Thus, despite the different magnetic field geometry (i.e., a vertical magnetic field versus an azimuthal field) and the different nature of the ideal MHD processes involved (shear, the Balbus–Hawley instability and the Parker instability versus shear only), the Tout and Pringle model gives an expression for the Shakura–Sunyaev parameter which is almost the same as the expression [Eq. (12.29)] obtained by Coroniti (i.e., $0.6 \, M_A$ versus $M_A^{2/3}$).

In order to produce enough accretion to be consistent with observations of various astrophysical phenomena such as quasars, M_A has to be on the order of 0.1. Both Tout and Pringle and Coroniti assume that such rapid reconnection is not a problem, and they refer to Petschek's (1964) work as justification for values of M_A on the order of 0.1. Although Petschek's mechanism shows that fast reconnection is possible, in principle, it has not been demonstrated that it will apply to the turbulent MHD regime occurring in the disk. To obtain a realistic picture of the connection between accretion and reconnection will probably require numerical simulations that treat the turbulent dynamo and reconnection processes self-consistently.

Several alternatives to the Coroniti and Tout–Pringle models have been proposed. Ichimaru (1976) has developed a turbulence model based upon both MHD and kinetic theory which avoids treating the reconnection rate as a free parameter. Galeev et al. (1979) proposed an MHD model which uses a dynamo mechanism driven by convection to amplify the magnetic field at a much faster rate than occurs in Coroniti's model. Because of this faster rate, magnetic flux cells are transported out of the disk by buoyancy so quickly that little reconnection occurs within the disk. Reconnection does occur once the cells have been ejected from the disk into the surrounding corona, but this reconnection is much more rapid because of the higher Alfvén speed there (see also Burm and Kupcrus, 1988). Also, Heyvaerts et al. (1996a,b) have developed a procedure for determining self-consistently the level of MHD turbulence in a disk, which is essential for determining the efficiency of the radial transport of material.

12.2.2 Ambipolar Diffusion

The plasma environments found in accretion disks are often quite different from those occurring in stellar atmospheres or planetary magnetospheres, and new physical processes can become important. Exotic environments involving relativistic plasmas around black holes or electron-positron plasma around neutron stars are outside the scope of this book, but there is a process known as *ambipolar diffusion* which is important in T-Tauri accretion disks and which can be discussed within the framework of MHD (Zweibel 1988, 1989).

Ambipolar diffusion occurs in low-density plasmas with a low degree of ionization. When there are many more neutral particles than charged ones, the bulk flow of the plasma is dominated by the neutrals, which are not directly affected by the magnetic field. The charged particles, in contrast, are affected, and consequently their velocity tends to be different from the neutrals. If the rate of collisions between the neutrals and ions is sufficiently large, the MHD equations can still be used, provided that Faraday's equation is written as

$$\frac{\partial \mathbf{B}}{\partial t} = \nabla \times (\mathbf{V}_i \times \mathbf{B} - \eta \nabla \times \mathbf{B}), \tag{12.43}$$

where $\mathbf{V}_i$ is the velocity of the ions (Mestel and Spitzer, 1956). Defining the drift velocity ($\mathbf{V}_{DR}$) as

$$\mathbf{V}_{DR} = \mathbf{V}_i - \mathbf{V}_n,$$

where $\mathbf{V}_n$ is the velocity of the neutrals, Eq. (12.43) yields

$$\frac{\partial \mathbf{B}}{\partial t} = \nabla \times (\mathbf{V}_n \times \mathbf{B} + \mathbf{V}_{DR} \times \mathbf{B} - \eta \nabla \times \mathbf{B}). \tag{12.44}$$

Since collisions act to make $\mathbf{V}_i$ and $\mathbf{V}_n$ the same, a continual Lorentz force is required to keep $\mathbf{V}_{DR}$ from going to zero. Therefore, the drift velocity is approximately

$$\mathbf{V}_{DR} \approx \frac{(\nabla \times \mathbf{B}) \times \mathbf{B}}{\mu \, \rho_i \, \overline{\nu}_{in}}, \tag{12.45}$$

where ρ_i is the ion density and $\bar{\nu}_{in}$ is the ion-neutral collision frequency. Substituting Eq. (12.45) into Eq. (12.44), we obtain

$$\frac{\partial \mathbf{B}}{\partial t} = \nabla \times \left[\mathbf{V}_n \times \mathbf{B} + \frac{(\nabla \times \mathbf{B}) \cdot \mathbf{B}}{\mu \, \rho_i \, \bar{\nu}_{in}} \mathbf{B} - (\eta + \eta_{AD}) \nabla \times \mathbf{B} \right], \qquad (12.46)$$

where the ambipolar-diffusion coefficient (η_{AD}) is

$$\eta_{AD} = \frac{B^2}{\mu \, \rho_i \, \bar{\nu}_{in}}. \qquad (12.47)$$

From Eq. (12.46) we see that the effect of ambipolar diffusion is to introduce a nonuniform magnetic diffusion coefficient which is independent of the standard magnetic diffusivity coefficient (η).

For ambipolar diffusion to be significant, the ion density (ρ_i) and the ion-neutral collision rate ($\bar{\nu}_{in}$) must have values which are smaller than those typically found in stellar envelopes. Therefore, ambipolar diffusion is not important in the solar photosphere, even though the ratio of ions to neutrals there is on the order of 10^6. The main regions where ambipolar diffusion is thought to be important are molecular clouds (Kulsrud and Anderson, 1992; Mestel, 1999) and accretion disks around newly formed stars (Shu et al., 1994).

At first glance, one might be tempted to conclude from the form of the last term on the right-hand side of Eq. (12.46) that ambipolar diffusion can replace the role of ordinary magnetic diffusion in magnetic reconnection. However, this is not the case because the ambipolar diffusion coefficient (η_{AD}) is zero at a magnetic null [cf. Eq. (12.47)]. Nevertheless, ambipolar diffusion can enhance the effectiveness of ordinary magnetic diffusion by sharpening the gradient of the magnetic field (Brandenburg and Zweibel, 1994, 1995). The sharpening occurs at magnetic nulls points because, as field lines try to diffuse through the plasma, they tend to hang up on the null points. It may seem somewhat surprising that ambipolar diffusion could lead to a sharpening of the field gradients, but it is able to do this because of its anisotropic and inhomogeneous nature. Thus, in those environments where it is important, ambipolar diffusion is likely to increase the rate of magnetic reconnection by increasing the field gradients.

12.2.3 Disk Flares

An accretion disk is surrounded by a corona which is distinct from that of the central star and which can, in principle, produce its own flares (Galeev et al., 1979; Heyvaerts et al., 1996b). Direct evidence that disk flares actually occur is not easy to come by. Emissions from disk flares must be distinguished from other sources of variability, such as flares on the central star itself, episodic infalls of accreting material, and fluctuations in outflow jets and stellar winds.

One of the most favourable systems for distinguishing disk flares from other phenomena is the dwarf nova (Livio and Pringle, 1992). A dwarf nova occurs in a close-binary pair consisting of a main-sequence star (spectral type G, typically) and a white dwarf. The two stars orbit each other with a period of less than 8 h and are sufficiently close that matter is continually transferred via an accretion disk onto the white dwarf. Since white dwarfs are inherently dim, emissions from the accretion disk are relatively easy to observe. For this reason, dwarf novae provide an ideal opportunity to study the dynamics of the accretion disk without interference from activity in the central star.

Dwarf novae undergo periodic outbursts in which the brightness of the disk emissions increases by a factor of 10 over a period of 5–20 days (Ostlie and Carroll, 1996). The outbursts are repeated at intervals of 30–300 days and are known to be due to an increase in the accretion rate by a factor of 100. Why the accretion rate increases remains a mystery, but it may be due to an instability in the accretion disk or in the transfer rate between the two stars (Livio and Pringle, 1992).

The outbursts caused by the enhanced accretion rate are not themselves considered as flares. Flare-like activity is instead associated with fluctuations in the disk brightness by a factor of two or more on times-scales of a few minutes, but variations at this level occur almost continuously and are often referred to as *flickering* (la Dous, 1993). Flickering occurs in the inner region of the accretion disk close to the white dwarf (Horne et al., 1994) and it appears to be caused by some kind of turbulence in the disk (Bruch, 1992). In many respects, flickering appears to be the accretion-disk analog of Parker's (1988) proposed nanoflares (see §§2.7, 11.3.3).

An example of a possible flare mechanism is illustrated by Fig. 12.10, which shows a simulation by Hayashi et al. (1996) of the dynamic effect on the field of the central star produced by the shearing motions in the disk (see also Tajima and Gilden, 1987; Lynden-Bell and Boily, 1994). As the shear in the field lines connected to the disk increases, the field expands outwards to form a bulge. Eventually, a current sheet appears behind the bulge, and reconnection in the current sheet creates a plasmoid which is ejected from the system. The process is very reminiscent of the solar simulation of coronal mass ejections by Mikić and Linker (1994) discussed in

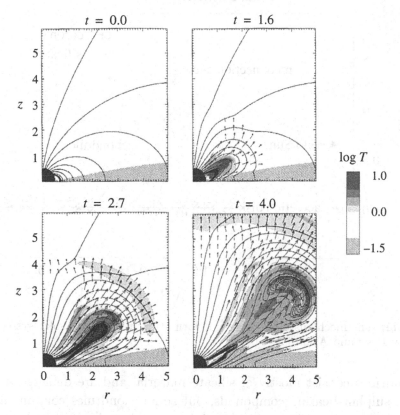

Fig. 12.10. Simulation of an accretion disk around a protostar having a dipole field aligned with its spin axis (from Hayashi et al., 1996).

Section 11.1.1 (see Fig. 11.3), except that in the Mikic and Linker simulation the shear is produced by photospheric convection rather than a rotating accretion disk.

Strange as it may seem, evidence for accretion-disk flares can also be found in certain types of stony meteorite known as *chondrites*, which are named after the small spherical objects they contain called *chondrules*. Radioactive dating shows that all chondrules formed approximately 4.6×10^9 yr ago, during a time span of less than 10^6 yr (Taylor, 1992), which is the epoch when the Sun would have been in its T-Tauri phase and surrounded by an accretion disk. Therefore, chondrules are considered as time capsules containing primitive material from the early solar accretion disk. Chondrules have a narrow range of sizes from 0.05 to 0.15 cm and contain a remnant magnetic field, indicating formation in an environment where the ambient magnetic field was about 10^{-5} tesla (0.1 G).

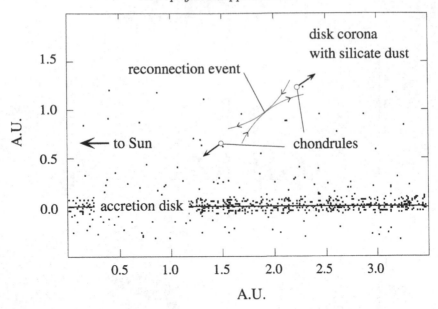

Fig. 12.11. The mechanism for chondrule formation in the early solar system proposed by Levy and Araki (1989).

Chondrules consist mostly of silicate material and are depleted in metals and sulphur-bearing compounds, but some chondrules contain carbon-bearing compounds. Figure 1.4(c) shows an example of a carbonaceous chondrite which fell in Chihuahua, Mexico, in 1969. Known as the Allende meteorite, it was found to contain amino acids, one of the building blocks of life. The chemical and isotopic composition of chondrules suggests that they were formed by flash heating events, which partially melted the silicate dust that was already depleted in metals and sulphur-bearing compounds (Taylor, 1992). These single-stage events lasted less than 1 s and did not heat the chondrules to more than 1,300 K.

At the present time, accretion-disk flares are still the most favoured explanation for the heating events which formed the chondrules (Levy and Araki, 1989; Taylor, 1992). To account for the compositional properties of chondrules, these disk flares should occur at a height of about 1 AU above the midplane of the early solar accretion disk, as shown in Fig. 12.11. At this height the density of the disk corona would have been low enough to allow rapid cooling of the chondrule after it had been heated by the flare. The depletion of metals and other heavy elements within chondrules is accounted for by gravitational separation of material with respect to height above the plane of the disk, and the remnant magnetic field in chondrules is explained by the magnetic field in the disk corona. The limited size range of the chondrules is usually attributed to the physics of the condensation process, which may be similar to the condensation of hailstones from super-cooled water in storm clouds.

Many other theories have been proposed for chondrule formation, such as impacts on planetary surfaces or collisions between planetesimals, but none of them come close to explaining the age and composition of chondrules (for a critique of other theories, see Taylor, 1992). Nevertheless, there are several problems remaining with the disk-flare model (see Cameron, 1994; Clayton and Jin, 1995), and there is still very little direct evidence that disk flares actually occur in T-Tauri systems.

13

Particle Acceleration

This book is devoted almost entirely to magnetohydrodynamic theories of reconnection and does not review the extensive literature on collisionless processes. Such processes are critical for the production of energetic particles by reconnection, but they constitute an enormous topic on which a whole text could easily be written. However, since the production of fast charged particles is one of the main consequences of reconnection in the cosmos, we feel it is important to devote some space to a brief discussion of this subject.

As we shall see, many particle acceleration theories rely, either implicitly or explicitly, on MHD concepts to supply information about the large-scale distribution of magnetic and electric fields. In the simplest theories (§13.1), the energetic particles are directly accelerated by an electric field whose behaviour is assumed to be known. In this case MHD models are often used to justify the particular choices made for the fields. However, even in more complex theories, MHD concepts such as turbulence (§13.2) and shock waves (§13.3) are often invoked. MHD theory is complemented by kinetic theory, which provides a great deal of additional information about the local plasma behaviour.

Energetic particles are common throughout the universe and reveal at once that it is not as a whole in thermodynamic equilibrium. They show up, for example, as cosmic rays incident on our atmosphere; they produce synchrotron emission from distant radio galaxies; and they are detected in the solar wind upstream of shocks produced by the Earth's magnetosphere and by coronal mass ejections. The basic reason that nature avoids equipartition for such particles is that they encounter other particles extremely rarely (e.g., Kirk, 1994): for example, in the million years, say, that a cosmic ray particle wanders around the disk of our galaxy, it has only a one-in-eight chance of colliding with another particle.

Furthermore, large-scale electric fields parallel to magnetic fields are extremely rare in the high-magnetic-Reynolds-number plasma of the universe, so for most of their lifetime charged particles move around under the effect of only the Lorentz force ($q\,\mathbf{v} \times \mathbf{B}$, where q is the charge). This force acts in a direction perpendicular to the particle velocity ($\mathbf{v}$) and thus does no work on the particle, thereby allowing it to conserve its energy. Although this property makes it easy for a particle to keep hold of its energy, it also makes it hard for the particle to acquire the energy in the first place. Indeed, in the absence of collisions, the only way for a particle to increase its energy is by the action of an electric field. (Note that, whereas in the rest of the book we have used $\mathbf{v}$ to denote plasma velocity, in this chapter $\mathbf{v}$ refers to a particle velocity and $\mathbf{u}$ refers to plasma velocity.)

To what energies are particles accelerated by reconnection? How many can be accelerated? What is the spectrum? These apparently simple questions have not yet been answered fully and are a matter for current research. The answers are highly complex, since, as we have seen in the previous chapters, there are many different types of reconnection and many different parameter regimes under which it operates. These include Sweet–Parker reconnection (§4.2), Petschek reconnection (§4.3), almost-uniform reconnection (§5.1), non-uniform reconnection (§5.2), impulsive bursty reconnection, and tearing-mode instability (§6.2) in two dimensions, as well as spine reconnection (§8.6.1), fan reconnection (§8.6.2), separator reconnection, (§8.6.3), quasi-separatrix layer reconnection (§8.7), and other types of singular magnetic field-line reconnection (§8.1.3) and general magnetic reconnection (§8.1.1) in three dimensions. So, clearly, the resulting acceleration of fast particles is likely to vary widely from one regime to another.

The role of magnetic reconnection in accelerating particles is complex not only because of the different reconnection regimes and parameter values under which it occurs, but also because of the complex magnetic environment that is usually associated with reconnection. This environment possesses three distinct types of region that are particularly conducive to accelerating particles. First of all, strong electric fields produced at the reconnection site itself may accelerate the particles directly (§13.1). Such fields may be present along the X-line of two-dimensional models and along the singular magnetic field lines of three-dimensional models (including spine, fan, and separator reconnection). Secondly, reconnection often produces a highly turbulent environment, both near the reconnection site and in the outflowing jets, where particles may be accelerated stochastically (§13.2). Thirdly, we have seen that MHD shock waves are often an integral part of a reconnecting field (§13.3). Slow-mode shocks tend to stand in the flow and are attached to the reconnection region. Fast-mode shocks are often generated where outflowing reconnection jets meet the ambient magnetic field, or they may propagate outwards from the reconnection site after the impulsive onset of reconnection. A brief survey (§13.4) of the operation of these three basic acceleration mechanisms in astrophysics, solar flares, coronal mass ejections, planetary magnetospheres, and the heliosphere concludes the chapter.

13.1 Direct Acceleration by Electric Fields

Since a magnetic field never does work on a particle, all acceleration involves the motion of the particle along an electric field. However, we use the term *direct acceleration* to indicate processes in which the acceleration is provided by an average, or mean, electric field rather than a fluctuating one. Thus, an auroral electron falling through a potential drop along a field line is directly accelerated, while a cosmic ray being scattered by magnetic fluctuations is not. The latter process is a classic example of *stochastic acceleration*, but processes also exist which blur the distinction between the terms direct and stochastic. For example, in Section 13.3 we shall discuss diffusive shock acceleration, which is a direct process when one looks at a single encounter between a particle and the shock, but it becomes a stochastic process when one considers the particle's multiple encounters with the shock and the turbulent magnetic fluctuations upstream of it.

The non-relativistic equation of motion for a particle of constant mass m, velocity $\mathbf{v}$ and charge q moving in an electric ($\mathbf{E}$) and magnetic ($\mathbf{B}$) field is in MKS units

$$m\frac{d\mathbf{v}}{dt} = q(\mathbf{E} + \mathbf{v} \times \mathbf{B}), \tag{13.1}$$

which may be separated for a unidirectional magnetic field into a component,

$$m\frac{dv_\parallel}{dt} = qE_\parallel, \tag{13.2}$$

along the magnetic field and a component,

$$m\frac{d\mathbf{v}_\perp}{dt} = q(\mathbf{E}_\perp + \mathbf{v}_\perp \times \mathbf{B}), \tag{13.3}$$

perpendicular to the field. When the field lines are not unidirectional, $v_\parallel$ is affected by $\mathbf{E}_\perp$ and $\mathbf{v}_\perp$ by $\mathbf{E}_\parallel$. The relativistic version of Eq. (13.1) replaces $md\mathbf{v}/dt$ by $d/dt(\bar{\gamma}m\mathbf{v})$, where $\bar{\gamma} = (1 - v^2/c^2)^{-1/2}$ is the Lorentz factor. It should be noted that in this chapter, in order to avoid confusion with other variables, we use an overbar for several variables, such as $\bar{\gamma}$ (here), $\bar{\nu}$ (§13.1.1), $\overline{\Omega}$ (§13.1.7), $\bar{p}$ (§13.2.3), $\bar{\mu}$ (§13.2.3), $\bar{\alpha}$ (§13.2.3), and $\bar{\kappa}$ (§13.3.2).

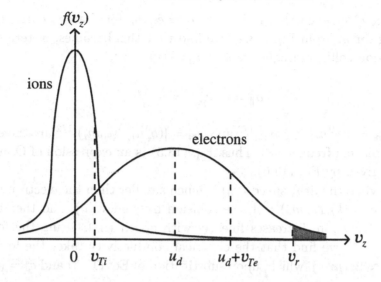

Fig. 13.1. Typical velocity distributions $f(v_z)$ of ions (with thermal speed v_{Ti}) and electrons (with thermal speed v_{Te}), shown in the rest frame of the ions. The electrons drift at a speed of u_d relative to the ions. Electrons beyond v_r run away (after Benz, 1993).

Let us consider first the simplest case of a particle moving along a uniform magnetic field under the action of an electric field (e.g., Benz, 1993) and leave until later the more complex cases of non-aligned electric fields and current sheets. We present only the basics of the classical processes, and so the reader should be aware that in reality there are many more aspects than we have room for here.

13.1.1 Acceleration along the Magnetic Field

According to Eq. (13.2), an electric field will accelerate particles indefinitely along a magnetic field, with ions and electrons going in opposite directions. This may continue until the electric force is balanced by collisional friction. For ions of mass m_i and charge q_i moving with a mean drift speed u_d relative to the electrons (Fig. 13.1), this occurs when

$$m_i \frac{u_d}{\tau} \simeq q_i E, \tag{13.4}$$

where τ is the slowing-down time due to interactions with the electrons. The parallel electrical conductivity ($\sigma_{\parallel}$) is then defined as the ratio

$$\sigma_{\parallel} = \frac{j}{E}, \tag{13.5}$$

where the electric current is $j \approx q_i n_i u_d \approx e n_e u_d$ for charge neutrality. Substituting for u_d from Eq. (13.4), we find that this becomes, in terms of the electron-ion collision frequency $\bar{\nu} = m_i n_i / (m_e n_e \tau)$,

$$\sigma_\parallel = \epsilon_0 \omega_{pi}^2 \tau = \frac{\epsilon_0 \omega_{pe}^2}{\bar{\nu}}, \tag{13.6}$$

where $\omega_{pe} = [(e^2 n_e)/(\epsilon_0 m_e)]^{1/2}$ and $\omega_{pi} = [(q_i^2 n_i)/(\epsilon_0 m_i)]^{1/2}$ are the electron and ion plasma frequencies. Thus, Eq. (13.5) is an expression of Ohm's Law with $\sigma_\parallel$ given by Eq. (13.6).

Now, when the drift speed (u_d) is much smaller than the electron thermal speed $[v_{Te} = (k_B T_e / m_e)^{1/2}]$, the collision frequency ($\bar{\nu}$), and therefore the collisional friction, increases linearly with speed (u_d). Using the formulae in Chapter 1, we find that the electrical conductivity takes the form $\sigma_\parallel = 1/(\mu \eta_\parallel) \approx 2/(\mu \eta_\perp)$, which, upon substitution of Eq. (1.13) and $c^2 = (\mu \epsilon_0)^{-1}$, yields

$$\sigma_\parallel = \frac{6(2\pi)^{3/2} m_e v_{Te}^3 \epsilon_0^2}{e^2 \ln \Lambda} \approx 0.0152 \frac{T_e^{3/2}}{\ln \Lambda} \text{mho m}^{-1}, \tag{13.7}$$

where $\ln \Lambda$ is the Coulomb logarithm [Eq. (1.14)]. The conductivity therefore increases as $T_e^{3/2}$ and so a hotter plasma is a better conductor. The corresponding drift speed from Eq. (13.4) may then be cast in the form

$$\frac{u_d}{v_{Te}} = 3\sqrt{\frac{\pi}{2}} \frac{E}{E_D}, \tag{13.8}$$

where

$$E_D = \frac{q_i \ln \Lambda}{4\pi \epsilon_0 \lambda_D^2}$$

is known as the *Dreicer field* [Eq. (1.66)] and $\lambda_D = \sqrt{[\epsilon_0 k_B T/(n_e e^2)]}$ is the Debye length [Eq. (1.64)].

13.1.2 Runaway Particles

As the drift speed (u_d) between ions and electrons approaches the electron thermal speed (v_{Te}), the linear relation (13.5) between E and j breaks down, with the result that an equilibrium between the electric force and the frictional force due to collisions is no longer possible. When this happens the particles are accelerated indefinitely, unless wave-particle instabilities are triggered. This runaway phenomenon was first analysed by Dreicer (1959), using the following equation of motion for an ion moving at speed u_d relative to an electron background:

$$m_l \frac{du_d}{dt} = eE - eE_D G(\bar{u}_d),$$

where

$$\bar{u}_d = \frac{u_d}{\sqrt{2}v_{Te}},$$

and

$$G(\bar{u}_d) = \frac{\text{erf}(\bar{u}_d)}{2\bar{u}_d^2} - \frac{e^{-\bar{u}_d^2}}{\bar{u}_d\sqrt{\pi}}.$$

The ion is accelerated by the electric force (eE), but slowed by the frictional force $eE_D G$, which is plotted in Fig. 13.2. Dreicer obtained this frictional force by considering the electrons to be Maxwellian, assuming $m_e \ll m_i$, ignoring the current-associated disturbances in the Maxwellian distribution, and assuming the electric field to be on the order of E_D or stronger. The most important aspect of this force is that it decreases when the drift speed (u_d) exceeds $\sqrt{2}v_{Te}$, behaving like u_d^{-2} when $u_d \gg v_{Te}$. Thus, once the drift speed enters the range where the force starts to decrease, a runaway occurs. The critical electric field (E_c) for which $u_d = \sqrt{2}v_{Te}$ is $E_c = 0.214\,E_D$; it is the maximum field for equilibrium and occurs at the maximum of the function $G(\bar{u}_d)$. More refined analyses, based primarily on the Fokker–Plank equation (Risken, 1989), have since been carried out by Dreicer (1960), Lebedev (1965), and Cohen (1976). When combined, these analyses provide a quantitative description for the complete transition from the weak electric field in the Spitzer regime to the strong electric field in the Dreicer regime.

The runaway phenomenon also exists for some particles at lower thresholds than this, especially electrons in the far wings of the velocity distribution, since the collisional friction acting on them is smaller than the electric force. Such electrons may be accelerated until they leave the system, and so runaway is an attractive acceleration process even when the electric field is smaller than the Dreicer field. The role of the runaway process in reconnection is not yet well understood, but, as we discussed in Section 1.7, the Dreicer electric field is typically smaller than the reconnection electric field. For example, in the Earth's plasma sheet it is four orders of magnitude smaller than the electric field associated with slow Sweet–Parker type of reconnection (see Table 1.2a).

The value of the runaway speed (v_r) for an electron (Fig. 13.1) may be estimated by equating the electric force (eE) to the frictional force. For

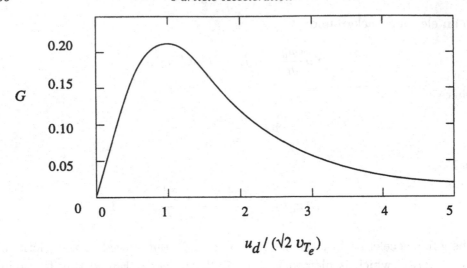

Fig. 13.2. The normalized frictional force (G) due to collisions as a function of the normalized drift speed (u_d) of ions relative to electrons.

$E \ll E_D$, the result is

$$v_r = v_{Te} \left(\frac{E_D}{E} \right)^{1/2}.$$

Ions and electrons have the same runaway velocity; but, because the ion thermal speed is generally much less than the electron thermal speed, the ion distribution function is much narrower than the electron distribution function (Fig. 13.1). There are therefore very much fewer ions beyond v_r than electrons, and so usually ion runaway can be neglected by comparison with electron runaway. The fraction of runaway electrons is

$$\frac{n_r}{n_0} \approx \exp\left(-\frac{v_r^2}{2v_{Te}^2} \right) = \exp\left(-\frac{E_D}{2E} \right),$$

which is much smaller than unity unless the electric field approaches the Dreicer field, when by Eq. (13.8) the drift speed is roughly the electron thermal speed.

13.1.3 *Micro-Instabilities and Anomalous Conductivity*

As we mentioned above, the drift speed (u_d) and current density (j) may exceed thresholds for various current-driven micro-instabilities well before the critical Dreicer field limit for total runaway is reached. These instabilities generate waves that are able to scatter the current-carrying particles, and thus impede the flow of current. The resulting anomalously low effective collision-time (τ_{eff}), or high collision-frequency ($\bar{\nu}_{\text{eff}}$), leads to an anomalously low conductivity (σ_{eff}) according to Eq. (13.6), or equivalently an anomalous resistivity ($1/\sigma_{\text{eff}}$) that is higher than the classical value (Galeev and Sagdeev, 1984).

When the electron and ion temperatures are similar ($T_e \approx T_i$), the ion-cyclotron instability has the lowest threshold, namely,

$$u_d \gtrsim 15 \frac{T_i}{T_e} v_{Ti},$$

and its main effect is to grow electrostatic ion-cyclotron waves and heat the ions. The resulting anomalous conductivity (σ_{eff}) is about $0.2(4\pi\epsilon_o\Omega_i)$ at saturation, where $\Omega_i = q_i B/m$ is the ion gyro-frequency.

When the drift speed satisfies

$$u_d \gtrsim 1.7 \left(v_{Te} + v_{Ti} \right),$$

so that the ion and electron distributions are displaced by more than the thermal speeds, they may be considered as two cold streams of particles. The resulting Buneman instability (or two-stream instability) produces electrostatic beam-mode waves and preferentially heats the electrons up to the drift speed ($v_{Te} \approx u_d$).

When the electron temperature far exceeds the ion temperature ($T_e \gg T_i$) and the drift speed exceeds the ion sound speed

$$u_d > c_{is} = \sqrt{\left(\frac{k_B T_e}{m_i} \right)}, \tag{13.9}$$

a powerful mode sets in called the *ion-acoustic instability*, which causes ion-acoustic waves to grow. The resulting effective collision frequency at saturation has been calculated by weak-turbulence theory to be

$$\bar{\nu}_{\text{eff}} = \frac{\omega_{pe}}{32\pi} \frac{T_e}{T_i} \frac{u_D}{v_{Te}},$$

and so the resulting anomalous conductivity, known as the *Sagdeev conductivity*, is

$$\sigma_{\text{eff}} = 32\pi\epsilon_o \frac{T_i}{T_e} \frac{v_{Te}}{u_d} \omega_{pe}.$$

Numerical and laboratory experiments find a temperature of $T_e \approx 10 T_i$ and $\bar{\nu}_{\text{eff}} \approx 0.2\,\omega_{pi}$ and a minimum anomalous conductivity of about $\sigma_{\text{eff}} \approx 80\pi\omega_{pe}$, which is typically about a million times smaller than the classical value.

Another mechanism which has been suggested as a mechanism for anomalous resistivity is the electrostatic double layer (Block, 1973; Hasan and Ter Haar, 1978; Raadu and Rasmussen, 1988). These are charge-separation layers that are commonly observed in laboratory plasmas and have thicknesses on the order of the Debye length. They are similar to the charge-separation sheaths that occurs at cathodes (see the discussion of the LCD experiment in Section 9.2, for example), except that they occur far away from any physical surface. The voltage drop within such a layer is on the order of $k_B T/e$ where T is the thermal temperature of the plasma. Thus, a double layer can act as a mechanism for anomalous resistivity because it allows a large electric field to exist parallel to the magnetic field. The kind of large-scale double layer that can exist in a laboratory plasma is not thought to occur in most astrophysical applications, but numerous, weak layers resulting as a secondary product of the ion-acoustic, or other, instabilities have been proposed as an alternative possibility (Raadu, 1989). The main difference is that in the laboratory an external potential difference may be imposed, but in an astrophysical plasma electric fields parallel to the magnetic field tend to occur only in extremely thin current concentrations (such as reconnecting current sheets or shock fronts) where ideal MHD breaks down.

In an astrophysical current sheet or current filament, it should be noted that anomalous conditions will only occur in extremely thin structures. For example, if $u_d = 10\ c_{is}$, the equations $j = neu_d = B/(\mu l)$ imply a sheet thickness (l) of only 40 m for typical solar coronal conditions of $B = 0.01$ tesla (100 G), $n = 10^{15}\,\mathrm{m}^{-3}$, $T_e = T_i = 2$ MK.

The onset of one of the above micro-instabilities is likely to lead to a burst or a series of bursts of enhanced ohmic heating and acceleration, rather than the anomalous conditions being maintained in a steady manner for a long time. The reason is that current-induced micro-instabilities tend to grow much faster than the magnetic diffusion-time over which the total current of a sheet tends to change. The current density therefore tends to remain constant while the conductivity rapidly falls from σ to σ_{eff} and the electric field correspondingly rises by a factor of $\sigma/\sigma_{\mathrm{eff}}$ (Fig. 13.3). The current sheet then expands over a longer time-scale (l^2/η_{eff}) by a factor of $\sigma/\sigma_{\mathrm{eff}}$, and at the same time the current density falls in value below threshold, so that the micro-instability switches off and the process repeats after the current density grows again. An alternative scenario is that the micro-instability reacts back on the plasma and keeps it close to the condition for marginal stability.

Runaway acceleration is another process that is affected by the onset of anomalous conditions. For ion-acoustic waves the effective collision frequency

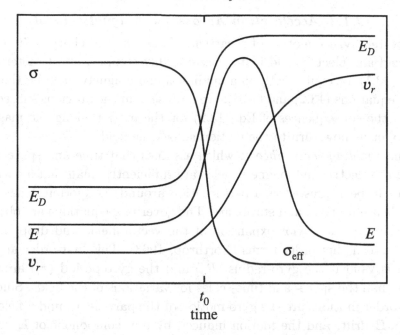

Fig. 13.3. A schematic variation in time of the plasma parameters (in arbitrary units) in a current sheet when the electrical conductivity (σ) falls to an anomalous value σ_{eff} (after Benz, 1993).

($\overline{\nu}_{\text{eff}}$) scales with electron velocity in just the same way as coulomb collisions, and so the above results for classical runaway simply scale with conductivity. E_D increases like $\sigma/\sigma_{\text{eff}}$ and so does E initially, so that the runaway speed (v_r) does not change. When E falls to its initial value, v_r increases like $\sqrt{(\sigma/\sigma_{\text{eff}})}$, as sketched in Fig. 13.3. Although the region of runaway beyond v_r decreases, the runaway electrons feel a higher electric field, and so a sudden onset of anomalous conductivity may produce a surge of high-energy runaway electrons. The total energy ($\frac{1}{2}I_{\text{run}}EL$) released in the form of runaways is always smaller than the ohmic heating (IEL), where L is the sheet length and I_{run} is the runaway current (always less than the total sheet current, I).

When strong electric currents flow across a magnetic field, as in a slow-mode or fast-mode shock wave (§1.5), the modified two-stream instability may occur provided $T_e \geq T_i$ and $u_d > c_{is}$. It drives lower-hybrid waves, which may accelerate non-relativistic electrons by a resonance to typically 1–40 keV. This mechanism has been proposed, for example, for solar flares and for reconnection in the Earth's magnetosphere (Huba et al., 1977; Sundaram et al., 1980; Malkov and Sotnikov, 1985; Das, 1992; Winske et al., 1995).

13.1.4 Acceleration Across the Magnetic Field

The non-relativistic motion of a particle of mass m and charge q in a magnetic field and electric field is in general highly complex and is determined by Eq. (13.1), or equivalently in a unidirectional magnetic field by the component equations (13.2) and (13.3). We have seen in the previous subsections some of the consequences of Eq. (13.2) for the motion along the magnetic field, so let us now turn to the motion across the field.

We use *guiding-centre theory*, which assumes that time and space variations of the electric and magnetic fields are sufficiently small that a particle's motion can be expressed as a perturbation around the motion it has when the fields are uniform and stationary. The governing equations are obtained by carrying out a Taylor expansion of the vector fields and dropping the second- and higher-order terms (Northrop, 1963). This first-order approximation is valid if the gyro-radius (R_g) and the gyro-period (τ_g) are much smaller than the space- and time-scales for variations of the fields. Since the zeroth-order motions are the gyro-motion of the particle around a field line, the $\mathbf{E} \times \mathbf{B}$ drift, and the motion induced by any component of $\mathbf{E}$ parallel to $\mathbf{B}$ (see Jackson, 1975), we can represent a particle's vector location ($\mathbf{x}$) as

$$\mathbf{x} = \mathbf{R} + \mathbf{R}_g,$$

where $\mathbf{R}$ is the vector location of the guiding centre, i.e., the centre of the gyro-orbit, and $\mathbf{R}_g$ is the vector stretching from the guiding centre to the particle. Substituting $\mathbf{x}$ into the equations of motion, expanding the fields, and then averaging over a gyro-period yields the guiding-centre equation:

$$\ddot{\mathbf{R}} = \frac{q}{m}\mathbf{E} + \frac{q}{m}\dot{\mathbf{R}} \times \mathbf{B} - \frac{\mu_m}{m}\nabla B, \qquad (13.10)$$

where

$$\mu_m = \frac{mv_g^2}{2B}$$

is the *magnetic moment* (or first adiabatic invariant) and $\mathbf{v}_g$ is the *gyro-velocity*. The magnitude of the gyro-velocity and radius are related by

$$v_g = \Omega R_g,$$

where

$$\Omega = \frac{qB}{m}$$

is the *gyro-frequency* (and $2\pi/\Omega = \tau_g$ is the gyro-period).

It is convenient to separate Eq. (13.10) into its components parallel and perpendicular to $\mathbf{B}$ by introducing the parallel and perpendicular velocities ($\mathbf{v}_\parallel$) and ($\mathbf{V}_\perp$). We use capital $\mathbf{V}_\perp$ to distinguish the perpendicular guiding-centre velocity from the perpendicular particle velocity ($\mathbf{v}_\perp$), where $\mathbf{v}_\perp = \mathbf{v}_g + \mathbf{V}_\perp$, but there is no need to distinguish between parallel velocities, since, by definition, $v_{g\parallel} = 0$. Thus, the guiding-centre velocity ($\dot{\mathbf{R}}$) can be written as

$$\dot{\mathbf{R}} = v_\parallel \hat{\mathbf{B}} + \mathbf{V}_\perp = \left[v_\parallel^{(0)} + v_\parallel^{(1)} + \cdots \right] \hat{\mathbf{B}} + \mathbf{V}_\perp^{(0)} + \mathbf{V}_\perp^{(1)} \cdots,$$

which to zeroth order is

$$\dot{\mathbf{R}} = v_\parallel^{(0)} \hat{\mathbf{B}} + \mathbf{V}_E^{(0)}. \tag{13.11}$$

Here

$$\mathbf{V}_E = \frac{\mathbf{E} \times \mathbf{B}}{B^2} \tag{13.12}$$

is the electric drift velocity and is the only zeroth-order velocity perpendicular to $\mathbf{B}$. All other drifts, such as gradient and curvature drifts, are of first order. In some applications it is assumed that the electric field is sufficiently weak that $\mathbf{V}_E$ is also of first order, but this assumption is not a general requirement of guiding-centre theory, and, as we shall shortly see, such an assumption eliminates the terms which lead to acceleration in the absence of a parallel electric field.

Let us now assume that $E_\parallel = 0$ and obtain the equation for motion along the field by taking the dot product of Eq. (13.10) with $\hat{\mathbf{B}}$. This gives

$$\hat{\mathbf{B}} \cdot \ddot{\mathbf{R}} = -\frac{\mu_m}{m}(\hat{\mathbf{B}} \cdot \nabla)B,$$

since all the other terms vanish. To lowest order, the left-hand side of this equation is

$$\hat{\mathbf{B}} \cdot \ddot{\mathbf{R}} = \dot{v}_\parallel + v_\parallel \hat{\mathbf{B}} \cdot \dot{\hat{\mathbf{B}}} + \hat{\mathbf{B}} \cdot \dot{\mathbf{V}}_E = \dot{v}_\parallel + \hat{\mathbf{B}} \cdot \dot{\mathbf{V}}_E,$$

because $\hat{\mathbf{B}}$ is a unit vector with constant amplitude and is therefore always perpendicular to $\dot{\hat{\mathbf{B}}}$. We also have that $\hat{\mathbf{B}} \cdot \dot{\mathbf{V}}_E = -\dot{\hat{\mathbf{B}}} \cdot \mathbf{V}_E$ because the product $\hat{\mathbf{B}} \cdot \mathbf{V}_E$ is zero and thus always has a zero derivative. Therefore,

$$\dot{v}_\parallel = -\frac{\mu_m}{m}(\hat{\mathbf{B}} \cdot \nabla)B + \frac{\mathbf{V}_E}{B} \cdot \left[\frac{\partial \mathbf{B}}{\partial t} + (\mathbf{V}_E \cdot \nabla)\mathbf{B} + v_\parallel (\hat{\mathbf{B}} \cdot \nabla)\mathbf{B} \right], \tag{13.13}$$

where the first term on the right-hand side is the derivative of B along a field line, and the square brackets give the changes in $\mathbf{B}$ as seen in the frame of a moving particle. Although $v_\parallel$ itself can be of zeroth order, the acceleration ($\dot{v}_\parallel$) is necessarily a first-order quantity, since all the terms on the right-hand side of Eq. (13.13) involve space and time variations which are never larger than first order in the guiding-centre theory.

Taking the cross product of Eq. (13.10) with $\hat{\mathbf{B}}$ yields an equation for the perpendicular component $(\mathbf{V}_\perp)$ of the guiding-centre velocity (Northrop, 1963), namely,

$$
\mathbf{V}_\perp = \frac{\mathbf{E} \times \hat{\mathbf{B}}}{B} + \frac{mv_\perp^2}{2qB^2}\hat{\mathbf{B}} \times \nabla B + \frac{mv_\parallel^2}{qB}\hat{\mathbf{B}} \times (\hat{\mathbf{B}} \cdot \nabla)\hat{\mathbf{B}}
$$

$$
+ \frac{m}{qB^2}\frac{\partial \mathbf{E}}{\partial t} + \frac{v_\parallel m}{qB}\hat{\mathbf{B}} \times \frac{\partial \hat{\mathbf{B}}}{\partial t} + \frac{v_\parallel m}{qB}\hat{\mathbf{B}} \times (\mathbf{V}_E \cdot \nabla)\hat{\mathbf{B}}
$$

$$
+ \frac{m}{qB}\hat{\mathbf{B}} \times \left[\mathbf{E} \times \frac{\partial(\hat{\mathbf{B}}/B)}{\partial t} \right] + \frac{v_\parallel m}{qB}\hat{\mathbf{B}} \times (\hat{\mathbf{B}} \cdot \nabla)\mathbf{V}_E
$$

$$
+ \frac{m}{qB}\hat{\mathbf{B}} \times (\mathbf{V}_E \cdot \nabla)\mathbf{V}_E. \tag{13.14}
$$

The drifts in the first line are the well-known electric field, gradient, and curvature drifts, but the other six drifts are not so well known. All of these other drifts can be neglected if the fields are quasi-steady and if the electric field is weak enough to make $(\mathbf{V}_E)$ the same size as the other drifts. However, if we make such an assumption, then all of the terms except for the first one on the right-hand side of Eq. (13.13) are also zero. Consequently, if temporal variations are negligible and the electric field is weak enough to eliminate all the additional drifts, then the only effect of the electric field is to cause an $\mathbf{E} \times \mathbf{B}$ drift, which does not lead to any change in the total energy of the particle.

13.1.5 *Example of Acceleration Across a Magnetic Field*

To illustrate how a relatively strong perpendicular electric field can lead to particle acceleration, we present a model due to Speiser (1968, 1969) which uses the above equations to analyze the qualitative behaviour of energetic particles in the polar caps of an open magnetosphere. The model fields are

$$
\mathbf{B} = B_0 \left(\frac{r_0}{r} \right) \hat{\mathbf{r}} \text{ and } \mathbf{E} = E_0 \hat{\mathbf{z}},
$$

where r, θ, and z are cylindrical coordinates and B_0, r_0, and E_0 are constants. The magnetic field $(\mathbf{B})$ corresponds to a line monopole along $r = 0$ with straight field lines radiating outwards as shown in Fig. 13.4. The electric field $(\mathbf{E})$ is uniform and parallel to the line monopole, so $\mathbf{E}$ is everywhere perpendicular to $\mathbf{B}$. Although this field provides a rather crude description of

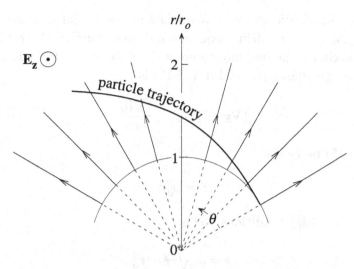

Fig. 13.4. The trajectory for a particle accelerated by an electric field $E_0\hat{z}$ in the magnetic field $B_0(r_0/r)\hat{r}$ of a line monopole.

the magnetic and electric fields over the poles of the Earth, it is sufficient to illustrate many of the features seen in numerical models using more accurate representations (e.g., Delcourt et al., 1989).

For convenience we shall also assume that the magnetic moment (μ_m) of the charged particle is negligible, but Speiser (1969) provides an analysis which does not make this assumption. All that is required for such an assumption to be valid in this particular example is for the initial gyro-speed (v_{g0}) to be much smaller than the electric drift speed (E_0/B_0). Physically, the assumption means that the mirror force $[-(\mu_m/m)\,\hat{\mathbf{B}}\cdot\nabla B]$ is negligible compared to the force exerted by the electric field.

Let us assume the particle is initially at $r=r_0, \theta=0, z=0$, which we may think of as being in the Earth's ionosphere, and let us also assume that the particle does not have an initial velocity component in the direction of the magnetic field. Therefore, equation (13.13) for the particle's parallel motion becomes

$$\dot{v}_r = \ddot{r} = \frac{V_E}{B}[(\mathbf{V}_E\cdot\nabla)\mathbf{B}]_\theta, \tag{13.15}$$

while the perpendicular drift motions, from terms 1, 6, and 8 in Eq. (13.14), are

$$V_\theta = V_E = \frac{E_0}{B}, \tag{13.16}$$

$$V_z = \frac{2v_r m E_0 r}{q B_0^2 r_0^2} = 2v_r \frac{\omega_0}{\Omega_0} \frac{r}{r_0}, \tag{13.17}$$

where $\Omega_0 = qB_0/m$ is the gyro-frequency at the initial starting point and $\omega_0 = E_0/(B_0 r_0)$ is the drift frequency corresponding to the rate at which the particle circles the line monopole in the θ-direction. In cylindrical coordinates the right-hand side of Eq. (13.15) is

$$[(\mathbf{V}_E \cdot \nabla)\mathbf{B}]_\theta = \frac{V_E B_r}{r},$$

and so (13.15) becomes

$$\ddot{r} = \omega_0^2 r.$$

Applying the initial conditions yields

$$\dot{r} = \omega_0 \sqrt{r^2 - r_0^2}$$

and

$$r = r_0 \cosh(\omega_0 t).$$

Therefore, the three velocity components vary with time as

$$v_r = V_{E0} \sinh(\omega_0 t), \tag{13.18}$$

$$V_\theta = V_{E0} \cosh(\omega_0 t), \tag{13.19}$$

$$V_z = V_{E0} \frac{\omega_0}{\Omega_0} \sinh(2\omega_0 t), \tag{13.20}$$

where $V_{E0} = E_0/B_0$ is the particle's initial drift speed. Note that, while v_r and V_θ are zeroth-order quantities, V_z is a first-order quantity because it is smaller than V_{E0} by the factor (ω_0/Ω_0) – an expansion parameter of the guiding-centre theory. [The other expansion parameter (R_g/r_0) is even smaller since $R_g/r_0 = (v_{g0}/V_{E0})(\omega_0/\Omega_0)$ and we have assumed that v_{g0}/V_{E0} is small.] Thus, to first order, the total particle energy is

$$W_p = \tfrac{1}{2}m\left(v_r^2 + V_\theta^2\right) = \tfrac{1}{2}mV_{E0}^2 \cosh(2\omega_0 t),$$

which is exactly the same expression obtained if we just consider the energy gained by the particle's motion along the electric field. Setting $V_z = \dot{z}$ in Eq. (13.20) and integrating yields

$$z = \frac{\omega_0}{2\Omega_0} r_0 \cosh(2\omega_0 t),$$

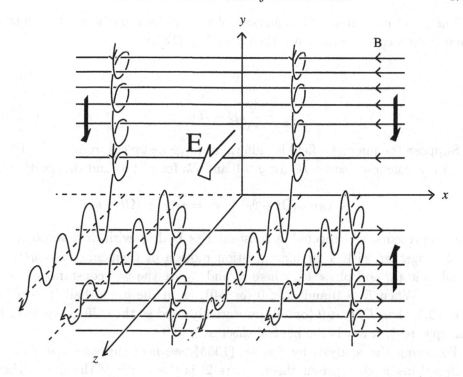

Fig. 13.5. The motion of charged particles in a neutral current sheet having $\mathbf{B} = B(y)\,\hat{\mathbf{x}}$. The particles first drift inwards across the field lines towards the xz-plane as they gyrate around the field lines in planes parallel to the yz-plane. Once they reach the xz-plane, they are accelerated in the z-direction, while they continue to oscillate up and down in the y-direction.

so the energy gain is again

$$W_p = qE_0z = \tfrac{1}{2}mV_{E0}^2\cosh(2\omega_0 t). \tag{13.21}$$

From this example we see that the particle gains energy because of its drift motion along the electric field and that the gain increases as the particle moves outwards into a weaker magnetic field. As we shall discuss in subsequent sections, the same effect occurs, and is even more pronounced, if the particle nears a null point in the magnetic field. This is one of the main reasons why magnetic reconnection is so closely associated with particle acceleration.

13.1.6 Particle Acceleration in Current Sheets

Consider the motion of a particle of mass m and charge q in a simple neutral current sheet (Fig. 13.5) with magnetic field $\mathbf{B} = B(y)\hat{\mathbf{x}}$ and under the action of a static and uniform electric field $\mathbf{E} = E\,\hat{\mathbf{z}}$.

The exact non-relativistic equations of motion for a particle with velocity components $v_y = \dot{y}, v_z = \dot{z}$ are then from Eq. (13.1)

$$\dot{v}_y = \frac{qB}{m} v_z, \tag{13.22}$$

$$\dot{v}_z = \frac{q}{m}(E - B v_y). \tag{13.23}$$

Suppose the magnetic field is uniform at large distances from $y = 0$. Then $B(y)$ is a constant (say, $-B_0$ for $y > 0$ and B_0 for $y < 0$) and the solution is

$$v_y = \frac{E}{B} + v_0 \cos(\Omega t + \theta_0), \quad v_z = -v_0 \sin(\Omega t + \theta_0).$$

This represents, as expected, a constant $\mathbf{E} \times \mathbf{B}$ drift with drift speed $v_{dr} = E/B$, together with a circular gyration motion of frequency $\Omega = qB/m$, amplitude v_0, and phase θ_0, where v_0 and θ_0 are the two constants of integration. When $E > 0$ and $B < 0$ (or >0), as in the upper half ($y > 0$) of Fig. 13.5, then $E/B < 0$ (or >0, respectively) and so the drift is downwards (or upwards), towards the neutral sheet at $y = 0$.

Following the analysis by Speiser (1965), we now suppose that $B(y) = -B_0 y/l$ inside the current sheet, where $2l$ is the width of the sheet. Then Eq. (13.23) may be integrated once to give

$$v_z = \frac{l}{\tau_g \tau_{dr}} t + \frac{y^2}{2\tau_g l} + \text{constant} \tag{13.24}$$

in terms of the two natural time-scales of the problem, namely the *gyration-time*

$$\tau_g = \frac{1}{\Omega_0} = \frac{m}{qB_0} \tag{13.25}$$

and the *drift-time* of particles into the sheet

$$\tau_{dr} = \frac{l}{v_{dr}} = \frac{lB_0}{E}. \tag{13.26}$$

For sufficiently large times, the first term on the right of Eq. (13.24) dominates, so

$$v_z \approx \frac{l\,t}{\tau_g \tau_{dr}}. \tag{13.27}$$

Substituting this into Eq. (13.22) with $B = -B_0 y/l$ gives

$$\ddot{y} + \frac{t}{\tau_{dr}\tau_g^2} y = 0,$$

namely an Airy equation of negative argument. It has the exact solution

$$y = c_1 A_i \left[\frac{-t}{\left(\tau_g^2 \tau_{dr} \right)^{1/3}} \right] + c_2 B_i \left[\frac{-t}{\left(\tau_g^2 \tau_{dr} \right)^{1/3}} \right],$$

where A_i and B_i are the Airy functions and c_1 and c_2 are constants of integration. Since this solution is only valid for large times, these constants cannot be evaluated in terms of the initial conditions, so no generality is lost if we arbitrarily assume $c_1 = y_0$ and $c_2 = 0$ at some point in time. Furthermore, since t is large, we can replace the Airy function A_i with its expansion for large negative argument (Spanier and Oldham, 1987) to obtain

$$y \approx \frac{y_0}{\pi^{1/2}} \left[\frac{\left(\tau_g^2 \tau_{dr} \right)^{1/3}}{t} \right]^{1/4} \sin \left(\frac{2t^{3/2}}{3\tau_g \sqrt{\tau_{dr}}} + \frac{\pi}{4} \right). \tag{13.28}$$

Thus, the particle oscillates back and forth across the field reversal with an amplitude that decreases in time as $t^{-1/4}$. Substituting the amplitude of Eq. (13.28) back into Eq. (13.24) gives the condition for the neglect of the second term as

$$t \gg (2\pi)^{-2/3} (y_0/l)^{4/3} \tau_g^{2/9} \tau_{dr}^{7/9}.$$

From integration of Eq. (13.27) the motion in the z-direction is

$$z \approx \frac{1}{2} \frac{l\, t^2}{\tau_g \tau_{dr}} + z_0 \approx \frac{1}{2} \frac{qE}{m} t^2 + z_0, \tag{13.29}$$

where z_0 is the constant of integration. Therefore, since the gain in kinetic energy is just $q E z$, the particle gains energy at the same rate as when there is no magnetic field. We should keep in mind, though, that this is only the case at large times when the motion due to acceleration along the electric field dominates the gyro-motion of the particle. At earlier times when the second term on the right-hand side of Eq. (13.24) is not negligible, Sonnerup (1971b) has shown that acceleration is indeed inhibited by the magnetic field as we would expect.

Runaway acceleration is prevented if there is a transverse magnetic field component $(\lambda_\perp B_0 \hat{y})$ in the y-direction. In this case there is no region where the magnetic field is zero, and a particle initially located at $y = 0$ will tend to gyrate about the transverse magnetic component. In fact, all the particles entering the sheet turn about the y-axis (with a frequency $\lambda_\perp/\tau_g$) and are ejected from it as shown in Fig. 13.6 (Speiser, 1965). This greatly limits the time that particles undergo acceleration in the z-direction. The resulting speed from Eq. (13.27) with which the particle leaves the sheet after a time $\tau_g/(\pi\lambda_\perp)$ is $l\pi\lambda_\perp/\tau_{\mathrm{dr}}$.

Thus, in total, a particle entering a current sheet feels three forces: the $\mathbf{E} \times \mathbf{B}$ force gives a drift in the y-direction; the parallel electric field produces acceleration; and the transverse magnetic field causes gyrations about the y-direction (Speiser, 1965; Sonnerup, 1971b; Cowley, 1978).

Many authors have considered test-particle acceleration. For instance, Bulanov and Sasorov (1976) and Moses et al. (1993) have modelled particle ejection from an X-line in detail. Also, Vekstein and Browning (1997) have considered the adiabatic acceleration of particles in an X-type configuration. They find substantial departures from a simple $\mathbf{E} \times \mathbf{B}$ drift, with jets of particles being accelerated along the separatrices with a power-law energy distribution. Also, it is interesting to note that in some applications such as the MRX laboratory experiment (§9.2) the ion gyro-radius is comparable with the sheet thickness while the electron gyro-radius is much smaller, so we enter the regime of electron MHD. Extending these ideas, test-particle acceleration has been studied in a two-dimensional turbulent magnetohydro-dynamic numerical experiment on reconnection by Matthaeus et al. (1984) and Ambrosiano et al. (1988). They find that particles can be accelerated to high energies. Some are trapped in the reconnection region for approx-imately an Alfvén transit time, while others are accelerated stochastically. They suggest this mechanism is capable of accelerating solar flare protons to several GeV (§13.4.2) and magnetospheric substorm ions to several hundred keV (§13.4.4).

The Speiser-type trajectories represent only one of four different types of particle motion that can occur in a current sheet. The other types are guiding-centre, resonant (or trapped), and chaotic motion (Ashour-Abdalla et al., 1990, 1991). Which type occurs depends primarily on the ratio of the radius of curvature of the field line to the gyro-radius of the particle. If the ratio is large (typically > 4 or more), the particles behave adiabatically and their acceleration can be understood in terms of guiding-centre theory (Büchner and Zelenyi, 1989). On the other hand, if the ratio is very small ($\ll \lambda_\perp$ in the above example), then the particles have Speiser-type motion. In between these two extremes lie resonant and chaotic motions (Speiser, 1988; Whipple et al., 1990). The resonant particles have trajectories similar in appearance to a Speiser-type trajectory, except that the amplitude of the oscillation in y is much smaller, and there is an integral relation between their oscillatory motion across the sheet and their oscillatory motion in the other two dimensions (Pellat and Francfort, 1976; Kaufmann and Lu, 1993).

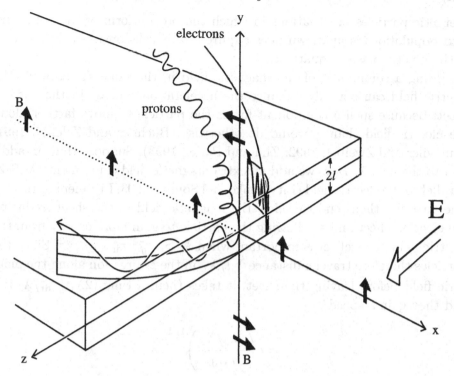

Fig. 13.6. Particle trajectories in a current sheet of width $2l$ with a magnetic field $B_x = -B_0 y/l$, $B_y = \lambda_\perp B_0$, and electric field $E\,\hat{z}$ (after Speiser, 1965).

Particles which do not satisfy this resonant condition, but still have small amplitude in y, are chaotic (Chen and Palmadesso, 1986).

In general, all types of trajectory are required for self-consistency between the field and currents (Kaufmann et al., 1993), but it is the particles undergoing Speiser-type motion which gain the most energy during a single encounter with the sheet. Full self-consistency between fields and particles also requires the introduction of charge-separation electric fields, which arise because of the different masses of the ions and electrons (Eastwood, 1972; Cowley, 1973a; Stern, 1990). Often self-consistency is not thought of as an issue when modelling energetic particles, since the term "high energy" refers only to those particles lying in the extreme upper range of the velocity distribution in phase space. This is the main reason that energetic particles are usually treated as test particles travelling in prescribed fields. However, a test-particle approach does not allow an estimate for the total number of accelerated particles at any given energy unless the phase-space distribution of the particles prior to their acceleration (i.e., the seed population) is known (e.g., Martens, 1988). Thus, to calculate the distribution of

energetic particles in situations in which the precise form of the fields and
seed populations is unknown does require a fully self-consistent calculation
of the kinetic plasma equations.

Adding a component of the magnetic field in the same direction as the
electric field can actually enhance the acceleration process in the current
sheet, because such a component forces the particle to move farther along
the electric field than it would do otherwise (Büchner and Zelenyi, 1991;
Bruhwiler and Zweibel, 1992; Zhu and Parks, 1993). Suppose that, in addi-
tion to the field $\lambda_\perp B_0 \hat{\mathbf{y}}$, we add an extra magnetic field component $(\lambda_\| B_0 \hat{\mathbf{z}})$
parallel to the electric field (Litvinenko and Somov, 1993; Litvinenko, 1996a).
The particles then tend to follow the magnetic field in the sheet, reducing
the ejection effect and so allowing the particle to gain more energy from the
electric field. This effect is important when $\lambda_\| > (\tau_g^{-1}\, \tau_{dr}\, \lambda_\perp)^{-1/2}$. Since the
particles therefore travel a distance $\lambda_\| l/\lambda_\perp$ in the z-direction along the mag-
netic field before leaving the sheet, it takes them a time $(2\lambda_\| \tau_g \tau_{dr}/\lambda_\perp)^{1/2}$
and they gain a speed

$$
v_z = l \left(\frac{2\lambda_\|/\lambda_\perp}{\tau_g \tau_{dr}} \right)^{1/2}.
\tag{13.30}
$$

Time-dependent fields also enhance acceleration by increasing the electric
field temporarily above its steady-state value (Galeev and Zelenyi, 1975;
Pellinen and Heikkila, 1978; Galeev, 1984; Matthaeus et al., 1984; Am-
brosiana et al., 1988; Scholer and Jamitzky, 1987; Lewis et al., 1990). Current
sheets containing neutral lines or points have been studied by Åström (1958),
Rusbridge (1971), Martin (1986), and Vekstein and Browning (1997). If a
neutral line or point lies within a thin sheet, then acceleration is enhanced,
as one would expect because of the weaker field. However, in the absence of
a thin current sheet, no substantial enhancement occurs, because most particles undergo chaotic motion, which randomizes their distribution in phase
space and causes many particles to lose energy (Martin, 1986).

So far we have demonstrated only that energetic particles are created
in regions where the magnetic field is weak or nearly parallel to the electric
field. Thus, one might be tempted to conclude that magnetic O-lines are just
as good a location for particle acceleration as magnetic X-lines. Stern (1979)
has argued that O-lines would be even better locations for particle acceleration than X-lines, because the particles are trapped in the weak magnetic
field near the O-type neutral line by the circular geometry of the field lines
there. By contrast, as in the case of the Speiser trajectories discussed above,
particles moving near an X-type neutral line are deflected away from the

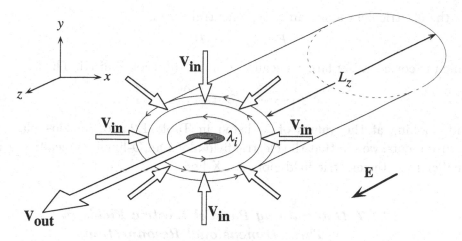

Fig. 13.7. Convective flow pattern in the vicinity of an O-type neutral line. The shaded region indicates the diffusion region where ideal MHD flow breaks down.

X-line into regions of stronger magnetic field where acceleration is more difficult. However, this conclusion presupposes that the electric fields along the X- and O-type neutral lines are the same, but Vasyliunas (1980) has pointed out that this is extremely unlikely in a plasma with a large magnetic Reynolds number.

According to Vasyliunas, the pattern of convective flow in the vicinity of an O-type neutral line, as shown in Fig. 13.7, greatly limits the strength of the electric field. For simplicity, let us consider the steady-state case with an electric field in the z-direction parallel to the O-line. Everywhere along the magnetic flux surfaces there is an inflow ($\mathbf{V}_{\text{in}}$) in the xy-plane which can only be balanced by an outflow ($\mathbf{V}_{\text{out}}$) in the z-direction. However, such a flow can exist only in the diffusion region around the O-line where ideal MHD breaks down. The fact that the diffusion region is necessarily quite small in a plasma with a high magnetic Reynolds number greatly limits the convective inflow into the neutral line and hence the convective electric field. Assuming mass conservation, a maximum possible convection speed of v_A, and a diffusion region thickness on the order of the ion-gyro radius, Vasyliunas found that the electric field ($E_{\text{O-type}}$) along the O-line has an upper limit on the order of

$$E_{\text{O-type}} \leq v_A B \frac{\lambda_i}{L_z},$$

where λ_i is the ion-inertial length (equal to the ion-gyro radius when the particle's perpendicular speed equals v_A). By comparison, the upper limit for the electric field along an X-type neutral line is

$$E_{X\text{-type}} \leq v_A B,$$

which is considerably larger because it is not diffusion-limited. Thus,

$$E_{O\text{-type}} \approx E_{X\text{-type}} \frac{\lambda_i}{L_z},$$

and, looking at the values of λ_i listed in Table 1.2a for various plasma environments, we see that the electric field along an O-line is typically much smaller than the electric field along an X-line.

13.1.7 *Determining Parallel Electric Fields in Three-Dimensional Reconnection*

The dialogue between Stern (1979) and Vasyliunas (1980), as to whether O-lines or X-lines are more likely to be sites of particle acceleration, highlights the importance of determining where electric fields – and particularly parallel electric fields – occur. As we have seen in Section 8.1.1, parallel electric fields are a general feature of three-dimensional reconnection, but it is not always immediately obvious where they will occur or what their strength will be. In this section we present an analysis by Schindler et al. (1991) which addresses this issue.

Let us consider a non-zero potential (Ψ) associated with three-dimensional reconnection as

$$\Psi = -\int_{s_1}^{s_2} E_{\parallel} \, ds, \tag{13.31}$$

evaluated along a magnetic field line with length coordinate s. Here $\Psi = \Psi(\alpha, \beta, t)$ is a function of the Euler potentials (α, β), and the points P$_1$ and P$_2$ at s_1 and s_2 are located in ideal regions on either side of a non-ideal region (D_R) through which the field line passes. Schindler et al. (1991) then show from Eqs. (8.5)–(8.7) that in general

$$\frac{\partial \Psi}{\partial \alpha} = \left[\frac{d\beta}{dt} \right], \quad -\frac{\partial \Psi}{\partial \beta} = \left[\frac{d\alpha}{dt} \right], \tag{13.32}$$

where the square brackets denote the jump in a value from P$_1$ to P$_2$.

The parallel potential Ψ is therefore determined (up to a constant of integration) entirely by the magnetic field and flow outside the non-ideal region, as prescribed by $d\beta/dt$ and $d\alpha/dt$. Of course, the resulting electric current and (possibly anomalous) conductivity are determined by microprocesses inside the non-ideal region (§13.1.3), so that part of the potential may accelerate particles (especially runaways) and part may drive microturbulence.

However, Eq. (13.32) has been used to estimate potential drops in several phenomena and to deduce the parallel electric fields associated with non-ideal plasma flows – an essential step for determining where particle acceleration occurs.

Some of the order-of-magnitude examples given by Schindler et al. (1991) are as follows. Consider an axisymmetric state with $\mathbf{v}, \mathbf{B}$, and $\mathbf{E}$ as functions of r and θ alone in spherical polar coordinates (r, θ, ϕ), and $\mathbf{B}$ consisting of poloidal $(\mathbf{B}_p)$ and toroidal (B_ϕ) parts. As Euler potentials, it is convenient to pick

$$\alpha = rA \sin \theta, \quad \beta = \phi + \beta_0(\alpha, s'),$$

where α is the flux function, A is the ϕ-component of the vector potential and s' is the arc-length along $\mathbf{B}_p$. Then $\partial \Psi / \partial \beta = d\alpha/dt = 0$, so that the second member of Eq. (13.32) is satisfied identically. Following Schindler et al. (1991) we now suppose that, in addition to the rotational symmetry of the field, the bulk flow velocity is similarly symmetric, but consists of a part undergoing rigid rotation at angular frequency $\overline{\Omega}$ and a part which deviates from rigid rotation at the angular frequency $\tilde{\Omega}\,(\alpha, s)$ due to poloidal motions. Then Eq. (13.32) may be integrated to give

$$\Psi = -\int_{\alpha_1}^{\alpha_2} [\tilde{\Omega}\,(\alpha, s_2) - \tilde{\Omega}\,(\alpha, s_1)]\, d\alpha, \tag{13.33}$$

where

$$\tilde{\Omega} = \overline{\Omega} - \frac{(\mathbf{v} \cdot \mathbf{B}_p)\, B_\phi}{r \sin \theta\, B_p^2}$$

and $\overline{\Omega} = v_\phi/(r \sin \phi)$. The function $\tilde{\Omega}$ measures the extent to which the flow violates the frozen-flux condition, and Eq. (13.33) therefore determines the potential drop along magnetic field lines as a function of this deviation. As we have stressed in previous chapters, a deviation is likely to occur wherever there is a large gradient in the magnetic field such as in a current sheet. However, Eq. (13.33) also implies that a small deviation over an extended range of field lines can create a significant parallel electric field.

As a specific example of where an extended, but small, deviation of the frozen-flux condition could lead to the creation of a parallel electric field, Schindler et al. (1991) considered the Earth's magnetosphere during quiet conditions. As we discussed in §10.6, it possesses a closed-field region near the equator, called the *plasmasphere* (see Fig. 10.1), which corotates at an angular velocity $\overline{\Omega}_E$ with the Earth and slips relative to the open field regions

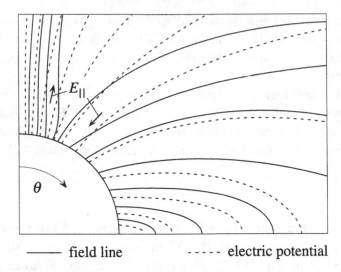

—— field line   - - - - - electric potential

Fig. 13.8. Schematic diagram of the electric potential and parallel electric field produced by small deviations from ideal MHD at an altitude of 1 R_E in the region between the open field lines of the polar cap and the corotating plasmasphere. The angle θ is the co-latitude.

near the poles. The field lines in the open regions are not corotating, but are instead convecting in the anti-sunward direction as they are dragged far back into the geomagnetic tail by the solar wind. In between a polar cap and the plasmasphere there must be a transition zone where co-rotation gives way to the solar-wind induced convection. Schindler et al. (1991) supposed that this transition would, at least at low altitudes, be extended over the latitudinal region separating the plasmasphere, at approximately a colatitude (θ) of 25° during quiet times, from the open field line boundary at about $\theta = 21°$ during quiet times (Hargreaves, 1993). In order of magnitude, Eq. (13.33) becomes

$$\Psi \approx \overline{\Omega}_E\,[\alpha_E],$$

where the jump $[\alpha_E]$ in α at one Earth-radius ($r = 1\,R_E$) between the colatitudes (θ) of 21° and 25° is about 6×10^7 weber. The resulting potential (Ψ) is about 4 kV, which is of the same order as the permanent large-scale auroral potential that exists at about 1 R_E above the Earth's surface. Figure 13.8 shows the qualitative behaviour of the electric potential and the parallel electric field associated with it.

Consider next a compact object such as a neutron star rotating with angular speed $\overline{\Omega}_0$, for which the low-lying field lines rotate at $\overline{\Omega}_0$ and the region beyond the light cylinder ($r\sin\theta = c/\overline{\Omega}_0$) does not corotate. The resulting slippage drives a parallel electric potential whose magnitude is independent of the plasma environment. It may be estimated as $\Psi \approx \overline{\Omega}_0 \, \alpha_0$, where α_0 is the range of the Euler potential α associated with field lines that extend beyond the light cylinder. For a dipole field (of moment m_0) $\alpha = \mu m_0 \sin^2 \theta / (4\pi r)$ and so with $\sin \theta \approx 1$, we find a potential of

$$\Psi \approx \frac{\mu \, m_0 \, \overline{\Omega}_0}{4\pi r}, \tag{13.34}$$

with $r \approx c/\overline{\Omega}_0$. For a rotational period of 0.01 s and a magnetic dipole moment of 5×10^{26} amp m^2 (corresponding to a field strength of 10^8 tesla at the surface of an object of radius 10 km), this gives a potential of 7×10^{16} V, high enough possibly to accelerate cosmic rays.

Another example would be corotation of the central region of our galaxy, which is thought to extend out to a radius (r) of order 300 pc. Adopting a magnetic moment corresponding to a field strength there of 3×10^{-9} tesla and an angular velocity corresponding to a speed $(v_0 = \overline{\Omega} r)$ of 300 km s^{-1}, Eq. (13.34) gives a potential of 4×10^{15} V, which may be important for accelerating galactic cosmic rays.

The basic result [Eq. (13.33)] may also be applied to non-axisymmetric examples. For instance, consider a substorm in the Earth's magnetosphere (§10.5), for which we may select as Euler potentials the values of $\alpha = \mu m_E \sin^2 \theta / (4\pi R_E)$ and ϕ with which a field line intersects the Earth's surface. Suppose that, before the substorm, a plasma element is located on a field line (α_a, ϕ_a), and after the substorm, when the field is more dipolar, it is located on a different field line (α_a, ϕ_b). Equation (13.33) then gives

$$\Psi = \int_{\alpha_a}^{\alpha_b} \left[\frac{d\phi}{dt} \right] d\alpha \approx \frac{\phi_b - \phi_a}{t_b - t_a} \delta\alpha, \tag{13.35}$$

where $\delta\alpha$ is the range of α associated with an auroral arc. Choosing a length-scale of 100 km at an auroral colatitude of, say, $\theta = 20°$ gives $\delta\alpha = 1.3 \times 10^7$ weber. Taking a range $(\phi_b - \phi_a)$ for ϕ of 25° and a time-scale $(t_b - t_a)$ of 30 min, we find a potential of 3 kV, which is the same as the observed peak in the auroral electron spectrum.

A similar approach may be adopted for solar flares. Choosing 9×10^{14} weber for the magnetic flux of an active region and 1% of that as the flux (F_m) associated with a magnetic reconfiguration, we find an appropriate value for $\delta\alpha$ in Eq. (13.35) is $\delta\alpha = F_m/(2\pi) = 1.4 \times 10^{12}$ weber. Taking 30° as the range of ϕ and 5 min as the time-scale, we find the resulting potential is 2×10^9 V, which corresponds to the highest proton energies (1–10 GeV) that are observed in flares (§13.4.2).

The above examples provide only rough estimates of the parallel electric fields based on an assumption about the deviation of the flow from ideal MHD. The deviation itself must be determined by a detailed analysis of the microphysics and plasma processes that are operating in a given environment (e.g., Hesse and Birn, 1990). Nevertheless, the examples provide insight into the origin of field-aligned electric fields associated with three-dimensional magnetic reconnection and non-ideal plasma flows – in other words, the electric fields which accelerate particles either directly or stochastically by means of the resulting turbulent states.

13.2 Stochastic Acceleration

Many different types of wave may grow in a reconnecting magnetic field and produce a turbulent environment, in which energy can be transferred to particles by a process of stochastic acceleration. The precursor of this idea came from Fermi (1949), when he suggested that cosmic rays are accelerated by the scattering of charged particles from randomly moving plasma clouds. The clouds act as magnetic mirrors that reflect the particles elastically. In a head-on collision a particle gains energy, but in an overtaking collision it loses energy.

This suggestion was developed in two ways. For what is called *first-order Fermi acceleration*, the two mirrors continually approach each other head on, and so the particles bounce to and fro many times, gaining energy at each reflection. If U is the cloud speed and v the particle speed (which is assumed to be much faster), then the energy gain is proportional to U/v.

For *second-order* (or *stochastic*) *Fermi acceleration*, the clouds move in random directions. This means that many of the collisions with a cloud cause a particle to lose energy rather than gain it. However, energy-depleting collisions are less frequent than their opposites, so there is still a net gain, but at a smaller rate proportional to U^2/v^2. This is therefore much less efficient than first-order Fermi acceleration. Second-order Fermi acceleration can still work even if energy-gaining (head-on) and energy-depleting (overtaking) collisions are equally likely, since there is a diffusion in the momentum part of phase space. Far more phase-space volume is available at high energies than at low energies, so a random walk in momentum space is most unlikely to lead towards zero velocity.

Although Fermi's original idea for cosmic-ray acceleration by interstellar clouds has been superseded, both his first- and second-order mechanisms have been greatly developed and retain a prominent place in astrophysics. The first-order mechanism can occur between oppositely directed MHD pulses (Parker, 1958) or astrophysical accretion flows (Cowsik and Lee, 1982; Schneider and Bogdan, 1989). However, the most effective configuration in which it operates is a *shock wave* (§13.3), which provides a head-on collision for particles that traverse the shock and scatter on magnetic irregularities located upstream and downstream of the shock front (Axford et al., 1977). We discuss this mechanism, called *diffusive shock acceleration*, in Section 13.3.2. First, we follow Lee (1994) and focus on stochastic Fermi acceleration. We give a short history (§13.2.1) and then describe a simple example of stochastic acceleration (§13.2.2) before discussing briefly the particle transport equation (§13.2.3) and shock-drift acceleration (§13.3.1).

13.2.1 Fermi's Formulation

Fermi's (1949) suggestion was that cosmic rays may be scattered by turbulent cloud motion in the interstellar medium, in such a way that each collision with a cloud of speed U increases the particle energy (W_p) by UW_p/v. Since the collision-rate between a particle and a cloud is proportional to their relative speed, head-on collisions are more probable than overtaking ones by an amount U/v. Averaged over many collisions, therefore, the energy increases at a rate

$$\frac{dW_p}{dt} = \left(\frac{U}{v}\right)^2 \frac{W_p v}{\lambda_{\mathrm{mfp}}} \tag{13.36}$$

per particle, where λ_{mfp} is the particle mean-free path.

Fermi suggested the following transport equation:

$$\frac{\partial N}{\partial t} + \frac{\partial}{\partial W_p}\left(\frac{dW_p}{dt}N\right) + \frac{N}{\tau} = Q(W_p), \tag{13.37}$$

in which the number density (N) of cosmic rays at a given energy (W_p) changes in time as a result of collisions with clouds (the second term); collisions with interstellar gas nuclei on a time-scale τ (the third term); and a source $[Q(W_p)]$ of some kind. He neglected escape from the Galaxy and, for a stationary distribution and injection at low energies, he derived from Eqs. (13.36) and (13.37) the solution

$$N(W_p) \sim W_p^{-1-\lambda_{\mathrm{mfp}}v/(\tau U^2)}.$$

Such a power-law distribution is a prime observational characteristic of cosmic rays, and so this result was a major achievement at the time. Stochastic Fermi acceleration is, however, no longer considered a viable mechanism for accelerating cosmic rays in the interstellar medium for several reasons including the fact that it is in general too slow (see Lee, 1994), but it has been invoked at other locations such as near compact objects or supernovae remnants.

Shock waves (§1.5) have many advantages over stochastic acceleration: they occur commonly in astrophysical plasmas, especially in reconnecting locations; they tend to accelerate particles rapidly and with a high efficiency; and they often have a large energy content. They are therefore now generally regarded as the most likely acceleration mechanism for galactic cosmic rays (Blandford and Eichler, 1987) and the cosmic ray anomalous component in the outer heliosphere (Fisk, 1976; Pesses et al., 1981; Jokipii, 1986), as well as many of the high-energy ion populations found in the heliosphere (Lee, 1992). Nevertheless, stochastic acceleration may well play an important role in interstellar and cometary pick-up ions in the solar wind (Ip and Axford, 1986; Bogdan et al., 1991) and solar cosmic rays (§13.4.2).

13.2.2 *One-Dimensional Stochastic Acceleration*

Perhaps the simplest situation that produces stochastic acceleration is the motion of a particle under the action of a one-dimensional electric field $[E(x,t)]$ that is fluctuating in space and time about a zero average value (Sturrock, 1994), so that $\langle E \rangle = 0$, where angular brackets represent an average over the fluctuations. As an illustrative example, let us consider how to set up the *particle transport equation* for this process.

The one-dimensional equation of motion of a single particle of mass m and charge q is

$$\frac{dv}{dt} = \frac{d^2 x}{dt^2} = \frac{q}{m} E(x,t). \tag{13.38}$$

Also, the distribution function $[f(x,v,t)]$ of a collection of particles is the number density of particles in (x,v) phase space at time t. It satisfies a *Fokker–Planck equation* (Risken, 1989) of the form

$$\frac{\partial f}{\partial t} + v \frac{\partial f}{\partial x} + \frac{q \langle E \rangle}{m} \frac{\partial f}{\partial v} = -\frac{\partial}{\partial v} \left(\left\langle \frac{\Delta v}{\Delta t} \right\rangle f \right) + \frac{1}{2} \frac{\partial^2}{\partial v^2} \left(\left\langle \frac{(\Delta v)^2}{\Delta t} \right\rangle f \right),$$

$$\tag{13.39}$$

where the left-hand side is simply the rate of change of f in time along a particle trajectory in (x,v) phase space. The collision term on the right-hand side represents the accumulated effects of small-angle collisions and the angled brackets represent mean values, in which the speed changes by an amount Δv in time Δt determined by Eq. (13.38). In higher dimensions the corresponding forms of the Fokker–Planck equation are natural extensions of Eq. (13.39), and the analysis of this section generalises in a natural way. The object here is then to evaluate the right-hand side of Eq. (13.39) and so derive the transport equation.

We can make progress by regarding a typical electric field $(E_0 = \langle E^2 \rangle^{1/2})$ as weak and expanding the particle trajectory in powers of $\bar{\epsilon} = qE_0x_0/(mv_0^2)$ as

$$x = x_0 + v_0t + X^I(t) + X^{II}(t) + \cdots,$$

where x_0 and v_0 are the particle's initial position and speed and X^I and X^{II} are of order $\bar{\epsilon}x_0$ and $\bar{\epsilon}^2x_0$, respectively. Substituting into Eq. (13.38) gives

$$\ddot{X}^I = \frac{q}{m}E(x_0 + v_0t, t), \quad \ddot{X}^{II} = \frac{q}{m}X^I\frac{\partial E}{\partial x}(x_0 + v_0t, t), \tag{13.40}$$

from which the change in first-order speed over a time Δt is

$$\Delta\dot{X}^I = \dot{X}^I(\Delta t) - \dot{X}^I(0) = \frac{q}{m}\int_0^{\Delta t} E(x_0 + v_0t', t')\,dt', \tag{13.41}$$

so that

$$\left\langle (\Delta\dot{X}^I)^2 \right\rangle = \frac{q^2}{m^2}\int_0^{\Delta t}\int_0^{\Delta t} \langle E(x_0 + v_0t', t')E(x_0 + v_0t'', t'') \rangle \, dt''dt'. \tag{13.42}$$

Now assume that the fluctuating electric field is statistically uniform in time and space, so that

$$\langle E(x, t)E(x + \xi, t + \tau) \rangle = \langle E^2 \rangle R(\xi, \tau),$$

where $R(\xi, \tau)$ is the correlation function for shifts of ξ and τ in space and time, respectively. Assume further that the width of R, known as the correlation time, is much shorter than the time Δt. Then Eq. (13.42) reduces to the required expression for the second coefficient in the Fokker–Planck equation, namely (on dropping the zero from v_0),

$$\left\langle \frac{(\Delta v)^2}{\Delta t} \right\rangle = \frac{q^2}{m^2}\langle E^2 \rangle \int R(v\tau, \tau)\, d\tau.$$

The first coefficient in Eq. (13.39) may be deduced by first integrating the second equation in Eq. (13.40) to give

$$\Delta\dot{X}^{II} \equiv \dot{X}^{II}(\Delta t) - \dot{X}^{II}(0) = \frac{q}{m}\int_0^{\Delta t} X^I(t')\frac{\partial E}{\partial x}(x_0 + v_0t', t')\, dt'. \tag{13.43}$$

Then, after integrating Eq. (13.41) to find $X^I(t')$, substituting into Eq. (13.43), and taking averages, we find

$$\langle \Delta v \rangle = \frac{q^2}{m^2}\langle E^2 \rangle \int_0^{\Delta t}\int_0^{\Delta t'} (t' - t'')\frac{\partial R}{\partial \xi}[v_0(t' - t''), t' - t'']\, dt''dt'.$$

Finally, assuming again that Δt is much longer than the correlation time gives the required coefficient as

$$\left\langle \frac{\Delta v}{\Delta t} \right\rangle = \frac{q^2}{2m^2} \langle E^2 \rangle \frac{\partial}{\partial v} \int R(v\tau, \tau)\, d\tau.$$

The Fokker–Planck equation, Eq. (13.39), therefore takes the form of a diffusion equation in velocity space, namely

$$\frac{\partial f}{\partial t} = \frac{\partial}{\partial v}\left[D(v)\frac{\partial f}{\partial v} \right], \tag{13.44}$$

whose diffusion coefficient is

$$D(v) = \frac{q^2}{2m^2} \langle E^2 \rangle \int R(v\tau, \tau)\, d\tau. \tag{13.45}$$

In deriving Eq. (13.44) from Eq. (13.39) we can omit the terms in $\partial f/\partial x$ and $\partial f/\partial v$ on the left-hand side of Eq. (13.39) if the distribution is spatially homogeneous (i.e., independent of x).

13.2.3 The Generalised Particle Transport Equation

Fermi's transport equation, Eq. (13.37), neglects the random nature of the stochastic acceleration process, which, as we have seen in Eq. (13.44), is described by a diffusive second-order derivative. This was first recognised by Davis (1956) and derived correctly by Parker and Tidman (1958). They considered elastic scattering from massive spheres [with distribution $f_s(\mathbf{v}_1, t)$] of particles with a spatially averaged distribution function $f(\overline{\mathbf{p}}, t)$ described by the Boltzmann equation

$$\frac{\partial f}{\partial t} = \int v_{\mathrm{rel}} f_s(\mathbf{v}_1)[f(\overline{\mathbf{p}}') - f(\overline{\mathbf{p}})]d^3\mathbf{v}_1 dS. \tag{13.46}$$

Here dS is an element of cross-section, $\overline{p}$ and $\overline{p}'$ are the pre-collision and post-collision momenta of the particle, and v_{rel} is the relative speed of a particle and sphere. The spheres are assumed to be infinitely massive with speed $v_1 \ll c$.

If the distributions are isotropic [$f(\overline{\mathbf{p}}) = f(\overline{p})$, $f_s(\mathbf{v}_1) = f_s(v_1)$] and $f(\overline{\mathbf{p}}')$ is expanded in a Taylor series about $f(\overline{\mathbf{p}})$, Eq. (13.46) reduces to

$$\frac{\partial f}{\partial t} = \frac{1}{\overline{p}^2}\frac{\partial}{\partial \overline{p}}\left[\overline{p}^2 D(\overline{p})\frac{\partial f}{\partial \overline{p}} \right], \tag{13.47}$$

where the diffusion coefficient is

$$D(\overline{p}) = \frac{\pi a^2 V^2 \overline{N} \overline{p}^2}{v} = \frac{V^2 \overline{p}^2}{3\lambda_s v}, \tag{13.48}$$

in which $\overline{N} = \int f_s(\mathbf{v}_1)d^3\mathbf{v}_1$ is the number density of spheres, a is the sphere radius, $V^2 = \overline{N}^{-1} \int v^2 f_s(\mathbf{v}_1)d^3\mathbf{v}_1$, and λ_s is the scattering mean-free path (Kulsrud and Ferrari, 1971). Note that Eq. (13.47) is the natural generalisation of Eq. (13.44) from Cartesian to spherical geometry. In terms of the energy (W_p), with $FdW_p = 4\pi\overline{p}^2 f d\overline{p}$ and $\overline{\alpha} = vD/\overline{p}^2 = V^2/(3\lambda_s)$, we can rewrite Eq. (13.47) as

$$\frac{\partial F}{\partial t} = \overline{\alpha} \left[\frac{\partial^2}{\partial W_p^2}(\overline{p}^2 vF) - \frac{\partial}{\partial W_p}(4\overline{p}F) \right].$$

For reference, we also give the relativistic version, which is

$$\frac{\partial F}{\partial t} = \frac{\overline{\alpha}}{c} \left(W_p'^2 \frac{\partial^2 F}{\partial W_p^2} - 2F \right).$$

A useful non-relativistic solution to Eq. (13.47) for impulsive mono-energetic injection may be derived by adding a source term $[\delta(t)\delta(\overline{p} - \overline{p}_0)]$ about a mean momentum $\overline{p}_0$ to the right-hand side, namely

$$f = \frac{\overline{p}_0}{\overline{p}\overline{\alpha}mt} I_2\left(\frac{2\sqrt{\overline{p}_0\overline{p}}}{\overline{\alpha}mt}\right) \exp\left(-\frac{\overline{p}_0 + \overline{p}}{\overline{\alpha}mt}\right), \tag{13.49}$$

where m is the particle mass and $I_2(x)$ is the modified Bessel function of the first kind. This equation has been used to describe the evolution of the energy spectrum of solar cosmic rays (Tverskoi, 1967), and its relativistic version is

$$f = \frac{\overline{p}_0^2}{\overline{p}^3} \left(\frac{c}{4\pi\overline{\alpha}t}\right)^{\frac{1}{2}} \exp\left\{ -\frac{[\ln(\overline{p}/\overline{p}_0) - 3\overline{\alpha}t/c]^2}{4\pi\overline{\alpha}t/c} \right\},$$

which has been invoked to describe electron transport in supernova remnants (Cowsik, 1979).

When a mono-energetic injection term and a loss term (over a time-scale τ) are added to Eq. (13.47), it becomes (in a stationary state)

$$0 = \frac{1}{\overline{p}^2}\frac{\partial}{\partial \overline{p}}\left[\overline{p}^2 D(\overline{p})\frac{\partial f}{\partial \overline{p}}\right] - \frac{f}{\tau} + \delta(\overline{p} - \overline{p}_0) \tag{13.50}$$

and possesses a non-relativistic solution (when $\overline{p} > \overline{p}_0$) of

$$f = \frac{2\overline{p}_0}{\overline{\alpha}m\overline{p}} I_2\left[\frac{2\overline{p}_0^{1/2}}{(\overline{\alpha}m\tau)^{1/2}}\right] K_2\left[\frac{2\overline{p}^{1/2}}{(\overline{\alpha}m\tau)^{1/2}}\right],$$

where $K_2(x)$ is the modified Bessel function of the second kind. The Bessel function form gives an excellent fit to solar cosmic-ray energy spectra

(Ramaty, 1979; Forman et al., 1986). In the relativistic limit, the corresponding solution is

$$f = \frac{c\bar{p}_0^2}{\bar{p}^3\alpha}\left(9 + \frac{4c}{\alpha\tau}\right)^{-1/2}\left(\frac{\bar{p}_0}{\bar{p}}\right)^{\Gamma},$$

where $\Gamma = \frac{1}{2}[9 + 4c/(\overline{\alpha}T)]^{1/2} - \frac{3}{2}$. This is more appropriate than Fermi's solution of Eq. (13.37) for cosmic rays. More general versions of Eq. (13.50) involving extra effects have been presented by Schlickeiser (1986a,b).

The basic equation [Eq. (13.47)] may also be derived by considering a quasilinear evolution of particles in the presence of a broad spectrum of waves. The form of the diffusion coefficient $[D(\bar{p})]$ depends on the types of waves that are considered [of intensity $I_0(k)$ such that $\langle|\delta B|^2\rangle = \int dk\, I_0(k)$], and of course there are many different possibilities. For example, Schlickeiser (1989) and Bogdan et al. (1991) have considered Alfvén waves propagating parallel $[I_+(k)]$ or anti-parallel $[I_-(k)]$ to the magnetic field $(B\,\hat{z})$. Ions with $v \gg v_A$ rapidly scatter in pitch-angle to isotropy in the mean wave-frame, and then they slowly diffuse according to Eq. (13.47) with

$$D(\bar{p}) = \frac{\pi q^2 v_A^2}{c^2 v}\int_{-1}^{1}\frac{1-\bar{\mu}^2}{|\bar{\mu}|}I_+\left(\frac{\Omega_i}{v\bar{\mu}}\right)I_-\left(\frac{\Omega_i}{v\bar{\mu}}\right)\left[I_+\left(\frac{\Omega_i}{v\bar{\mu}}\right)+I_-\left(\frac{\Omega_i}{v\bar{\mu}}\right)\right]^{-1}d\bar{\mu},$$

$$(13.51)$$

where Ω_i is the ion-cyclotron frequency and $\bar{\mu}$ is the cosine of the particle pitch-angle (the angle between the particle velocity and the magnetic field). Whereas the original stochastic acceleration mechanism involves nonresonant interactions between a particle and the magnetic field, Eq. (13.51) describes pitch-angle scattering of ions by a resonant set of waves. The ion-cyclotron resonance shows up in the argument $\Omega_i/(v\bar{\mu})$ of the functions I_+ and I_- in Eq. (13.51). The particles gain energy by being scattered elastically by scattering centres (namely the waves in this case), which move relative to each other without converging. (A net convergence would instead produce first-order Fermi acceleration.)

The quasi-linear transport equation [Eq. (13.47) or (13.54) below] has a detailed history, but a few of the key developments are as follows. A quasi-linear theory for MHD waves was developed in the 1960s (e.g., Vedenov et al., 1961; Drummond and Pines, 1962; Kennel and Engelmann, 1966; Hall and Sturrock, 1967). A diffusion-convection equation for cosmic-ray modulation was then proposed by Parker (1965) and by Gleeson and Axford (1967), and a derivation with a quasi-linear theory for magnetostatic turbulence was given by Jokipii (1966) and by Hasselmann and Wibberenz (1968). This was extended to electromagnetic turbulence by Tystovich et al.

(1973) and Melrose (1980). The diffusion-convection equation was later applied to shock-wave acceleration by Axford et al. (1977), Krymsky (1977), Bell (1978), and Blandford and Ostriker (1978). In particular, cosmic-ray dynamics in relativistic flows were modelled by Kirk et al. (1988) and Webb (1989), while a detailed quasi-linear theory for undamped Alfvén waves and fast magnetoacoustic waves was developed more recently by Schlickeiser and Miller (1998).

In general, the distribution function depends on three variables in momentum space. But, if we neglect the particle drifts that arise from inhomogeneities in the electric and magnetic fields and adopt a frame of reference in which the electric field vanishes, then the distribution function may be independent of the phase of gyration about the magnetic field. Thus, it depends on only two momentum coordinates (e.g., Kirk, 1994). The particles are then said to be *gyrotropic* and, in the absence of scattering, the magnetic moment (§13.1.4)

$$\mu_m = \frac{\frac{1}{2}\overline{\gamma}^2 m v_\perp^2}{B} \tag{13.52}$$

is invariant, where $\overline{\gamma}$ is the Lorentz factor. For the two momentum coordinates, it is convenient to choose the magnitude ($\overline{p}$) of the momentum and the pitch-angle cosine ($\overline{\mu}$). Then, since in quasi-linear theory the scattering centres each produce a small deflection of the particle, the transport equation becomes

$$\frac{df}{dt} = \frac{\partial}{\partial \overline{\mu}}\left(D_{\overline{\mu}\overline{\mu}}\frac{\partial f}{\partial \overline{\mu}} + D_{\overline{\mu}\overline{p}}\frac{\partial f}{\partial \overline{p}}\right) + \frac{1}{\overline{p}^2}\frac{\partial}{\partial \overline{p}}\left(\overline{p}^2\left(D_{\overline{\mu}\overline{p}}\frac{\partial f}{\partial \overline{\mu}} + D_{\overline{p}\overline{p}}\frac{\partial f}{\partial \overline{p}}\right)\right), \tag{13.53}$$

in which the Fokker–Planck collision term on the right represents pitch-angle scattering and the time-derivative is taken along a particle trajectory (Schlickeiser, 1989). When the distribution is gyrotropic and the electric field vanishes in the rest frame of the plasma, this time-derivative reduces to $d/dt = \partial/\partial t + \mathbf{v}\cdot\nabla$.

The pitch-angle scattering term tends to make the particles isotropic *in the frame of the scattering centres*, and, for scattering by Alfvén waves, that frame is essentially the plasma frame. In such a frame the scattering term has the simple form of Eq. (13.53). However, it is often more convenient to use a laboratory frame in which the plasma is in motion. In order to preserve the simple form of the right-hand side of Eq. (13.53), it is therefore common to use a mixed system of phase-space coordinates with the momentum ($\overline{p}$) measured in the plasma frame and the position variable ($\mathbf{x}'$) measured in the laboratory frame in which the plasma velocity is $\mathbf{u}$, say. In such a frame, the full nonrelativistic transport equation is (Kirk et al., 1988)

$$\frac{\partial f}{\partial t} + v_i \frac{\partial f}{\partial x_i'} - \frac{\partial u_j}{\partial x_i'} \bar{p}_i \frac{\partial f}{\partial \bar{p}_j} - m \frac{\partial u_i}{\partial t} \frac{\partial f}{\partial \bar{p}_i} - m u_i \frac{\partial u_j}{\partial x_i'} \frac{\partial f}{\partial \bar{p}_j} + \frac{\bar{p}_i}{m} \frac{\partial f}{\partial x_i'}$$

$$= \frac{\partial}{\partial \bar{\mu}} \left(D_{\bar{\mu}\bar{\mu}} \frac{\partial f}{\partial \bar{\mu}} + D_{\bar{\mu}\bar{p}} \frac{\partial f}{\partial \bar{p}} \right) + \frac{1}{\bar{p}^2} \frac{\partial}{\partial \bar{p}} \left(\bar{p}^2 \left(D_{\bar{\mu}\bar{p}} \frac{\partial f}{\partial \bar{\mu}} + D_{\bar{p}\bar{p}} \frac{\partial f}{\partial \bar{p}} \right) \right).$$

$$(13.54)$$

For relativistic particles this is accurate up to the order of u^2/c^2, provided the fluid flow is non-relativistic.

13.2.4 Stochastic Acceleration and Magnetic Reconnection

A variety of proposals have been made as to how stochastic acceleration and magnetic reconnection can act together to produce energetic particles. One possibility is that reconnection acts as a pre-acceleration stage for stochastic acceleration, either for electrons (Smith, 1977) or ions (Melrose, 1983). Preacceleration by the reconnection electric field in a current sheet or X-line helps overcome the inefficiency of stochastic acceleration, allowing it to accelerate particles to higher energy more promptly.

Reconnection has also been proposed as an energy source for generating and maintaining the magnetic fluctuations that drive stochastic acceleration. This process is particularly attractive for solar flares because the strong Alfvénic flows generated by reconnection in the corona can create a turbulent wave field very quickly (e.g., Moore et al., 1995; La Rosa et al., 1996). Also, energization over many hours in large flares cannot be explained unless there is a mechanism, such as reconnection, which can replenish the magnetic energy drained from the fluctuations by the accelerated particles (Ryan and Lee, 1991).

Hybrid processes have been proposed which introduce magnetic fluctuations into a current sheet or X-line. Ambrosiano et al. (1988) examined test-particle acceleration with fluctuations added to a sheet pinch. The dominant effect of these fluctuations is to create many small magnetic islands which help to trap particles within the sheet where they can be most easily accelerated. However, fluctuation of the electric field in the sheet also plays a significant role because it diffuses the particles in momentum space. The combination of these two effects far outweighs the additional acceleration produced by the scattering of particles from the random wave fluctuations outside the sheet. Similar results have been found by Kobak and Ostrowski (1997), who added magnetic fluctuations to the three-dimensional models of Craig et al. (1995) for spine and fan reconnection (see §3.5), and then considered particle acceleration.

Finally, Gray and Matthaeus (1991) have analyzed particle acceleration in two-dimensional MHD turbulence with randomly oriented current sheets. The turbulence was numerically generated from an initial configuration containing random variations of magnetic field in a periodic computational domain (Matthaeus and Lamkin, 1986). In addition, the system was un-driven by any external energy input, so that the initial fluctuations decayed with time by means of a turbulent cascade. Figure 13.9 shows contours of the magnetic flux function and the absolute magnitude of the current density after about 10 Alfvén scale-times for a simulation by Matthaeus and Montgomery (1981) that is similar to the one used by Gray and Matthaeus (1991). Randomly orientated current sheets are present, some with current flowing into the plane of the figure and others with current flowing out of the plane. This configuration differs substantially from the one used by Ambrosiano et al. (1988), which has a monolithic current sheet with an electric field everywhere in the same direction.

Gray and Matthaeus (1991) analyzed the acceleration properties of a configuration similar to that shown in Fig. 13.9, using an initial distribution of test particles with a Gaussian profile in energy, as shown in Fig. 13.9(c). After about 4 Alfvén scale-times, a high-energy tail developed, as indicated by the curve labeled "final". Comparing this distribution with a similar one calculated by Ambrosiano et al. (1988), Gray and Matthaeus concluded that the acceleration process is slower by a factor of about two but is still much more efficient than stochastic acceleration in a fluctuating field without current sheets.

13.3 Shock-Wave Acceleration

Shock waves of both the slow-mode and fast-mode types (§1.5) are naturally produced during the process of magnetic reconnection. They are potentially fertile grounds for accelerating energetic particles in two main ways. In *shock-drift acceleration* (§13.3.1), the motion of individual particles is followed in the electromagnetic field of the shock front and their interaction with fluctuating electric and magnetic fields is neglected. Since they may enter the shock with different pitch angles, they may emerge with a range of energies. In *diffusive shock acceleration* the particles bounce many times to and fro between irregularities upstream and downstream of the shock front, passing many times through the shock front and gaining energy each time. Many comprehensive reviews have been compiled on this extensive subject

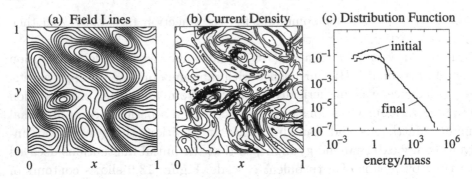

Fig. 13.9. (a) Magnetic field and (b) absolute value of the current density for a two-dimensional simulation of MHD turbulence (after Matthaeus and Lamkin, 1986). (c) Results of a test-particle analysis show the development of a high-energy tail (curve labeled "final") from an initially Gaussian distribution (curve labeled "initial"). All units are dimensionless (after Gray and Matthaeus, 1991).

(e.g., Axford, 1981; Drury, 1983; Blandford and Eichler, 1987; Jones and Ellison, 1991). Here we follow a particularly clear account by Kirk (1994).

13.3.1 Shock-Drift Acceleration

Before considering the motion of a particle near a shock wave, let us consider what is the most convenient frame of reference in which to analyse the properties of the shock. It is preferable to sit in a frame of reference moving with the shock front and watch the plasma approaching the shock from the upstream side and leaving it on the downstream side. In general, the magnetic field will be inclined to the plasma velocity. The flow will therefore carry the magnetic field lines in from one side and out on the other, so that the point of intersection of the magnetic field lines and the shock front moves along the shock with speed u_I, say. However, there are many frames of reference moving with the shock front, and, provided $u_I < c$ (so that the shock front is *subluminal*), the simplest is the *de Hoffmann–Teller frame*, namely the one that moves along the shock front at the speed u_I. In this frame the electric field vanishes both upstream and downstream, and the plasma moves everywhere along the magnetic field lines (Fig. 13.10). (For *superluminal shocks*, i.e., $u_I > c$, it is not possible to transform away the electric field, but there is a useful so-called *perpendicular shock frame*, for which **E** and **B** are normal to each other and both are parallel to the plane of the shock front, while the plasma velocity is oblique to the shock front.)

In highly relativistic flows, such as the wind of the Crab nebula, shock waves are usually superluminal, but superluminal shocks can also occur in

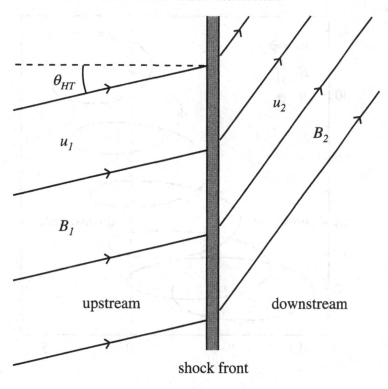

Fig. 13.10. A (subluminal) MHD fast-mode shock in the de Hoffmann–Teller frame of reference.

nonrelativistic flows if the magnetic field is aligned at a small enough angle to the shock. For example, a perpendicular shock (in which the magnetic field is parallel to the shock front; §1.5) is always superluminal. However, in the solar wind, say, only a small part of a shock front is superluminal, since the magnetic field of such a shock in practice has to be inclined to the shock front by less than approximately a tenth of a degree to be superluminal.

Consider now for simplicity the process of shock-drift acceleration at the perpendicular shock shown as a dashed line in Fig. 13.11, in which the plasma is approaching the shock front at a speed (u_1) that is much smaller than the particle speed (v). In this upstream region the particle gyrates about the magnetic field $(B_1 \hat{\mathbf{y}})$, while its guiding centre drifts with the speed $\mathbf{E} \times \mathbf{B}_1/B_1^2 = u_1 \hat{\mathbf{x}}$. When the particle penetrates the shock front into the downstream region, its orbit has a smaller radius of curvature, since the magnetic field strength (B_2) there is higher. The net effect for a positively charged particle is to find its orbit moved in the z-direction parallel to the electric field, and so the electric field does work on the particle and

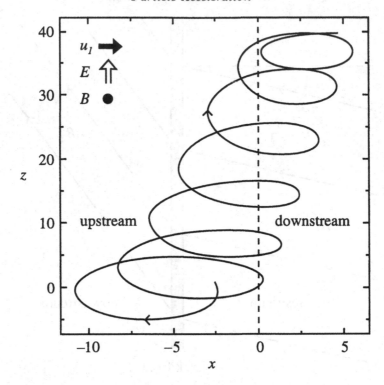

Fig. 13.11. The trajectory (solid curve) of a positively charged particle crossing a nonrelativistic perpendicular shock front (dashed line) with compression ratio 3 and upstream speed (u_1) of 0.1 c. $\mathbf{B}$ is in the negative y-direction, $\mathbf{E}$ is in the z-direction, and the plasma flows from left to right in the x-direction. Different scales are used in the x- and z-directions, with the unit of length being 0.1 times the gyro-radius in the upstream region (after Kirk, 1994).

increases its energy. (An electron would move in the opposite direction.) Since the guiding-centre motion is very similar to the gradient drift in an inhomogeneous magnetic field (§13.1.4), this process is referred to as *shock-drift acceleration*.

In spite of the sudden increase in magnetic field at the shock front, the magnetic moment [Eq. (13.52)] is conserved, at least when $v \gg u_1$. An immediate consequence is that, even for strong shocks, the energy gain is modest. For example, a typical maximum compression (B_2/B_1) of a factor of 4 in the magnetic field (for $\gamma = 5/3$) produces an energy increase by a factor of $B_2/B_1 = 4$ for a non-relativistic particle and of 2 for a relativistic particle.

An additional possibility is that an incident particle may be reflected back into the upstream region (unless the shock is superluminal). Since the magnitude of the momentum is conserved in the de Hoffmann–Teller frame, the conservation of magnetic moment across the shock may be written in terms of the pitch-angles (α_{p1}) and (α_{p2}) as

$$\frac{1 - \cos^2 \alpha_{p1}}{B_1} = \frac{1 - \cos^2 \alpha_{p2}}{B_2}.$$

Thus, reflection occurs when there is no solution for $\cos \alpha_{p2}$, namely when the pitch-angle (α_{p1}) is large enough that

$$\cos^2 \alpha_{p1} < 1 - \frac{B_2}{B_1}. \tag{13.55}$$

Furthermore, the condition for there to be an adiabatic invariant is that the orbit intersects the shock many times, namely

$$\tan \alpha_p \tan \theta_{HT} \gg 1,$$

where θ_{HT} is the inclination of the upstream magnetic field to the shock front (Fig. 13.10). This condition fails for low pitch-angles (α_p) or low magnetic field inclinations (θ_{HT}), but even in this case numerical calculations show the conservation of magnetic moment to be surprisingly good (Hudson, 1965; Terasawa, 1981; Decker, 1988). For non-relativistic particles at shocks that are well within the subluminal regime, the maximum mean energy gain is $4(B_2/B_1) - 3$, which, for $B_2/B_1 = 4$, is 13 (Toptyghin, 1980). For relativistic particles it is smaller than this at a subluminal shock [i.e., only $2(B_2/B_1) - 1$], but at shocks that are almost superluminal the maximum increase is slightly larger. An interesting additional feature of shock-drift acceleration, arising from the shock potential at quasi-perpendicular shocks, is *shock surfing acceleration*, which may be an effective injection mechanism at quasi-perpendicular shocks (e.g., Lee et al., 1996).

Effective shock-drift acceleration is more important for electrons than ions, since most (i.e., thermal) electrons have a gyro-radius much smaller than the shock thickness and so their motion is adiabatic, whereas thermal ions tend to have a gyro-radius similar to the shock thickness and so are not reflected. Also, thermal electrons can satisfy Eq. (13.55) and so be reflected, whereas the slower-moving thermal ions with a narrower distribution function (Fig. 13.1) generally have their pitch-angles inside the loss cone [that is, they do not satisfy Eq. (13.55)] and so are not reflected. Various attempts have been made to extend the basic theory to try and see how the energy gain can be increased (e.g., Decker, 1988). However, in most cases it does not exceed a factor on the order of 10, and so another mechanism (such as diffusive shock acceleration, see the next section) is required to explain more dramatic acceleration such as occurs in cosmic rays.

Shock-drift acceleration may be important for particle acceleration at sites of magnetic reconnection. Shimada et al. (1997) have demonstrated that the shock-drift mechanism can accelerate particles trapped between the slow-mode shocks of a Petschek-type configuration, but the process is not simple. By itself, a slow shock is not particularly favourable for particle acceleration, because the magnetic field decreases across it, which means that particles upstream of the shock are not reflected by a stronger downstream magnetic field as they are in the case of a fast-mode shock. Reflection does occurs for the downstream particles, but when such particles collide with the shock they lose energy since the shock, acting as a magnetic mirror, is moving away from the particle in the rest frame of the plasma. A detailed analysis by Webb et al. (1983) has shown that particle acceleration can occur when particles penetrate the shock and are not reflected by it. This acceleration is produced solely by the curvature drift, which competes with the gradient drift acting in the opposite sense. Thus, it is perhaps not surprising that the energy gained upon penetration is relatively small and typically less than a factor of two.

Shimada et al. (1997) managed to overcome the unpromising acceleration properties of a slow-mode shock by invoking two new features. The first is the Petschek-shock configuration, with its pairs of slow-mode shocks flanking the reconnection outflow. In concert, these shocks trap the particles within the outflow region and force them to undergo many shock encounters. However, this mechanism acting alone does not lead to any acceleration because, as stated above, each encounter with the shock from the downstream side causes the particle to lose energy. The second new feature is turbulence upstream of the shocks. As we shall see in the next section (§13.3.2), accelerated particles upstream of fast shocks can generate turbulent fluctuations. Although Isenberg (1986) has shown that this is not true for slow-mode shocks, the reconnection process itself is very likely to generate the required turbulence when it is impulsive and bursty (Priest, 1986). The upstream turbulence is important because it allows those particles which are accelerated when they penetrate the shock to be reflected back through it and undergo further acceleration. In order for a net acceleration to occur, the energy gained by the particles when they penetrate into the upstream region must be greater than the energy they lose when they are reflected at the shocks.

Figure 13.12 shows the acceleration modelled by Shimada et al. (1997) for particles in the outflow jet of a Petschek-type configuration. The net gain a particle experiences after half a dozen scatterings is about 5. Even though many scatterings are needed to gain a significant amount of energy, the

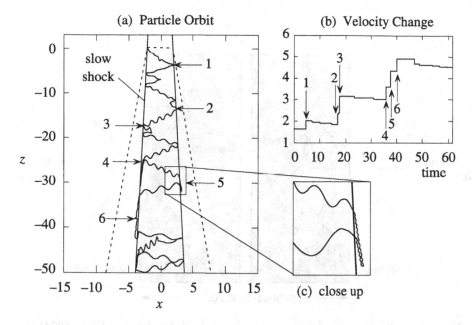

Fig. 13.12. (a) Particle orbit and (b) velocity change for a Petschek-type configuration containing two slow-mode shocks. The curvy line is the particle orbit, and the dashed line is a magnetic field line. The close-up (c) shows the drift orbit of the particle as it passes through the shock before being scattered back into the jet by the upstream turbulence (from Shimada et al., 1997).

process is still quite fast because of the very short time between the particle encounters with the shock waves. Shimada et al. (1997) estimate that, for upstream magnetic fluctuations of normalized amplitude $\delta B/B = 0.04$, the acceleration time-scale is more than 300 times faster than the diffusive shock acceleration mechanism (§13.3.2) at a single fast-mode shock. Furthermore, they also find that shock-drift acceleration at Petschek-type shocks provides a plausible explanation for the 1-MeV particles observed in the geomagnetic tail of the Earth's magnetosphere. Such particles are difficult to explain by direct acceleration because the total potential drop across the X-line in the tail is less than 100 kV.

13.3.2 Diffusive Shock Acceleration

Particle motion in an electromagnetic field is reversible, and so a potential problem with the shock-drift mechanism is that deceleration at a rarefaction wave may counteract acceleration at a shock. However, if pitch-angle scattering occurs far upstream or downstream of the shock, it will isotropize the distributions and introduce irreversibility into the process. In this section

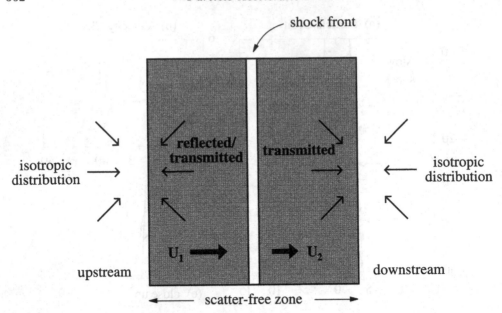

Fig. 13.13. The scatter-free zone around a shock front (after Kirk, 1994).

we consider the effect of scattering on shock acceleration and set up a simple model for particle diffusion on both sides of a shock discontinuity.

Suppose that the mean-free path for scattering is much longer than the thickness of a shock front, and consider a shock front surrounded by a scatter-free zone in which pitch-angle scattering is absent (Fig. 13.13). Further upstream and downstream there exist regions in which scattering creates isotropic distributions. Calculating the resulting distribution functions is in general highly complex. Even though we may ensure that particles incident from the left on the left-hand boundary in Fig. 13.13 form part of an isotropic distribution, some particles at that boundary may have been reflected at the shock front while others may have travelled through the shock from the right-hand boundary. (Note that diffusive acceleration also occurs when the skin depth rather than the scattering mean-free path is much larger than the shock thickness.)

Progress may be made by assuming small departures from isotropy, which leads to spatial diffusion of the particles described by the transport equation for pitch-angle diffusion [Eq. (13.53) or (13.54)]. Consider a simple case in which the fluid velocity $(u\,\hat{\mathbf{x}})$ is constant, much smaller than the particle velocity $(\mathbf{v})$, and is directed along the magnetic field, so that Eq. (13.53) or (13.54) reduces to

$$\frac{\partial f}{\partial t} + (u + \overline{\mu}v)\frac{\partial f}{\partial x} = \frac{\partial}{\partial \overline{\mu}}\left(D_{\overline{\mu}\overline{\mu}}\frac{\partial f}{\partial \overline{\mu}}\right), \tag{13.56}$$

where $\bar{\mu}$ is the cosine of the pitch-angle between the particle velocity and the magnetic field. Suppose also that f is stationary in the laboratory frame and almost isotropic in the plasma frame, so that the distribution consists of an isotropic part $[f^{(0)}]$ and a small anisotropic part $[f^{(1)}]$:

$$f(\mathbf{x}, \bar{\mu}, \bar{p}) = f^{(0)}(x, \bar{p}) + f^{(1)}(x, \bar{\mu}, \bar{p}),$$

where $f^{(1)}$ is of order u/v and

$$\int_{-1}^{1} f^{(1)}(x, \bar{\mu}, \bar{p}) \, d\bar{\mu} = 0. \tag{13.57}$$

Substituting into Eq. (13.56) gives

$$(u + \bar{\mu}v) \left[\frac{\partial f^{(0)}}{\partial x} + \frac{\partial f^{(1)}}{\partial x} \right] = \frac{\partial}{\partial \bar{\mu}} \left[D_{\bar{\mu}\bar{\mu}} \frac{\partial f^{(1)}}{\partial \bar{\mu}} \right] \tag{13.58}$$

or, to lowest order in u/v,

$$\bar{\mu}v \frac{\partial f^{(0)}}{\partial x} = \frac{\partial}{\partial \bar{\mu}} \left[D_{\bar{\mu}\bar{\mu}} \frac{\partial f^{(1)}}{\partial \bar{\mu}} \right].$$

Integrating this twice gives

$$f^{(1)} = K - \frac{v}{2} \frac{\partial f^{(0)}}{\partial x} \int_{-1}^{\bar{\mu}} \frac{1 - \bar{\mu}^2}{D_{\bar{\mu}\bar{\mu}}} d\bar{\mu} \tag{13.59}$$

as the solution for the anisotropic part of the distribution, where K is a constant determined by Eq. (13.57).

The resulting particle flux $[\mathcal{F}(\bar{p}, x) d\bar{p}]$ per unit area along the magnetic field in the momentum interval $d\bar{p}$ is

$$\mathcal{F}(\bar{p}, x) \equiv 2\pi \bar{p}^2 v \int_{-1}^{1} \bar{\mu} f^{(1)} d\bar{\mu} = -\bar{\kappa} \frac{\partial \mathcal{N}}{\partial x}, \tag{13.60}$$

where $\mathcal{N}(\bar{p}, x) d\bar{p}$ is the particle density in the interval $d\bar{p}$ defined as

$$\mathcal{N} = 4\pi \bar{p}^2 f^{(0)} \tag{13.61}$$

and the spatial diffusion coefficient is

$$\bar{\kappa} = \frac{v^2}{8} \int_{-1}^{1} \frac{(1 - \bar{\mu}^2)^2}{D_{\bar{\mu}\bar{\mu}}} d\bar{\mu}.$$

The linear relation (13.60) between particle flux and density gradient is known as *Fick's law* and is a fundamental property of diffusion processes.

Next, the isotropic part $[f^{(0)}]$ of the distribution (f) may be found by integrating Eq. (13.58) over $\bar{\mu}$ from -1 to $+1$ to give the *cosmic-ray transport equation* for stationary transport in the diffusion approximation at constant plasma speed (u),

$$\frac{\partial}{\partial x}\left[\bar{\kappa}\frac{\partial f^{(0)}}{\partial x}\right] - u\frac{\partial f^{(0)}}{\partial x} = 0, \tag{13.62}$$

where the first term represents spatial diffusion and the second term spatial advection. A generalisation of this (which is also known as the *diffusion–convection equation*) is

$$\frac{\partial f}{\partial t} + \mathbf{u}\cdot\nabla f = \nabla\cdot(\bar{\kappa}\nabla f) + \tfrac{1}{3}\nabla\cdot\mathbf{u}\,\bar{p}\frac{\partial f}{\partial\bar{p}}$$

$$+ \frac{1}{\bar{p}^2}\frac{\partial}{\partial\bar{p}}\left(\bar{p}^2 D\frac{\partial f}{\partial\bar{p}}\right) + I - L, \tag{13.63}$$

where D is the diffusion coefficient and the terms describe (in order) the time-dependence of f, spatial advection, spatial diffusion, adiabatic expansion or compression, momentum diffusion, injection sources (I), and loss processes (L) (Parker, 1965). An alternative elegant derivation is given by Gleeson and Axford (1967). Often a drift term is included for nearly isotropic distributions, which adds a term $\mathbf{v}_D\cdot\nabla f$ (Jokipii and Levy, 1977). The drift term actually includes shock drift in a shock acceleration configuration (Jokipii, 1982).

Consider now a *parallel shock*, namely one in which both the plasma velocity and magnetic field are parallel to the shock normal. The solutions to Eq. (13.62) in the upstream $(x<0)$ and downstream $(x>0)$ regions (denoted by subscripts 1 and 2, respectively) are therefore

$$f_1^{(0)} = A_1(\bar{p}) + C_1(\bar{p})\exp\int_0^x\frac{u_1}{\bar{\kappa}_1}dx', \quad f_2^{(0)} = f_2^{(0)}(\bar{p}), \tag{13.64}$$

where A_1 and C_1 are integration constants and the exponential term is absent from $f_2^{(0)}$ since it would grow indefinitely with x. A parallel shock does not affect a particle at all. It neither accelerates nor reflects it, and so the transverse momenta are conserved $(\bar{p}_{z1}=\bar{p}_{z2}, \bar{p}_{y1}=\bar{p}_{y2})$, while the longitudinal momentum in the plasma frame only changes because the plasma slows down (from u_1 to u_2). Therefore $\bar{p}_{x1} = \bar{p}_{x2} - m(u_1 - u_2)$ and

$$\bar{p}_2 \approx \bar{p}_1\left(1 + \frac{\Delta u}{v_1}\bar{\mu}_1\right), \tag{13.65}$$

where $\Delta u = u_1 - u_2$.

Across the scatter-free zones, *Liouville's theorem* (the constancy of f along a particle trajectory) implies that $f_1(\bar{p}_1, \bar{\mu}_1) = f_2(\bar{p}_2)$, where f_2 is independent of $\bar{\mu}_2$ by Eq. (13.64). Using Eq. (13.65) with $\Delta u/v_1 \ll 1$, we therefore find

$$f_1(\bar{p}_1, \bar{\mu}_1) \approx f_2^{(0)}(\bar{p}_1) + \bar{\mu}_1 \frac{\Delta u}{v_1} \bar{p}_1 \left[\frac{df_2^{(0)}}{d\bar{p}} \right]_{\bar{p}_1}. \tag{13.66}$$

Integrating this over $\bar{\mu}_1$ gives $N_1 = N_2$ at $x = 0$ and so by Eqs. (13.61) and (13.64)

$$A_1 + C_1 = f_2^{(0)}. \tag{13.67}$$

Thus, the change in plasma velocity through the shock produces an anisotropy in the distribution function proportional to $\bar{\mu}_1$ but does not affect the particle density.

A second relation comes from applying Fick's law [Eq. (13.60)] to the left of the shock front, where it yields

$$\tfrac{1}{2} v_1 \int_{-1}^{1} \bar{\mu}_1 f_1^{(1)} d\bar{\mu}_1 = -\kappa_1 \frac{\partial f_1^{(0)}}{\partial x}.$$

Substituting for $f_1^{(1)}$ from Eq. (13.59) and $f_1^{(0)}$ from Eq. (13.64) gives

$$\frac{\Delta u}{3} \bar{p}_1 \left[\frac{df_2^{(0)}}{d\bar{p}} \right]_{\bar{p}_1} = -u_1 C_1. \tag{13.68}$$

Eliminating C_1 between Eqs. (13.67) and (13.68) then produces a differential equation for $f_2^{(0)}(\bar{p})$ in terms of $A_1(\bar{p})$, which represents the particle spectrum in the region far upstream ($x \rightarrow -\infty$) where the exponential term in Eq. (13.64) vanishes. The solution is

$$f_2^{(0)}(\bar{p}) = \frac{a}{\bar{p}^s} + s \int_0^{\bar{p}} \left(\frac{\bar{p}'}{\bar{p}} \right)^s \frac{A_1(\bar{p}')}{\bar{p}} d\bar{p}', \tag{13.69}$$

where a is a constant and the power-law index is given by

$$s = \frac{3u_1}{\Delta u} = \frac{3u_1}{u_1 - u_2}. \tag{13.70}$$

If there are no particles entering the system far upstream, then $A_1 = 0$ and, in order to obtain a nontrivial solution, we need some injection process to boost a small number of particles up to an initial momentum, at which the acceleration process takes over. If particles far upstream possess a power-law spectrum ($A_1 = A_0 \bar{p}^{-q}$, say, with $q < s$), then the effect of the shock is to produce a spectrum

$$f_2^{(0)} = \frac{s}{s-q} A_0 \bar{p}^{-q},$$

which has the same shape but is amplified.

It is interesting to note that the spectrum of accelerated particles (when $q > s$) is a power law of index s that depends only on the compression ratio (u_1/u_2) and is independent of the diffusion coefficient ($\bar{\kappa}$), although this result can be modified by the effect of boundaries. Furthermore, the increase in energy on crossing a shock once is of the order of $\Delta u/v_1$ by Eq. (13.65), and so particles with an amplification $\bar{p}/\bar{p}_0$, say, must have crossed the shock $\bar{p}v_1/(\Delta u\,\bar{p}_0)$ times. In addition, the assumption of isotropy fails for particles just above thermal energy (as observed at the Earth's bow shock, for example) or at a relativistic shock.

A microscopic picture of diffusive shock acceleration from Bell (1978) is equivalent to the above transport-equation approach, and it shows more clearly the physics of the process. Consider the change in momentum as a particle crosses a shock and then crosses back. Equation (13.65) gives the change in momentum ($\bar{p}_2 - \bar{p}_1$) in going from upstream to downstream for $-u_1/v_1 < \bar{\mu}_1 < 1$. Suppose the same particle is scattered downstream, so that only its pitch angle changes. Then, when it returns upstream, it has a new momentum of $\bar{p}_1 + \Delta\bar{p}_1 = \bar{p}_2(1 - \bar{\mu}_2\Delta u/v_2)$ for $-1 < \bar{\mu}_2 < -u_2/v_2$. The probability of a particle crossing the shock is proportional to the relative velocity between it and the front (i.e., $|\bar{\mu}v + u|$) if the distribution is isotropic, and therefore the average momentum gain over the cycle becomes to first order in $\Delta u/v$

$$\left\langle \frac{\Delta\bar{p}}{\bar{p}_1} \right\rangle = \frac{4\Delta u}{3v}. \tag{13.71}$$

The gain in momentum per interaction is therefore of first order in $\Delta u/v$, which is why it was called a *first-order* Fermi process, whereas randomly moving scattering centres would give a second-order gain (§13.2). In order to gain an appreciable energy a particle has to cross the shock and cross back many times, and, at large distances from the shock, it is partly advected with the plasma and partly scattered. Indeed, from the macroscopic solution [Eq. (13.64)] we know that far upstream (where the distribution falls off exponentially) all the particles are advected back to the shock front, whereas far downstream (where the distribution is spatially uniform) some particles are being scattered back towards the shock while others are being advected away.

Let us estimate the fraction of particles crossing the shock which subsequently escape without returning. The flux of particles across the shock front from upstream to downstream per second is

$$\dot{n}_{\text{cross}} = 2\pi\bar{p}_2^2 \int_{-u_2/v_2}^{1} |\bar{\mu}_2 v_2 + u_2| \, f_2(\bar{p}_2) \, d\bar{\mu}_2 = \frac{v}{4} \mathcal{N}_2, \qquad (13.72)$$

whereas the rate at which particles escape across a distant downstream boundary is

$$\dot{n}_{\text{esc}} = 2\pi\bar{p}_2^2 \int_{-1}^{1} (\bar{\mu}_2 v_2 + u_2) \, f_2(\bar{p}_2) \, d\bar{\mu}_2 = u_2 \mathcal{N}_2.$$

The escape probability (P_{esc}) per cycle is therefore

$$P_{\text{esc}} = \frac{\dot{n}_{\text{esc}}}{\dot{n}_{\text{cross}}} = \frac{4u_2}{v}. \qquad (13.73)$$

Then the steady-state spectrum can be calculated from a balance between the total number of particles with momentum greater than $\bar{p} + \Delta\bar{p}$ crossing the shock into the downstream region and the number crossing with momentum greater than $\bar{p}$ minus those which escape, namely

$$u_2 \int_{\bar{p}+\Delta\bar{p}}^{\infty} \dot{n}_{\text{cross}} \, d\bar{p}' = (1 - P_{\text{esc}}) u_2 \int_{\bar{p}}^{\infty} \dot{n}_{\text{cross}} \, d\bar{p}'.$$

After substituting for $\Delta\bar{p}_2$, $\dot{n}_{\text{cross}}$, and P_{esc} from Eqs. (13.71)–(13.73), this becomes

$$\Delta u \, \mathcal{N}_2(\bar{p}) = 3u_2 \int_{\bar{p}}^{\infty} \mathcal{N}_2(\bar{p}') \, d\bar{p}',$$

with solution

$$\mathcal{N}_2 \sim \bar{p}^{-1-P_{\text{esc}}/(\Delta\bar{p}/\bar{p})} = \bar{p}^{2-s}, \qquad s = \frac{3u_1}{u_1 - u_2}, \qquad (13.74\text{a,b})$$

in agreement with Eqs. (13.61) and (13.70).

The above theory for diffusive shock acceleration at a parallel shock also applies to oblique shocks (where the flow and magnetic field are no longer both aligned with the shock normal), provided the particle transport is diffusive – in other words, if the plasma speed is much smaller than the particle speed and the anisotropy is small [$f^{(1)} \ll f^{(0)}$]. The first condition is satisfied at non-relativistic shocks, but the second can be violated if shock-drift acceleration is sufficiently strong.

When the anisotropy is large the eigenfunction expansion method of the previous section can easily be adapted, provided cross-field diffusion is negligible, the distribution function is gyrotropic and the magnetic moment is conserved across the shock (Kirk and Heavens, 1989). Surprisingly, oblique shocks tend to produce flatter spectra than parallel shocks. Shock-drift acceleration can also be effective at oblique shocks (Jokipii, 1982).

Diffusive shock acceleration does not work for slow-mode shocks because the magnetic field decreases across them, and, therefore, they cannot act as magnetic mirrors for the upstream particles. Consequently, one might suppose that diffusive shock acceleration is not important for fast magnetic reconnection that involves slow-mode shocks. However, as we discussed in Section 11.1.2, the outflow from a Petschek-type reconnection process is supermagnetosonic with respect to the fast-mode wave speed (Podgorny and Syrovatsky, 1981; Soward and Priest, 1982), so any obstacle which obstructs the outflow will create a fast-mode shock. Although suggestions have been made in the past that diffusive particle acceleration might be important at such shocks (Forbes, 1986), it is only recently that a quantitative analysis has been carried out by Tsuneta and Naito (1998) of just how effective such a process might be.

In a perfectly symmetric configuration the fast-mode shock which terminates a reconnection jet is a perpendicular shock. Since the geometry of perpendicular shocks is not favourable for diffusive shock acceleration, Tsuneta and Naito supposed that in general the termination shock is oblique (rather than perpendicular) in most practical situations. This supposition seems quite reasonable since any gradient in the external magnetic field in the direction of the outflow jet will cause the field lines in the outflow to be curved (Vasyliunas, 1975). Numerical simulations often show curvature to be present (see, e.g., Figs. 7.9, 11.16 and 11.17). Furthermore, the fast shock will also be oblique if the reconnecting fields themselves are asymmetric as is often the case (e.g., reconnection at the dayside magnetopause).

Tsuneta and Naito (1998) also noted that the problem of obtaining a high enough injection energy for diffusive shock acceleration to be effective is generally not a problem for the termination shock because the plasma flowing into it has already been energized by its passage through the slow shocks. Though not mentioned by Tsuneta and Naito (1998), it is also likely that the shock-drift mechanism of Shimada et al. (1997) for reconnection slow shocks (§13.3.1) provides an additional injection source. In fact, it is possible that acceleration at the fast shock may be geometrically coupled to that occurring near the slow shocks because particles reflected from the fast shock can in turn be reflected by the slow shocks. Tsuneta and Naito (1998) concluded that diffusive shock acceleration could be effective at reconnection sites in the solar corona and could also explain hard X-ray sources observed in these sites.

13.4 Particle Acceleration in the Cosmos

Particles are being accelerated in many different environments through-out the cosmos, and in some cases magnetic reconnection may be directly responsible. In the previous sections of this chapter we have described the three main mechanisms for acceleration (direct, stochastic, and shock) and have at the same time provided illustrative examples of how these mechanisms work in reconnection processes. Here we present a broad survey of some of the research areas where these mechanisms have been utilized.

13.4.1 Cosmic Rays

The Earth's atmosphere is bombarded by cosmic-ray particles at a rate of about $1,000 \, \text{m}^{-2} \, \text{s}^{-1}$ (e.g., Gaisser, 1990). They are ionized nuclei, about 90% of them being protons, 9% alpha particles, and the rest heavier nuclei; their most notable feature is their high energy. Most cosmic rays are relativistic and some are ultrarelativistic with energies up to at least 10^{20} eV (20 J). Where do they come from and what mechanisms are accelerating them to such high energies?

Most cosmic rays seem to come from outside the solar system, but from within the Milky Way. They are anti-correlated with solar activity, since they are more effectively excluded from the solar neighbourhood when the solar wind is strongest. The low-energy particles ($\lesssim 100$ MeV per nucleon) are mainly produced within the solar system, and some below 10^{10} eV are ac-celerated during solar flares and coronal mass ejections. The highest-energy cosmic rays, however, have gyro-radii in the galactic magnetic field that are larger than the size of our Galaxy and so are probably of extra-galactic origin (perhaps from relativistic termination shocks at the tips of galactic jets).

Most cosmic rays up to 10^{15} eV, however, are generally thought to be produced by supernovae blast waves, which are the explosions of dying stars that release about 10^{44} J (10^{51} ergs) of energy into the interstellar medium of the Galaxy approximately every 50 yr (e.g., Axford, 1994). Approximately 10% of this energy has to be channelled into galactic cosmic rays in order to explain the observed cosmic-ray energy density. The cosmic rays have kinetic energies from 10^8 to 10^{20} eV per nucleon. The proton intensity is isotropic and steady over geological time-scales ($\approx 10^9$ yr) from cosmogenic nucleii observations. It has a power-law spectrum between 10^{10} and 10^{15} eV per nucleon with a differential density [Eq. (13.61)] of

$$\mathcal{N}(\bar{p}) \sim \bar{p}^{2-s},$$

where $s = 4.71 \pm 0.05$. Below 10^{10} eV it is modulated by the Sun, whereas above 10^{16} eV it steepens to $s = 5$ and flattens again at 10^{19}eV. The energy density is about 10^{-13} J m^{-3}(10^{-12} ergs cm$^{-3} \approx$ eV cm^{-3}), which is larger than that of the microwave background or starlight but comparable with the thermal and magnetic energy densities.

The most convincing explanation of the way in which the energy is extracted from a supernova and channelled into cosmic rays below about 10^{14} eV is in terms of diffusive shock acceleration (§13.3.2). In order to explain the observations, the theory has been extended to include the effect of time-dependence from a supernova shock (e.g., Axford, 1981; Schlickeiser, 1986a, 1986b; Drury, 1991). Coupling the cosmic-ray transport equation with realistic hydrodynamics was initially performed with a self-similar hydrodynamic solution but is now being accomplished numerically. In addition, a realistic spectrum of expected cosmic rays has been calculated by assuming acceleration at the shock during a so-called self-similar Sedov phase (during which the supernova energy is conserved and remains inside the shock front) followed by adiabatic expansion after the particles leave the neighbourhood of the shock (e.g., Bogdan and Volk, 1983). Furthermore, the effects of the cosmic rays in the momentum and energetics of the shock front have been incorporated.

Energetic particles may be produced too in astrophysical jets and active galactic nuclei (AGNs), as evidenced by a variety of observations. Bright hot spots in jets of double radio sources exhibit synchrotron radiation from spiralling electrons having Lorentz factors up to 10^6. Gamma rays between 100 MeV and 10 GeV have been detected from AGNs by the Compton Gamma-Ray Observatory (and with TeV energies by ground-based Cherenkov telescopes, Watson, 1993). These gamma rays imply the presence in the source of either electrons with somewhat larger energy or protons with much larger energy.

The acceleration mechanism for ultra-high-energy cosmic rays up to 3×10^{20} eV is at present an enigma. Some authors (e.g., Ostrowski, 1998) suggest that diffusive shock acceleration at shocks produced by relativistic galactic jets can be modified to give much higher energies than previously supposed. Others propose more speculative mechanisms, such as the decay of super-heavy relic particles or black holes (Hillier, 1984). In the latter case it is conceivable that magnetic reconnection might play a role, although, at the present time, there is no evidence that it does.

13.4.2 Solar Flares and Coronal Mass Ejections

Solar flares occur in two major types: gradual, eruptive flares (§11.1), with most ions observed as solar cosmic rays probably accelerated by a shock wave driven by a coronal mass ejection, and impulsive flares (§11.2), which tend to be compact and produce hard X-ray and gamma-ray emission (e.g., Priest, 1982). This latter type of flare can produce a wide range of radiation from radio to gamma rays in excess of 1 GeV, including: microwave and hard X-ray/gamma-ray continuum due to gyrosynchrotron emission and bremsstrahlung, respectively, from subrelativistic and relativistic electrons; lower-frequency radio emission from plasma radiation; soft X-rays from thermal bremsstrahlung continuum and atomic lines; visible and EUV line emission from hot thermal plasma; gamma-ray lines from excited nuclei, neutrons and positrons produced by interactions of MeV ions with ambient nuclei; and gamma-ray continuum (>10 MeV) due to the decay of pions into gamma rays or ultra-relativistic electrons and positrons.

As we have described in Section 11.1, large-scale magnetic eruptions from the Sun are associated with coronal mass ejections and erupting prominences. When the eruption is from an active region, a two-ribbon solar flare is produced, but when it is from a quiet region with a much weaker magnetic field there is generally no solar flare. Most geo-effective particles appear to be accelerated in the shock wave that is driven ahead of a coronal mass ejection (Reames, 1990).

A comprehensive review of impulsive flare observations and models for the copious particle acceleration in them has been given by Miller et al. (1997). The standard simple picture for the close association between hard and soft X-ray emission is given by the nonthermal thick-target model (e.g., Emslie et al., 1981), in which electrons are accelerated to more than 20 keV by magnetic energy release in the coronal part of active-region loops or arcades arching above the solar surface. The accelerated electrons stream through the corona down towards the cooler chromosphere, emitting gyrosynchrotron microwaves as they spiral in the coronal magnetic field. In the chromosphere they produce hard X-ray bremsstrahlung by means of interactions with the ambient protons (and also give up much more energy to the electrons). The heated plasma is then driven upwards by chromospheric evaporation and fills the coronal loop, where it emits in thermal soft X-rays for many minutes or hours (e.g., Antonucci et al., 1982). The observation of simultaneous impulsive soft and hard X-ray emission from chromospheric footpoints of coronal magnetic loops is strong support for the model (e.g., Hudson et al., 1994).

The energies to which the electrons are accelerated are (from hard X-ray spectra) at least 100 keV (Dennis, 1988) and sometimes (from gamma-ray emission) up to 10 MeV (when the spectral index, s, is as low as 1.5). The time-scale for acceleration can be as short as 0.15 s for these 100-keV electrons that produce spikes in hard X-rays and a few seconds for the higher energies (10 MeV) from gamma-ray time-profiles.

The calculated numbers of particles that are accelerated are model-dependent. It is now believed that, although many of the electrons are non-thermal (as in the thick-target model), some of them may be superhot ($\gtrsim 10^8$ K) and so emit hard X-rays thermally in a much more efficient manner. To produce each hard X-ray spike nonthermally, approximately 5×10^{34} electrons/s must be accelerated with a total energy of 5×10^{19} J s^{-1} (5×10^{26} ergs s^{-1}), whereas for the whole flare the nonthermal thick-target model would give 10^{37} s^{-1} and 3×10^{24} J (3×10^{31} ergs). By comparison, there are only 10^{37} electrons in a typical coronal loop of area 10^{14} m^2, length 10^7 m, and density 10^{16} m^{-3}. Thus, there may have to be a replenishment of electrons from the chromosphere during the acceleration. The existence of a rapidly varying nonthermal component responsible for the higher-energy X-rays (and rapidly varying low-energy X-rays) together with a more slowly increasing superhot thermal component at low-energy X-rays ($\lesssim 30$ keV) is suggested by hard X-ray balloon observations at a high spectral resolution (Lin et al., 1981). It is also supported by Yohkoh images that exhibit hard X-ray sources both at the foot-points and above the summits of coronal loops during some flares viewed near the limb of the Sun (Masuda, 1994).

Energetic nonthermal ions in a solar flare or coronal mass ejection are revealed directly by in-situ space measurements and indirectly by the γ-ray lines produced in a thick-target interaction region (the chromosphere and photosphere). Many γ-ray flares have 100-MeV ions, and some accelerate protons up to 10 GeV (Chupp, 1984; Ramaty and Murphy, 1987). The acceleration time is about 1 s. The number of protons above 30 MeV is approximately 3×10^{30} s^{-1}, or in total 10^{32} for a typical burst duration of 30 s. The corresponding values above 1 MeV are 10^{35} s^{-1} and 3×10^{36} (Ramaty et al., 1995). Their energy content is approximately 3×10^{24} J (3×10^{31} ergs), which is comparable with the nonthermal electron energy (based on the thick-target model) and also with the magnetic energy contained in a cube of side 10^7 m and containing a magnetic field of 0.1 tesla (1 kG).

Aschwanden et al. (1994, 1995, 1996, 1997) have obtained invaluable clues about the properties of acceleration mechanisms from hard X-ray observations and from comparison with radio bursts. They used energy-dependent hard X-ray observations to deduce time-of-flight distances of electrons between the acceleration site and the hard X-ray footpoint emission. From these they inferred that the acceleration site is probably located in the cusp region above the flare loop at an altitude of about 10^4 km. Type III radio burst observations suggest that the electron density in the acceleration site is 10^{15}–10^{16} m^{-3}, which is lower than the density in the soft X-ray flare loops observed by Yohkoh (0.2–2.5×10^{17} m^{-3}). They also found many hard X-ray pulses to be correlated with type III bursts, suggesting the presence of bidirectional electron beams. They deduced an upper limit of 2,000 km for the length of the acceleration site and an acceleration time of about 4 ms for the ($\gtrsim 5$ keV) radio-emitting electrons. Furthermore, in some cases oscillations in the radio emission suggests acceleration in a pulsed mode.

Various acceleration mechanisms have been considered to account for energetic particles from solar flares (Heyvaerts, 1981; Scholer, 1988b). The two most important mechanisms are: direct electric field acceleration in the reconnection region, which accelerates particles very quickly in a small region to high energies and so is a natural explanation for γ-ray spikes and hard X-ray spikes; and shock acceleration, which tends to accelerate many more particles more slowly and so is a likely explanation for the main phase of a flare. Let us consider these and other mechanisms briefly as follows.

First of all, stochastic acceleration by low-amplitude waves (§13.2) has been proposed to be caused by a resonant wave-particle interaction in which the Doppler-shifted wave frequency ($\omega - k_\parallel v_{\mathrm{dr}}$) in the particle's guiding centre frame equals a multiple of the gyro-frequency (Ω). For example, so-called *transit-time damping* (Lee and Völk, 1975) due to a resonance with fast-mode MHD waves is extremely efficient for electron acceleration under flare conditions ($B_0 \approx 0.05$ tesla $= 500$ G, $n \approx 10^{16}$ m^{-3}, $T \approx 3$ MK), where the Alfvén speed is comparable with the electron thermal speed (Miller et al., 1996). It is a promising mechanism for explaining hard X-ray spikes.

Another flare particle mechanism is stochastic Fermi acceleration with large-amplitude fast-mode waves produced by many small reconnection sites (La Rosa et al., 1996). Alfvén waves generated by reconnection and cascading to short wavelengths have been invoked for ion acceleration (Miller and Ramaty, 1992; Miller and Roberts, 1995). Ion abundance enhancements may be caused by a gyro-resonance with electrostatic or electromagnetic waves that are close to the cyclotron frequency of the ion (Fisk, 1978). For example, the enhancement of the ^{3}He/^{4}He ratio from the normal coronal value of 5×10^{-4} to 0.1–10 has been explained by Temerin and Roth (1992) and Litvinenko (1996b), using electromagnetic ion-cyclotron waves. Furthermore, lower-hybrid waves may account for radio emission but do not appear to accelerate enough electrons for hard X-ray bursts (Vlahos et al., 1982; McClements et al., 1993).

Shock waves are also natural particle accelerators in solar flares and coronal mass ejections. They are present at the reconnection region (as slow-mode shocks) and where the reconnection jet meets the ambient field (as a fast-mode shock), and they also propagate away from the flare site (as fast-mode shocks) where they show up as Moreton waves and type II radio bursts. Electron-drift acceleration is probably important in type II radio bursts (Holman and Pesses, 1983). Diffusive acceleration is more likely at fast-mode shocks than slow-mode shocks, since the scattering centres tend to converge towards the shock in the shock frame of a fast shock but not a slow shock (Isenberg, 1986). It is a viable mechanism for ion acceleration up to 100 MeV in less than 1 s (Ellison and Ramaty, 1985). Furthermore, Tsuneta and Naito (1998) have suggested the acceleration of 20–100 keV nonthermal electrons in about 0.3–0.6 s at the fast shock in the reconnection jet below the reconnection site: as in many mechanisms, such electrons could create the commonly observed double-source hard X-ray structure at the chromospheric footpoints of the reconnected field lines (Fig. 11.14).

Direct electric field acceleration of electrons by super-Dreicer fields is a very natural and efficient mechanism, since a typical reconnection field is $10^3\,\mathrm{V\,m^{-1}}$ by comparison with the Dreicer field of about $10^{-2}\,\mathrm{V\,m^{-1}}$ (see Table 1.2). The energy gain depends on the magnetic field components in the reconnecting current sheet. A transverse component of say 10^{-4} tesla (1 G) by comparison with a longitudinal component of 10^{-2} tesla (100 G) would produce acceleration up to 100 keV, while a lower transverse field would allow higher energies to be reached before the electrons spiral out of the sheet (Martens, 1988; Martens and Young, 1990; Litvinenko, 1996a). Sub-Dreicer fields have also been considered by Holman (1995).

In Section 13.1.2 we presented a simple argument that has been commonly used to infer that acceleration in a simple solar-flare current sheet is limited to only about 10^{30} electrons/s. However, Martens (1988) realised that this is an underestimate. Although the number of electrons that is drifting into the sheet with the field lines is typically 4×10^{35} electrons/s, the actual number accelerated depends on how long a particular particle stays in the current sheet. As we have seen (§13.1.6), the transverse magnetic field tends to turn the particles and eject them from it before they travel the whole length of the sheet. Thus, if the electric field is, say, $300\,\mathrm{V\,m^{-1}}$, a typical 30 keV electron would to be accelerated for only 100 m along the sheet (compared with the typical sheet dimension of 100 Mm in a direction along the ribbons of a two-ribbon flare). Hence, at any cross-section of the current sheet, the fraction of the total number of accelerated particles that is passing by and providing the current is only 10^{-6} of the total available. As a consequence, Martens (1988) found that the estimates for the electric current, magnetic energy conversion and particle number in a flare are all reasonable.

So far, there have been few detailed studies of the subtle relation between the large-scale macroscopic magnetohydrodynamics of a solar flare (§§11.1, 11.2) and the microscopic plasma physics of the acceleration processes. The global environment within which the acceleration occurs is created by the MHD processes, but they are in turn affected by the microscopic physics – for example, the onset and value of the anomalous transport coefficients. Macrophysics and microphysics also come together in the process of fragmentation, which was first pointed out for hard X-rays and microwaves by de Jager and Jonge (1978) and later considered in, for instance, spikes (Benz, 1985), in type III radio bursts (Aschwanden et al., 1990), and in metric spikes (Benz et al., 1996).

Three other global macroscopic scenarios have been proposed within which the microscopics have to be calculated in more detail in future. The first is the eruption of a magnetic arcade and its close-down by reconnection to create the flaring loops (Tsuneta, 1995, 1996) and chromospheric ribbons seen in a typical eruptive two-ribbon flare (§11.1). Here the promising idea by Tsuneta and Naito (1998) of first-order Fermi acceleration of nonthermal electrons at the fast-mode shock below the reconnection site has to be developed and compared with direct electric field acceleration and betatron acceleration by loop shrinkage (§11.1.2). The second scenario is the interaction by reconnection of separate coronal loops, as in the Emerging Flux Model (§11.2.1). The third is a fragmentation of the energy release in many small current sheets (Benz, 1985; Vlahos, 1989; Vlahos et al., 1995), although this could equally well occur in either of the other two scenarios.

In each case, there is: some direct electric field acceleration associated with the reconnection; some shock-wave acceleration (both near the reconnection site and in the fast-mode shock that propagates out from the site of the initial energy release); and some stochastic acceleration from the turbulence at the reconnection site (or sites) and in the jets that are accelerated from it. Important issues to be addressed in the process of developing more quantitative models are: the mechanisms for and the extent of current filamentation; the nature and details of return currents; and the efficiency of acceleration and the fraction of magnetic energy release that goes into particles.

13.4.3 Shocks in the Solar Wind

Strong shock waves are formed where the supersonic solar wind meets a planetary magnetosphere (such as that of the Earth) or when solar-wind streams of different speeds collide with one another, and thus both are natural particle accelerators in the heliosphere. At the orbit of Earth the solar wind typically has a density of about 10^{-20} kg m^{-3} (10^{-23} g cm^{-3}), a number density of 10^7 m^{-3} (10 cm^{-3}), a temperature of 10^5 K, a speed of 400 km s^{-1}, and a magnetic field of 3×10^{-5} G. The size of the Earth's bow shock is about 10^5 km, which is only 30 times the Larmor radius of a 10-keV proton and 10^{-10} the size of a supernova remnant, so that particles can easily escape sideways before they can be accelerated up to relativistic energies.

Detailed measurements have been made of the particle and wave properties in the environment of the Earth's bow shock by spacecraft. Energetic beams with speeds of about $1{,}000$ km s^{-1} in the upstream region are thought to be accelerated by the shock-drift mechanism (§13.3.1) on the quasi-perpendicular part of the shock, where the magnetic field is nearly tangential to the shock front (Krauss-Varban and Burgess, 1991).

Upstream of the quasi-parallel part of the shock there is a fairly isotropic component of 100-keV particles with Larmor radii comparable with the wavelength of Alfvén waves. In addition, Langmuir and whistler waves are driven by electron beams and ion-acoustic modes by supra-thermal protons. According to Kennel et al. (1986), the best theory for the 10–100 keV diffuse ions upstream of the Earth's bow shock is diffusive shock acceleration, as worked out by Lee (1982).

Supra-thermal particles have also been detected in interplanetary space with energies of 0.5–10 MeV and are frequently associated with shock waves. Some of these are at corotating interaction regions, where fast solar-wind streams catch up with slower streams. Others are at interplanetary travelling shock waves (energetic storm particle events) produced by solar flares or coronal mass ejections (Lee, 1983).

Finally, the solar wind is thought to be decelerated by the interstellar medium in a termination shock at about 85 AU from the Sun. It may be responsible for accelerating by the diffusive shock mechanism (§13.3.2) the anomalous cosmic ray component, which has an energy up to 100 MeV and lies between the energies of normal heliospheric particles and galactic cosmic rays.

13.4.4 Planetary Magnetospheres

Detailed observations have also been made during magnetospheric substorms (§10.5) of particles accelerated in the auroral, polar-cap, and cusp regions and in the geomagnetic tail (Lopez and Baker, 1994; Williams, 1997; Onsager and Lockwood, 1997; Baker et al., 1998). Direct electric field acceleration by field-aligned electric fields associated with three-dimensional magnetic reconnection may play an important role here (§13.1). Particle acceleration is also a prominent feature of magnetic storms (§10.5), during which ions of both solar-wind and terrestrial origin are accelerated and form an energetic ring current in the inner magnetosphere (e.g., Daglis et al., 1997).

The magnetosphere is thus a highly complex system in which a variety of acceleration mechanisms are at work, including shock waves, reconnection, neutral-sheet electric fields, convection electric fields, cross-field diffusion driven by fluctuating electric and magnetic fields, auroral parallel electric fields, and various local wave-particle processes (Cornwall, 1986; Bryant, 1999). Furthermore, magnetospheres of other planets such as Jupiter show clear evidence for acceleration in auroral regions, magnetotails, and equatorial regions by electric fields, wave-particle interactions, and magnetic pumping (Möbius, 1994).

A major problem, namely the acceleration of particles in a magnetospheric substorm (§10.5), has recently been resolved by Birn et al. (1996, 1997, 1998). They first of all developed a three-dimensional computational model for the overall resistive MHD behaviour in the geomagnetic tail, which shows how the tail goes unstable and forms a thin current sheet followed by reconnection at about 20 Earth radii (R_E) down the tail. It is this thin current sheet which produces an anomalous resistivity by current-driven microinstabilities and so allows the subsequent reconnection to take place. Then a plasmoid forms and is ejected along the tail in the way that is sketched in Fig. 10.12. At the same time, a substorm current-wedge is formed in the inner tail (Fig. 10.13) and a key feature occurs, called *dipolarization*, in which the reconnected field lines that had been stretched out in the growth-phase rapidly spring back towards the Earth to a more dipolar configuration. Such a process is also observed in solar flares, where it is called *loop shrinkage* (§11.1.2). As far as nonthermal particles are concerned, it was natural to suggest that they are accelerated at the reconnection line situated at 20 R_E. However, there are two difficulties with this suggestion. Firstly, the fast particles above 20 keV are observed closer to the Earth than this, namely, at 7–10 R_E. Secondly, the particles are dispersionless – i.e., all the energies are observed simultaneously – whereas, if they had been accelerated at 20 R_E, they would show dispersion, with the faster particles arriving at 7–10 R_E first. The solution to this dilemma was provided by three calculations: (i) the MHD simulations, which showed that the strongest cross-tail electric fields (up to 20 mV/m) occur not at the reconnection line but in the dipolarization region by magnetic induction where the field lines are moving rapidly; (ii) test proton calculations, which reproduced the observed dispersionless ion fluxes above 20 keV; and (iii) electron acceleration calculations in the time-dependent guiding-centre approximation (§13.1.4), which explained the observed properties of electrons from 100 eV to 1 MeV. The mechanism that accelerates the particles to higher energies than is possible with the $\mathbf{E} \times \mathbf{B}$ drift associated with the bulk MHD flow is partly direct electric field acceleration at the reconnection line but more importantly *betatron acceleration* by mirroring of the particles in the field lines that are springing back towards Earth.

References

Acton, L.W. (1996). Comparison of Yohkoh X-ray and other solar activity parameters for November 1991 to November 1995, in *Cool Stars, Stellar Systems and the Sun*, eds. R. Pallavacini and A.K. Dupree **109**, 45–54.

Acton, L.W., Feldman, U., Bruner, M.E., Doschek, G.A., Hirayama, T., Hudson, H.S., Lemen, J.R., Ogawa, Y., Strong, K.T., and Tsuneta, S. (1992). The morphology of 20 MK plasma in large non-impulsive solar flares, *Publ. Astron. Soc. Japan* **44**, L71–L75.

Akasofu, S.-I. (1968). *Polar and Magnetospheric Substorms* (Reidel, Dordrecht).

Alekseyev, I.I., and Belen'kaya, Y.S. (1983). Electric field in an open model of the magnetosphere, *Geomagn. Aeron.* **23**, 57–61.

Alfvén, H. (1943). On sunspots and the solar cycle, *Ark. f. Mat. Ast. Fys.* **29A**, 1–17.

Altyntsev, A.T., Bardakov, V.M., Krasov, V.I., Lebedev, N.V., and Paperni, V.L. (1986). Laboratory simulation of energy release in solar flares, *Solar Phys.* **106**, 131–145.

Altyntsev, A.T., Krasov, V.I., Lebedev, N.V., Paperni, V.L., and Strokin, N.A. (1989). Dissipation of magnetic energy on merging of current fibers, *Sov. Phys. JETP* **96**, 1243–1251.

Altyntsev, A.T., Lebedev, N.V., and Strokin, N.A. (1990). Ion acceleration in a quasi-neutral current sheet, *Planet. Space Sci.* **38**, 751–763.

Aly, J.J. (1984). On some properties of force-free magnetic fields in infinite regions of space, *Astrophys. J.* **283**, 349–362.

Aly, J.J. (1991). How much energy can be stored in a three-dimensional force-free field?, *Astrophys. J.* **375**, L61–L64.

Aly, J.J. (1994). Asymptotic formation of a current sheet in an indefinitely sheared force-free field: an analytical example, *Astron. Astrophys.* **288**, 1012–1020.

Aly, J.J., and Amari, T. (1989). Current sheets in two-dimensional potential magnetic fields. 1. General properties, *Astron. Astrophys.* **221**, 287–294.

Amari, T., and Aly, J.J. (1990). Extended massive current sheets in a two-dimensional constant-α force-free field: a model for quiescent prominences, *Astron. Astrophys.* **231**, 213–220.

Amari, T., Luciani, J.F., Aly, J.J., and Tagger, M. (1996). Plasmoid formation in a single sheared arcade and application to coronal mass ejections, *Astron. Astrophys.* **306**, 913–923.

Ambrosiano, J.J., Matthaeus, W.H., Goldstein, M.L., and Plante, D. (1988). Test particle acceleration in turbulent reconnecting magnetic fields, *J. Geophys. Res.* **93**, 14383–14400.

Andersen, O.A., and Kunkel, W.B. (1969). Tubular pinch and tearing instability, *Phys. Fluids* **12**, 2009–2018.

Anderson, C., and Priest, E.R. (1993). Time-dependent magnetic annihilation at a stagnation point, *J. Geophys. Res.* **98**, 19395–19407.

Antiochos, S.K., and Sturrock, P.A. (1982). The cooling and condensation of flare coronal plasma, *Astrophys. J.* **254**, 343–348.

Antonucci, E., Gabriel, A.H., Acton, L.W., Leibacher, J.W., Culhane, J.L., Rapley, C.G., Doyle, J.G., Machado, M.E., and Orwig, L.E. (1982). Impulsive phase of flares in soft X-ray emission, *Solar Phys.* **78**, 107–123.

Appenzeller, I. (1990). T-Tauri stars and flare stars – Common properties and differences, in *Flare Stars in Star Clusters, Associations and the Solar Vicinity*, IAU Symp. 137, eds. L.V. Mirzoyan, B.R. Pettersen, and M.K. Tsvetkov (Kluwer, Dordrecht) pp. 209–212.

Arendt, U., and Schindler, K. (1988). On the existence of three-dimensional magnetohydrostatic equilibria, *Astron. Astrophys.* **204**, 229–234.

Arnol'd, V.I. (1974). The asymptotic Hopf invariant and its applications, *Sel. Math. Sov.* **5**, 327–345.

Arnoldy, R.L. (1971). Signatures in the interplanetary medium for substorms, *J. Geophys. Res.* **76**, 5189–5201.

Arnoldy, R.L., Kane, S.R., and Winkler, J.R. (1968). The observation of 10–50 keV solar flare particles, in *Structure and Development of Solar Active Regions*, IAU Symp. 35, ed. K.O. Kiepenheuer, (Reidel, Dordrecht) pp. 490–509.

Arrowsmith, D.K., and Place, C.M. (1990). *An Introduction to Dynamical Systems* (Cambridge Univ. Press, Cambridge).

Aschwanden, M.J., and Benz, A.O. (1997). Electron densities in solar flare loops, chromospheric evaporation upflows and acceleration sites, *Astrophys. J.* **480**, 825–839.

Aschwanden, M.J., Schwartz, R.A., Benz, A.O., Lin, R.P., and Pelling, R. (1990). Flare fragmentation and type III productivity in the 1980 June 27 flare, *Solar Phys.* **130**, 39–55.

Aschwanden, M.J., Benz, A.O., Dennis, B.R., and Kundu, M.R. (1994). Pulsed acceleration in solar flares, *Astrophys. J. Suppl.* **90**, 631–638.

Aschwanden, M.J., Benz, A.O., Dennis, B.R., and Schwartz, R.A. (1995). Solar electron beams detected in hard X-rays and radio waves, *Astrophys. J.* **455**, 347–365.

Aschwanden, M.J., Wills, M.J., Hudson, H.S., Kosugi, T., and Schwartz, R.A. (1996). Electron time-of-flight distances and flare loop geometries compared from CGRO and YOHKOH observations, *Astrophys. J.* **468**, 398–417.

Ashour–Abdalla, M., Berchem, J., Büchner, J., and Zelenyi, L.M. (1990). Chaotic scattering and acceleration of ions in the earth's magnetotail, *Geophys. Res. Lett.* **17**, 2317–2320.

Ashour–Abdalla, M., Berchem, J., and Büchner, J. (1991). Large- and small-scale structures in the plasma sheet: A signature of chaotic motion and resonance effects, *Geophys. Res. Lett.* **18**, 1603–1606.

Åström, E. (1958). Electron orbits in hyperbolic magnetic fields, *Tellus* **8**, 260–267.

Aubrey, M.P., Russell, C.T., and Kivelson, M.G. (1970). Inward motions of the magnetopause before a substorm, *J. Geophys. Res.* **75**, 7018–7031.

Axford, W.I. (1967). Magnetic storm effects associated with the tail of the magnetosphere, *Space Sci. Rev.* **7**, 149–157.

Axford, W.I. (1981). Acceleration of cosmic rays by shock waves, *Proc. 17th Int. Conf. Cosmic Rays* **12**, 155–203.

Axford, W.I. (1984). Magnetic field reconnection, in *Magnetic Reconnection in Space and Laboratory Plasmas*, Geophys. Monograph 30, ed. E.W. Hones, Jr. (Amer. Geophys. Union, Washington, DC) pp. 1–8.

Axford, W.I. (1994). The origins of high-energy cosmic rays, *Astrophys. J. Suppl.* **90**, 937–944.

Axford, W.I., and Hines, C.O. (1961). A unifying theory of high-latitude geophysical phenomena and magnetic storms, *Can. J. Phys.* **39**, 1433–1464.

Axford, W.I., and McKenzie, J.F. (1996). Implications of observations of the solar wind and corona for solar wind models, *Astrophys. Space Sci.* **243**, 1–3.

Axford, W.I., Leer, E., and Skadron, G. (1977). The acceleration of cosmic rays by shock waves, *Proc. 15th Int. Cosmic Ray Conf.* **11**, 132–137.

Aydemir, A.Y., Wiley, J.C., and Ross, D.W. (1989). Toroidal studies of sawtooth oscillations in tokamaks, *Phys. Fluids B* **1**, 774–787.

Bajer, K. (1990). *Flow Kinematics and Magnetic Equilibria*, PhD thesis (Cambridge Univ., Cambridge).

Baker, D.A. (1984). The role of magnetic reconnection phenomena in the reversed field pinch, in *Magnetic Reconnection in Space and Laboratory Plasmas*, Geophys. Monograph 30, ed. E.W. Hones, Jr. (Amer. Geophys. Union, Washington, DC) pp. 332–340.

Baker, D.N., Pulkkinen, T.I., Hesse, M., and McPherron, R.L. (1997). A quantitative assessment of energy storage and release in the magnetotail, *J. Geophys. Res.* **102**, 7159–7168.

Baker, D.N., Li, X., Blake, J.B., and Kanekal, S. (1998). Strong electron acceleration in the Earth's magnetosphere, *Adv. Space Res.* **21**, 609–613.

Baldwin, P., and Roberts, P.H. (1972). On resistive instabilities, *Phil. Trans. Roy. Soc. Lond.* **272**, 303–330.

Barnes, C.W., and Sturrock, P.A. (1972). Force-free magnetic field structures and their role in solar activity, *Astrophys. J.* **174**, 659–670.

Batchelor, G.K. (1967). *An Introduction to Fluid Dynamics* (Cambridge Univ. Press, Cambridge).

Bateman, G. (1978). *MHD Instabilities* (MIT Press, Cambridge, USA).

Baty, H., Luciani, J.F., and Bussac, M.N. (1992). Asymmetric reconnection and stochasticity in tokamaks, *Nuclear Fusion* **32**, 1217–1223.

Baum, P.J., and Bratenahl, A. (1976). Laboratory solar flare experiments, *Solar Phys.* **47**, 331–344.

Baum, P.J., and Bratenahl, A. (1977). On reconnection experiments and their interpretation, *J. Plasma Phys.* **18**, 257–272.

Baum, P.J., and Bratenahl, A. (1980). Magnetic reconnection experiments, *Adv. Electronic Phys.* **54**, 1–67.

Bell, A.R. (1978). The acceleration of cosmic rays in shock fronts, *Mon. Not. Roy. Astron. Soc.* **182**, 147–156.

Benedetti, M.D., and Pegoraro, F. (1995). Resistive modes at a magnetic X-point, *Plasma Phys. Control. Fusion* **37**, 103–116.

Bentley, R.D., Doschek, G.A., Simnett, G.M., Rilee, M.L., Mariska, J.T., Culhane, J.L., Kosugi, T., and Watanabe, T. (1994). The correlation of solar flare hard X-ray bursts with doppler blue-shifted soft X-ray flare emission, *Astrophys. J.* **421**, L55–L58.

Benz, A.O. (1985). Radio spikes and the fragmentation of flare energy release, *Solar Phys.* **96**, 357–370.

Benz, A.O. (1993). *Plasma Astrophysics* (Kluwer, Dordrecht).

Benz, A.O., Csillaghy, A., and Aschwanden, M.J. (1996). Metric spikes and electron acceleration in the solar corona, *Astron. Astrophys.* **309**, 291–300.

Berger, M.A. (1984). Rigorous new limits on magnetic helicity dissipation in the solar corona, *Geophys. Astrophys. Fluid Dyn.* **30**, 79–104.

Berger, M.A. (1986). Topological invariants of field lines rooted to planes, *Geophys. Astrophys. Fluid Dynamics* **34**, 265–281.

Berger, M.A. (1988). An energy formula for non-linear force-free magnetic fields, *Astron. Astrophys.* **201**, 355–361.

Berger, M.A. (1991a). Generation of coronal magnetic fields by random surface motions, *Astron. Astrophys.* **252**, 369–376.

Berger, M.A. (1991b). Third-order braid invariants, *J. Phys. A.* **24**, 4027–4036.

Berger, M.A. (1993). Energy-crossing number relations for braided magnetic fields, *Phys. Rev. Lett.* **70**, 705–708.

Berger, M.A. (1998). Magnetic helicity and filaments, in *New Perspectives on Solar Prominences*, IAU Colloq. 167, eds. D. Webb, B. Schmieder, and D. Rust (Astron. Soc. Pacific, San Francisco) pp. 102–111.

Berger, M.A., and Field, G.B. (1984). The topological properties of magnetic helicity, *J. Fluid Mech.* **147**, 133–148.

Berghmans, D., Clette, F., and Moses, D. (1998). Quiet-Sun EUV transient brightenings and turbulence, in *A Cross-roads for European Solar and Heliospheric Physics*, eds. E.R. Priest, F. Moreno–Insertis, and R.A. Harris (ESA SP-417, Noordwijk) pp. 229–234.

Besser, P.B., Biernat, H.I.C., and Rijnbeek, R.P. (1990). Planar MHD stagnation-point flows with velocity shear, *Planet. Space Sci.* **38**, 411–418.

Bhattacharjee, A., and Dewar, R.L. (1982). Energy principle with global invariants, *Phys. Fluids* **25**, 887–897.

Bhattacharjee, A., Dewar, R.L., and Monticello, D. (1980). Energy principle with global invariants for toroidal plasmas, *Phys. Rev. Lett.* **45**, 347–350.

Bhattacharjee, A., Brunel, F., and Tajima, T. (1983). Magnetic reconnection driven by the coalescence instability, *Phys. Fluids* **26**, 3332–3337.

Biermann, L. (1951). Kometenschweife und solare Korpuskularstrahlung, *Z. Astrophys.* **29**, 274–286.

Biernat, H.K., Heyn, M.F., and Semenov, V.S. (1987). Unsteady Petschek reconnection, *J. Geophys. Res.* **92**, 3392–3396.

Biernat, H.K., Semenov, V.S., and Rijnbeek, R.P. (1998). Time-dependent 3D Petschek-type reconnection: A case study for magnetopause conditions, *J. Geophys. Res.* **103**, 4693–4706.

Birn, J. (1991). The geomagnetic tail, *Rev. Geophys. Suppl.* **29**, 1049–1065.

Birn, J., and Hesse, M. (1990). The magnetic topology of a plasmoid flux rope in an MHD simulation of magnetotail reconnection, in *Physics of Magnetic Flux Ropes*, Geophys. Monograph 58, eds. C.T. Russell, E.R. Priest, and L.C. Lee (Amer. Geophys. Union, Washington, DC) pp. 655–661.

Birn, J., and Hones, E.W., Jr. (1981). 3D computer modelling of dynamic reconnection in the geomagnetic tail. *J. Geophys. Res.* **86**, 6802–6808.

Birn, J., Hesse, M., and Schindler, K. (1989). Filamentary structure of a three-dimensional plasmoid, *J. Geophys. Res.* **94**, 241–251.

Birn, J., Hesse, M., and Schindler, K. (1996). MHD simulations of magnetotail dynamics, *J. Geophys. Res.* **101**, 12939–12954.

Birn, J., Thomsen, M.F., Borovsky, J.E., Reeves, G.D., McComas, D.J., and Belian, R.D. (1997). Substorm ion injections: Geosynchronous observations and test particle orbits in three-dimensional dynamic MHD fields, *J. Geophys. Res.* **102**, 2325–2341.

Birn, J., Thomsen, M.F., Borovsky, J.E., Reeves, G.D., McComas, D.J., and Belian, R.D. (1998). Substorm electron injections: geosynchronous observations and test particle simulations, *J. Geophys. Res.* **103**, 9235–9248.

Biskamp, D. (1982). Effect of secondary tearing instability on the coalescence of magnetic islands, *Phys. Lett.* **87A**, 357–360.

Biskamp, D. (1986). Magnetic reconnection via current sheets, *Phys. Fluids* **29**, 1520–1531.

Biskamp, D. (1994). *Nonlinear Magnetohydrodynamics* (Cambridge Univ. Press, Cambridge).

Biskamp, D., and Welter, H. (1989). Dynamics of decaying 2D MHD turbulence *Phys. Fluids B* **1**, 1964–1979.

Blanchard, G.T., Lyons, L.R., de la Beaujardière, O., Doe, R.A., and Mendillo, M. (1996). Measurement of the magnetotail reconnection rate, *J. Geophys. Res.* **101**, 15265–15276.

Blanchard, G.T., Lyons, L.R., and de la Beaujardière, O. (1997). Magnetotail reconnection rate during magnetospheric substorms, *J. Geophys. Res.* **102**, 14303–14312.

Blandford, R.D., and Eichler, D. (1987). Particle acceleration at astrophysical shocks: a theory of cosmic ray origin, *Phys. Rep.* **154**, 1–75.

Blandford, R.D., and Ostriker, J.P. (1978). Supernova shock acceleration of the cosmic rays in the galaxy, *Astrophys. J.* **237**, 793–808.

Block, L.P. (1973). The magnetosphere, in *Cosmical Geophysics*, eds. A. Egeland, Ø. Holter, and A. Omholt (Scandinavian Univ. Books, Oslo) pp. 103–119.

Bobrova, N., and Syrovatsky, S.I. (1979). Singular lines of one-dimensional force-free magnetic fields, *Solar Phys.* **61**, 379–387.

Bobrova, N., and Syrovatsky, S.I. (1980). Dissipative instability of a 1D force-free magnetic field, *Sov. J. Plasma. Phys.* **6**, 1–3.

Bogdan, T.J., and Volk, H.J. (1983). Onion-shell model of cosmic ray acceleration in supernova remnants, *Astron. Astrophys.* **122**, 129–136.

Bogdan, T.J., Lee, M.A., and Schneider, P. (1991). Coupled quasi-linear wave damping and stochastic acceleration of pickup ions in the solar wind, *J. Geophys. Res.* **96**, 161–178.

Braginsky, S.I. (1965). Transport processes in a plasma, *Rev. Plasma Phys.* **1**, 205–311.

Brandenburg, A., and Zweibel, E.G. (1994). The formation of sharp structures by ambipolar diffusion, *Astrophys. J.* **427**, L91–L94.

Brandenburg, A., and Zweibel, E.G. (1995). Effects of pressure and resistivity on the ambipolar diffusion singularity: too little, too late, *Astrophys. J.* **448**, 734–741.

Brandenburg, A., Nordlund, A., Stein, R.F., and Torkelsson, U. (1996). The disk accretion rate for dynamo-generated turbulence, *Astrophys. J.* **458**, L45–L48.

References

Bratenahl, A., and Baum, P.J. (1976). On flares, substorms and the theory of impulsive transfer events, *Solar Phys.* **47**, 345–360.

Bratenahl, A., and Yeates, C.M. (1970). Experimental study of magnetic flux transfer at the hyperbolic neutral point, *Phys. Fluids* **13**, 2696–2709.

Brown, D.S., and Priest, E.R. (1999). Topological bifurcations of 3D magnetic fields, *Proc. Roy. Soc. Lond.*, in press.

Brown, M.R., Canfield, R.C., and Pevtsov, A.A. (1999). *Magnetic Helicity in Space and Laboratory Plasmas*, Geophys. Monograph 111 (Amer. Geophys. Union, Washington, DC).

Browning, P.K. (1992). Energy relations in reconnection, in *Magnetic Reconnection in Physics and Astrophysics*, ed. P. Maltby (Univ. of Oslo, Oslo) pp. 177–190.

Browning, P.K., Sakurai, T., and Priest, E.R. (1986). Coronal heating in closely-packed flux tubes: a Taylor–Heyvaerts relaxation theory, *Astron. Astrophys.* **158**, 217–227.

Browning, P.K., Cunningham, G., Gee, S.J., Gibson, K.J., al-Karkhy, A., Kitson, D.A., Martin, R., and Rusbridge, M.G. (1992). Power flow in a gun-injected spheromak plasma, *Phys. Rev. Lett.* **68**, 1718–1721.

Bruch, A. (1992). Flickering in cataclysmic variables: its properties and origins, *Astron. Astrophys.* **266**, 237–265.

Brueckner, G.E., and Bartoe, J.D.F. (1983). Observations of high-energy jets in the corona above the quiet Sun, the heating of the corona, and the acceleration of the solar wind, *Astrophys. J.* **272**, 329–348.

Bruhwiler, D.L., and Zweibel, E.G. (1992). Energy spectrum of particles accelerated near a magnetic X-line, *J. Geophys. Res.* **97**, 10825–10830.

Brunel, F., Tajima, T., and Dawson, J.M. (1982). Fast magnetic reconnection processes, *Phys. Rev. Lett.* **49**, 323–326.

Bruning, D.H., Chenoweth, R.E., Jr., and Marcy, G.W. (1987). Magnetic fields on K and M dwarfs, in *Cool Stars, Stellar Systems, and the Sun*, eds. J.L. Linsky and R.E. Stencel (Springer-Verlag, Berlin) pp. 36–37.

Brushlinskii, K.V., Zaborov, A.M., and Syrovatsky, S.I. (1980). Numerical analysis of the current sheet near a magnetic null line, *Sov. J. Plasma Phys.* **6**, 165–173.

Bryant, D.A. (1999). *Electron Acceleration in the Aurora and Beyond* (Inst. of Phys., Bristol).

Büchner, J., and Zelenyi, L.M. (1989). Regular and chaotic charged particle motion in magnetotail-like field reversals. 1. Basic theory of trapped motion, *J. Geophys. Res.* **94**, 11821–11842.

Büchner, J., and Zelenyi, L.M. (1991). Regular and chaotic particle motion in sheared magnetic field reversals, *Adv. Space Res.* **11**, 177–182.

Bulanov, S.V., and Frank, A.G. (1992). An approach to the experimental study of magnetic reconnection in three-dimensional magnetic configurations, *Sov. J. Plasma Phys.* **18**, 797–799.

Bulanov, S.V., and Olshanetsky, M.A. (1984). Magnetic collapse near zero points of the magnetic field, *Phys. Lett.* **100A**, 35–38.

Bulanov, S.V., and Sasorov, P.V. (1976). Energy spectrum of particles accelerated in the neighborhood of a neutral line, *Sov. Astron.* **19**, 464–468.

Bulanov, S.V., Syrovatsky, S.I., and Sakai, J. (1978). Stabilising influence of plasma flow on dissipative tearing instability, *Sov. Phys. JETP Lett.* **28**, 177–179.

Bulanov, S.V., Sakai, J., and Syrovatsky, S.I. (1979). Tearing-mode instability in approximately steady MHD configurations, *Sov. J. Plasma. Phys.* **5**, 157–163.

Bulanov, S.V., Shasharaina, S.G., and Pegoraro, F. (1990). MHD modes near the X-line of a magnetic configuration, *Plasma Phys. Control. Fusion* **32**, 377–389.

Bungey, T.N., and Priest, E.R. (1995). Current sheet configurations in potential and force-free fields, *Astron. Astrophys.* **293**, 215–224.

Bungey, T.N., Titov, V.S., and Priest, E.R. (1996). Basic topological elements of coronal magnetic fields, *Astron. Astrophys.* **308**, 233–247.

Burm, H., and Kuperus, M. (1988). Accretion disk coronae and the contribution to angular momentum transport in the disk, *Astron. Astrophys.* **192**, 165–169.

Byrne, P.B. (1995). Flares in late-type stars, UV, in *Flares and Flashes*, IAU Colloq. 151, eds. J. Greiner, H.W. Duerbeck, and R.E. Gershberg (Springer, Berlin) pp. 137–145.

Byrne, P.B., Eibe, M.T., and Rolleston, W.R.J. (1996). Cool prominences in the corona of the rapidly rotating dMe star HK Aquarii, *Astron. Astrophys.* **311**, 651–660.

Cameron, A.C. (1995). New limits on starspot lifetimes for AB Doradus, *Mon. Not. Roy. Astron. Soc.* **275**, 534–544.

Cameron, A.C. (1996). Stellar prominences, in *Stellar Surface Structure*, IAU Symp. 176, eds. K.G. Strassmeier and J.L. Linsky (Kluwer, Dordrecht) pp. 449–460.

Cameron, A.G.W. (1994). Astrophysical processes contributing to the formation of meteoritic components, *Meteoritics* **29**, 454.

Canfield, R.C., Cheng, C.-C., Dere, K.P., Dulk, G.A., McLean, D.J., Robinson, R.D., Jr., Schmahl, E.J., and Schoolman, S.A. (1980). Radiative energy output of the 5 September 1973 flare, in *Solar Flares: A Monograph from the Skylab Solar Workshop II*, ed. P.A. Sturrock (Colorado Assoc. Univ. Press, Boulder) pp. 451–469.

Canfield, R.C., Reardon, K.P., Leka, K.D., Shibata, K., Yokoyama, T., and Shimojo, M. (1996). Hα surges and X-ray jets in AR7260, *Astrophys. J.* **464**, 1016–1029.

Caramana, E.J., Nebel, R.A., and Schnack, D.D. (1983). Nonlinear, single helicity magnetic reconnection in the reversed-field pinch, *Phys. Fluids* **26**, 1305–1319.

Cargill, P.J., and Priest, E.R. (1982). Slow shock heating and the Kopp-Pneuman model for post-flare loops, *Solar Phys.* **76**, 357–375.

Cargill, P.J., Mariska, J.T., and Antiochos, S.K. (1995). Cooling of solar flare plasmas. I. Theoretical considerations, *Astrophys. J.* **439**, 1034–1043.

Carmichael, H. (1964). A process for flares, in *Physics of Solar Flares*, ed. W.N. Hess (NASA SP-50, Washington, DC) pp. 451–456.

Carreras, B., Hicks, H.R., and Lee, D.K. (1981). Effects of coupling on the stability of tearing modes, *Phys. Fluids* **24**, 66–77.

Catalano, S. (1996). Flares on active binary systems, in *Magnetohydrodynamic Phenomena in the Solar Atmosphere, Prototypes of Stellar Magnetic Activity*, IAU Colloq. 153, eds. Y. Uchida, T. Kosugi, and H.S. Hudson (Kluwer, Dordrecht) pp. 227–234.

Chae, J., Schühle, U., and Lemaire, P. (1998). SUMER measurements of nonthermal motions: constraints on coronal heating mechanisms, *Astrophys. J.* **505**, 957–973.

Chandrasekhar, S. (1961). *Hydrodynamic and Hydromagnetic Stability* (Dover, New York).

Chapman, S., and Ferraro, V.C.A. (1931). A new theory of magnetic storms, *Magn. Atmos. Elec.* **36**, 77–97.

Chapman, S., and Kendall, P.C. (1963). Liquid instability and energy transformation near a magnetic neutral line: A soluble non-linear hydromagnetic problem, *Proc. Roy. Soc. Lond.* **271**, 435–448.

Chapman, S., and Kendall, P.C. (1966). Comment on "Some exact solutions of magnetohydrodynamics," *Phys. Fluids* **9**, 2306–2307.

Chen, J., and Palmadesso, P.J. (1986). Chaos and nonlinear dynamics of single-particle orbits in a magnetotail-like magnetic field, *J. Geophys. Res.* **91**, 1499–1508.

Cheng, A.F. (1979). Unsteady magnetic merging in one dimension, *J. Geophys. Res.* **84**, 2129–2134.

Cheng, C.-C. (1983). Numerical simulations of loops heated to solar flare temperatures. I. Gas dynamics, *Astrophys. J.* **265**, 1090–1102.

Cheng, C.-C., and Pallavicini, R. (1991). Numerical simulations of flares on M dwarf stars: Hydrodynamics and coronal X-ray emission, *Astrophys. J.* **381**, 234–249.

Chew, G.F., Goldberger, M.L., and Low, F.E. (1956). The Boltzmann equation and the one-fluid hydromagnetic equations in the absence of particle collisions, *Proc. Roy. Soc. Lond.* **236**, 112–118.

Chodura, R., and Schlüter, A. (1981). A 3D code for MHD equilibrium and stability, *J. Comp. Phys.* **41**, 68–88.

Chupp, E.L. (1984). High energy neutral radiations from the Sun, *Ann. Rev. Astron. Astrophys.* **22**, 359–387.

Clark, A. (1964). Production and dissipation of magnetic energy by differential fluid motions, *Phys. Fluids* **7**, 1299–1305.

Clauer, C.R., and McPherron, R.L. (1974). Mapping the local time – universal time development of magnetospheric substorms using mid-latitude magnetic observations, *J. Geophys. Res.* **79**, 2811–2820.

Clayton, D.D., and Jin, L. (1995). Interpretation of 26Al in meteoritic inclusions, *Astrophys. J.* **451**, L87–L91.

Cliver, E., and Kahler, S. (1991). High coronal flares and impulsive acceleration of solar energetic particles, *Astrophys. J.* **366**, L91–L94.

Cohen, R.H. (1976). Runaway electrons in an impure plasma, *Phys. Fluids* **19**, 239–244.

Cook, J.W., Keenan, F.P., Dufton, P., Kingston, A.E., Pradhan, A.K., Zhang, H. L., Doyle, J.G., and Hayes, M.A. (1995). The OIV and SIV intercombination lines in solar and stellar ultraviolet spectra, *Astrophys. J.* **444**, 936–942.

Coppi, B., Galvao, R., Pellat, R., Rosenbluth, M.N., and Rutherford, P.M. (1976). Resistive internal kink modes, *Sov. J. Plasma Phys.* **2**, 533–535.

Cornwall, J.M. (1986). Magnetospheric ion acceleration processes, in *Ion Acceleration in the Magnetosphere and Ionosphere*, Geophys. Monograph 38, ed. T. Chang (Amer. Geophys. Union, Washington, DC) pp. 3–16.

Coroniti, F.V. (1981). On the magnetic viscosity in Keplerian accretion disks, *Astrophys. J.* **244**, 587–599.

Coroniti, F.V., and Kennel, C.F. (1973). Can the ionosphere regulate magnetospheric convection?, *J. Geophys. Res.* **78**, 2837–2851.

Coroniti, F.V., and Kennel, C.F. (1979). Magnetospheric reconnection, substorms, and energetic particle acceleration, in *Particle Acceleration in Planetary Magnetospheres*, eds. J. Arons, C. Max, and C. McKee (Amer. Inst. Phys., New York) pp. 169–178.

Cowley, S.W.H. (1973a). A self-consistent model of a simple magnetic neutral sheet system surrounded by a cold, collisionless plasma, *Cosmic Electrodynamics* **3**, 448–501.

Cowley, S.W.H. (1973b). A qualitative study of the reconnection between the Earth's magnetic field and an interplanetary field of arbitrary orientation, *Radio Science* **8**, 903–913.

Cowley, S.W.H. (1974a). Convection region solutions for the reconnection of antiparallel magnetic fields of unequal magnitude, *J. Plasma Phys.* **12**, 341–352.

Cowley, S.W.H. (1974b). On the possibility of magnetic fields and fluid flows parallel to the X-line in a reconnection geometry, *J. Plasma Phys.* **12**, 319–339.

Cowley, S.W.H. (1978). A note on the motion of charged particles in one-dimensional magnetic current sheets, *Planet. Space Sci.* **26**, 539–545.

Cowley, S.W.H. (1980). Plasma populations in a simple open model magnetosphere, *Space Sci. Rev.* **26**, 217–275.

Cowley, S.W.H. (1984). The distant geomagnetic tail in theory and observation, in *Magnetic Reconnection in Space and Laboratory Plasmas*, ed. E.W. Hones, Jr. (Amer. Geophys. Union, Washington, DC) pp. 228–239.

Cowley, S.W.H. (1985). Magnetic reconnection, in *Solar System Magnetic Fields*, ed. E.R Priest (D. Reidel, Dordrecht) pp. 121–155.

Cowley, S.W.H., and Southwood, D.J. (1980). Some properties of a steady state geomagnetic tail, *Geophys. Res. Lett.* **7**, 833–836.

Cowling, T.G. (1953). Solar electrodynamics, in *The Sun*, ed. G.P. Kuiper (Univ. Chicago Press, Chicago) pp. 532–591.

Cowling, T.G. (1965). General aspects of stellar and solar magnetic fields, in *Solar and Stellar Magnetic Fields*, ed. R. Lüst (North-Holland Publishing Co., Amsterdam) pp. 405–417.

Cowsik, R. (1979). Evolution of the radio spectrum of Cassiopeia A, *Astrophys. J.* **227**, 856–862.

Cowsik, R., and Lee, M.A. (1982). Transport of neutrinos, radiation and energetic particles in accretion flows, *Proc. Roy. Soc. Lond. A* **383**, 409–437.

Cox, D.P. (1972). Theoretical structure and spectrum of a shock wave in the interstellar medium: The Cygnus loop, *Astrophys. J.* **178**, 143–157.

Craig, I.J.D., and Fabling, R.B. (1996). Exact solutions for steady-state spine and fan magnetic reconnection, *Astrophys. J.* **462**, 969–976.

Craig, I.J.D., and Henton, S.M. (1995). Exact solutions for steady-state incompressible magnetic reconnection, *Astrophys. J.* **450**, 280–288.

Craig, I.J.D., and McClymont, A.N. (1991). Dynamic magnetic reconnection at an X-type neutral point, *Astrophys. J.* **371**, L41–L44.

Craig, I.J.D., and McClymont, A.N.M. (1993). Linear theory of fast reconnection at an X-type neutral point, *Astrophys. J.* **405**, 207–215.

Craig, I.J.D., and Rickard, G.J. (1994). Linear models of steady state, incompressible magnetic reconnection, *Astron. Astrophys.* **287**, 261–267.

Craig, I.J.D., and Watson, P.G. (1992). Fast dynamic reconnection at X-type neutral points, *Astrophys. J.* **393**, 385–395.

Craig, I.J.D., Fabling, R.B., Henton, S.M., and Rickard, G.J. (1995). An exact
    solution for steady-state magnetic reconnection in 3D, *Astrophys. J.*
    455, L197–L199.
Crooker, N.U. (1979). Dayside merging and cusp geometry, *J. Geophys. Res.*
    84, 951–959.
Crooker, N.U., Lyon, J.G., and Fedder, J.A. (1998). MHD model merging with
    IMF B_y: Lobe cells, sunward polar cap convection, and overdraped lobes,
    J. Geophys. Res. **103**, 9143–9151.
Crooker, N.U., Siscoe, G.L., and Toffoletto, F.R. (1990). A tangent subsolar
    merging line, *J. Geophys. Res.* **95**, 3787–3793.
Cross, M.A., and Van Hoven, G. (1976). High-conductivity magnetic tearing
    instability, *Phys. Fluids* **19**, 1591–1595.
Daglis, I.A., Axford, W.I., Sarris, E.T., Livi, S., and Wilken, B. (1997). Particle
    acceleration in geospace and its association with solar events, *Solar Phys.*
    172, 287–296.
Dahlburg, R.B., Antiochos, S.K., and Zang, T.A. (1991). Dynamics of solar
    coronal magnetic fields, *Astrophys. J.* **383**, 420–430.
Dahlburg, R.B., Antiochos, S.K., and Zang, T.A. (1992). Secondary instability in
    three-dimensional magnetic reconnection, *Phys. Fluids B* **4**, 3902–3914.
Dahlburg, R.B., Antiochos, S.K., and Norton, D. (1997). Magnetic flux-tube
    tunneling, *Phys. Rev. E* **56**, 2094–2103.
Das, A.C. (1992). Lower hybrid turbulence and tearing-mode instability in
    magnetospheric plasma, *J. Geophys. Res.* **97**, 12275–12277.
Davis, L. (1956). Modified Fermi mechanism for the acceleration of cosmic rays,
    Phys. Rev. **101**, 351–358.
Decker, R.B. (1988). Role of drifts in diffusive shock acceleration, *Astrophys. J.*
    324, 566–573.
de Jager, C., and de Jonge, G. (1978). Properties of elementary flare bursts, *Solar
    Phys.* **58**, 127–137.
de Jager, C., Heise, J., and van Genderen, A.M. (1989). Coordinated observations
    of a large impulsive flare on UV Ceti, *Astron. Astrophys.* **211**, 157–172.
de la Beaujardière, O., Lyons, L.R., and Friis–Christensen, E. (1991).
    Sondrestrom radar measurements of reconnection electric fields, *J. Geophys.
    Res.* **96**, 13907–13912.
Delcourt, D.C., Chappell, C.R., Moore, T.E., and Waite, J.H., Jr. (1989).
    A three-dimensional numerical model of ionospheric plasma in the
    magnetosphere, *J. Geophys. Res.* **94**, 11893–11920.
Delva, M., Schwingenschuh, K., Niedner, M.B., and Gringauz, K.I. (1991). Comet
    Halley remote plasma tail observations and in situ solar wind properties:
    Vega-1/2 IMF/plasma observations and ground-based optical observations
    from 1 December 1985 to 1 May 1986, *Planet. Space Sci.* **39**, 697–708.
Démoulin, P., and Priest, E.R. (1997). The importance of photospheric intense
    flux tubes for coronal heating, *Solar Phys.* **175**, 123–155.
Démoulin, P., Priest, E.R., and Ferreira, J. (1991). Instability of a prominence
    supported in a linear force-free field, *Astron. Astrophys.* **245**, 289–298.
Démoulin, P., Priest, E.R., and Lonie, D.P. (1996a). 3D magnetic reconnection
    without null points. 2. Application to twisted flux tubes, *J. Geophys.
    Res.* **101**, 7631–7646.
Démoulin, P., Hénoux, J.C., Priest, E.R., and Mandrini, C.H. (1996b). Quasi-
    separatrix layers in solar flares. I. Method, *Astron. Astrophys.* **308**, 643–655.

Dennis, B.R. (1988). Solar flare hard X-ray observations, *Solar Phys.* **118**, 49–94.

Dere, K.P., Bartoe, J.D.F., Brueckner, G.E., Ewing, J., and Lund, P. (1991). Explosive events and magnetic reconnection in the solar atmosphere, *J. Geophys. Res.* **96**, 9399–9407.

Desch, M.D., Farrell, W.M., Kaiser, M.L., Lepping, R.P., Steinberg, J.T., and Villanueva, L.A. (1991). The role of solar wind reconnection in driving the Neptune radio emission, *J. Geophys. Res.* **96**, 19111–19116.

Diamond, P.H., Hazeltine, R.D., An, Z.G., Carreras, B.A., and Hicks, H.R. (1984). Theory of anomalous tearing mode growth and the major tokamak disruption, *Phys. Fluids* **27**, 1449–1462.

Dixon, A.M., Browning, P.K., and Priest, E.R. (1988). Coronal heating by relaxation in a sunspot magnetic field, *Geophys. Astrophys. Fluid Dynamics* **40**, 293–327.

Dobrowolny, M., Veltri, P., and Mangeney, A. (1983). Dissipative instabilities of magnetic neutral layers with velocity shear, *J. Plasma Phys.* **29**, 393–407.

Dodson, H.W., and Hedeman, E.R. (1968). The proton flare of August 28, 1966, *Solar Phys.* **4**, 229–239.

Donati, J.-F., and Cameron, A.C. (1997). Differential rotation and magnetic polarity patterns on AB Doradus, *Mon. Not. Roy. Astron. Soc.* **291**, 1–19.

Doschek, G.A., Cheng, C.C., Oran, E.S., Boris, J.P., and Mariska, J.T. (1983). Numerical simulations of loops heated to solar flare temperatures. II. X-ray and UV spectroscopy, *Astrophys. J.* **265**, 1103–1119.

Doschek, G.A., Strong, K.T., and Tsuneta, S. (1995). The bright knots at the tops of soft X-ray loops: quantitative results from YOHKOH, *Astrophys. J.* **440**, 370–385.

Drake, J.F., Biskamp, D., and Zeiler, A. (1997). Breakup of the electron current layer during 3D collisionless magnetic reconnection, *Geophys. Res. Lett.* **24**, 2921–2924.

Dreicer, H. (1959). On the theory of run-away electrons, in *Plasma Physics and Thermonuclear Research* 1, eds. C. Longmire, J.L. Tuck, and W.B. Thompson (Pergamon, New York) pp. 491–507.

Dreicer, H. (1960). Electron and ion runaway in a fully ionized gas. II., *Phys. Rev.* **117**, 329–342.

Drummond, W.E., and Pines, D. (1962). Nonlinear stability of plasma oscillations, *Nuclear Fusion Suppl.* **3**, 1049–1057.

Drury, L.O'C. (1983). An introduction to the theory of diffusive shock acceleration of energetic particles in tenuous plasmas, *Rep. Prog. Phys.* **46**, 973–1027.

Drury, L.O'C. (1991). Time-dependent diffusive acceleration of test particles at shocks, *Mon. Not. Roy. Astron. Soc.* **251**, 340–350.

Dubois, M.A., and Samain, A. (1980). Evolution of magnetic islands in tokamaks, *Nuclear Fusion* **20**, 1101–1109.

Dungey, J.W. (1953). Conditions for the occurrence of electrical discharges in astrophysical systems, *Phil. Mag.* **44**, 725–738.

Dungey, J.W. (1958). The neutral point discharge theory of solar flares. A reply to Cowling's criticism, in *Electromagnetic Phenomena in Cosmical Physics*, IAU Symp. 6, ed. B. Lehnert (Cambridge Univ. Press, Cambridge) pp. 135–140.

Dungey, J.W. (1961). Interplanetary magnetic field and the auroral zones, *Phys. Rev. Lett.* **6**, 47–48.

Dungey, J.W. (1965). The length of the magnetospheric tail, *J. Geophys. Res.* **70**, 1753.

Dungey, J.W. (1994). Memories, maxims and motives, *J. Geophys. Res.* **99**, 19189–19197.

Eardley, D.M., and Lightman, A.P. (1975). Magnetic viscosity in relativistic accretion disks, *Astrophys. J.* **200**, 187–203.

Eastwood, J.W. (1972). Consistency of fields and particle motion in the Speiser model of the current sheet, *Planet. Space Sci.* **20**, 1555–1568.

Edenstrasser, J.W., and Schuurman, W. (1983). Axisymmetric finite-beta minimum energy equilibria of weakly toroidal discharges, *Phys. Fluids* **26**, 500–507.

Einaudi, G., and Rubini, F. (1989). Resistive instabilities in a flowing plasma. III. Effects of viscosity, *Phys. Fluids B* **1**, 2224–2228.

Einaudi, G., and Van Hoven, G. (1983). The stability of coronal loops: finite-length and pressure-profile limits, *Solar Phys.* **88**, 163–177.

Einaudi, G., Lionello, R., and Velli, M. (1997). Magnetic reconnection in solar coronal loops, *Adv. Space. Res.* **19**, 1875–1878.

Ejiri, A., and Miyamoto, K. (1995). Ion-heating model during magnetic reconnection in reversed-field pinch plasmas, *Plasma Phys. Control. Fusion* **37**, 43–56.

Elliott, J.A. (1993). Plasma kinetic theory, in *Plasma Physics*, ed. R. Dendy (Cambridge Univ. Press, Cambridge) pp. 29–53.

Ellison, D.C., and Ramaty, R. (1985). Shock acceleration of electrons and ions in solar flares, *Astrophys. J.* **298**, 400–408.

Emslie, A.G., Brown, J.C., and Machado, M.E. (1981). Discrepancies between theoretical and empirical models of the flaring solar chromosphere and their possible resolution, *Astrophys. J.* **246**, 337–343.

Erickson, G.M., and Wolf, R.A. (1980). Is steady convection possible in the Earth's magnetotail?, *Geophys. Res. Lett.* **7**, 897–900.

Fabling, R.B., and Craig, I.J.D. (1996). Exact solutions for steady-state planar magnetic reconnection in an incompressible viscous plasma, *Phys. Plasmas* **3**, 2243–2247.

Fairfield, D.H., and Cahill, L.J., Jr. (1977). Transition-region magnetic field and polar magnetic disturbances, *J. Geophys. Res.* **71**, 155–170.

Falconer, D.A., Moore, R.L., and Porter, J.G. (1999). Micro-coronal bright points observed in the quiet magnetic network by SOHO/EIT, in *High-Resolution Solar Atmospheric Dynamics*, eds. J.A. Bookbinder, E.F. DeLuca, and L. Golub (Astron. Soc. Pacific, Provo Utah) in press.

Fedder, J.A., Mobarry, C.M., and Lyon, J.G. (1991). Reconnection voltage as a function of IMF clock angle, *Geophys. Res. Lett.* **18**, 1047–1050.

Feldman, W.C. et al. (1984). Evidence for slow-mode shocks in the deep geomagnetic tail, *Geophys. Res. Lett.* **11**, 599–602.

Fermi, E. (1949). On the origin of the cosmic radiation, *Phys. Rev.* **75**, 1169–1174.

Finn, J.M. (1975). The destruction of magnetic surfaces in tokamaks by current perturbations, *Nuclear Fusion* **15**, 845–854.

Finn, J.M., and Antonsen, T.M. (1983). Turbulent relaxation of compressible plasmas with flow, *Phys. Fluids* **26**, 3540–3552.

Finn, J.M., and Antonsen, T.M. (1985). Magnetic helicity: what is it and what is it good for?, *Comments Plasma Phys. Control. Fusion* **9**, 111–120.

Finn, J.M., and Guzdar, P.N. (1991). Formation of a flux-core spheromak, *Phys. Fluids B* **3**, 1041–1051.

Finn, J.M., and Kaw, P.K. (1977). Coalescence instability of magnetic islands, *Phys. Fluids* **20**, 72–78.

Finn, J.M., and Manheimer, W.M. (1982). Resistive interchange modes in reversed-field pinches, *Phys. Fluids* **25**, 697–701.

Finn, J.M., and Sovinec, C.R. (1998). Nonlinear tearing modes in the presence of resistive wall and rotation, *Phys. Plasmas* **5**, 461–480.

Finn, J.M., Guzdar, P.N., and Chen, J. (1992). Fast plasmoid formation in double arcades, *Astrophys. J.* **393**, 800–814.

Fisher, G.H., Canfield, R.C., and McClymont, A.N. (1985). Flare loop radiative hydrodynamics, *Astrophys. J.* **289**, 414–441.

Fisk, L.A. (1976). The acceleration of energetic particles in the interplanetary medium by transit time damping, *J. Geophys. Res.* **81**, 4633–4640.

Fisk, L.A. (1978). *He*-rich flares: a possible explanation, *Astrophys. J.* **224**, 1048–1055.

Fontenla, J.M. (1993). Generation of electric currents in two-dimensional magnetic nulls, *Astrophys. J.* **419**, 837–854.

Forbes, T.G. (1982). Implosion of a uniform current sheet in a low-beta plasma, *J. Plasma Phys.* **27**, 491–505.

Forbes, T.G. (1986). Fast-shock formation in line-tied magnetic reconnection models of solar flares, *Astrophys. J.* **305**, 553–563.

Forbes, T.G. (1988). Magnetohydrodynamic boundary conditions for global models, in *Modeling Magnetospheric Plasma*, eds. T.E. Moore and J.J. Waite, Jr. (Amer. Geophys. Union, Washington, DC) pp. 319–328.

Forbes, T.G. (1990). Numerical simulation of a catastrophe model for coronal mass ejections, *J. Geophys. Res.* **95**, 11919–11931.

Forbes, T.G. (1991). Magnetic reconnection in solar flares, *Geophys. Astrophys. Fluid Dynamics* **62**, 15–36.

Forbes, T.G., and Acton, L.W. (1996). Reconnection and field line shrinkage in solar flares, *Astrophys. J.* **459**, 330–341.

Forbes, T.G., and Isenberg, P.A. (1991). A catastrophe mechanism for a coronal mass ejection, *Astrophys. J.* **373**, 294–307.

Forbes, T.G., and Malherbe, J.M. (1986). A shock-condensation mechanism for loop prominences, *Astrophys. J.* **302**, L67–L70.

Forbes, T.G., and Malherbe, J.M. (1991). A numerical simulation of magnetic reconnection and radiative cooling in line-tied current sheets, *Solar Phys.* **135**, 361–391.

Forbes, T.G., and Priest, E.R. (1982a). Neutral line motion due to reconnection in two-ribbon solar flares and geomagnetic substorms, *Planet. Space Sci.* **30**, 1183–1197.

Forbes, T.G., and Priest, E.R. (1982b). A numerical study of line-tied magnetic reconnection, *Solar Phys.* **81**, 303–324.

Forbes, T.G., and Priest, E.R. (1984a). Numerical simulation of reconnection in an emerging magnetic flux region, *Solar Phys.* **94**, 315–340.

Forbes, T.G., and Priest, E.R. (1984b). Reconnection in solar flares, in *Solar Terrestrial Physics: Present and Future*, eds. D.M. Butler and K. Papadopoulous (NASA RP-1120, Washington, DC) pp. 1–35.

Forbes, T.G., and Priest, E.R. (1987). A comparison of analytical and numerical models for steadily driven reconnection, *Rev. Geophys.* **25**, 1583–1607.

Forbes, T.G., and Priest, E.R. (1995). Photospheric magnetic field evolution and eruptive flares, *Astrophys. J.* **446**, 377–389.

Forbes, T.G., and Speiser, T.W. (1979). Temporal evolution of magnetic reconnection in the vicinity of a magnetic neutral line, *J. Plasma Phys.* **21**, 107–126.

Forbes, T.G., Hones, E.W., Jr., Bame, S.J., Asbridge, J.R., Paschmann, G., Sckopke, N., and Russell, C.T. (1981). Evidence for the tailward retreat of a magnetic neutral line in the magnetotail during substorm recovery, *Geophys. Res. Lett.* **8**, 261–264.

Forbes, T.G., Priest, E.R., and Hood, A.W. (1982). Evolution of current sheets following the onset of enhanced resistivity, *J. Plasma. Phys.* **27**, 157–176.

Forbes, T.G., Malherbe, J.M., and Priest, E.R. (1989). The formation of flare loops by magnetic reconnection and chromospheric ablation, *Solar Phys.* **120**, 285–307.

Forman, M.A., Ramaty, R., and Zweibel, E.G. (1986). The acceleration and propagation of solar flare energetic particles, in *Physics of the Sun*, ed. P.A. Sturrock (Reidel, Dordrecht) pp. 249–289.

Formisano, V., and Amata, E. (1975). Evidence for magnetic field-line reconnection in the solar wind, in *The Magnetospheres of the Earth and Jupiter*, ed. V. Formisano (Reidel, Dordrecht) pp. 205–217.

Foster, J.C., Fairfield, D.H., Ogilvie, K.W., and Rosenberg, T.J. (1971). Relationship of interplanetary parameters and occurrence of magnetospheric substorms, *J. Geophys. Res.* **76**, 6971–6975.

Fredrickson, E.D., McGuire, K.M., Bell, M.G., Bush, C.E., Budny, R.V., Janos, A.C., Mansfield, D.K., Nagayama, Y., Park, H.K., Schivell, J.F., Taylor, G., Zarnstorff, M.C., Drake, J.F., and Kleva, R. (1993). Phenomenology of high-density disruptions in the TFTR-tokamak, *Nuclear Fusion* **33**, 141–146.

Frisch, U., Pouquet, A., Leorat, J., and Mazure, A. (1975). Field opening and reconnection, *J. Fluid Mech.* **68**, 769–778.

Furth, H.P., Killeen, J., and Rosenbluth, M.N. (1963). Finite-resistivity instabilities of a sheet pinch, *Phys. Fluids* **6**, 459–484.

Furth, H.P., Rutherford, P.M., and Selberg, H. (1973). Tearing mode in the cylindrical tokamak, *Phys. Fluids* **16**, 1054–1063.

Gahm, G.F. (1994). Flares in T-Tauri stars, in *Flares and Flashes*, IAU Colloq. 151, eds. J. Greiner, H.W. Duerbeck, and R.E. Gershberg (Springer-Verlag, Berlin) pp. 203–211.

Gaisser, T.K. (1990). *Cosmic Ray Particles and Particle Physics* (Cambridge Univ. Press, Cambridge).

Galeev, A.A. (1984). Spontaneous reconnection of magnetic field lines in a collisionless plasma, in *Handbook of Plasma Physics* **2**, eds. M.N. Rosenbluth and R.Z. Sagdeev (North-Holland, Amsterdam) pp. 305–336.

Galeev, A.A., and Sagdeev, R.Z. (1984). Current instabilities and anomalous resistivity of plasma, in *Handbook of Plasma Physics* **2**, eds. M.N. Rosenbluth and R.Z. Sagdeev (North-Holland, Amsterdam) pp. 271–303.

Galeev, A.A., and Zelenyi, L.M. (1975). Metastable states of diffuse neutral sheet and the substorm explosive phase, *Sov. Phys. JETP Lett.* **22**, 170–172.

Galeev, A.A., Rosner, R., and Vaiana, G.S. (1979). Structured coronae of accretion disks, *Astrophys. J.* **229**, 318–326.

Galsgaard, K., and Nordlund, Å. (1996). Heating and activity of the solar corona. 1. Boundary shearing of an initially homogeneous magnetic field, *J. Geophys. Res.* **101**, 13445–13460.

Galsgaard, K., and Nordlund, Å. (1997a). Heating and activity of the solar corona. 2. Kink instability in a flux tube, *J. Geophys. Res.* **102**, 219–230.

Galsgaard, K., and Nordlund, Å. (1997b). Heating and activity of the solar corona. 3. Dynamics of a low-beta plasma with three-dimensional null points, *J. Geophys. Res.* **102**, 231–248.

Galsgaard, K., MacKay, D.H., Priest, E.R., and Nordlund, Å. (1999). On the location of energy release and temperature profiles along coronal loops, *Solar Phys.*, in press.

Garren, D.A., and Chen, J. (1994). Lorentz self-forces on curved current loops, *Phys. Plasmas* **1**, 3425–3436.

Gekelman, W., and Pfister, H. (1988). Experimental observations of the tearing of an electron current sheet, *Phys. Fluids* **31**, 2017–2025.

Gekelman, W., and Stenzel, R.L. (1984). Magnetic field line reconnection experiments. 6. Magnetic turbulence, *J. Geophys. Res.* **89**, 2715–2733.

Gekelman, W., Stenzel, R.L., and Wild, N. (1982). Magnetic field-line reconnection experiments. 3. Ion acceleration, flows and anomalous scattering, *J. Geophys. Res.* **87**, 101–110.

Gershberg, R.E. (1983). On activities of UV Ceti-type flare stars and of T-Tau-type stars, in *Activity in Red-Dwarf Stars*, IAU Colloq. 71, eds. P.B. Byrne and M. Rodonò (Reidel, Dordrecht) pp. 487–495.

Giovanelli, R.G. (1946). A theory of chromospheric flares, *Nature* **158**, 81–82.

Gleeson, L.J., and Axford, W.I. (1967). Cosmic rays in the interplanetary medium, *Astrophys. J. Lett.* **149**, L115–L118.

Goedbloed, J.P. (1983). *Lecture Notes on Ideal MHD*, Rijnhuizen Report 83–145, Niewegein, Netherlands.

Goedheer, W.J., and Westerhof, E. (1988). Sawtooth, transport and cyclotron heating in T-10, *Nuclear Fusion* **28**, 565–576.

Gold, T. (1959). Motions in the magnetosphere of the Earth, *J. Geophys. Res.* **64**, 1219–1224.

Golub, L., Zirin, H., and Wang, H. (1994). The roots of coronal structure in the Sun's surface, *Solar Phys.* **153**, 179–198.

Gomez, D.O., and Ferro Fontan, C. (1988). Coronal heating by selective decay of MHD turbulence, *Solar Phys.* **116**, 33–44.

Gomez, D.O., Martens, P.C.H., and Golub, L. (1993). NIXT power spectra of X-ray emission from solar active regions. 1. Observations, *Astrophys. J.* **405**, 767–772.

Goossens, M. (1991). MHD waves and wave heating in nonuniform plasmas, in *Advances in Solar System MHD*, eds. E.R. Priest and A.W. Hood (Cambridge Univ. Press, Cambridge) pp. 137–172.

Gosling, J.T. (1975). Large-scale inhomogeneities in the solar wind of solar origin, *Rev. Geophys. Space Phys.* **13**, 1053–1076.

Gosling, J.T. (1990). Coronal mass ejections and magnetic flux ropes in interplanetary space, in *Physics of Magnetic Flux Ropes*, eds. C.T. Russell, E.R. Priest, and L.C. Lee (Amer. Geophys. Union, Washington, DC) pp. 343–364.

Gosling, J.T. (1993). The solar flare myth, *J. Geophys. Res.* **98**, 18937–18949.

Gosling, J.T. (1996). Magnetic topologies of coronal mass ejection events: Effects of three-dimensional reconnection, in *Solar Wind Eight*, eds. D. Winterhalter, J.T. Gosling, S.R. Habbal, W.S. Kurth, and M. Neugebauer (Amer. Inst. Phys., New York) pp. 438–444.

References

Gosling, J.T., Birn, J., and Hesse, M. (1995). Three-dimensional magnetic reconnection and the magnetic topology of coronal mass ejection events, *Geophys. Res. Lett.* **22**, 869–872.

Gratton, F.T., Heyn, M.F., Biernat, H.K., Rijnbeek, R.P., and Gnavi, G. (1988). MHD stagnation-point flows in the presence of resistivity and viscosity, *J. Geophys. Res.* **93**, 7318–7324.

Gray, P.C., and Matthaeus, W.H. (1991). MHD turbulence, reconnection, and test-particle acceleration, in *Particle Acceleration in Cosmic Plasmas*, eds. G.P. Zank and T.K. Gaisser (Amer. Inst. Phys., New York) pp. 261–266.

Green, R.M. (1965). Modes of annihilation and reconnection of magnetic fields, in *Solar and Stellar Magnetic Fields*, ed. R. Lüst (North-Holland Publishing Co., Amsterdam) pp. 398–404.

Greene, J.M. (1988). Geometrical properties of three-dimensional reconnecting magnetic fields with nulls, *J. Geophys. Res.* **93**, 8583–8590.

Greene, J.M. (1993). Reconnection of vorticity lines and magnetic lines, *Phys. Fluids B* **5**, 2355–2362.

Habbal, S.R., and Grace, E. (1991). The connection between coronal bright points and the variability of the quiet-Sun extreme-ultraviolet emission, *Astrophys. J.* **382**, 667–676.

Hada, T., and Kennel, C.F. (1985). Nonlinear evolution of slow waves in the solar wind, *J. Geophys. Res.* **90**, 531–535.

Haerendel, G., Paschmann, G., Sckopke, N., and Rosenbauer, H. (1978). The frontside boundary layer of the magnetosphere and the problem of reconnection, *J. Geophys. Res.* **82**, 3195–3216.

Hahm, T.S., and Kulsrud, R.M. (1985). Forced magnetic reconnection, *Phys. Fluids* **28**, 2412–2418.

Haisch, B.M. (1983). X-ray observations of stellar flares, in *Activity in Red-Dwarf Stars*, IAU Colloq. 71, eds. P.B. Byrne and M. Rodonò (Reidel, Dordrecht) pp. 255–268.

Haisch, B.M. (1989). An overview of solar and stellar flare research, *Solar Phys.* **121**, 3–18.

Haisch, B.M. (1996). Stellar X-ray flares, in *Magnetodynamic Phenomena in the Solar Atmosphere, Prototypes of Stellar Magnetic Activity*, IAU Colloq. 153, eds. Y. Uchida, T. Kosugi, and H.S. Hudson (Kluwer, Dordrecht) pp. 235–242.

Haisch, B.M., and Schmitt, J.H.M.M. (1996). Advances in solar-stellar astrophysics, *Pub. Astron. Soc. Pacific* **108**, 113–129.

Haisch, B.M., Strong, K.T., and Rodonò, M. (1991). Flares on the Sun and other stars, *Ann. Rev. Astron. Astrophys.* **29**, 275–324.

Hall, D.E., and Sturrock, P.A. (1967). Stochastic magnetic pumping, *Phys. Fluids* **10**, 1593–1595.

Hameiri, E., and Hammer, J.H. (1982). Turbulent relaxation of compressible plasmas, *Phys. Fluids* **25**, 1855–1862.

Hammer, J.H. (1984). Reconnection in spheromak formation and sustainment, in *Magnetic Reconnection in Space and Laboratory Plasmas*, ed. E.W. Hones, Jr. (Amer. Geophys. Union, Washington, DC) pp. 319–331.

Hammond, C.M., Feldman, W.C., and Phillips, J.L. (1996). Ulysses observations of double ion beams associated with coronal mass ejections, *Adv. Space Res.* **17**, 303–306.

Hargreaves, J.K. (1993). *The Solar-Terrestrial Environment* (Cambridge Univ. Press, Cambridge).

Harrison, R.A. (1995). The nature of solar flares associated with coronal mass ejections, *Astron. Astrophys.* **304**, 585–594.

Harrison, R.A. (1997). EUV Blinkers: The significance of variations in the extreme ultraviolet quiet Sun, *Solar Phys.* **175**, 467–485.

Harrison, R.A., Waggett, P.W., Bentley, R.D., Phillips, K.J.H., Bruner, M., Dryer, M., and Simnett, G.M. (1985). The X-ray signature of solar coronal mass ejections, *Solar Phys.* **97**, 387–400.

Hartmann, L. (1987). Stellar magnetic fields: Optical observations and analysis, in *Cool Stars, Stellar Systems and the Sun,* eds. J.L. Linsky and R.E. Stencel (Springer-Verlag, Berlin) pp. 1–9.

Harvey, K.L. (1985). The relationship between coronal bright points as seen in He I 10830 and the evolution of the photospheric network magnetic fields, *Australian. J. Phys.* **38**, 875–883.

Harvey, K.L., and Recely, F. (1984). He I 10830 observations of the 3N/M4.0 flare of 4 September 1982, *Solar Phys.* **91**, 127–139.

Hasan, S.S., and Ter Haar, D. (1978). The Alfvén–Carlquist double-layer theory of solar flares, *Astrophys. Space Sci.* **56**, 89–107.

Hassam, A.B. (1984). Collisional tearing in field-reversed configurations, *Phys. Fluids* **27**, 2877–2880.

Hassam, A.B. (1992). Reconnection of stressed magnetic fields, *Astrophys. J.* **399**, 159–163.

Hasselmann, K., and Wibberenz, G. (1968). Scattering of charged particles by random electromagentic fields. *Z. Geophys.* **34**, 353–88.

Hawley, J.F., and Balbus, S.A. (1991). A powerful local shear instability in weakly magnetized disks. II. Nonlinear evolution, *Astrophys. J.* **376**, 223–233.

Hawley, J.F., Gammie, C.F., and Balbus, S.A. (1996). Local three-dimensional simulations of an accretion disk hydromagnetic dynamo, *Astrophys. J.* **464**, 690–703.

Hayashi, M.R. (1981). Numerical simulations of forced coalescence, *J. Phys. Soc. Japan* **50**, 3124–3130.

Hayashi, M.R., Shibata, K., and Matsumoto, R. (1996). X-ray flares and mass outflows driven by magnetic interaction between a protostar and its surrounding disk, *Astrophys. J.* **468**, L37–L40.

Heikkila, W.J. (1978). Electric field topology near the dayside magnetopause, *J. Geophys. Res.* **83**, 1071–1078.

Hesse, M., and Birn, J. (1990). Parallel electric fields in a simulation of magnetotail reconnection and plasmoid evolution, in *Physics of Magnetic Flux Ropes*, eds. C.T. Russell, E.R. Priest, and L.C. Lee (Amer. Geophys. Union, Washington, DC) pp. 679–685.

Hesse, M., and Birn, J. (1993). Three-dimensional magnetotail equilibria by numerical relaxation techniques, *J. Geophys. Res.* **98**, 3973–3982.

Hesse, M., and Schindler, K. (1988). A theoretical foundation of general magnetic reconnection, *J. Geophys. Res.* **93**, 5539–5567.

Hesse, M., Birn, J., and Schindler, K. (1990). On the topology of flux transfer events, *J. Geophys. Res.* **95**, 6549–6560.

Heyn, M.F. (1997). Rapid reconnection and compressible plasma waves, *Astrophys. Space Sci.* **256**, 343–348.

Heyn, M.F., and Pudovkin, M.L. (1993). A time-dependent model of magnetic field annihilation, *J. Plasma Phys.* **49**, 17–27.

Heyn, M.F., and Semenov, V.S. (1996). Rapid reconnection in compressible plasma, *Phys. Plasmas* **3**, 2725–2741.

Heyn, M.F., Biernat, H.K., Rijnbeek, R.P., and Semenov, V.S. (1988). The structure of reconnection layers, *J. Plasma Phys.* **40**, 235–252.

Heyvaerts, J. (1981). Particle acceleration in solar flares, in *Solar Flare MHD*, ed. E.R. Priest (Gordon and Breach, New York) pp. 429–555.

Heyvaerts, J., and Kuperus, M. (1978). The triggering of plasma turbulence during fast flux emergence in the solar corona, *Astron. Astrophys.* **64**, 219–234.

Heyvaerts, J., and Priest, E.R. (1976). Thermal evolution of current sheets and the flash phase of solar flares, *Solar Phys.* **47**, 223–231.

Heyvaerts, J., and Priest, E.R. (1983). Coronal heating by phase-mixed Alfvén waves, *Astron. Astrophys.* **117**, 220–234.

Heyvaerts, J., and Priest, E.R. (1984). Coronal heating by reconnection in DC current systems: a theory based on Taylor's hypothesis, *Astron. Astrophys.* **137**, 63–78.

Heyvaerts, J., and Priest, E.R. (1989). A model for a non-Keplerian magnetic accretion disc with a magnetically heated corona, *Astron. Astrophys.* **216**, 230–244.

Heyvaerts, J., and Priest, E.R. (1992). A self-consistent turbulent model for solar coronal heating, *Astrophys. J.* **390**, 297–308.

Heyvaerts, J., Priest, E.R., and Rust, D.M. (1977). An emerging flux model for the solar flare phenomenon, *Astrophys. J.* **216**, 123–137.

Heyvaerts, J., Bardou, A., and Priest, E.R. (1996a). Interaction of turbulent accretion disks with embedded magnetic fields, in *Solar and Astrophysical Magnetohydrodynamic Flows*, ed. K.C. Tsinganos (Kluwer, Dordrecht) pp. 659–672.

Heyvaerts, J., Priest, E.R., and Bardou, A. (1996b). Magnetic field diffusion in self-consistently turbulent accretion disks, *Astrophys. J.* **473**, 403–421.

Hide, R. (1979). On the magnetic flux linkage of an electrically conducting fluid, *Geophys. Astrophys. Fluid Dyn.* **12**, 171–176.

Hiei, E., and Hundhausen, A.J. (1996). Development of coronal helmet streamer of 24 January 1992, in *Magnetodynamic Phenomena in the Solar Atmosphere, Prototypes of Stellar Magnetic Activity*, eds. Y. Uchida, T. Kosugi, and H.S. Hudson (Kluwer, Dordrecht) pp. 125–126.

Hiei, E., Hundhausen, A.J., and Sime, D.G. (1993). Reformation of a coronal helmet streamer by magnetic reconnection after a coronal mass ejection, *Geophys. Res. Lett.* **20**, 2785–2788.

Hill, T. (1983). Magnetospheric models, in *Physics of the Jovian Magnetosphere*, ed. A.J. Dessler (Cambridge Univ. Press, Cambridge) pp. 353–394.

Hillier, R. (1984). *Gamma Ray Astronomy* (Clarendon Press, Oxford).

Hirayama, T. (1974). Theoretical model of flares and prominences. I. Evaporating flare model, *Solar Phys.* **34**, 323–338.

Hollweg, J.V. (1983). Coronal heating by waves, in *Solar Wind 5*, ed. M. Neugebauer (NASA CP-2280, Washington, DC) pp. 1–21.

Hollweg, J.V. (1986). Viscosity and the Chew–Goldberger–Low equations in the solar corona, *Astrophys. J.* **306**, 730–739.

Holman, G.D. (1995). DC electric field acceleration of ions in solar flares, *Astrophys. J.* **452**, 451–456.

Holman, G.D., and Pesses, M.E. (1983). Solar type II radio emission and shock drift acceleration of electrons, *Astrophys. J.* **267**, 837–843.

Holt, E.H., and Haskell, R.E. (1965). *Plasma Dynamics* (Macmillan, New York).

Holzer, T.E., and Reid, G.C. (1975). The response of the dayside magnetopause-ionosphere system to time-varying field line reconnection at the magnetopause. 1. Theoretical model, *J. Geophys. Res.* **80**, 2041–2049.

Hones, E.W., Jr. (1973). Plasma flow in the plasma sheet and its relation to substorms, *Radio Sci.* **8**, 979–990.

Hood, A.W. (1990). Structure and stability of solar and stellar coronae, *Computer Phys. Reports* **12**, 177–203.

Hood, A.W., and Priest, E.R. (1981). Critical conditions for magnetic instabilities in force-free coronal loops, *Geophys. Astrophys. Fluid Dynamics* **17**, 297–318.

Horne, K., Marsh, T.R., Cheng, F.H., Hubney, I., and Lanz, T. (1994). HST eclipse mapping of dwarf nova on Carinae in quiescence: An "Fe II curtain" with Mach 6 velocity dispersion veils the white dwarf, *Astrophys. J.* **426**, 294–307.

Hornig, G., and Rastätter, L. (1997). The role of helicity in the reconnection process, *Adv. Space Res.* **19**, 1789–1792.

Hornig, G., and Rastätter, L. (1998). The magnetic structure of $B \neq 0$ reconnection, *Physica Scripta* **T74**, 34–39.

Hornig, G., and Schindler, K. (1996). Magnetic topology and the problem of its invariant definition, *Phys. Plasmas* **3**, 781–791.

Houdebine, E.R., Foing, B.H., Doyle, J.G., and Rodonò, M. (1993). Dynamics of flares on late type dMe stars. III. Kinetic energy and mass momentum budget of a flare on AD Leonis, *Astron. Astrophys.* **278**, 109–128.

Hoyle, F. (1949). *Some Recent Researches in Solar Physics* (Cambridge Univ. Press, Cambridge).

Huba, J.D., Gladd, N.T., and Papadopoulos, K. (1977). The lower-hybrid-drift instability as a source of anomalous resistivity for magnetic field line reconnection, *Geophys. Res. Lett.* **4**, 125–128.

Hudson, H.S., Strong, K.T., Dennis, B.R., Zarro, D., Inda, M., Kosugi, T., and Sakao, T. (1994). Impulsive behaviour in solar soft X-radiation. *Astrophys. J.* **422**, L25–L27.

Hudson, P.D. (1965). Reflection of charged particles by plasma shocks, *Mon. Not. Roy. Astron. Soc.* **131**, 23–50.

Hughes, W.J. (1995). The magnetopause, the magnetotail and magnetic reconnection, in *Introduction to Space Physics*, eds. M.G. Kivelson and C.T. Russell (Cambridge Univ. Press, Cambridge) pp. 227–287.

Hughes, W.J., and Sibeck, D.G. (1987). On the three-dimensional structure of plasmoids, *Geophys. Res. Lett.* **14**, 636–639.

Hundhausen, A.J. (1988). The origin and propagation of coronal mass ejections, in *Proceedings of the Sixth International Solar Wind Conference*, eds. V. Pizzo, T. Holzer, and D.G. Sime (NCAR/TN 306, Boulder) pp. 181–241.

Ichimaru, S. (1976). Magnetohydrodynamic turbulence in disk plasmas and magnetic field fluctuations in the Galaxy, *Astrophys. J.* **208**, 701–705.

Illing, R.M.E., and Hundhausen, A.J. (1983). Possible observation of a disconnected magnetic structure in a coronal transient, *J. Geophys. Res.* **88**, 10210–10214.

Imshennik, V.S., and Syrovatsky, S.I. (1967). Two-dimensional flow of an ideally conducting gas in the vicinity of the zero line of a magnetic field, *Sov. Phys. JETP* **25**, 656–664.

Inhester, B., Birn, J., and Hesse, M. (1992). The evolution of line-tied coronal arcades including a convergent footpoint motion, *Solar Phys.* **138**, 257–281.

Innes, D.E., Inhester, B., Axford, W.I., and Wilhelm, K. (1997). Bi-directional plasma jets produced by magnetic reconnection on the Sun, *Nature* **386**, 811–813.

Inverarity, G.W., and Priest, E.R. (1995a). Turbulent coronal heating. II. Twisted flux tube, *Astron. Astrophys.* **296**, 395–404.

Inverarity, G.W., and Priest, E.R. (1995b). Turbulent coronal heating. III. Wave heating in coronal loops, *Astron. Astrophys.* **302**, 567–577.

Ip, W.-H., and Axford, W.I. (1986). The acceleration of particles in the vicinity of comets, *Planet. Space. Sci.* **34**, 1061–1065.

Ip, W.-H., and Mendis, D.A. (1978). The flute instability as the trigger mechanism for disruption of cometary plasma tails, *Astrophys. J.* **223**, 671–675.

Irby, J.H., Drake, J.F., and Grien, H.R. (1979). Observation and interpretation of magnetic field line reconnection and tearing in a theta pinch, *Phys. Rev. Lett.* **42**, 228–231.

Isenberg, P.A. (1986). On the difficulty with accelerating particles at slow-mode shocks, *J. Geophys. Res.* **91**, 1699–1700.

Jackson, J.D. (1975). *Classical Electrodynamics* (Wiley and Sons, New York).

Jafelice, L.C., and Friaca, A.C.S. (1996). The role of magnetic reconnection in emission-line filaments in cooling flows, *Mon. Not. Roy. Astron. Soc.* **280**, 438–446.

Jakimiec, J., Sylwester, B., Sylwester, J., Lemen, J.R., Mewe, R., Bentley, R.D., Peres, G., Serio, S., and Schrijver, C.J. (1987). High-temperature plasma diagnostics of solar flares and comparison with model calculations, in *Solar Maximum Analysis*, eds. V.E. Stepanov and V.N. Obridko (VNU Science Press, Utrecht) pp. 91–101.

Jamitzky, F., and Scholer, M. (1995). Steady-state magnetic reconnection at high magnetic Reynolds number: A boundary layer analysis, *J. Geophys. Res.* **100**, 19277–19285.

Jardine, M. (1994). Three-dimensional steady-state magnetic reconnection, *J. Plasma Phys.* **51**, 399–422.

Jardine, M., Allen, H.R., and Grundy, R.E. (1993). Three-dimensional magnetic field annihilation, *J. Geophys. Res.* **98**, 19409–19417.

Jardine, M., Allen, H.R., Grundy, R.E., and Priest, E.R. (1992). A family of two-dimensional nonlinear solutions for magnetic field annihilation, *J. Geophys. Res.* **97**, 4199–4207.

Jardine, M., and Priest, E.R. (1988a). Weakly nonlinear theory of fast steady-state magnetic reconnection, *J. Plasma Phys.* **40**, 143–161.

Jardine, M., and Priest, E.R. (1988b). Global energetics of fast magnetic reconnection, *J. Plasma Phys.* **40**, 505–515.

Jardine, M., and Priest, E.R. (1988c). Reverse currents in fast magnetic reconnection, *Geophys. Astrophys. Fluid Dynamics* **42**, 163–168.

Jardine, M., and Priest, E.R. (1989). Compressible models of fast steady-state magnetic reconnection, *J. Plasma Phys.* **42**, 111–132.

Jardine, M., and Priest, E.R. (1990). Energetics of compressible fast steady-state magnetic reconnection, *J. Plasma. Phys.* **43**, 141–150.

Jeffries, R.D. (1993). Prominence activity on the rapidly rotating field star HD 197890, *Mon. Not. Roy. Astron. Soc.* **262**, 369–376.

Jin, S.P., and Ip, W.-H. (1991). Two-dimensional compressible MHD simulation of a driven reconnection process, *Phys. Fluids B* **3**, 1927–1936.

Johnson, F.S. (1960). The gross character of the geomagnetic field in the solar wind, *J. Geophys. Res.* **65**, 3049–3051.

Jokipii, J.R. (1966). Cosmic-ray propagation. I. Charged particles in a random magnetic field, *Astrophys. J.* **146**, 480–487.

Jokipii, J.R. (1982). Particle drift, diffusion and acceleration at shocks, *Astrophys. J.* **255**, 716–720.

Jokipii, J.R. (1986). Cosmic rays in interplanetary magnetic fields, *Science* **233**, 483.

Jokipii, J.R., and Levy, E.R. (1977). Effects of particle drifts on the solar modulation of galactic cosmic rays, *Astrophys. J.* **213**, L85–88.

Jones, F.C., and Ellison, D.C. (1991). The plasma physics of shock acceleration, *Space Sci. Rev.* **58**, 259–346.

Jordan, C. (1992). Modelling of solar coronal loops, *Memorie Societa Astronomica Italiana* **63**, 3–4.

Kadomtsev, B.B. (1975). Disruptive instability in tokamaks, *Sov. J. Plasma Phys.* **1**, 389–391.

Kahler, S.W. (1992). Solar flares and coronal mass ejections, *Ann. Rev. Astron. Astrophys.* **30**, 113–141.

Kahler, S.W., and 30 coauthors (1982). Coordinated X-ray, optical and radio observations of flaring activity on YZ Canis Minoris, *Astrophys. J.* **252**, 239–349.

Kahn, F.D., and Brett, L. (1993). Magnetic reconnection in the disc halo, *Mon. Not. Roy. Astron. Soc.* **263**, 37–48.

Kan, J.R. (1988). A theory of patchy and intermittent reconnection for magnetospheric flux transfer events, *J. Geophys. Res.* **95**, 5613–5623.

Kaplan, W. (1952). *Advanced Calculus* (Addison-Wesley, Reading).

Karpen, J.T., Antiochos, S.K., De Vore, C.R., and Golub, L. (1998). Dynamic responses to magnetic reconnection in solar arcades, *Astrophys. J.* **495**, 491–501.

Katsova, M.M., Boiko, A.Y., and Livshits, M.A. (1997). The gas-dynamic model of impulsive flares, *Astron. Astrophys.* **321**, 549–556.

Kaufmann, R.L., and Lu, C. (1993). Cross-tail current: Resonant orbits, *J. Geophys. Res.* **98**, 15447–15465.

Kaufmann, R.L., Larson, D.J., and Lu, C. (1993). Mapping and energization in the magnetotail. 2. Particle acceleration, *J. Geophys. Res.* **98**, 9321–9333.

Kennel, C.F. (1988). Shock structure in classical magnetohydrodynamics, *J. Geophys. Res.* **93**, 8545–8557.

Kennel, C.F., and Coroniti, F.V. (1975). Is Jupiter's magnetosphere like a pulsar's or Earth's?, *Space Sci. Rev.* **17**, 857–883.

Kennel, C.F., and Engelman, F. (1966). Velocity space diffusion from weak plasma turbulence, *Phys. Fluids* **9**, 2377–2388.

Kennel, C.F., Coroniti, F.V., Scarf, F.L., Livesey, W.H., Russell, C.F., Smith, E.J., Wenzel, K.P., and Scholer, M. (1986). A test of Lee's quasi-linear theory of ion acceleration by interplanetary travelling shocks, *J. Geophys. Res.* **91**, 11917–11928.

Kiendl, M.T., Semenov, V.S., Kubyshkin, I.V., Biernat, H.K., Rijnbeek, R.P., and

Besser, B.P. (1997). MHD analysis of Petschek-type reconnection in nonuniform field and flow geometries, *Space. Sci. Rev.* **67**, 1–47.

Kippenhahn, R., and Schlüter, A. (1957). Eine Theorie der Solaren Filamente, *Zs. Ap.* **43**, 36–62.

Kirk, J.G. (1994). Particle acceleration, in *Plasma Astrophysics* eds. J.G. Kirk, D.B. Melrose, and E.R. Priest (Springer-Verlag, Berlin) pp. 225–314.

Kirk, J.G., and Heavens, A.F. (1989). Particle acceleration at oblique shock fronts, *Mon. Not. Roy. Astron. Soc.* **239**, 995–1011.

Kirk, J.G., Schlickeiser, R., and Schneider, P. (1988). Cosmic-ray transport in accelerating flows, *Astrophys. J.* **328**, 269–274.

Kirkland, K.B., and Sonnerup, B.U.Ö. (1979). Self-similar resistive decay of a current sheet in a compressible plasma, *J. Plasma Phys.* **22**, 289–302.

Kivelson, M.G., and Russell, C.T. (1995). *Introduction to Space Physics* (Cambridge Univ. Press, Cambridge).

Kivelson, M.G., Khuvana, K.K., Russell, C.T., Walker, R.J., Warnecke, J., Coroniti, F.V., Polanskey, C., Southwood, D.J., and Schubert, G. (1996). Discovery of Ganymede's magnetic field by the Galileo spacecraft, *Nature* **384**, 537–541.

Klapper, I. (1998). Constraints on finite-time current sheet formation at null points in two-dimensional ideal incompressible MHD, *Phys. Plasmas* **5**, 910–914.

Kleva, R.G., and Drake, J.F. (1991). Density limit disruptions in tokamaks, *Phys. Fluids B* **3**, 372–383.

Kliem, B., and Seehafer, N. (1991). Plasma heating by current sheets in solar active regions, in *Mechanisms of Chromospheric and Coronal Heating*, eds. P. Ulmschneider, E.R. Priest, and R. Rosner (Springer-Verlag, Berlin) pp. 564–566.

Klimchuk, J.A., Sturrock, P.A., and Yang, W.H. (1988). Coronal magnetic fields produced by photospheric shear, *Astrophys. J.* **335**, 456–467.

Kobak, T.Z., and Ostrowski, M. (1997). Process of energetic particle acceleration in a three-dimensional magnetic field reconnection model, in *Relativistic Jets in Active Galactic Nuclei*, eds. M. Ostrowski et al. (Uniwersytet Jagiellonski, Obserwatorium Astronomiczne, Krakow) pp. 185–188.

Kohl, J.L., and 25 authors (1997). First results from the SOHO Ultraviolet Coronagraph Spectrometer, *Solar Phys.* **175**, 613–644.

Kopp, R.A., and Pneuman, G.W. (1976). Magnetic reconnection in the corona and the loop prominence phenomenon, *Solar Phys.* **50**, 85–98.

Krauss-Varban, D., and Burgess, D. (1991). Electron acceleration at nearly perpendicular collisionless shocks. 2. Reflection at curved shocks, *J. Geophys. Res.* **96**, 143–154.

Krucker, S., Benz, A.O., Bastian, T.S., and Acton, L.W. (1998). X-ray network flares of the quiet Sun, *Astrophys. J.* **488**, 499–505.

Krymsky, G.F. (1977). A regular mechanism for the acceleration of charged particles on the front of a shock wave, *Sov. Phys. Dokl.* **22**, 327–328.

Kulsrud, R.M., and Anderson, S.W. (1992). The spectrum of random magnetic fields in the mean field dynamo theory of the Galactic magnetic field, *Astrophys. J.* **396**, 606–630.

Kulsrud, R.M., and Ferrari, A. (1971). The relativistic quasilinear theory of particle acceleration by hydromagnetic turbulence, *Astrophys. Space Sci.* **12**, 302–318.

Kulsrud, R.M., and Hahm, T.S. (1982). Forced magnetic reconnection, *Physica Scripta* **T2**, 525–528.

Kusano, K., Suzuki, Y., and Nishikawa, K. (1995). A solar flare triggering mechanism based on the Woltjer–Taylor minimum energy principle, *Astrophys. J.* **441**, 942–951.

La Belle-Hamer, A.L., Fu, Z.F., and Lee, L.C. (1987). A mechanism for patchy reconnection at the dayside magnetopause, *Geophys. Res. Lett.* **15**, 152–155.

La Dous, C. (1993). Dwarf novae and nova-like stars, in *Cataclysmic Variables and Related Objects*, eds. M. Hack and C. La Dous (NASA SP-507, Washington, DC) pp. 15–258.

Lakhina, G.S., and Schindler, K. (1988). The effects of plasma sheet boundary flow and plasma mantle flow on the ion tearing instability, *J. Geophys. Res.* **93**, 8591–8601.

Landau, L.D., and Lifshitz, E.M. (1984). *Electrodynamics of Continuous Media* (Pergamon Press, Oxford).

Lang, K.R. (1992). *Astrophysical Data: Planets and Stars* (Springer-Verlag, Berlin).

La Rosa, T.N., Moore, R.L., Miller, J.A., and Shore, S.N. (1996). New promise for electron bulk energization in solar flares: preferential acceleration of electrons over protons in reconnection-driven MHD turbulence, *Astrophys. J.* **467**, 454–464.

Lau, Y.-T., and Finn, J.M. (1990). Three-dimensional kinematic reconnection in the presence of field nulls and closed field lines, *Astrophys. J.* **350**, 672–691.

Lau, Y.-T., and Finn, J.M. (1991). Three-dimensional kinematic reconnection of plasmoids, *Astrophys. J.* **366**, 577–599.

Lau, Y.-T., and Finn, J.M. (1992). Dynamics of a three-dimensional incompressible flow with stagnation points, *Physica D* **57**, 283–310.

Lau, Y.-T., and Finn, J.M. (1996). Magnetic reconnection and the topology of interacting twisted flux tubes, *Phys. Plasmas* **3**, 3983–3997.

Laval, G., and Gressilon, G. (1979). *Intrinsic Stochasticity in Plasmas* (Ed. de Physique, Orsay).

Laval, G., Pellat, R., and Vuillemin, M. (1966). Instabilités électromagnetiques des plasma sans collisions, in *Proceedings of the 2nd Conference on Plasma Physics and Controlled Nuclear Fusion Research*, **2** (Int. Atomic Energy Agency, Vienna) pp. 259–277.

Lebedev, A.N. (1965). Contribution to the theory of runaway electrons, *Sov. Phys. JETP* **21**, 931–933.

Lee, L.C., and Fu, Z.F. (1985). A theory of magnetic flux transfer at the Earth's magnetopause, *Geophys. Res. Lett.* **12**, 105–108.

Lee, L.C., and Fu, Z.F. (1986). Multiple X-line reconnection. 1. A criterion for the transition from a single X-line to a multiple X-line reconnection, *J. Geophys. Res.* **91**, 6807–6815.

Lee, L.C., Lin, Y., Shi, Y., and Tsurutani, B.T. (1989). Slow shock characteristics as a function of distance from the X-line in the magnetotail, *Geophys. Res. Lett.* **16**, 903–906.

Lee, M.A. (1982). Coupled hydromagnetic wave excitation and ion acceleration upstream of the Earth's bow shock, *J. Geophys. Res.* **87**, 5063–5080.

Lee, M.A. (1983). Coupled hydromagnetic wave excitation and ion acceleration at interplanetary travelling shocks, *J. Geophys. Res.* **88**, 6109–6119.

Lee, M.A. (1992). Particle acceleration in the heliosphere, in *Particle Acceleration in Cosmic Plasmas*, eds. G.P. Zank and T.K. Gaisser (Amer. Inst. Phys., New York) pp. 27–44.

Lee, M.A. (1994). Stochastic Fermi acceleration and solar cosmic rays, in *High-Energy Solar Phenomena – A New Era of Spacecraft Measurements*, eds. J.M. Ryan and W.T. Vestrand (Amer. Inst. Phys., New York) pp. 134–142.

Lee, M.A., and Ryan, J.M. (1986). Time-dependent coronal shock acceleration of energetic solar flare particles, *Astrophys. J.* **303**, 829–842.

Lee, M.A., and Volk, H.J. (1975). Hydromagnetic waves and cosmic-ray diffusion theory, *Astrophys. J.* **198**, 485–492.

Lee, M.A., Shapiro, V.D., and Sagdeev, R.Z. (1996). Pickup ion acceleration by shock surfing, *J. Geophys. Res.* **101**, 4777–4789.

Lesch, H. (1996). Magnetic reconnection in accretion disc coronae, in *Solar and Astrophysical Magnetohydrodynamic Flows*, ed. K.C. Tsinganos (Kluwer, Dordrecht) pp. 673–682.

Lesch, H., and Reich, W. (1992). The origin of monoenergetic electrons in the arc of the galactic center – particle acceleration by magnetic reconnection, *Astron. Astrophys.* **264**, 493–499.

Levy, E.H., and Araki, S. (1989). Magnetic reconnection flares in the protoplanetary nebula and the possible origin of meteorite chondrules, *Icarus* **81**, 74–91.

Levy, R.H., Petschek, H.E., and Siscoe, G.L. (1964). Aerodynamic aspects of the magnetospheric flow, *AIAA J.* **2**, 2065–2076.

Lewis, Z.V., Cowley, S.W.H., and Southwood, D.J. (1990). Impulsive energization of ions in the near-earth magnetotail during substorms, *Planet. Space Sci.* **38**, 491–505.

Lighthill, J. (1978). *Waves in Fluids* (Cambridge Univ. Press, Cambridge).

Lin, J., and Forbes, T.G. (1999). The effect of reconnection on the CME process, *J. Geophys. Res.*, in press.

Lin, R.P., Schwartz, R.A., Pelling, R.M., and Hurley, K.C. (1981). A new component of hard X-rays in solar flares, *Astrophys. J. Lett.* **251**, L109–L114.

Linardatos, D. (1993). Magnetostatic equilibria and analogous Euler flows, *J. Fluid Mech.* **246**, 569–591.

Linker, J.A., Van Hoven, G., and Schnack, D.D. (1990). MHD simulations of coronal mass ejections: importance of the driving mechanism, *J. Geophys. Res.* **95**, 4229–4238.

Linker, J.A., Van Hoven, G., and McComas, D.J. (1992). Simulations of coronal disconnection events, *J. Geophys. Res.* **97**, 13733–13740.

Linsky, J.L. (1989). Solar and stellar magnetic fields and structures: Observations, *Solar Phys.* **121**, 187–196.

Linsky, J.L., Neff, J.E., Brown, A., Gross, B.D., Simon, T. Andrews, A.D., Rodonò, M., and Feldman, P.A. (1989). Rotational modulation and flares on RS Canum Venaticorum and BY Draconis stars. X. The 1981 October 3 flare on V711 Tauri (HR 1099), *Astron. Astrophys.* **211**, 173–186.

Lionello, R., Velli, M., Einaudi, G., and Mikić, Z. (1998). Nonlinear MHD evolution of line-tied coronal loops, *Astrophys. J.* **494**, 840–850.

Litvinenko, Y.E. (1994). An explanation for the flare-frequency energy dependence, *Solar Phys.* **151**, 195–198.

Litvinenko, Y.E. (1996a). Particle acceleration in reconnecting current sheets with a nonzero magnetic field, *Astrophys. J.* **462**, 997–1004.

Litvinenko, Y.E. (1996b). On the formation of the Helium-3 spectrum in impulsive solar flares, in *High-Energy Solar Physics*, eds. R. Ramaty, N. Mandzhavidze, and X.-M. Hua (Amer. Inst. Phys., New York) pp. 498–503.

Litvinenko, Y.E., and Somov, B.V. (1993). Particle acceleration in reconnecting current sheets, *Solar Phys.* **146**, 127–133.

Litvinenko, Y.E., Forbes, T.G., and Priest, E.R. (1996). A strong limitation on the rapidity of flux-pile-up reconnection, *Solar Phys.* **167**, 445–448.

Liu, W.W., and Hill, T.W. (1990). Effect of plasma mantle injection on the dynamics of the distant magnetotail, *J. Geophys. Res.* **95**, 18849–18860.

Livio, M., and Pringle, J.E. (1992). Dwarf nova outbursts – the ultraviolet delay and the effect of a weakly magnetized white dwarf, *Mon. Not. Roy. Astron. Soc.* **259**, 23–26.

Lockwood, M. (1991). Flux transfer events at the dayside magnetopause: Transient reconnection or magnetosheath dynamic pressure pulses?, *J. Geophys. Res.* **96**, 5497–5509.

Lockwood, M., Cowley, S.W.H., and Sandholt, P.E. (1990). Transient reconnection search for ionospheric signatures, *EOS, Trans. Amer. Geophys. Union* **71**, 719–720.

Longcope, D.W. (1996). Topology and current ribbons: A model for current, reconnection and flaring in a complex, evolving corona, *Solar Phys.* **169**, 91–121.

Longcope, D.W. (1998). Current sheet formation and reconnection of separator field lines, in *Observational Plasma Physics: Five Years of Yohkoh and Beyond*, ed. Watanabe, T. (Kluwer, Dordrecht) pp. 179–184.

Longcope, D.W., and Cowley, S.C. (1996). Current sheet formation along three-dimensional magnetic separatrices, *Phys. Plasmas* **3**, 2885–2897.

Longcope, D.W., and Silva A.V.R. (1998). A current ribbon model for energy storage and release with application to the flare of 7 Jan. 1992, *Solar Phys.* **179**, 349–377.

Longcope, D.W., and Strauss, H.R. (1993). The coalescence instability and the development of current sheets in two-dimensional MHD, *Phys. Fluids B* **5**, 2858–2869.

Longcope, D.W., and Strauss, H.R. (1994). Spontaneous reconnection of line-tied flux tubes, *Astrophys. J.* **426**, 742–757.

Lopez, R.E., and Baker, D.N. (1994). Evidence for particle acceleration during magnetospheric substorms, *Astrophys. J.* **90**, 531–539.

Lopez, R.E., Lui, A.T.Y., Sibeck, D.G., Takahashi, K., McEntire, R.W., Zanetti, L.J., and Krimigis, S.M. (1989). On the relationship between the energetic particle flux morphology and the change in the magnetic field during substorms, *J. Geophys. Res.* **94**, 17105–17119.

Lottermoser, R.-F., Scholer, M., and Matthews, A.P. (1998). Ion kinetic effects in magnetic reconnection: hybrid simulations, *J. Geophys. Res.* **103**, 4547–4559.

Low, B.C. (1981). Eruptive solar magnetic fields, *Astrophys. J.* **251**, 352–363.

Low, B.C. (1982). Magnetostatic atmospheres with variations in three dimensions, *Astrophys. J.* **263**, 952–969.

Low, B.C. (1986). Models of partially open magnetospheres with and without magnetodisks, *Astrophys. J.* **310**, 953–965.

Low, B.C. (1987). Electric current sheet formation in magnetic field induced by continuous magnetic footpoint displacements, *Astrophys. J.* **323**, 358–367.

Low, B.C., and Smith, D.F. (1993). The free energies of partially open coronal magnetic fields, *Astrophys. J.* **410**, 412–425.

Low, B.C., and Wolfson, R. (1988). Spontaneous formation of electric current sheets and the origin of solar flares, *Astrophys. J.* **324**, 574–581.

Lu, E.T., Hamilton, R.J., McTiernan, J.M., and Bromund, K.R. (1993). Solar flares and avalanches in driven dissipative systems, *Astrophys. J.* **412**, 841–852.

Luhmann, J.G., Walker, R.J., Russell, C.T., Crooker, N.U., Spreiter, J.R., and Stahara, S.S. (1984). Patterns of potential magnetic field merging sites on the dayside magnetopause, *J. Geophys. Res.* **89**, 1739–1742.

Lui, A.T.Y., Lopez, R.E., Anderson, B.J., Takahashi, K., Zanetti, L.J., McEntire, R.W., Potemra, T.A., Klumpar, D.M., Greene, E.M., and Strangeway, R. (1992). Current disruptions in the near-Earth neutral sheet, *J. Geophys. Res.* **97**, 1461–1480.

Lundquist, S. (1951). On the stability of magneto-hydrostatic fields, *Phys. Rev.* **83**, 307–311.

Lynden–Bell, D., and Boily, C. (1994). Self-similar solutions up to flashpoint in highly wound magnetostatics, *Mon. Not. Roy. Astron. Soc.* **267**, 146–152.

Lyons, L.R., and Speiser, T.W. (1985). Ohm's law for a current sheet, *J. Geophys. Res.* **90**, 8543–8546.

Lyons, L.R., and Williams, D.J. (1984). *Quantitative Aspects of Magnetospheric Physics* (Reidel, Dordrecht).

Macek, W.M. (1989). Reconnection at the heliopause, *Adv. Space Res.* **9**, 257–261.

Macek, W.M. (1990). Reconnection pattern at the magnetopause, in *Physics of the Outer Heliosphere*, eds. S. Grzedzielski and D.E. Page (Pergamon Press, Oxford) pp. 399–402.

Machado, M.E., Moore, R.L., Henandez, A.M., Rovira, M.G., Hagyard, M.J., and Smith, J.B., Jr. (1988). The observed characteristics of flare energy release. I. Magnetic structure at the energy release site, *Astrophys. J.* **326**, 425–450.

Malara, F., Einaudi, G., and Mangeney, A. (1989). Stability properties of a cometary plasma, *J. Geophys. Res.* **94**, 11813–11819.

Malherbe, J.M., and Priest, E.R. (1983). Current sheet models for solar prominences, *Astron. Astrophys.* **123**, 80–88.

Malkov, M.A., and Sotnikov, V.I. (1985). Lower hybrid drift instability and reconnection of magnetic lines of force, *Sov. J. Plasma Phys.* **11**, 626–631.

Mandrini, C.H., Démoulin, P., Rovira, M.G., de la Beaujardière, J.-F., and Hénoux, J.C. (1995). Constraints on flare models set by the active region magnetic topology of AR 6233, *Astron. Astrophys.* **303**, 927–939.

Mansurov, S.M. (1969). New evidence of a relationship between magnetic fields in space and on Earth, *Geomagn. Aeron.* **9**, 622–623.

Martens, P.C.H. (1988). The generation of proton beams in two-ribbon flares, *Astrophys. J. Lett.* **330**, L131–L133.

Martens, P.C.H., and Kuin, N.P.M. (1989). A circuit model for filament eruptions and two-ribbon flares, *Solar Phys.* **122**, 263–302.

Martens, P.C.H., and Young, A. (1990). Neutral beams in two-ribbon flares and in the geomagnetic tail, *Astrophys. J.* **73**, 333–342.

Martin, R.F., Jr. (1986). Chaotic particle dynamics near a two-dimensional magnetic neutral point with application to the geomagnetic tail, *J. Geophys. Res.* **91**, 11985–11992.

Martin, S.F., and Švestka, Z. (1988). Flaring arches. I. The major events of 1980 November 6 and 12, *Solar Phys.* **116**, 91–118.

Martin, S.F., Livi, S.H.B., and Wang, J. (1985). The cancellation of magnetic flux. II. In a decaying active region, *Australian J. Phys.* **38**, 929–959.

Martinell, J.J. (1990). Expansion and stability of a magnetic arcade during a solar flare, *Astrophys. J.* **365**, 342–353.

Masuda, S. (1994). *Hard X-ray Sources and the Primary Energy Release Site in Solar Flares*, PhD thesis (Univ. of Tokyo, Tokyo).

Masuda, S., Kosugi, T., Hara, H., Tsuneta, S., and Ogawara, Y. (1994). A loop-top hard X-ray source in a compact solar flare, *Nature* **371**, 495–497.

Matsumoto, R., Tajima, T., Shibata, K., and Kaisig, M. (1993). Three-dimensional magnetohydrodynamics of emerging magnetic flux in the solar atmosphere, *Astrophys. J.* **414**, 357–371.

Matthaeus, W.H., and Lamkin, S.L. (1986). Turbulent magnetic reconnection, *Phys. Fluids* **29**, 2513–2534.

Matthaeus, W.H., and Montgomery, D. (1981). Nonlinear evolution of the sheet pinch, *J. Plasma Phys.* **25**, 11–41.

Matthaeus, W.H., Ambrosiano, J.J., and Goldstein, M.L. (1984). Particle acceleration by turbulent MHD reconnection, *Phys. Rev. Lett.* **53**, 1449–1452.

Matthaeus, W.H., Ghosh, S., Oughton, S., and Roberts, D.A. (1996). Anisotropic three-dimensional MHD turbulence, *J. Geophys. Res.* **101**, 7619–7629.

Matthews, P.C., Proctor, M.R.E., and Weiss, N.O. (1995). Compressible magnetoconvection in three dimensions: planforms and nonlinear behaviour, *J. Fluid Mech.* **305**, 281–305.

Maynard, N.C., Burker, W.J., Basinska, E.M., Erickson, G.M., Hughes, W.J., Singer, M.J., Yahnin, A.G., Hardy, D.A., and Mozer, F.S. (1996). Dynamics of the inner magnetosphere near times of substorm onset, *J. Geophys. Res.* **101**, 7705–7736.

McClements, K.G., Bingham, R., Su, J.J., Dawson, J.J., and Spicer, D.S. (1993). Lower hybrid resonance acceleration of electrons and ions in solar flares and the associated microwave emission, *Astrophys. J.* **409**, 465–475.

McClymont, A.N., and Craig, I.J.D. (1996). Dynamical finite-amplitude magnetic reconnection at an X-type neutral point, *Astrophys. J.* **466**, 487–495.

McComas, D.J., Gosling, J.T., Phillips, J.L., Bame, S.J., Luhman, J.G., and Smith, E.J. (1989). Electron heat flux dropouts in the solar wind – Evidence for interplanetary magnetic field reconnection?, *J. Geophys. Res.* **94**, 6907–6917.

McComas, D.J., Phillips, J.L., Hundhasuen, A.J., and Burkepile, J.T. (1991). Observations of disconnection of open coronal magnetic structures, *Geophys. Res. Lett.* **18**, 73–76.

McComas, D.J., Gosling, J.T., and Phillips, J.L. (1992). Regulation of the interplanetary magnetic field, in *Solar Wind Seven*, eds. E. Marsch and R. Schwenn (Pergamon Press, Oxford) pp. 643–646.

McComas, D.J., Gosling, J.T., Hammond, C.M., Moldwin, M.B., Phillips, J.L., and Forsyth, R.J. (1995). Reconnection of open field lines ahead of coronal mass ejections, *Space Sci. Rev.* **72**, 129–132.

McKenzie, J.F., Banaskiewicz, M., and Axford, W.I. (1995). Acceleration of the high-speed solar wind, *Astron. Astrophys.* **303**, L45–L48.

McKenzie, J.F., Axford, W.I., and Banaskiewicz, M. (1997). The fast solar wind, *Geophys. Res. Lett.* **24**, 2877–2880.

McKenzie, J.F., Sukhorakova, G.W., and Axford, W.I. (1998). The source region of the fast solar wind, *Astron. Astrophys.* **330**, 1145–1148.

McPherron, R.L. (1995). Magnetospheric dynamics, in *Introduction to Space Physics*, eds. M.G. Kivelson and C.T. Russell (Cambridge Univ. Press, Cambridge) pp. 400–458.

McTiernan, J.M., Kane, S.R., Loran, J.M., Lemen, J.R., Acton, L.W., Hara, H., Tsuneta, S., and Kosugi, T. (1993). Temperature and density structure of the 1991 November 2 flare observed by the Yohkoh soft X-ray telescope and hard X-ray telescope, *Astrophys. J.* **416**, L91–L93.

Melrose, D.B. (1980). *Plasma Astrophysics* (Gordon and Breach, New York).

Melrose, D.B. (1983). Prompt acceleration of greater than 30 MeV per nucleon ions in solar flares, *Solar Phys.* **89**, 149–162.

Mestel, L. (1999). *Stellar Magnetism* (Oxford Univ. Press, Oxford).

Mestel, L., and Ray, T.P. (1985). Disk-like magnetogravitational equilibria, *Mon. Not. Roy. Astron. Soc.* **212**, 275–300.

Mestel, L., and Spitzer, L. (1956). Star formation in magnetic dust clouds, *Mon. Not. Roy. Astron. Soc.* **116**, 503–514.

Mestel, L., and Spruit, H.C. (1987). On magnetic braking of late-type stars, *Mon. Not. Roy. Astron. Soc.* **26**, 57–66.

Mestel, L., and Strittmatter, P.A. (1967). The magnetic field of a contracting gas cloud. II. Finite diffusion effects – an illustrative example, *Mon. Not. Roy. Astron. Soc.* **137**, 95–105.

Meyer, J.P. (1993). Elemental abundances in active regions, flares and the interplanetary medium, *Adv. Space Res.* **13**, 377–390.

Mikić, Z., and Linker, J.A. (1994). Disruption of coronal magnetic field arcades, *Astrophys. J.* **430**, 898–912.

Mikić, Z., Barnes, D.C., and Schnack, D.D. (1988). Dynamical evolution of a solar coronal magnetic field arcade, *Astrophys. J.* **328**, 830–847.

Mikić, Z., Schnack, D.D., and Van Hoven, G. (1990). Dynamical evolution of twisted magnetic flux tubes. I. Equilibrium and linear stability, *Astrophys. J.* **361**, 690–700.

Miller, J.A., and Ramaty, R. (1992). Stochastic acceleration in impulsive solar flares, in *Particle Acceleration in Cosmic Plasmas*, eds. G.P. Zank and T.K. Gaisser (Amer. Inst. Phys., New York) pp. 223–228.

Miller, J.A., and Roberts, D.A. (1995). Stochastic proton acceleration by cascading Alfvén waves in impulsive solar flares, *Astrophys. J.* **452**, 912–932.

Miller, J.A., La Rosa, T.N., and Moore, R.L. (1996). Stochastic electron acceleration by cascading fast-mode waves in impulsive solar flares, *Astrophys. J.* **461**, 445–464.

Miller, J.A., Cargill, P.J., Emslie, A.G., Holman, G.D., Dennis, B.R., La Rosa, T.N., Winglee, R.M., Benka, S.G., and Tsuneta, S. (1997). Critical issues for understanding particle acceleration in impulsive solar flares, *J. Geophys. Res.* **102**, 14631–14659.

Milne–Thomson, L.M. (1960). *Theoretical Hydrodynamics* (Macmillan, London).

Milne, A.M., and Priest, E.R. (1981). Internal structure of reconnecting current sheets and the emerging flux model for solar flares, *Solar Phys.* **73**, 157–181.

Milroy, R.D. (1984). Reconnection during the formation of field-reversed configurations, in *Magnetic Reconnection in Space and Laboratory Plasmas*, ed. E.W. Hones, Jr. (Amer. Geophys. Union, Washington, DC) pp. 305–312.

Mitchell, H.G., and Kan, J.R. (1978). Merging of magnetic fields with field-aligned plasma flow components, *J. Plasma Phys.* **20**, 31–45.

Mitchell, D.G., Williams, D.J., Huang, C.Y., Frank, L.A., and Russell, C.T. (1990). Current carriers in the near-Earth cross-tail current sheet during the substorm growth phase, *Geophys. Res. Lett.* **17**, 583–586.

Möbius, E. (1994). Sources and acceleration of energetic particles in planetary magnetospheres, *Astrophys. J.* **90**, 521–530.

Moffatt, H.K. (1969). The degree of knottedness of tangled vortex lines, *J. Fluid Mech.* **35**, 117–129.

Moffatt, H.K. (1978). *Magnetic Field Generation in Electrically Conducting Fluids* (Cambridge Univ. Press, Cambridge).

Moffatt, H.K. (1985). Magnetostatic equilibria and analogous Euler flows of arbitrarily complex topology. Part 1. Fundamentals, *J. Fluid Mech.* **159**, 359–378.

Moffatt, H.K. (1990). Structure and stability of solutions of the Euler equations: a Lagrangian approach, *Phil. Trans. Roy. Soc. Lond. A* **333**, 321–342.

Montgomery, D. (1983). Theory of hydromagnetic turbulence, in *Solar Wind 5*, ed. M. Neugebauer (NASA CP-2280, Washington, DC) pp. 107–130.

Montgomery, D., and Phillips, L. (1988). Minimum dissipation rates in magnetohydrodynamics, *Phys. Rev. A* **38**, 2953–2964.

Montgomery, D., Turner, L., and Vahala, G. (1978). Three-dimensional magnetohydrodynamic turbulence in cylindrical geometry, *Phys. Fluids* **21**, 757–764.

Montmerle, T., and Casanova, S. (1996). X-ray flares and variability of young stellar objects, in *Magnetohydrodynamic Phenomena in the Solar Atmosphere, Prototypes of Stellar Magnetic Activity*, IAU Colloq. 153, eds. Y. Uchida, T. Kosugi, and H.S. Hudson (Kluwer, Dordrecht) pp. 247–258.

Moore, R.L. et al. (1980). The thermal X-ray flare plasma, in *Solar Flares*, ed. P.A. Sturrock (Colorado Assoc. Univ. Press, Boulder) pp. 341–409.

Moore, R.L., LaRosa, T.N., and Orwig, L.E. (1995). The wall of reconnection-driven magnetohydrodynamic turbulence in a large solar flare, *Astrophys. J.* **438**, 985–966.

Moses, R.W., Finn, J.M., and Laing, K.H. (1993). Plasma heating by collisionless magnetic reconnection: analysis and computations, *J. Geophys. Res.* **98**, 4013–4040.

Mullan, D.J. (1986). M dwarfs: Theoretical work, in *The M-Type Stars*, eds. J.R. Johnson and F.R. Querci (NASA SP-492, Washington, DC) pp. 455–479.

Muskelishvilli, N. (1953). *Singular Integral Equations* (Noordhoff, Groningen, Holland).

Nagai, F. (1980). A model of hot loops associated with solar flares. I. Gasdynamics in the loops, *Solar Phys.* **68**, 351–379.

Nagai, T. et al. (1998). Structure and dynamics of magnetic reconnection of substorm onsets with Geotail observations, *J. Geophys. Res.* **103**, 4419–4440.

Neff, J.E., Linsky, J.L., Walter, F.M., and Rodonò, M. (1989). Rotational modulation and flares on RS Canum Venaticorum and BY Draconis stars. XI. Ultraviolet spectral images of AR Lacertae in September 1985, *Astron. Astrophys.* **215**, 79–91.

Neidig, D.F. (1989). The importance of solar white-light flares, *Solar Phys.* **121**, 261–269.

Nerney, S., Suess, S.T., and Schmahl, E.J. (1993). Flow downstream of the heliospheric terminal shock – the magnetic field on the heliopause, *J. Geophys. Res.* **98**, 15169–15176.

Neugebauer, M.M., and Snyder, C.W. (1962). Solar plasma experiment, *Science*, **138**, 1095–1096.

Neugebauer, M.M., Alexander, C.J., Schwenn, R., and Richter, A.K. (1986). Tangential discontinuities in the solar wind – correlated field and velocity changes and the Kelvin–Helmholtz instability, *J. Geophys. Res.* **91**, 13694–13698.

Newcomb, W.A. (1958). Motion of magnetic lines of force, *Ann. Phys.* **3**, 347–385.

Niedner, M.B., Jr. (1984). Magnetic reconnection in comets, in *Magnetic Reconnection in Space and Laboratory Plasmas*, ed. E.W. Hones, Jr. (Amer. Geophys. Union, Washington, DC) pp. 79–89.

Niedner, M.B., Jr., and Brandt, J.C. (1978). Interplanetary gas. XXIII. Plasma tail disconnection events in comets: evidence for magnetic field line reconnection at interplanetary sector boundaries?, *Astrophys. J.* **223**, 655–670.

Nishida, A., and Nagayama, N. (1973). Synoptic survey of the neutral line in the magnetotail during the substorm expansion phase, *J. Geophys. Res.* **78**, 3782–3798.

Norman, C., and Heyvaerts, J. (1985). Anomalous magnetic field diffusion during star formation, *Astron. Astrophys.* **147**, 247–256.

Northrop, T.G. (1963). *The Adiabatic Motion of Charged Particles* (Interscience Publishers, New York).

O'Brien, M.R., and Robinson, M.R. (1993). Tokamak experiments, in *Plasma Physics*, ed. R. Dendy (Cambridge Univ. Press, Cambridge) pp. 189–208.

Ofman, L., Chen, X.L., Morrison, P.J., and Steinolfson, R.S. (1991). Resistive tearing mode instability with shear flow and viscosity, *Phys. Fluids B* **3**, 1364–1373.

Ofman, L., Morrison, P.J., and Steinolfson, R.S. (1993a). Nonlinear evolution of resistive tearing mode instability with shear flow and viscosity, *Phys. Fluids B* **5**, 376–387.

Ofman, L., Morrison, P.J., and Steinolfson, R.S. (1993b). Magnetic reconnection at stressed X-type neutral points, *Astrophys. J.* **417**, 748–756.

Ogino, T., Walker, R.J., and Ashour–Abdalla, M. (1986). An MHD simulation of the interaction of the solar wind with the outflowing plasma from a comet, *Geophys. Res. Lett.* **13**, 929–932.

Ogino, T., Walker, R.J., and Ashour–Abdalla, M. (1989). An MHD simulation of the formation of magnetic flux tubes at the Earth's dayside magnetopause, *Geophys. Res. Lett.* **16**, 155–158.

Ohyabu, N., Okamura, S., and Kawashima, N. (1974). Strong ion heating in a magnetic neutral point discharge, *Phys. Fluids* **17**, 2009–2013.

Omidi, N., and Winske, D. (1992). Kinetic structure of slow shocks – effects of the electromagnetic ion/ion cyclotron instability, *J. Geophys. Res.* **97**, 14801–14821.

Onsager, T.G., and Lockwood, M. (1997). High-latitude particle precipitation and its relationship to magnetospheric source regions, *Space. Sci. Rev.* **80**, 77–107.

Onsager, T.G., and Mukai, T. (1995). Low-altitude signature of the plasma sheet boundary layer: Observations and model, *Geophys. Res. Lett.* **22**, 855–858.

Onsager, T.G., Thomsen, M.F., Elphic, R.C., and Gosling, J.T. (1991). Model of electron and ion distributions in the plasma sheet boundary layer, *J. Geophys. Res.* **96**, 20999–21011.

Ortolani, S. (1987). Reversed-field pinch confinement physics, *Plasma Phys. Control Fusion* **31**, 1665–1683.

Ostlie, D.A., and Carroll, B.W. (1996). *Modern Stellar Astrophysics* (Addison–Wesley, Wokingham, England).

Ostrowski, M. (1998). Acceleration of ultra-high energy cosmic ray particles in relativistic jets in extragalactic radio sources, *Astron. Astrophys.* **335**, 134–144.

Otto, A. (1995). Forced three-dimensional magnetic reconnection due to linkage of magnetic flux tubes, *J. Geophys. Res.* **100**, 11863–11874.

Otto, A., Schindler, K., and Birn, J. (1990). Quantitative study of the nonlinear formation and acceleration of plasmoids in the Earth's magnetotail, *J. Geophys. Res.* **95**, 15023–15037.

Overskei, D., and Politzer, P.A. (1976). Plasma turbulence in the vicinity of a magnetic neutral line, *Phys. Fluids* **19**, 683–689.

Pallavicini, R., Peres, G., Serio, S., Vaiana, G., Acton, L., Leibacher, J., and Rosner, R. (1983). Closed coronal structures. V. Gasdynamic models of flaring loops and comparison with SMM observations, *Astrophys. J.* **270**, 27–287.

Pan, H.C., Jordan, C., Makishima, K., Stern, R.A., Hayashida, K., and Inda–Koide, M. (1995). X-ray observations of the dMe star EQ1839.6+8002 in 1980–1993, in *Flares and Flashes*, eds. J. Greiner, H.W. Duerbeck, and R.E. Gershberg (Springer, Berlin) pp. 171–174.

Papaloizou, J.C.B., and Lin, D.N.C. (1995). Theory of accretion disks. I. Angular momentum transport processes, *Ann. Rev. Astron. Astrophys.* **33**, 505–540.

Papamastorakis, I., Paschmann, G., Baumjohann, W., and Sonnerup, B.U.Ö. (1989). Orientation, motion and other properties of flux transfer event structures on September 4, 1984, *J. Geophys. Res.* **94**, 8852–8866.

Paré, V.K. (1984). Reconnection in tokamaks, in *Magnetic Reconnection in Space and Laboratory Plasmas*, ed. E.W. Hones, Jr. (Amer. Geophys. Union, Washington, DC) pp. 341–346.

Park, W., Monticello, D.A., and White, R.B. (1984). Reconnection rates of magnetic fields including the effects of viscosity, *Phys. Fluids* **27**, 137–149.

Parker, E.N. (1953). Instability of thermal fields, *Astrophys. J.* **117**, 431–436.

Parker, E.N. (1957). Sweet's mechanism for merging magnetic fields in conducting fluids, *J. Geophys. Res.* **62**, 509–520.

Parker, E.N. (1958). Dynamics of the interplanetary gas and magnetic field, *Astrophys. J.* **128**, 664–676.

Parker, E.N. (1963). The solar flare phenomenon and theory of reconnection and annihilation of magnetic fields, *Astrophys. J. Supp.* **8**, 177–211.

Parker, E.N. (1965). The passage of energetic charged particles through interplanetary space, *Planet. Space Sci.* **13**, 9–49.

Parker, E.N. (1972). Topological dissipation and the small-scale fields in turbulent gases, *Astrophys. J.* **174**, 499–510.

Parker, E.N. (1973). Comments on the reconnection rate of magnetic fields, *J. Plasma Phys.* **9**, 49–63.

Parker, E.N. (1974). The dynamical properties of twisted ropes of magnetic field and the vigor of new active regions on the Sun, *Astrophys. J.* **191**, 245–254.

Parker, E.N. (1979). *Cosmical Magnetic Fields* (Clarendon Press, Oxford).

Parker, E.N. (1983). Magnetic neutral sheets in evolving fields. II. formation of the solar corona, *Astrophys. J.* **264**, 642–647.

Parker, E.N. (1988). Nanoflares and the solar X-ray corona, *Astrophys. J.* **330**, 474–479.

Parker, E.N. (1989). Spontaneous tangential discontinuities and the optical analogy for static magnetic fields, *Geophys. Astrophys. Fluid Dynamics* **45**, 159–168.

Parker, E.N. (1990). Tangential discontinuities and the optical analogy for stationary fields. V. Formal integration of the force-free field equations, *Geophys. Astrophys. Fluid Dynamics* **52**, 183–210.

Parker, E.N. (1996). The alternative paradigm for magnetospheric physics, *J. Geophys. Res.* **101**, 10587–10625.

Parker, E.N., and Tidman, D.A. (1958). Suprathermal particles, *Phys. Rev.* **111**, 1206–1211.

Parks, G.K. (1991). *Physics of Space Plasmas* (Addison–Wesley, Reading, Massachusetts).

Parnell, C.E., and Jupp, P.E. (1999). Statistical analysis of the energy distribution of sub-microflares in the quiet Sun, *Astrophys. J.*, in press.

Parnell, C.E., Priest, E.R., and Golub, L. (1994). The three-dimensional structures of X-ray bright points, *Solar Phys.* **151**, 57–74.

Parnell, C.E., Smith, J., Neukirch, T., and Priest, E.R. (1996). The structure of three-dimensional magnetic neutral points, *Phys. Plasmas* **3**, 759–770.

Paschmann, G., Sonnerup, B.U.Ö., Papamastorakis, I., Sckopke, N., Haerendel, G., Bame, S.J., Asbridge, J.R., Gosling, J.T., Russell, C.T., and Elphic, R.C. (1979). Plasma acceleration at the Earth's magnetopause: Evidence for reconnection, *Nature* **282**, 243–246.

Paschmann, G., Haerendel, G., Papamastorakis, I., Sckopke, N., Bame, S.J., Gosling, J.T., and Russell, C.T. (1982). Plasma and magnetic field characteristics of magnetic flux transfer events, *J. Geophys. Res.* **87**, 2159–2168.

Pease, R.S. (1993). Survey of fusion plasma physics, in *Plasma Physics*, ed. R. Dendy (Cambridge Univ. Press, Cambridge) pp. 475–507.

Pellat, R., and Francfort, P. (1976). Magnetic merging in collisionless plasmas, *Geophys. Res. Lett.* **3**, 433–436.

Pellinen, R.J., and Heikkila, W.J. (1978). Energization of charged particles to high energies by an induced substorm electric field within the magnetotail, *J. Geophys. Res.* **83**, 1544–1550.

Pesses, M., Jokipii, J.R., and Eichler, D. (1981). Cosmic-ray drift, shock-wave acceleration and the anomalous components of cosmic rays, *Astrophys. J.* **246**, L85–L88.

Petschek, H.E. (1964). Magnetic field annihilation, in *Physics of Solar Flares*, ed. W.N. Hess (NASA SP-50, Washington, DC) pp. 425–439.

Petschek, H.E., and Thorne, R.M. (1967). The existence of intermediate waves in neutral sheets, *Astrophys. J.* **147**, 1157–1163.

Pettersen, B.R. (1989). A review of stellar flares and their characteristics, *Solar Phys.* **121**, 299–312.

Phan, T.D., and Sonnerup, B.U.Ö. (1990). MHD stagnation-point flows at a current sheet including viscous and resistive effects, *J. Plasma Phys.* **44**, 525–546.

Pilipp, W.G., and Morfill, G. (1978). The formation of the plasma sheet resulting from plasma mantle dynamics, *J. Geophys. Res.* **83**, 5670–5678.

Pike, C.D., and Mason, H.E. (1998). Rotating transition-region features observed with the SOHO Coronal Diagnostic Spectrometer, *Solar Phys.* **182**, 333–348.

Platt, U., and Neukirch, T. (1994). Theoretical study of onset conditions for solar eruptive processes – influence of the boundaries, *Solar Phys.* **153**, 287–306.

Podgorny, A.I., and Syrovatsky, S.I. (1981). Formation and development of a current sheet for various magnetic viscosities and gas pressures, *Sov. J. Plasma Phys.* **7**, 580–585.

Poletto, G., and Kopp, R.A. (1986). Macroscopic electric fields during two-ribbon flares, in *The Lower Atmosphere of Solar Flares*, ed. D.F. Neidig (NSO, Sunspot, NM) pp. 453–465.

Poletto, G., Pallavicini, R., and Kopp, R.A. (1988). Modelling of long-duration two-ribbon flares on M dwarf stars, *Astron. Astrophys.* **201**, 93–99.

Politano, H., Pouquet, A., and Sulem, P.L. (1995). Current and vorticity dynamics in three-dimensional magnetohydrodynamic turbulence, *Phys. Plasmas* **2**, 2931–2939.

Poston, T., and Stewart, I. (1978). *Catastrophe Theory and Its Applications* (Pitman, San Francisco).

Pouquet, A. (1978). On two-dimensional MHD turbulence, *J. Fluid Mech.* **88**, 1–16.

Priest, E.R. (1972a). A modification and criticism of Petschek's mechanism, *Mon. Not. Roy. Astron. Soc.* **159**, 389–402.

Priest, E.R. (1972b). Sweet's mechanism for the destruction of magnetic flux, *Quart. Journal Mech. and App. Math.* **25**, 319–332.

Priest, E.R. (1981). Theory of loop flows and instability, in *Solar Active Regions*, ed. F.Q. Orall (Colo. Assoc. Univ. Press, Boulder) pp. 213–275.

Priest, E.R. (1982). *Solar Magnetohydrodynamics* (Reidel, Dordrecht).

Priest, E.R. (1985). The MHD of current sheets, *Rep. Prog. Phys.* **48**, 955–1090.

Priest, E.R. (1986). Magnetic reconnection on the Sun, *Mit. Astron. Ges.* **65**, 41–51.

Priest, E.R. (1987). Appearance and disappearance of magnetic flux at the solar surface, in *Role of Fine-Scale Magnetic Fields on Structure of the Solar Atmosphere*, eds. E. Schroter, M. Vazquez, and A. Wyller (Cambridge Univ. Press, Cambridge) pp. 297–316.

Priest, E.R. (1992). Basic magnetic configuration and energy supply processes for an Interacting Flux Model of eruptive solar flares, in *Eruptive Solar Flares*, eds. Z. Svestka, B. Jackson, and M. Machado (Springer-Verlag, Berlin) pp. 15–32.

Priest, E.R. (1993). Coronal heating mechanisms, in *Physics of Solar and Stellar Coronae*, eds. J. Linsky and S. Serio (Kluwer, Dordrecht) pp. 515–532.

Priest, E.R. (1996). Reconnection of magnetic lines of force, in *Solar and Astrophysical MHD Flows*, ed. K. Tsinganos (Kluwer, Dordrecht) pp. 151–170.

Priest, E.R., and Cowley, S.W.H. (1975). Some comments on magnetic field reconnection, *J. Plasma Phys.* **14**, 271–282.

Priest, E.R., and Démoulin, P. (1995). Three-dimensional reconnection without null points, *J. Geophys. Res.* **100**, 23443–23463.

Priest, E.R., and Forbes, T.G. (1986). New models for fast steady-state magnetic reconnection, *J. Geophys. Res.* **91**, 5579–5588.

Priest, E.R., and Forbes, T.G. (1989). Steady magnetic reconnection in three dimensions, *Solar Phys.* **119**, 211–214.

Priest, E.R., and Forbes, T.G. (1990). Magnetic field evolution during prominence eruptions and two-ribbon flares, *Solar Phys.* **126**, 319–350.

Priest, E.R., and Forbes, T.G. (1992a). Magnetic flipping – reconnection in three dimensions without null points, *J. Geophys. Res.* **97**, 1521–1531.

Priest, E.R., and Forbes, T.G. (1992b). Does fast magnetic reconnection exist?, *J. Geophys. Res.* **97**, 16757–16772.

Priest, E.R., and Lee, L.C. (1990). Nonlinear magnetic reconnection models with separatrix jets, *J. Plasma Phys.* **44**, 337–360.

Priest, E.R., and Raadu, M.A. (1975). Preflare current sheets in the solar atmosphere, *Solar Phys.* **43**, 177–188.

Priest, E.R., and Soward, A.M. (1976). On fast magnetic field reconnection, in *Basic Mechanisms of Solar Activity*, IAU Symp. 71, eds. V. Bumba and J. Kleczek (Reidel, Dordrecht) pp. 353–366.

Priest, E.R., and Titov, V.S. (1996). Magnetic reconnection at three-dimensional null points, *Phil. Trans. Roy. Soc. Lond.* A **354**, 2951–2992.

Priest, E.R., Parnell, C.E., and Martin, S.F. (1994a). A converging flux model of an X-ray bright point and an associated cancelling magnetic feature, *Astrophys. J.* **427**, 459–474.

Priest, E.R., Titov, V.S., and Rickard, G.K. (1994b). The formation of magnetic singularities by nonlinear time-dependent collapse of an X-type magnetic field, *Phil. Trans. Roy. Soc. Lond.* A **351**, 1–37.

Priest, E.R., Titov, V.S., Vekstein, G.E., and Rickard, G.J. (1994c). Steady linear X-point magnetic reconnection, *J. Geophys. Res.* **99**, 21467–21480.

Priest, E.R., Lonie, D.P., and Titov, V.S. (1996). Bifurcations of magnetic topology by the creation or annihilation of null points, *J. Plasma Phys.* **56**, 507–530.

Priest, E.R., Bungey, T.N., and Titov, V.S. (1997). The 3D topology and interaction of complex magnetic flux systems, *Geophys. Astrophys. Fluid Dynamics* **84**, 127–163.

Priest, E.R., Foley, C.R., Heyvaerts, J., Arber, T.D., Culhane, J.L., and Acton, L.W. (1998). Nature of the heating mechanism for the diffuse solar corona, *Nature* **393**, 545–547.

Pritchett, P.L., and Wu, C.C. (1979). Coalescence of magnetic islands, *Phys. Fluids* **22**, 2140–2146.

Pritchett, P.L., Lee, Y.C., and Drake, J.F. (1980). Linear analysis of the double tearing mode, *Phys. Fluids* **23**, 1368–1374.

Proctor, M.R.E., Matthews, P.C., and Rucklidge, A.M. (1993). *Solar and Planetary Dynamos* (Cambridge Univ. Press, Cambridge).

Pudovkin, M.I., and Semenov, V.S. (1985). Magnetic field reconnection theory and the solar wind – magnetosphere interaction: A review, *Space Sci. Rev.* **41**, 1–89.

Pudovkin, M.I., and Shukhova, L.Z. (1988). Certain manifestations of magnetic field reconnection processes in the solar wind, *Geomagn. Aeron.* **28**, 667–670.

Raadu, M.A. (1989). The physics of double layers and their role in astrophysics, *Phys. Reports* **178**, 25–97.

Raadu, M.A., and Rasmussen, J.J. (1988). Dynamical aspects of electrostatic double layers, *Astrophys. Space Sci.* **144**, 43–71.

Rädler, K.-H. (1986). On the effect of differential rotation on axisymmetric and nonaxisymmetric magnetic fields of cosmic bodies, in *Proceedings of the Joint Varenna-Abastumani International School and Workshop on Plasma Astrophysics*, ed. T.D. Guyenne (ESA SP-251, Noordwijk) pp. 569–574.

Ramaty, R. (1979). Energetic particles in solar flares, in *Particle Acceleration Mechanisms in Astrophysics*, eds. J. Arons, C. McKee, and C. Max (Amer. Inst. Phys., New York) pp. 135–154.

Ramaty, R., and Murphy, R.J. (1987). Nuclear processes and accelerated particles in solar flares, *Space Sci. Rev.* **45**, 213–268.

Ramaty, R., Mandzhavidze, N., Kozlovsky, B., and Murphy, R.J. (1995). Solar atmospheric abundances and energy content in flare-accelerated ions from gamma-ray spectroscopy, *Astrophys. J.* **455**, L193–L196.

Rastätter, L., and Neukirch, T. (1997). Magnetic reconnection in a magnetosphere–accretion–disc system: Axisymmetric stationary states and 2D reconnection simulations, *Astron. Astrophys.* **323**, 923–930.

Rastätter, L., Voge, A., and Schindler, K. (1994). On current sheets in 2D ideal MHD caused by pressure perturbations, *Phys. Plasmas* **1**, 3414–3424.

Reames, D.V. (1990). Acceleration of energetic particles by shock waves from large solar flares, *Astrophys. J.* **358**, L63–L67.

Rechester, A.B., and Rosenbluth, M.N. (1978). Electron heat transport in a tokamak with destroyed magnetic surfaces, *Phys. Rev. Lett.* **40**, 38–41.

Rechester, A.B., and Stix, T.H. (1976). Magnetic braiding due to weak asymmetry, *Phys. Rev. Lett.* **36**, 587–591.

Reid, G.C., and Holzer, T.E. (1975). The response of the dayside magnetosphere-ionosphere system to time-varying field line reconnection at the magnetopause. 2. Erosion event of March 27, 1968, *J. Geophys. Res.* **80**, 2050–2056.

Richardson, J.D., Belcher, J.W., Selesnick, R.S., Zhang, M., Siscoe, G.L., and Eviatar, A. (1988). Evidence for periodic reconnection at Uranus?, *Geophys. Res. Lett.* **15**, 733–736.

Richter, A.K., Hsieh, K.C., Luttrell, A.H., Marsch, E., and Schwenn, R. (1985). Review of interplanetary shock phenomena near and within 1 AU, in *Collisionless Shocks in the Heliosphere: Reviews of Current Research*, eds. B. Tsurutani and R. Stone (Amer. Geophys. Union, Washington, DC) pp. 33–50.

Rijnbeek, R.P., and Semenov, V.S. (1993). Features of a Petschek-type reconnection model, *Trends in Geophys. Res.* **2**, 247–268.

Rijnbeek, R.P., Cowley, S.W.H., Southwood, D.J., and Russell, C.T. (1982). Observations of reverse polarity flux transfer events at the Earth's dayside magnetopause, *Nature* **300**, 23–26.

Rijnbeek, R.P., Cowley, S.W.H., Southwood, D.J., and Russell, C.T. (1984). A survey of dayside flux transfer events observed by ISEE 1 and 2 magnetometers, *J. Geophys. Res.* **89**, 786–800.

Risken, H. (1989). *The Fokker-Planck Equation* (Springer-Verlag, Berlin).

Roache, P.J. (1982). *Computational Fluid Dynamics* (Hermosa Press, Albuquerque, New Mexico).

Roberts, B. (1984). Waves in inhomogeneous media, in *The Hydromagnetics of the Sun* (ESA SP-220, Noordwijk) pp. 137–145.

Roberts, B., and Priest, E.R. (1975). On the maximum rate of reconnection for Petschek's mechanism, *J. Plasma Phys.* **14**, 417–431.

Roberts, P.H. (1967). *An Introduction to Magnetodynamics* (Longmans, London).

Rodonò, M. (1986). The atmospheres of M-dwarfs: Observations, in *The M-Type Stars*, eds. J.R. Johnson and F.R. Querci (NASA SP-492, Washington, DC) pp. 409–453.

Rodonò, M., and Lanza, A.F. (1996). Active longitudes and rotation of active stars, in *Magnetodynamic Phenomena in the Solar Atmosphere, Prototypes of Stellar Magnetic Activity*, IAU Colloq. 153, eds. Y. Uchida, T. Kosugi, and H.S. Hudson (Kluwer, Dordrecht) pp. 375–386.

Romanova, M.M., and Lovelace, R.V.E. (1992). Magnetic field reconnection and particle acceleration in extragalactic jets, *Astron. Astrophys.* **262**, 26–36.

Rosenbluth, M.N., and Bussac, M.N. (1979). MHD stability of a spheromak, *Nuclear Fusion* **19**, 489–498.

Rosenbluth, M.N., Sagdeev, R.Z., Taylor, J.B., and Zaslavsky, G.M. (1966). Destruction of magnetic surfaces by magnetic field irregularities, *Nuclear Fusion* **6**, 297–300.

Rosner, R., and Knobloch, E. (1982). On perturbations of magnetic field configurations, *Astrophys. J.* **262**, 349–357.

Rosner, R., and Vaiana, G.S. (1978). Cosmic flare transients: constraints upon models for energy storage and release derived from the event frequency distribution, *Astrophys. J.* **222**, 1104–1108.

Rosner, R., Tucker, W.H., and Vaiana, G.S. (1978). Dynamics of the quiescent solar corona, *Astrophys. J.* **220**, 643–645.

Rossi, B., and Olbert, S. (1970). *Introduction to the Physics of Space* (McGraw-Hill, New York).

Roumeliotis, G., and Moore, R.L. (1993). A linear solution for magnetic reconnection driven by converging or diverging footpoint motion, *Astrophys. J.* **416**, 386–391.

Rusbridge, M.G. (1971). Non-adiabatic charged particle motion near a magnetic field zero line, *Plasma Phys.* **13**, 977.

Russell, C.T., and Elphic, R.C. (1978). Initial ISEE magnetometer results: magnetopause observations, *Space Sci. Rev.* **22**, 681–715.

Russell, C.T., and McPherron, R.L. (1973). The magnetotail and substorms, *Space Sci. Rev.* **15**, 205–266.

Russell, C.T., Saunders, M.A., Phillips, J.L., and Fedder, J.A. (1986). Near-tail reconnection as the cause of cometary tail disconnections, *J. Geophys. Res.* **91**, 1417–1423.

Rust, D.M., and Bar, V. (1973). Magnetic fields, loop prominences and the great flares of August, 1972, *Solar Phys.* **33**, 445–459.

Rutherford, P.H. (1973). Nonlinear growth of the tearing mode, *Phys. Fluids* **16**, 1903–1908.

Ryan, J.M., and Lee, M.A. (1991). On the transport and acceleration of solar flare particles in a coronal loop, *Astrophys. J.* **368**, 316–324.

Ryu, D., and Goodman, J. (1992). Convective instability in differentially rotating disks, *Astrophys. J.* **388**, 438–450.

Saar, S.H. (1996). Recent measurements of stellar magnetic fields, in *Magnetodynamic Phenomena in the Solar Atmosphere, Prototypes of Stellar Magnetic Activity*, IAU Colloq. 153, eds. Y. Uchida, T. Kosugi, and H.S. Hudson (Kluwer, Dordrecht) pp. 367–374.

Sakai, J.-I. (1983). Forced reconnection by fast magnetosonic waves in a current sheet with stagnation-point flows, *J. Plasma Phys.* **30**, 109–124.

Sakai, J.-I. (1990). Prompt particle acceleration and plasma jet formation during current loop coalescence in solar flares, *Astrophys. J. Suppl.* **73**, 321–332.

Sakao, T., et al. (1992). Hard X-ray imaging observations by Yohkoh of the 1991 November 15 solar flare, *Publ. Astron. Soc. Japan* **44**, L83–L87.

Sakimoto, P.J., and Coroniti, F.V. (1989). Buoyancy-limited magnetic viscosity in quasi-stellar accretion disk models, *Astrophys. J.* **342**, 49–63.

Sakurai, T., and Levine, R. (1981). Generation of coronal electric currents due to convective motions on the photosphere, *Astrophys. J.* **248**, 817–837.

Sato, T. (1979). Strong plasma acceleration by slow shocks resulting from magnetic reconnection, *J. Geophys. Res.* **84**, 7177–7190.

Sato, T. (1985). 3D reconnection between two colliding magnetized plasmas, *Phys. Rev. Lett.* **54**, 1502–1505.

Sato, T., Walker, R.J., and Ashour-Abdalla, M. (1984). Driven magnetic reconnection in three dimensions: energy conversion and field-aligned current generation, *J. Geophys. Res.* **89**, 9761–9769.

Saunders, M.A., Russell, C.T., and Sckopke, N. (1984a). A dual-satellite study of the spatial properties of FTE's, in *Magnetic Reconnection in Space and Laboratory Plasmas*, ed. E.W. Hones, Jr. (Amer. Geophys. Union, Washington, DC) pp. 145–152.

Saunders, M.A., Russell, C.T., and Sckopke, N. (1984b). Flux transfer events: Scale size and interior structure, *Geophys. Res. Lett.* **11**, 131–134.

Schatten, K.H., and Wilcox, J.M. (1967). Response of the geomagnetic activity index Kp to the interplanetary magnetic field, *J. Geophys. Res.* **72**, 5185–5191.

Schindler, K. (1974). A theory of the substorm mechanism, *J. Geophys. Res.* **79**, 2803–2810.

Schindler, K., and Birn, J. (1993), On the cause of thin current sheets in the near-Earth magnetotail and their possible significance for magnetospheric substorms, *J. Geophys. Res.* **98**, 15477–15485.

Schindler, K., Hesse, M., and Birn, J. (1988). General magnetic reconnection, parallel electric fields and helicity, *J. Geophys. Res.* **93**, 5547–5557.

Schindler, K., Hesse, M., and Birn, J. (1991). Magnetic field-aligned electric potentials in nonideal plasma flows, *Astrophys. J.* **380**, 293–301.

Schlickeiser, R. (1986a). An explanation of abrupt cutoffs in the optical-infrared spectra of nonthermal sources – a new pile-up mechanism for relativistic electron spectra, *Astron. Astrophys.* **136**, 227–236.

Schlickeiser, R. (1986b). Stochastic particle acceleration in cosmic objects, in *Cosmic Radiation in Contemporary Astrophysics*, ed. M. Shapiro (Reidel, Dordrecht) pp. 27–55.

Schlickeiser, R. (1989). Cosmic-ray transport and acceleration. 1. Derivation of the kinetic equation and application to cosmic rays in static cold media, *Astrophys. J.* **336**, 243–263.

Schlickeiser, R., and Miller, J.A. (1998). Quasi-linear theory of cosmic ray transport and acceleration: the role of oblique MHD waves and transit-time damping, *Astrophys. J.* **492**, 352–378.

Schmidt, G. (1966). *Physics of High-Temperature Plasmas* (Academic Press, London).

Schmieder, B., Forbes, T.G., Malherbe, J.M., and Machado, M.E. (1987).
Evidence for gentle chromospheric evaporation during the gradual phase of
large solar flares, *Astrophys. J.* **317**, 956–963.

Schmieder, B., Mein, P., Simnett, G.M., and Tandberg-Hanssen, E. (1988). An
example of the association of X-ray and UV emission with H-alpha surges,
Astron. Astrophys. **201**, 327–338.

Schmieder, B., Golub, L., and Antiochos, S.K. (1994). Comparison between cool
and hot plasma behaviours of surges, *Astrophys. J.* **425**, 326–330.

Schmieder, B., Heinzel, P., van Driel-Gesztelyi, L., Wiik, J.E., and Lemen, J.
(1996). Hot and cool post-flare loops: Formation and dynamics, in
*Magnetodynamic Phenomena in the Solar Atmosphere, Prototypes of Stellar
Magnetic Activity*, eds. Y. Uchida, T. Kosugi, and H.S. Hudson (Kluwer,
Dordrecht) pp. 211–212.

Schnack, D., and Killeen, J. (1978). Linear and nonlinear calculations of the
tearing mode, in *Theoretical and Computational Plasma Physics* (IAEA,
Vienna) pp. 337–360.

Schnack, D., and Killeen, J. (1979). Nonlinear saturation of the tearing mode in a
reversed-field pinch, *Nuclear Fusion* **19**, 877–887.

Schnack, D., Caramana, E.J., and Nebel, R.A. (1985). Three-dimensional MHD
studies of the reversed-field pinch, *Phys. Fluids* **28**, 321–333.

Schneider, P., and Bogdan, T.J. (1989). Energetic particle acceleration in
spherically symmetric accretion flows: Importance of a momentum-dependent
diffusion coefficient, *Astrophys. J.* **347**, 496–504.

Scholer, M. (1988a). Strong core magnetic field in magnetopause flux transfer
events, *Geophys. Res. Lett.* **15**, 748–751.

Scholer, M. (1988b). Acceleration of energetic particles in solar flares, in *Activity
in Cool Star Envelopes*, eds. O. Havnes, B.R. Pettersen, J.H.M.M. Schmitt,
and J.E. Solheim (Kluwer, Dordrecht) pp. 195–210.

Scholer, M. (1989). Undriven magnetic reconnection in an isolated current sheet,
J. Geophys. Res. **94**, 8805–8812.

Scholer, M., and Jamitzky, F. (1987). Particle orbits during the development of
plasmoids, *J. Geophys. Res.* **92**, 12181–12186.

Schrijver, C.J., Mewe, R., van den Oord, G.H.J., and Kaastra, J.S. (1995). EUV
spectroscopy of cool stars. II. Coronal structure of selected cool stars
observed with the EUVE, *Astron. Astrophys.* **302**, 438–456.

Schrijver, C.J., Title, A.M., Harvey, K.L., Sheeley, N.R., Wang, Y.-M., van den
Oord, G.H.J., Shine, R.A., Tarbell, T.D., and Hurlburt, N.E. (1997).
Large-scale coronal heating by the dynamic, small-scale magnetic field of the
Sun, *Nature* **48**, 424–425.

Schüller, F.C. (1995). Disruptions in tokamaks, *Phys. Control. Fusion* **37**,
A135–A162.

Schumacher, J., and Kliem, B. (1996). Dynamic current sheets with localized
anomalous resistivity, *Phys. Plasmas* **3**, 4703–4711.

Seehafer, N. (1985). An example of a solar flare caused by magnetic field
nonequilibrium, *Solar Phys.* **96**, 307–316.

Seehafer, N. (1986). On the magnetic field line topology in solar active regions,
Solar Phys. **105**, 223–235.

Semenov, V.S., Heyn, M.F., and Kubyshkin, I.V. (1983a). Reconnection of
magnetic field lines in a nonstationary case, *Sov. Astron.* **27**, 660–665.

Semenov, V.S., Kubyshkin, I.V., Heyn, M.F., and Biernat, H.K. (1983b). Field-line reconnection in the 2D asymmetric case, *J. Plasma Phys.* **30**, 321–344.

Semenov, V.S., Vasilyev, E.P., and Pudovkin, A.I. (1984). A scheme for the nonsteady reconnection of magnetic lines of force, *Geomagn. Aeron.* **24**, 370–373.

Semenov, V.S., Kubyshkin, I.V., Lebedeva, V.V., Rijnbeek, R.P., Heyn, M.F., Biernat, H.K., and Farrugia, C.J. (1992). A comparison and review of steady-state and time-varying reconnection, *Planet. Space Sci.* **40**, 63–87.

Sevillano, E., and Ribe, F.L. (1984). Driven magnetic reconnection during the formation of a two-cell field-reversed configuration, in *Magnetic Reconnection in Space and Laboratory Plasmas*, ed. E.W. Hones, Jr. (Amer. Geophys. Union, Washington, DC) pp. 313–318.

Shafranov, V.D. (1966). Plasma equilibrium in a magnetic field, *Revs. Plasma Phys.* **2**, 103–151.

Shakura, N.I., and Sunyaev, R.A. (1973). Black holes in binary systems. Observational appearance, *Astron. Astrophys.* **24**, 337–355.

Shibata, K., Nozawa, S., Matsumoto, R., Sterling, A.C., and Tajima, T. (1990). Emergence of solar magnetic flux from the convection zone into the photosphere and chromosphere, *Astrophys. J.* **351**, L25–L28.

Shibata, K., Ishido, Y., Acton, L.W., Strong, K.T., Hirayama, T., Uchida, Y., McAllister, A.H., Matsumoto, R., Tsuneta, S., Shimizu, T., Hara, H., Sakurai, T., Ichimoto, K., Nishino, Y., and Ogawara, Y. (1992). Observations of X-ray jets with the Yohkoh soft X-ray telescope, *Publ. Astron. Soc. Japan* **44**, L173–L179.

Shibata, K., Nitta, N., Strong, K.T., Matsumoto, R., Yokoyama, T., Hirayama, T., Hudson, H., and Ogawara, Y. (1994). A gigantic coronal jet ejected from a compact active region in a coronal hole, *Astrophys. J.* **431**, L51–L53.

Shibata, K., Shimojo, M., Yohoyama, T., and Ohyama, M. (1996). Theory and observations of X-ray jets, in *Magnetic Reconnection in the Solar Atmosphere*, eds. R.D. Bentley and J.T. Mariska, pp. 29–38.

Shimada, N., Terasawa, T., and Jokipii, J.R. (1997). Stochastic particle acceleration by a pair of slow shocks, *J. Geophys. Res.* **102**, 22301–22310.

Shivamoggi, B.K. (1985). MHD theories of magnetic field reconnection, *Phys. Rep.* **127**, 99–184.

Shu, F., Najita, J., Ostriker, E., and Wilkin, F. (1994). Magneto-centrifugally driven flows from young stars and disks. I. A generalized model, *Astrophys. J.* **429**, 781–796.

Simnett, G.M. (1992). Energy build-up, transport, and release in solar flares, in *Solar Physics and Astrophysics at Interferometric Resolution* (ESA SP-344, Noordwijk) pp. 73–80.

Simnett, G.M., et al. (1997). LASCO observations of disconnected structures out to beyond 28 solar radii during coronal mass ejections, *Solar Phys.* **175**, 685–698.

Simon, T., Linsky, J.L., and Schiffer, F.H. (1980). IUE spectra of a flare in the RS Canum Venaticorum-type system UX Arietis, *Astrophys. J.* **239**, 911–918.

Siscoe, G.L., and Sanchez, E. (1987). An MHD model for the complete open magnetotail boundary, *J. Geophys. Res.* **92**, 7405–7412.

Siscoe, G.L., Ness, N.F., and Yeates, C.M. (1975). Substorms on Mercury?, *J. Geophys. Res.* **80**, 4359–4363.

Smale, A.P., Charles, P.A., Corbet, R.H.D., Jordan, C., Brown, A., and Walter, F. (1986). X-ray and optical observations of a dMe star in the T-Tauri Field, *Mon. Not. Roy. Astron. Soc.* **221**, 77–92.

Smith, D.F. (1977). Particle acceleration by strong plasma turbulence. II – Acceleration of nonrelativistic electrons in solar flares, *Astrophys. J.* **217**, 644–656.

Smith, E.J., Slavin, J.A., Tsurutani, B.T., Feldman, W.C., and Bame, S.J. (1984). Slow mode shocks in the Earth's magnetotail ISEE-3, *Geophys. Res. Lett.* **11**, 1054–1057.

Somov, B.V. (1992). *Physical Processes in Solar Flares* (Kluwer Academic, Dordrecht).

Somov, B.V., and Syrovatsky, S.I. (1976). Hydrodynamic plasma flow in a strong magnetic field, *Proc. Lebedev. Phys. Inst.* **74**, 13–72.

Somov, B.V., and Verneta, A.I. (1994). Tearing instability of reconnecting current sheets in space plasmas, *Space Sci. Rev.* **65**, 253–288.

Somov, B.V., Sermulina, B.J., and Spektor, A.R. (1982). Hydrodynamic response of the solar chromosphere to elementary flare burst. 2. Thermal model, *Solar Phys.* **81**, 281–292.

Song, Y., and Lysak, R.L. (1989). Evaluation of twist helicity of FTE flux tubes, *J. Geophys. Res.* **94**, 5273–5281.

Sonnerup, B.U.Ö. (1970). Magnetic field reconnection in a highly conducting incompressible fluid, *J. Plasma Phys.* **4**, 161–174.

Sonnerup, B.U.Ö. (1971a). Magnetopause structure during the magnetic storm of September 24, 1961, *J. Geophys. Res.* **76**, 6717–6735.

Sonnerup, B.U.Ö. (1971b). Adiabatic particle orbits in a magnetic null sheet, *J. Geophys. Res.* **76**, 8211–8222.

Sonnerup, B.U.Ö. (1979). Magnetic field reconnection, in *Solar System Plasma Physics III*, eds. L.J. Lanzerotti, C.F. Kennel, and E.N. Parker (North Holland Pub. Co., Amsterdam) pp. 45–108.

Sonnerup, B.U.Ö. (1984). Magnetic field reconnection at the magnetopause: an overview, in *Magnetic Reconnection in Space and Laboratory Plasmas*, ed. E.W. Hones, Jr. (Amer. Geophys. Union, Washington, DC) pp. 92–103.

Sonnerup, B.U.Ö. (1987). On the stress balance in flux transfer events, *J. Geophys. Res.* **92**, 8613–8620.

Sonnerup, B.U.Ö. (1988). On the theory of steady-state reconnection, *Computer Phys. Commun.* **49**, 143–159.

Sonnerup, B.U.Ö., and Priest, E.R. (1975). Resistive MHD stagnation-point flows at a current sheet, *J. Plasma Phys.* **14**, 283–294.

Sonnerup, B.U.Ö., and Wang, D.J. (1987). Structure of reconnection boundary layers in incompressible MHD, *J. Geophys. Res.* **92**, 8621–8633.

Sonnerup, B.U.Ö., Paschmann, G., Papamastorakis, I., Sckopke, N., Haerendel, G., Bame, S.J., Asbridge, J.R., Gosling, J.T., and Russell, C.T. (1981). Evidence for magnetic field reconnection at the Earth's magnetopause, *J. Geophys. Res.* **86**, 10049–10067.

Southwood, D.J., Farrugia, C.J., and Saunders, M.A. (1988). What are flux transfer events?, *Planet. Space Sci.* **36**, 503–508.

Soward, A.M. (1982). Fast magnetic field-line reconnection in a compressible fluid. 2. Skewed field lines, *J. Plasma Phys.* **28**, 415–443.

Soward, A.M., and Priest, E.R. (1977). Fast magnetic field reconnection, *Phil. Trans. Roy. Soc. Lond. A* **284**, 369–417.

Soward, A.M., and Priest, E.R. (1982). Fast magnetic field-line reconnection in a compressible fluid. 1. Coplanar field lines, *J. Plasma Phys.* **28**, 335–367.

Soward, A.M., and Priest, E.R. (1986). Magnetic reconnection with jets, *J. Plasma Phys.* **35**, 333–350.

Spanier, J., and Oldham, K.B. (1987). *An Atlas of Functions* (Hemisphere Publishing, New York).

Speiser, T.W. (1965). Particle trajectories in model current sheets, part 1: analytical solutions, *J. Geophys. Res.* **70**, 4219–4226.

Speiser, T.W. (1968). Plasma density and acceleration in the tail from the reconnection model, in *Earth's Particles and Fields*, ed. B.M. McCormac (Reinhold Book Corp., New York) pp. 393–402.

Speiser, T.W. (1969). Some recent results using the Dungey model, in *Atmospheric Emissions*, eds. B.M. McCormac and A. Omholt (Van Nostrand Reinhold, New York) pp. 337–349.

Speiser, T.W. (1988). Plasma sheet theories, in *Modeling Magnetospheric Plasma*, Geophys. Monograph 44, eds. T.E. Moore and J.H. Waite, Jr. (Amer. Geophys. Union, Washington, DC) pp. 277–288.

Spitzer, L. (1962). *Physics of Fully Ionized Gases* (Interscience, New York).

Steele, C.D.C., Hood, A.W., Priest, E.R., and Amari, T. (1989). Nonequilibrium of a cylindrical magnetic arcade, *Solar Phys.* **123**, 127–141.

Steinolfson, R.S. (1991). Coronal evolution due to shear motion, *Astrophys. J.* **382**, 677–687.

Steinolfson, R.S., and Gurnett, D.A. (1995). Distances to the termination shock and heliopause from a simulation analysis of the 1992–93 heliospheric radio emission event, *Geophys. Res. Lett.* **22**, 651–654.

Steinolfson, R.S., and Van Hoven, G. (1983). The growth of the tearing mode: boundary and scaling effects, *Phys. Fluids* **26**, 117–123.

Steinolfson, R.S., and Van Hoven, G. (1984a). Radiative tearing: magnetic reconnection on a fast thermal instability time-scale, *Astrophys. J.* **276**, 391–398.

Steinolfson, R.S., and Van Hoven, G. (1984b). Nonlinear evolution of the resistive tearing mode, *Phys. Fluids* **27**, 1207–1214.

Stenzel, R.L. (1992). Three-dimensional driven reconnection in a laboratory plasma, in *Magnetic Reconnection in Physics and Astrophysics*, ed. P. Maltby (Univ. of Oslo, Oslo) pp. 3–18.

Stenzel, R.L., and Gekelman, W. (1979). Experiments on magnetic field line reconnection, *Phys. Rev. Lett.* **42**, 1055–1057.

Stenzel, R.L., and Gekelman, W. (1981a). Magnetic field line reconnection experiments. 1. Field topologies, *J. Geophys. Res.* **86**, 649–658.

Stenzel, R.L., and Gekelman, W. (1981b). Magnetic field line reconnection experiments. 2. Plasma parameters, *J. Geophys. Res.* **86**, 659–666.

Stenzel, R.L., and Gekelman, W. (1984). Particle acceleration during reconnection in laboratory plasmas, *Adv. Space Res.* **4**, 459–470.

Stenzel, R.L., and Gekelman, W. (1986). Lessons from laboratory experiments on reconnection, *Adv. Space Res.* **6**, 135–147.

Stenzel, R.L., Gekelman, W., and Wild, N. (1982a). Double-layer formation during current sheet disruptions in a reconnection experiment, *Geophys. Res. Lett.* **9**, 680–683.

Stenzel, R.L., Gekelman, W., and Wild, N. (1982b). Magnetic field line reconnection experiments. 4. Resistivity, heating and energy flow, *J. Geophys. Res.* **87**, 111–117.

Stenzel, R.L., Gekelman, W., and Wild, N. (1983). Magnetic field line reconnection experiments. 5. Current disruptions and double layers, *J. Geophys. Res.* **88**, 4793–4804.

Stern, D.P. (1979). The role of O-type neutral lines in magnetic merging during substorms and solar flares, *J. Geophys. Res.* **84**, 63–71.

Stern, D.P. (1989). A brief history of magnetospheric physics before the space flight era, *Revs. Geophys.* **27**, 103–114.

Stern, D.P. (1990). Substorm electrodynamics, *J. Geophys. Res.* **95**, 12057–12067.

Stix, M. (1989). *The Sun* (Springer-Verlag, New York).

Strachan, N.R., and Priest, E.R. (1994). A general family of non-uniform reconnection models with separatrix jets, *Geophys. Astrophys. Fluid Dynamics* **74**, 245–274.

Strauss, H.R. (1977). Dynamics of high-beta tokamaks, *Phys. Fluids* **20**, 1354–1360.

Strauss, H.R. (1981). Resistive ballooning modes, *Phys. Fluids* **24**, 2004–2009.

Strauss, H.R. (1986). Hyper-resistivity produced by tearing mode turbulence, *Phys. Fluids* **29**, 3668–3671.

Strauss, H.R. (1993). Fast three-dimensional driven reconnection, *Geophys. Res. Lett.* **20**, 325–328.

Sturrock, P.A. (1966). Model of the high-energy phase of solar flares, *Nature* **211**, 695–697.

Sturrock, P.A. (1968). A model of solar flares, in *Structure and Development of Solar Active Regions*, IAU Symp. 35, ed. K.O. Kiepenheuer (Reidel, Dordrecht) pp. 471–479.

Sturrock, P.A. (1991). Maximum energy of semi-infinite magnetic field configurations, *Astrophys. J.* **380**, 655–659.

Sturrock, P.A. (1994). *Plasma Physics* (Cambridge Univ. Press, Cambridge).

Sturrock, P.A., and Uchida, Y. (1981). Coronal heating by stochastic magnetic pumping, *Astrophys. J.* **246**, 331–336.

Sturrock, P.A., and Woodbury, E.T. (1967). Force-free magnetic fields and solar filaments, in *Plasma Astrophysics*, ed. P. Sturrock (Academic Press, London) pp. 155–167.

Sturrock, P.A., Kaufman, P., Moore, R.L., and Smith, D.F. (1984). Energy release in solar flares, *Solar Phys.* **94**, 341–357.

Suess, S.T. (1990). The heliopause, *Rev. Geophys.* **28**, 97–115.

Suess, S.T., and Nerney, S. (1993). The polar heliospheric magnetic field, *Geophys. Res. Lett.* **20**, 329–332.

Sundaram, A.K., Das, A.C., and Sen, A. (1980). A nonlinear mechanism for magnetic reconnection and substorm activity, *Geophys. Res. Lett.* **7**, 921–924.

Suzuki, Y., Kusaro, K., and Nishikawa, K. (1997). Three-dimensional simulation study of MHD relaxation processes in the solar corona, *Astrophys. J.* **474**, 782–791.

References

561

Svalgaard, L. (1973). Polar cap magnetic variations and their relationship with interplanetary magnetic sector structure, *J. Geophys. Res.* **78**, 2064–2078.

Švestka, Z. (1976). *Solar Flares* (D. Reidel, Dordrecht).

Švestka, Z., and Cliver, E.W. (1992). History and basic characteristics of eruptive flares, in *Eruptive Solar Flares*, eds. Z. Švestka, B.V. Jackson, and M.E. Machado (Springer-Verlag, New York) pp. 1–14.

Švestka, Z., and Simon, P. (1969). Proton flare project, *Solar Phys.* **10**, 3–59.

Švestka, Z., Martin, S.F., and Kopp, R.A. (1980). Particle acceleration in the process of the eruptive opening and reconnection of magnetic fields, in *Solar and Interplanetary Dynamics*, eds. M. Dryer and E. Tandberg-Hanssen (Reidel, Dordrecht) pp. 217–221.

Švestka, Z., Dodson-Prince, H.W., Martin, S.F., Mohler, O.C., Moore, R.L., Nolte, J.T., and Petrasso, R.D. (1982). Study of the post-flare loops on 29 July 1973, *Solar Phys.* **78**, 271–285.

Švestka, Z., Fontenla, J.M., Machado, M.E., Martin, S.F., Neidig, D.F., and Poletto, G. (1987). Multithermal observations of newly formed loops in a dynamic flare, *Solar Phys.* **108**, 237–250.

Švestka, Z., Farnik, F., and Tang, F. (1990). X-ray bright surges, *Solar Phys.* **127**, 149–163.

Sweet, P.A. (1958a). The neutral point theory of solar flares, in *Electromagnetic Phenomena in Cosmical Physics*, IAU Symp. 6, ed. B. Lehnert (Cambridge Univ. Press, London) pp. 123–134.

Sweet, P.A. (1958b). The production of high-energy particles in solar flares, *Nuovo Cimento Suppl.* **8**, Ser. X, 188–196.

Sylwester, B., Sylwester, J., Serio, S., Reale, F., Bentley, R.D., and Fludra, A. (1993). Dynamics of flaring loops. III. Interpretation of flare evolution in the emission measure temperature diagram, *Astron. Astrophys.* **267**, 586–594.

Syrovatsky, S.I. (1971). Formation of current sheets in a plasma with a frozen-in strong magnetic field, *Sov. Phys. JETP* **33**, 933–940.

Syrovatsky, S.I. (1978). On the time-evolution of force-free fields, *Solar Phys.* **58**, 89–94.

Syrovatsky, S.I. (1982). Model for flare loops, fast motions, and opening magnetic field in the corona, *Solar Phys.* **76**, 3–20.

Syrovatsky, S.I., Frank, A.G., and Khodzhaev, A.Z. (1973). Current distribution near the null line of a magnetic field and turbulent plasma resistance, *Sov. Phys. Tech. Phys.* **18**, 580–586.

Tajima, T., and Gilden, D. (1987). Reconnection-driven oscillations in dwarf novae disks, *Astrophys. J.* **320**, 741–745.

Takahashi, K., Zanetti, L.J., Lopez, R.E., McEntire, R.W., Potemra, T.A., and Yumoto, K. (1987). Disruption of the magnetotail current sheet observed by AMPTE/CCE, *Geophys. Res. Lett.* **14**, 1019–1022.

Tandberg-Hanssen, E., and Emslie, A.G. (1988). *The Physics of Solar Flares* (Cambridge Univ. Press, Cambridge).

Tataronis, J., and Grossman, W. (1973). Decay of MHD waves by phase mixing, *Zs. Physik* **261**, 203–216.

Taylor, J.B. (1974). Relaxation of toroidal plasma and generation of reverse magnetic fields, *Phys. Rev. Lett.* **33**, 1139–1141.

Taylor, J.B. (1986). Relaxation and magnetic reconnection in plasmas, *Revs. Modern Phys.* **58**, 741–763.

Taylor, S.R. (1992). *Solar System Evolution* (Cambridge Univ. Press, Cambridge).

Temerin, M., and Roth, I. (1992). The production of He-rich flares by electromagnetic hydrogen cyclotron waves, *Astrophys. J. Lett.* **391**, L105–L108.

Terasawa, T. (1981). Energy spectrum of ions accelerated through the Fermi process at the terrestrial bow shock, *J. Geophys. Res.* **86**, 7595–7606.

Thyagaraga, A. (1981). Perturbation analysis of a simple model of magnetic island structures, *Phys. Fluids* **24**, 1716–1724.

Ting, A.C., Mattaeus, W.H., and Montgomery, D. (1986). Turbulent relaxation processes in MHD, *Phys. Fluids* **29**, 3261–3274.

Titov, V.S. (1992). On the method of calculating two-dimensional potential magnetic fields with current sheets, *Solar Phys.* **139**, 401–404.

Titov, V.S., and Priest, E.R. (1993). The collapse of an X-type neutral point to form a reconnecting time-dependent current sheet, *Geophys. Astrophys. Fluid Dynamics* **72**, 249–276.

Titov, V.S., and Priest, E.R. (1997). Visco-resistive magnetic reconnection due to steady inertialess flows. Part 1. Exact analytical solutions, *J. Fluid Mech.* **348**, 327–347.

Titov, V.S., Priest, E.R., and Démoulin, P. (1993). Conditions for the appearance of 'bald patches' at the solar surface, *Astron. Astrophys.* **276**, 564–570.

Toptyghin, I.N. (1980). Acceleration of particles by shocks in a cosmic plasma, *Space Sci. Rev.* **26**, 157–213.

Tout, C., and Pringle, J.E. (1992). Accretion disk viscosity – a simple model for a magnetic dynamo, *Mon. Not. Roy. Astron. Soc.* **259**, 604–612.

Truemann, R.A., and Baumjohann,W. (1997). *Advanced Space Plasma Physics* (Imperial College Press, London).

Tsinganos, K.C. (1982). MHD equilibrium, *Astrophys. J.* **252**, 775–790.

Tsinganos, K.C., Distler, J., and Rosner, R. (1984). On the topological stability of magnetostatic equilibria, *Astrophys. J.* **278**, 409–419.

Tsuneta, S. (1993). Solar flares as an ongoing magnetic reconnection process, in *The Magnetic and Velocity Fields of Solar Active Regions*, eds. H. Zirin, G. Ai, and H. Wang (Astron. Soc. Pacific, San Francisco) pp. 239–248.

Tsuneta, S. (1995). Particle acceleration and magnetic reconnection in solar flares, *Pub. Astron. Soc. Japan.* **47**, 691–697.

Tsuneta, S. (1996). Structure and dynamics of magnetic reconnection in a solar flare, *Astrophys. J.* **456**, 840–849.

Tsuneta, S., Hara, H., Shimizu, T., Acton, L.W., Strong, K.T., Hudson, H.S., and Ogawara, Y. (1992). Observation of a solar flare at the limb with the Yohkoh Soft X-ray Telescope, *Publ. Astron. Soc. Japan* **44**, L63–L69.

Tsuneta, S., and Naito, T. (1998). Fermi acceleration at the fast shock in a solar flare and the impulsive loop-top hard X-ray source, *Astrophys. J.* **495**, L67–L70.

Tsytovich, V.N., Stenflo, L., Wilhelmsson, H., Gustavsson, H.-G., and Ostberg, K. (1973). One-dimensional model for nonlinear reflection of laser radiation by an inhomogeneous plasma layer, *Phys. Scr. (Sweden)* **7**, 241–249.

Tur, T.J. (1977). *Aspects of Current Sheet Theory*, PhD thesis (St., Andrews Univ., St. Andrews).

Tur, T.J., and Priest, E.R. (1976). The formation of current sheets during the emergence of new magnetic flux from below the photosphere, *Solar Phys.* **48**, 89–102.

Tur, T.J., and Priest, E.R. (1978). A trigger mechanism for the emerging flux model of solar flares, *Solar Phys.* **58**, 181–200.

Turner, L. (1984). Analytical solutions of $\nabla \times \mathbf{B} = \lambda \mathbf{B}$ having separatrices for geometries with one ignorable coordinate, *Phys. Fluids* **27**, 1677–1685.

Turner, L., and Christiansen, J.F. (1981). Incomplete relaxation of pinch discharges, *Phys. Fluids* **24**, 893–898.

Tverskoi, B.A. (1967). Contribution to the theory of Fermi statistical acceleration, *Sov. Phys. JETP* **25**, 317–325.

Uberoi, M.S. (1963). Some exact solutions of magnetohydrodynamics, *Phys. Fluids* **6**, 1379–1381.

Uberoi, M.S. (1966). Reply to comments by S. Chapman and P.C. Kendall, *Phys. Fluids* **9**, 2307.

Uchida, Y. (1986). Magnetohydrodynamic phenomena in the solar and stellar outer atmospheres, *Astrophys. Space Sci.* **118**, 127–148.

Ugai, M. (1984). Self-consistent development of fast magnetic reconnection with anomalous plasma resistivity, *Plasma Phys. Control. Fusion* **26**, 1549–1563.

Ugai, M. (1988). MHD simulations of fast reconnection spontaneously developing in a current sheet, *Comp. Phys. Commun.* **49**, 185–192.

Ugai, M. (1995a). Computer studies on powerful magnetic energy conversion by the spontaneous fast reconnection mechanism, *Phys. Plasmas* **2**, 388–397.

Ugai, M. (1995b). Computer studies on plasmoid dynamics associated with the spontaneous fast reconnection mechanism, *Phys. Plasmas* **2**, 3320–3328.

Ugai, M., and Tsuda, T. (1977). Magnetic field-line reconnection by localized enhancement of resistivity. I. Evolution in a compressible MHD fluid, *J. Plasma Phys.* **17**, 337–356.

Ugai, M., and Tsuda, T. (1979). Magnetic field-line reconnection by localized enhancement of resistivity. III. Controlling factors, *J. Plasma Phys.* **21**, 459–473.

Ugai, M., and Wang, W.B. (1998). Computer simulations on three-dimensional plasmoid dynamics by spontaneous fast reconnection, *J. Geophys. Res.* **103**, 4573–4585.

Uzdensky, D.A., and Kulsrud, R.M. (1998). On the viscous boundary layer near the center of the resistive reconnection region, *Phys. Plasmas* **5**, 3249–3256.

Van Allen, J.A., Fennel, J.F., and Ness, N.F. (1971). Asymmetric access of energetic solar protons to the Earth's north and south polar caps, *J. Geophys. Res.* **19**, 4262–4275.

van Ballegooijen, A.A. (1985). Electric currents in the solar corona and the existence of magnetostatic equilibrium, *Astrophys. J.* **298**, 421–430.

van Ballegooijen, A.A., and Martens, P.C.H. (1989). Formation and eruption of solar prominences, *Astrophys. J.* **343**, 971–984.

van den Oord, G.H.J., and Mewe, R. (1989). The X-ray flare and the quiescent emission from Algol as detected by EXOSAT, *Astron. Astrophys.* **213**, 245–260.

van den Oord, G.H.J., et al. (1996). Flare energetics: Analysis of a large flare on YZ CMi observed simultaneously in the ultraviolet, optical and radio, *Astron. Astrophys.* **310**, 908–922.

Van Hoven, G., and Hendrix, D.L. (1995). Diagnosis of general magnetic reconnection, *J. Geophys. Res.* **100**, 19819–19828.

Van Tend, W., and Kuperus, M. (1978). The development of coronal electric current systems in active regions and their relation to filaments and flares, *Solar Phys.* **59**, 115–127.

Vasyliunas, V.M. (1972). Nonuniqueness of magnetic field line reconnection, *J. Geophys. Res.* **77**, 6271–6274.

Vasyliunas, V.M. (1975). Theoretical models of magnetic field line merging, *Rev. Geophys. Space Phys.* **13**, 303–336.

Vasyliunas, V.M. (1980). Upper limit on the electric field along a magnetic O-line, *J. Geophys. Res.* **85**, 4616–4620.

Vasyliunas, V.M. (1984). Steady-state aspects of magnetic field line merging, in *Magnetic Reconnection in Space and Laboratory Plasmas*, ed. E.W. Hones, Jr. (Amer. Geophys. Union, Washington, DC) pp. 23–31.

Vedenov, A.A., Velikhov, E.P., and Sagdeev, R.Z. (1961). Quasi-linear theory of plasma oscillations, *Nuclear Fusion Supp.* **2**, 465–475.

Vekstein, G.E., and Browning, P.K. (1997). Electric-drift generated trajectories and particle acceleration in collisionless magnetic reconnection, *Phys. Plasmas* **4**, 2261–2268.

Vekstein, G.E., and Jain, R. (1998). Energy release and plasma heating by forced magnetic reconnection, *Phys. Plasmas* **5**, 1506–1513.

Vekstein, G.E., and Priest, E.R. (1992). Magnetohydrostatic equilibria and cusp formation at an X-type neutral line by footpoint shearing, *Astron. Astrophys.* **384**, 333–340.

Vekstein, G.E., and Priest, E.R. (1993). Magnetostatic equilibria and current sheets in a sheared magnetic field with an X-point, *Solar Phys.* **146**, 119–125.

Vekstein, G.E., Priest, E.R., and Amari, T. (1990). Formation of current sheets in force-free magnetic fields, *Astron. Astrophys.* **243**, 492–500.

Velhikhov, E.P. (1959). Stability of an ideally conducting liquid flowing between cylinders rotating in a magnetic fluid, *Sov. Phys. JETP* **9**, 995–998.

Velikanova, L.G., Kirii, N.P., Kiselev, D.T., Markov, V.S., Preobrazhenskii, N.G., and Frank, A.G. (1993). Evolution of a current layer from the results of spectro–tomographic studies, *Sov. J. Plasma Phys.* **18**, 800–806.

Velli, M., Grappin, R., and Mangeney, A. (1990). Physical consistency in modelling interplanetary MHD fluctuations, *Phys. Rev. Lett.* **64**, 2592.

Verbunt, F. (1982). Accretion disks in stellar X-ray sources, *Space Sci. Rev.* **32**, 379–404.

Verigin, M.I., Axford, W.I., Gringauz, K.I., and Richter, A.K. (1987). Acceleration of cometary plasma in the vicinity of comet Halley associated with an inter-planetary magnetic field polarity change, *Geophys. Res. Lett.* **14**, 987–990.

Vilhu, O., Muhli, P., Huovelin, J., Hakala, P., Rucinski, S.M., and Cameron, A.C. (1998). Ultraviolet spectroscopy of AB Doradus with the Hubble Space Telescope: Impulsive flares, bimodal profiles and Doppler images of the CIV 1549 line in a young star, *Astron. J.* **115**, 1610–1616.

Vlahos, L. (1989). Particle acceleration in solar flares, *Solar Phys.* **121**, 431–447.

Vlahos, L., Gergely, T.E., and Papadopoulos, K. (1982). Electron acceleration and radiation signatures in loop coronal transients, *Astrophys. J.* **258**, 812–822.

Vlahos, L., Georgoulis, M., Kluiving, R., and Paschos, P. (1995). The statistical flare, *Astron. Astrophys.* **299**, 897–911.

Vogt, S.S., and Penrod, G.D. (1983). Doppler imaging of spotted stars: Application to the RS Canum Venaticorum star HR1099, *Publ. Astron. Soc. Pacific* **95**, 565–576.

Voight, G.-H. (1978). A steady-state field line reconnection model for the Earth's magnetosphere, *J. Atmos. Terr. Phys.* **40**, 355–365.

Waddell, B.V., Rosenbluth, M.N., Monticello, D.A., and White, R.B. (1976). Nonlinear growth of the $m = 1$ tearing mode, *Nuclear Fusion* **16**, 528–532.

Waddell, B.V., Carreras, B., Hicks, H.R., Holmes, J.A., and Lee, D.K. (1978). Mechanism for major disruptions in tokamaks, *Phys. Rev. Lett.* **41**, 1386–1389.

Wagner, W.J., Hildner, E., House, L.L., Sawyer, C., Sheridan, K.V., and Dulk, G.A. (1981). Radio and visible light observations of matter ejected from the Sun, *Astrophys. J.* **244**, L123–L126.

Walker, R.J., Ogino, T., Raeder, J., and Ashour-Abdalla, M. (1993). A global magnetohydrodynamic simulation of the magnetosphere when the interplanetary magnetic field is southward: The onset of magnetotail reconnection, *J. Geophys. Res.* **98**, 17235–17251.

Wang, X., and Bhattacharjee, A. (1992). Forced reconnection and current sheet formation in Taylor's model, *Phys. Fluids B* **4**, 1795–1799.

Wang, X., and Bhattacharjee, A. (1996). A three-dimensional reconnection model of the magnetosphere: Geometry and kinematics, *J. Geophys. Res.* **101**, 2641–2653.

Watson, A.A. (1993). The search for the origin of the highest energy cosmic rays, *Inst. of Phys. Conf. Ser.* **133**, 135–148.

Webb, D.F., and Cliver, E.W. (1995). Evidence for magnetic disconnection of mass ejections in the corona, *J. Geophys. Res.* **100**, 5853–5870.

Webb, D.F., Cheng, C.-C., Dulk, G.A., Edberg, S.J., Martin, S.F., McKenna, S.L., and McLean, D.J. (1980). Mechanical energy output of the 5 September 1973 flare, in *Solar Flares: A Monograph from the Skylab Solar Workshop II*, ed. P.A. Sturrock (Colorado Assoc. Univ. Press, Boulder) pp. 471–499.

Webb, D.F., Forbes, T.G., Aurass, H., Chen, J., Martens, P., Rompolt, B., Rusin, V., Martin, S.F., and Gaizauskas, V. (1994). Material ejection: Report of the Flares 22 Workshop, *Solar Phys.* **153**, 73–89.

Webb, G.M. (1989). The diffusion approximation and transport theory for cosmic rays in relativistic flows, *Astrophys. J.* **340**, 1112–1123.

Webb, G.M., Axford, W.I., and Terasawa, T. (1983). On the drift mechanism for energetic charged particles at shocks, *Astrophys. J.* **270**, 537–553.

Weber, E.J., and Leverett, D., Jr. (1967). The angular momentum of the solar wind, *Astrophys. J.* **148**, 217–227.

Weiler, E.J., Owen, F.N., Bopp, B.W., Schmitz, M., Hall, D.S., Fraquelli, D.A., Pirola, V., Ryle, M., and Gibson, D.M. (1978). Coordinated ultraviolet, optical and radio observations of HR 1099 and UX Arietis, *Astrophys. J.* **225**, 919–931.

Wesson, J.A. (1966). Finite resistivity instabilites of a sheet pinch, *Nuclear Fusion* **6**, 130–134.

Wesson, J.A. (1978). Hydromagnetic stability of tokamaks, *Nuclear Fusion* **18**, 87–132.

Wesson, J.A. (1981). MHD stability theory, in *Plasma Physics and Nuclear Fusion Research*, ed. R. Gill (Academic Press, New York) pp. 191–233.

Wesson, J.A. (1986). Sawtooth oscillations, *Plasma Phys. Control. Fusion* **28**, 243–248.

Wesson, J.A. (1989). The sawtooth mystery, *Physics Today* **42**, 562–563.

Wesson, J.A. (1990). Sawtooth reconnection, *Nuclear Fusion* **30**, 2545–2549.

Wesson, J.A. (1997). *Tokamaks* (Clarendon Press, Oxford).

Whipple, E.C., Rosenberg, M., and Brittnacher, M. (1990). Magnetotail acceleration using generalized drift theory – A kinetic merging scenario, *Geophys. Res. Lett.* **17**, 1045–1048.

White, R.B., Monticello, D.A., Rosenbluth, M.N., and Waddell, B.V. (1977). Saturation of the tearing mode, *Phys. Fluids* **20**, 800–805.

White, W.W., Siscoe, G.L., Erickson, G.M., Kaymaz, Z., Maynard, N.C., Siebert, K.D., Sonnerup, B.U.Ö., and Weimer, D.R. (1998). The magnetospheric sash and the cross tail S, *Geophys. Res. Letts.* **25**, 1605–1608.

Wiegelmann, T., and Schindler, K. (1995). Formation of thin current sheets in a quasistatic magnetotail model, *Geophys. Res. Lett.* **22**, 2057–2060.

Wild, N., Gekelman, W., and Stenzel, R.L. (1981). Resistivity and energy flow in a plasma undergoing magnetic field line reconnection, *Phys. Rev. Lett.* **46**, 339–342.

Williams, D.J. (1997). Considerations of source, transport, acceleration, heating and loss processes responsible for geomagnetic tail particle populations, *Space Sci. Rev.* **80**, 369–389.

Winningham, J.D., Yasuhara, F., Akasofu, S.-I., and Heikkila, W.J. (1975). The latitudinal morphology of 10-eV to 10-keV electron fluxes during magnetically quiet and disturbed times in the 2100–0300 JLT sector, *J. Geophys. Res.* **80**, 3148–3171.

Winske, D., Thomas, V.A., and Omidi, N. (1995). Diffusion at the magnetopause: A theoretical perspective, in *Physics of the Magnetopause*, Geophys. Monograph 90, eds. P. Song, B.U.Ö. Sonnerup, and M.F. Thomsen (Amer. Geophys. Union, Washington, DC) pp. 321–330.

Wolfson, R., and Low, B.C. (1992). Energy buildup in sheared force-free magnetic fields, *Astrophys. J.* **391**, 353–358.

Woltjer, L. (1958). A theorem on force-free fields, *Proc. Nat. Acad. Sci. USA* **44**, 489–491.

Wright, A.N. (1987). The evolution of an isolated reconnected flux tube, *Planet. Space Sci.* **35**, 813–819.

Wright, A.N. (1999). The role of magnetic helicity in magnetospheric physics, in *Magnetic Helicity in Space and Laboratory Plasmas*, Geophys. Monograph 111, eds. M.R. Brown, R.C. Canfield, and A.A. Pevtsov (Amer. Geophys. Union, Washington, DC) pp. 267–276.

Wright, A.N., and Berger, M.A. (1989). The effect of reconnection upon the linkage and interior structure of magnetic flux tubes, *J. Geophys. Res.* **94**, 813–819.

Wu, F., and Low, B.C. (1987). Static current-sheet models of quiescent prominences, *Astrophys. J.* **312**, 431–443.

Wu, J.-P., and Lin, J.-Y. (1989). The probability of magnetic field reconnection in the solar wind and its observational evidence, *Chinese J. Space Sci.* **9**, 117–126.

Wu, S.T., Guo, W.P., and Wang, J.F. (1995). Dynamical evolution of a coronal streamer-bubble system, *Solar Phys.* **157**, 325–348.

Yamada, M., Ono, Y., Hayakawa, A., and Katsurai, M. (1990). Magnetic reconnection of plasma toroids with cohelicity and counterhelicity, *Phys. Rev. Lett.* **65**, 721–724.

Yamada, M., Perkins, F.W., and MacAulay, A.K. (1991). Initial results from investigation of three-dimensional magnetic reconnection in a laboratory plasma, *Phys. Fluids B* **3**, 2379–2385.

Yamada, M., Levinton, F.M., Pomphredy, N., Budny, R., Manickam, J., and Nagayama, Y. (1994). Investigation of magnetic reconnection during a sawtooth crash in a high-temperature plasma, *Phys. Plasmas* **1**, 3269–3276.

Yamada, M., Ji, H., Hsu, S., Carter, T., Kulsrud, R.M., Bretz, N., Jobes, F., Ono, Y., and Perkins, F.W. (1997a). Study of driven magnetic reconnection in a laboratory plasma, *Phys. Plasmas* **4**, 1937–1944.

Yamada, M., Ji, H., Hsu, S., Carter, T., Russell, K., Ono, Y., and Perkins, F. (1997b). Identification of Y-shaped and O-shaped diffusion regions during magnetic reconnection in laboratory plasmas, *Phys. Rev. Lett.* **78**, 3117–3120.

Yan, M., Lee, L.C., and Priest, E.R. (1992). Fast magnetic reconnection with small shock angles, *J. Geophys. Res.* **97**, 8277–8293.

Yan, M., Lee, L.C., and Priest, E.R. (1993). Magnetic reconnection with large separatrix angles, *J. Geophys. Res.* **98**, 7593–7602.

Yang, C.-K., and Sonnerup, B.U.Ö. (1977). Compressible magnetopause reconnection, *J. Geophys. Res.* **82**, 699–703.

Yeh, T. (1983). Diamagnetic force on a flux tube, *Astrophys. J.* **264**, 630–634.

Yeh, T., and Axford, W.I. (1970). On the reconnection of magnetic field lines in conducting fluids, *J. Plasma Phys.* **4**, 207–229.

Yi, Y., Caputo, F.M., and Brandt, J.C. (1994). Disconnection events (DEs) and sector boundaries: The evidence from comet Halley 1985–1986, *Planet. Space Sci.* **42**, 705–720.

Yokoyama, T., and Shibata, K. (1997). Magnetic reconnection coupled with heat conduction, *Astrophys. J.* **474**, L61–L64.

Yoshida, T., and Tsuneta, S. (1996). Temperature structure of solar active regions, *Astrophys. J.* **459**, 342–346.

Zhu, Z., and Parks, G. (1993). Particle orbits in model current sheet with a nonzero B(y) component, *J. Geophys. Res.* **98**, 7603–7608.

Zimbardo, G. (1989). A self-consistent picture of Jupiter's nightside magnetosphere, *J. Geophys. Res.* **94**, 8707–8719.

Zimbardo, G. (1993). Observable implications of tearing-mode instability in Jupiter's nightside magnetosphere, *Planet. Space Sci.* **41**, 357–361.

Zirin, H. (1988). *Astrophysics of the Sun* (Cambridge Univ. Press, Cambridge).

Zirin, H., and Lackner, D.R. (1969). The solar flares of August 28 and 30, *Solar Phys.* **6**, 86–103.

Zukakishvili, G.G., Kavartskhava, I.F., and Zukakishvili, L.M. (1978). Plasma behaviour near the neutral line between parallel currents, *Sov. J. Plasma Phys.* **4**, 405–410.

Zweibel, E.G. (1988). Ambipolar diffusion drifts and dynamos in turbulent gases, *Astrophys. J.* **329**, 384–391.

Zweibel, E.G. (1989). Magnetic reconnection in partially ionized gases, *Astrophys. J.* **340**, 550–557.

Zweibel, E.G., and Li, H. (1987). The formation of current sheets in the solar atmosphere, *Astrophys. J.* **312**, 423–430.

Zwingmann, W., Schindler, K., and Birn, J. (1985). On sheared magnetic-field structures containing neutral points, *Solar Phys.* **99**, 133–143.

Appendix 1
NOTATION

Latin Alphabet

Note that subscripts 0, 1, 2 denote values of variables that are particular to a given section. Also, the paragraph equation references in brackets below are to the first occurrence of a particular notation.

a	a constant that is particular to a given section
A	magnetic flux function (§1.3.1)
A_0	area of emitting region (§12.1)
A_{np}	nonpotential part of a flux function (§5.2.2)
A_p	potential part of a flux function (§5.2.2)
$\mathbf{A}$	magnetic vector potential (§1.2.2)
$\mathbf{A}_p$	magnetic vector potential in a Coulomb gauge (§8.5.2)
b	a constant or parameter that is particular to a given section
B_e	external magnetic field (§1.6)
B_i	inflow magnetic field to diffusion region (§4.2)
B_N	normal field component at a shock wave (§4.3)
B_{pc}	polar cap magnetic field (§10.1)
B_{sw}	solar wind magnetic field (§10.1)
B_ϕ	toroidal magnetic field component (§6.4)
$\mathbf{B}$	magnetic induction (§1.2.1)
$\mathbf{B}_{mp}$	magnetopause magnetic field (§10.3)
$\mathbf{B}_p$	poloidal magnetic field (§2.6.1)
$\mathcal{B}$	complex magnetic field, $B_y + iB_x$ (§2.2.4)
c	speed of light (§1.7)
c_{is}	ion sound speed (Eq. 13.9)
c_s	sound speed (Table 1.1)
c_v	specific heat at constant volume (§1.2.2)
$\bar{c}$	a current parameter (§5.2.2)

569

C	a curve (§1.4); or a field-line constant (§8.1.3)
d/dt	convective derivative (§1.2.1)
$\text{daw}(x)$	Dawson integral function (§3.2)
D	diffusion coefficient (§13.2.2)
$2\pi D/\mu$	dipole moment (§2.2.3)
D_R	the domain of a diffusion region (§8.1.1)
e	internal energy per unit mass (§1.2.1)
$\text{erf}(x)$	error function (§3.1.1)
$\mathbf{e}$	eigenvector (§8.7)
E_a	asymptotic electric field at large distances from a diffusion region (§8.1.1)
E_A	convective electric field based on the Alfvén speed (§1.7)
E_c	ionospheric electric field (§10.5)
E_D	Dreicer electric field (Eq. 1.66)
E_h	source of heating (§12.1.1)
E_m	volumetric emission measure (§12.1)
E_{rec}	reconnection electric field (§10.5)
E_{SP}	Sweet-Parker electric field (§1.7, §4.2)
E^*	electric field at two-dimensional X-line (i.e. an X-point) (§7.2)
$E_{\parallel}$	component of electric field parallel to the magnetic field (§8.1)
$\mathbf{E}$	electric field (§1.2.1)
$\mathbf{E}_{\perp}$	component of electric field normal to the magnetic field (§13.1)
f	distribution function (§13.2.2)
$f(x)$	a function of x, say, that is particular to a given section
f_s	distribution function of massive spheres (§13.2.3)
$\bar{f}$	filling factor (§12.1)
F	magnetic flux of a tube (§2.4); or a velocity function (§3.5); or a Hamiltonian function (§8.1.1)
F_B	polar cap magnetic flux (§10.1)
F_H	heating flux (§11.2.1)
$\mathbf{F}$	net force on particles (§1.7)
$\mathbf{F}_B$	flow of a magnetic field (§8.1.2)
$\mathbf{F}_g$	external force (§1.2.1)
$\mathbf{F}_{\text{kink}}$	kink force (§10.1)
$\mathbf{F}_w$	mapping across a magnetic field (§8.1.2)
$\mathcal{F}$	displacement gradient tensor (§8.7); or particle flux (§13.3.2)
$g(x)$	a function of x, say, that is particular to a given section
G	the potential of a non-ideal term in Ohm's law (§8.1.1); or gravitational constant (§12.2.1)

h	height of flare loops (§12.1)
h_d	half-thickness of accretion disc (§12.2.1)
H	magnetic helicity (Eq. 8.49)
H_2	cross helicity in two dimensions (§11.3.3)
H_3	cross helicity in three dimensions (§11.3.3)
H_g	gravitational scale-height (§11.1.1)
H_m	mutual magnetic helicity (§8.5.3)
H_s	self magnetic helicity (§8.5.3)
$\mathbf{H}$	magnetic field (§1.2.1)
$\mathcal{H}$	Heavyside function (§11.1.1)
I	electric current (§9.1.2)
$I_2(x)$	modified Bessel function (§13.2.3)
I_{run}	runaway electric current (§13.1.3)
j_N	current density at neutral point (§4.2)
$j_\parallel$	electric current parallel to the spine of a null (§8.2)
$j_\perp$	electric current normal to the spine of a null (§8.2)
$\mathbf{j}$	electric current density (§1.2.1)
J	electric current per unit length of a sheet (§2.3.1)
J_n	Bessel function of order n (§1.4)
$\mathcal{J}$	Jacobian (§8.7)
k	wave number (§2.3.2)
k_B	Boltzmann's constant (§1.2.1)
$\mathbf{k}$	vector wave number (§6.1)
K	a constant that is particular to a given section
$K_2(x)$	modified Bessel function of second kind (§13.2.3)
l	half-width of current sheet (§1.3.2)
l_{mp}	thickness of magnetopause (§10.3)
l_p	length-scale for photospheric motions (§11.3.3)
L	half-length of current sheet (§1.3.2); or half-length of coronal loop (§12.1.1)
L_0	a typical length-scale (§1.2.2)
$L_{collision}$	current sheet thickness when particle collisions are important (Eq. 1.63)
L_e	external or overall length-scale (§1.6)
$L_{inertia}$	current sheet thickness when inertia terms are important (Eq. 1.54)
L_{Hall}	current sheet thickness when Hall terms are important (Eq. 1.56)
L_{ij}	linking number of two curves (§8.5.3)

L_{pc}	length of Earth's polar cap (§10.1)
L_{stress}	current sheet thickness when electron-stress term is important (Eq. 1.58)
L_{tail}	length of geomagnetic tail (§10.1)
L_u	Lundquist number (Table 1.1, Eq. 1.65)
Lu	luminosity (§12.1)
m	particle mass (§1.1)
m_e	electron mass (§1.2.2)
m_p	proton mass (§1.2.1)
$\overline{m}$	mean particle mass (§1.2.1)
M	Alfvén Mach number (Table 1.1)
M_e	external Alfvén Mach number or reconnection rate (§1.6)
M_e^*	maximum reconnection rate (§4.3)
M_i	inflow Alfvén Mach number to diffusion region (§4.2)
$\overline{M}$	Mach number (Table 1.1)
M^*	Kummer function (§3.2)
M_*	mass of a star (§12.2.1)
$\dot{M}$	accretion rate (§12.2.1)
n	total number of particles per unit volume (§1.2.1)
n_e	electron number density (§1.2.1)
n_d	plasma number density based on line ratios (§12.1)
n_i	ion number density (§1.7)
n_r	number of runaway particles (§13.1.2)
$\mathbf{n}$	unit normal to a surface (§8.5.1)
N	norm of a mapping (§8.7); or number of flaring events (§11.2.1)
$\dot{N}$	flux of accelerated particles (§13.1.2)
$\overline{N}$	number density of spheres (§13.2.3)
$\mathcal{N}$	particle density (§13.3.2)
p	plasma pressure (§1.2.1)
p_e	external plasma pressure (§4.2)
p_i	plasma pressure at inflow to diffusion region (§4.2)
p_N	plasma pressure at neutral point (§4.2)
p_o	plasma pressure at outflow from diffusion region (§4.2)
p_s	plasma pressure at stagnation point (§3.2)
$p_\parallel$	plasma pressure along the magnetic field (§1.7)
$p_\perp$	plasma pressure normal to the magnetic field (§1.7)
$\overline{\mathbf{p}}$	particle momentum (§13.2.3)
$\underline{\mathbf{P}}_e$	electron stress tensor (§1.7)

P_m	magnetic Prandtl number (Table 1.1)
P_n	Legendre polynomial of degree n (§2.2.5)
P_r	Prandtl number (Table 1.1)
q	particle electric charge (§1.7); or safety factor (§6.4)
q_0	value of safety factor on axis of tube (§6.4)
q_a	value of safety factor at surface of tube (§6.4)
$\mathbf{q}$	energy flux (Eq. 1.24)
$Q(T)$	temperature dependence of optically thin radiative loss function (§1.2.1)
Q_ν	viscous heating (§1.2.1)
Q_r	radiative energy loss (§1.2.1)
r	radial distance in cylindrical polar coordinates (§1.4)
r_s	radius of resonant surface (§6.3.3)
R	radius of a current loop (§2.2.5); or major radius of torus (§6.4); correlation function (§13.2.2)
R_e	Reynolds number (Table 1.1)
R_E	radius of the Earth (§10.1)
R_g	particle gyro-radius (§13.1.4)
R_{ge}	electron gyro-radius (§13.1.4)
R_{gi}	ion gyro-radius (Eq. 1.60)
R_m	magnetic Reynolds number (§1.2.2)
R_{me}	magnetic Reynolds number associated with overall (external) length-scale (§1.6)
R_*	stellar radius (§12.1)
$\overline{R}$	$\log_e r$ (§4.4)
$\mathbf{R}$	general non-ideal term in Ohm's law (§8.1.1); or position of guiding centre of particle (§13.1.4)
$\mathcal{R}$	universal gas constant (§1.2.1)
Re	real part of a complex function (§2.2.4)
s	entropy (§1.2.2); or distance along a curve (§2.6.1); or power-law index (§13.3.2)
S_R	surface area of plasma processed through reconnection site (§12.1.2)
S	a surface (§1.4); or shear of an arcade (§8.5.3)
S_v	ratio of viscous time to Alfvén time (§6.3.1)
$\mathbf{S}$	Poynting flux (§1.4)
$\underline{\mathbf{S}}$	viscous stress tensor (§1.2.1)
t	time (§1.2.1)
T	temperature (§1.2.1)

T_e	electron temperature (§1.2.2)
T_i	ion temperature (§1.7)
u_d	particle drift speed (§13.1.1)
u_I	speed of intersection of magnetic field line with shock front (§13.3.1)
(u, v)	general orthogonal coordinates (§8.7)
$\mathbf{u}$	plasma velocity (ch. 13 only)
U	cloud speed (§13.2)
v_A	Alfvén speed (Table 1.1)
v_d	magnetic diffusion speed (Eq. 1.17)
v_D	viscous diffusion speed (§1.2.2)
v_e	external flow speed (§1.6)
v_g	group speed (§6.3.3); or gyro-speed (§13.1.4)
v_i	inflow speed to diffusion region (§4.2)
v_o	outflow speed from diffusion region (§4.2)
v_p	photospheric speed (§11.3.3)
v_r	particle runaway speed (§13.1.2)
v_{rel}	relative speed of particle and sphere (§13.2.3)
v_{rise}	rise speed of flare loops (§12.1.2)
v_R	reconnection speed (§11.1.1)
v_{Ti}	ion thermal speed (§13.1.3)
v_{Te}	electron thermal speed (§13.1.1)
$\mathbf{v}$	bulk plasma flow velocity (§1.2.1); or particle velocity (ch 13 only)
$\mathbf{V}_E$	electric drift velocity (§13.1.4)
$\mathbf{v}_g$	gyro-velocity (§13.1.4)
$\mathbf{v}_\perp$	component of plasma velocity (§8.1.2); or particle velocity (§13.1.4) normal to the magnetic field
V_0	a typical speed (§1.2.2)
V_c	polar-cap convection speed (§10.5)
V_{gun}	gun voltage (§9.1.3)
V_{pc}	field line speed in polar cap (§10.1)
V_R	volume of diffusion region (§8.5.4); or flare ribbon speed (§11.1.2)
V_{sw}	solar-wind speed (§10.1)
$\mathbf{V}_i$	velocity of ions (§12.2.2)
$\mathbf{V}_{\text{DR}}$	drift velocity between ions and neutrals (§12.2.2)
$\mathbf{V}_n$	velocity of neutral particles (§12.2.2)

$\mathbf{V}_\perp$	guiding centre velocity (§13.1.4)
w	total energy density (Eq. 1.23); or width of magnetic island (§6.5.1); or thickness of an arcade shell (§8.5.3)
$\mathbf{w}$	magnetic field line velocity (§1.4)
$\mathbf{w}_\perp$	component of $\mathbf{w}$ normal to the magnetic field (§8.1.2)
W	ribbon width (§11.1.2); or energy of flaring events (§11.2.1)
W_k	kinetic energy (§2.4)
W_m	magnetic energy (§2.4)
W_p	particle energy (§13.1.4)
W_r	radiative energy from a flare (§12.1)
W_{tail}	width of geomagnetic tail (§10.1)
$\mathbf{W}^{(4)}$	four-velocity of moving surface (§8.1.4)
X	density ratio across a shock wave (§1.5)
z	complex variable, $x + iy$ (§2.2.4); or the third cartesian direction

Greek Alphabet

α	constant or function of proportionality for a linear force-free field (§2.3.1); or Euler potential (§8.1.1)
α_p	pitch-angle (§13.3.1)
α^*	square of ratio of Sweet-Parker inflow speed to inflow speed on axis of diffusion region (§4.6.3); or radiative-loss parameter (§12.1.1)
$\bar{\alpha}$	null-point parameter (§1.3.2); or diffusion parameter (§13.2.3)
β	plasma beta (Table 1.1, Eq. 1.59); or Euler potential (§8.1.1)
β_p	plasma beta based on poloidal magnetic field component (§6.4)
γ	ratio of specific heats (§1.2.1)
$\bar{\gamma}$	Lorentz factor (§13.1)
Δ'	jump in magnetic gradient across singular layer in tearing theory (§6.2)
ϵ	a small parameter in an expansion that is particular to a given section
ϵ_0	electrical permittivity of free space (§1.2.1)
ζ	coefficient of bulk viscosity (§1.2.1); complex variable $x + iy$ (§11.1.1)
η	magnetic diffusivity (§1.2.2)

η_{AD}	ambipolar diffusion coefficient (§12.2.2)
η_e	scalar electrical resistivity (§1.2.1)
$\eta_\perp$	magnetic diffusivity normal to the magnetic field (Eq. 1.13)
$\eta_\parallel$	magnetic diffusivity along the magnetic field (§1.2.2)
$\tilde{\eta}$	anomalous diffusivity (§6.3.3)
η^*	turbulent magnetic diffusivity (§11.3.3)
$\boldsymbol{\eta}_e$	electrical resistivity tensor (§1.2.1)
θ	angle in polar coordinates (§1.4)
θ_{HT}	upstream magnetic field inclination to shock front in de Hoffmann-Teller frame (§13.3.1)
κ_0	coefficient of thermal conduction (§12.1.1)
$\overline{\kappa}$	spatial diffusion coefficient (§13.3.2)
$\boldsymbol{\kappa}$	thermal conductivity tensor (§1.2.1)
λ	a parameter or scalar function that is particular to a given section
λ_e	electron-inertial length or skin-depth (Eq. 1.55)
λ_i	ion-inertial length or skin-depth (Eq. 1.57)
λ_D	Debye length (Eq. 1.64)
λ_{mfp}	particle mean-free path (§1.2.2, Eq 1.62)
Λ	coulomb logarithm (Eq 1.14)
μ	magnetic permeability (§1.2.1)
μ_m	magnetic moment (§13.1.4)
$\overline{\mu}$	cosine of particle pitch angle (§13.2.3)
$\tilde{\mu}$	mean atomic weight (§1.7)
ν	coefficient of kinematic viscosity (§1.2.1)
$\overline{\nu}$	electron-ion collision frequency (§13.1.1)
$\overline{\nu}_{eff}$	effective collision frequency (§13.1.3)
$\overline{\nu}_{in}$	ion-neutral collision frequency (§12.2.2)
ν^*	turbulent viscosity (§11.3.3)
$\boldsymbol{\xi}$	a plasma displacement (§2.6.1)
ξ_a	perturbation at surface of tube of radius a (§6.4.1)
∇	gradient operator (§1.2.1)
ρ	mass density (§1.2.1)
ρ_c	electric charge density (§1.7)
σ	electrical conductivity (§1.2.1)
σ^*	anomalous electrical conductivity (§1.7)
σ_{eff}	effective electrical conductivity (§13.1.3)
τ	slowing-down time (§13.1.1); or confinement time (§9.1)

τ_A	Alfvén travel-time (§6.1)
τ_{BH}	time-scale for Balbus-Hawley instability (§12.2.1)
τ_c	coalescence time (§6.5.3)
τ_{cr}	cooling time due to conduction and radiation (§11.1.2); or corotation time (§12.2.1)
τ_d	magnetic diffusion time (Eq 1.1.6)
τ_{decay}	decay-time for a flare (§12.1)
τ_{dr}	drift time (Eq. 13.26)
τ_{eff}	effective particle collision-time (§13.1.3)
τ_g	gyro-period of particle (§13.1.4)
τ_G	gravitational time (§6.3)
τ_κ	conduction time (§11.2.2)
τ_l	time for acoustic waves to cross a current sheet (§11.2.2)
τ_L	time for acoustic waves to travel along a current sheet (§11.2.2)
τ_m	tearing time (§6.2)
τ_p	correlation time of footpoint motions (§11.3.3)
τ_P	time-scale for Parker instabilitiy (§12.2.1)
τ_r	radiative time-scale (§11.2.2)
τ_{rise}	rise-time for a flare (§12.1)
τ_R	reconnection time (§12.1.2)
τ_s	time-scale for shearing (§12.2.1)
ϕ	angle in polar coordinates measured from $y = x$ (§5.3.2); or angle measured in toroidal direction around torus (§6.4)
ϕ_s	angle between separatrices (§7.1.1)
Φ	electric scalar potential (§1.2.2)
Φ^*	electric potential (§1.4)
Φ_A	gauge function for magnetic vector potential (§8.5.1)
Φ_m	magnetic potential (§2.2.5)
Φ_{pc}	polar cap potential (§10.1)
Φ_{sw}	solar wind potential (§10.1)
Φ_T	twist in a magnetic flux tube (§6.4)
χ	optically thin radiative-loss parameter (§12.1.1)
χ_x, χ_y	velocity gradient functions (§7.1.1)
ψ	stream function (§3.3.1)
Ψ	potential along a magnetic field (§8.1.1)
ω	frequency of a time variation (§6.2)
ω_A	Alfvén rate (§6.2)
ω_d	diffusion rate (§6.2)

ω_i	angular speed of rotation of a flux tube (§8.5.2)
ω_{pe}	electron plasma frequency (§1.7)
ω_{pi}	ion plasma frequency (§1.7)
ω	vorticity (§1.2.2)
Ω	gyro-frequency (§13.1.6)
Ω_a	orbital frequency in accretion disc (§12.2.1)
$\overline{\Omega}_E$	Earth's angular frequency (§13.1.6)
Ω_i	ion gyro-frequency (§13.1.3)

Appendix 2
UNITS

In this book we have adopted the rationalised mks system of units. However, we have commonly quoted magnetic fields (**B**) in gauss (G), which is acceptable in the SI system. The relationship between Gaussian cgs and rationalised mks units is as follows.

Quantity	Gaussian	Rationalised mks
length	1 cm =	10^{-2} m
mass	1 g =	10^{-3} kg
time	1 s =	1 s
force	1 dyne =	10^{-5} N
energy	1 erg =	10^{-7} joule (J)
power	1 erg s^{-1} =	10^{-7} watt (W)
charge	1 statcoul =	$\frac{1}{3} \cdot 10^{-9}$ coulomb (C)
electric field	1 statvolt cm^{-1} =	3×10^4 V m^{-1}
electric current	1 statamp =	$\frac{1}{3} \times 10^{-9}$ ampère (A)
current density	1 statamp cm^{-2} =	$\frac{1}{3} \times 10^{-5}$ A m^{-2}
electrical conductivity	1 s^{-1} =	$\frac{1}{9} \times 10^{-9}$ siemens m^{-1}
magnetic induction (B)	1 G =	10^{-4} tesla (T)
magnetic field (H)	1 oersted =	$\frac{1}{4\pi} \times 10^3$ A m^{-1}
magnetic flux	1 maxwell (Mx) =	10^{-8} weber (Wb)

Equations in the Gaussian system can be transformed into the mks system by replacing each symbol in the equation as follows. Symbols for length and time and their derivatives remain unchanged.

Quantity	Gaussian	Rationalised mks
speed of light	c	$(\epsilon_0\mu_0)^{-1/2}$
charge	e	$e(4\pi\epsilon_0)^{-1/2}$
current	j	$j(4\pi\epsilon_0)^{-1/2}$
electric field	E	$E(4\pi\epsilon_0)^{1/2}$
magnetic induction	B	$B(4\pi/\mu_0)^{1/2}$
electric displacement	D	$D(4\pi/\epsilon_0)^{1/2}$
magnetic field	H	$H(4\pi\mu_0)^{1/2}$
dielectric constant	ϵ	ϵ/ϵ_0
magnetic permeability	μ	μ/μ_0
electrical conductivity	σ	$\sigma(4\pi\epsilon_0)^{-1}$

More details on the relationship between these systems of units can be found in Appendix I of Priest (1982).

Appendix 3

USEFUL EXPRESSIONS

Physical Constants

Speed of light	$c = 2.998 \times 10^8$	$\mathrm{m\,s^{-1}}$
Electron charge	$e = 1.602 \times 10^{-19}$	C
Electron Mass	$m_e = 9.109 \times 10^{-31}$	kg
Proton Mass	$m_p = 1.673 \times 10^{-27}$	kg
Mass ratio	$m_p/m_e = 1837$	
Electron volt	$1 \text{ eV} = 1.602 \times 10^{-19}\mathrm{J}$	
	$1 \text{ eV}/k_B = 11\,605 \text{ K}$	
Boltzmann constant	$k_B = 1.381 \times 10^{-23}$	$\mathrm{J\ deg^{-1}}$
Gravitational constant	$G = 6.672 \times 10^{-11}$	$\mathrm{N\ m^2\,kg^{-2}}$
Gas constant	$\tilde{R} = 8.3 \times 10^3$	$\mathrm{m^2\,s^{-2}\,deg^{-1}}$
Permeability of free space	$\mu_0 = 4\pi \times 10^{-7}$	
	$= 1.257 \times 10^{-6}$	$\mathrm{henry\ m^{-1}}$
Permittivity of free space	$\epsilon_0 = 8.854 \times 10^{-12}$	$\mathrm{farad\ m^{-1}}$

Plasma Properties (n_e in m^{-3}, B in Gauss, T in °K, v in m s^{-1}, l in m)

Sound speed	$c_s = 152\,T^{1/2}$	m s^{-1}
Alfvén speed	$v_A = 2.18 \times 10^{12} B\,n^{-1/2}$	m s^{-1}
Plasma beta	$\beta = 3.5 \times 10^{-21}\,n T B^{-2}$	
Magnetic Reynolds number	$R_m = 2 \times 10^{-9} l\,v\,T^{3/2}$	
Scale-height (for $\tilde{\mu} = 0.6$)	$H_g = 50\,T$	m
Electrical conductivity	$\sigma = 10^{-3}\,T^{3/2}$	siemens m^{-1}
Magnetic diffusivity	$\eta = 10^9\,T^{-3/2}$	m^2 s^{-1}
Thermal conductivity	$\kappa_{\parallel} = 10^{-11}\,T^{5/2}$	W m^{-1} deg^{-1}
Electron plasma frequency	$\omega_{pe} = 56.4\,n_e^{1/2}$	rad s^{-1}
Electron gyro-frequency	$\Omega_e = 1.76\,B \times 10^7$	rad s^{-1}
Proton gyro-frequency	$\Omega_p = 9.58\,B \times 10^3$	rad s^{-1}
Debye length	$\lambda_D = 6.9(T_e/n_e)^{1/2} \times 10$	m
Electron gyro-radius	$R_{ge} = 2.38 \times 10^{-2}(T_e/11605)^{1/2}B^{-1}$	m
Proton gyro-radius	$R_{gp} = 1.02(T_p/11605)^{1/2}B^{-1}$	m
Electron thermal speed	$v_{Te} = (k_B T_e/m_e)^{1/2} = \Omega R_{ge}$	
	$\quad = 4.19 \times 10^5 (T_e/11605)^{1/2}$	m s^{-1}
Proton thermal speed	$v_{Ti} = (k_B T_p/m_p)^{1/2} = \Omega_p R_{gp}$	
	$\quad = 9.79 \times 10^3 (T_p/11605)^{1/2}$	m s^{-1}

Cylindrical Polar Coordinates (r, θ, z)
(Note that in many writings the notation (R, ϕ, z) is adopted instead, but here we have used the letters R and ϕ for other purposes.)

$$\nabla A = \frac{\partial A}{\partial r}\,\hat{\mathbf{r}} + \frac{1}{r}\frac{\partial A}{\partial \theta}\,\hat{\boldsymbol{\theta}} + \frac{\partial A}{\partial z}\,\hat{\mathbf{z}}.$$

$$\nabla \cdot \mathbf{B} = \frac{1}{r}\frac{\partial}{\partial r}(rB_r) + \frac{1}{r}\frac{\partial B_\theta}{\partial \theta} + \frac{\partial B_z}{\partial z}.$$

$$\nabla \times \mathbf{B} = \left(\frac{1}{r}\frac{\partial B_z}{\partial \theta} - \frac{\partial B_\theta}{\partial z}\right)\hat{\mathbf{r}} + \left(\frac{\partial B_r}{\partial z} - \frac{\partial B_z}{\partial r}\right)\hat{\boldsymbol{\theta}}$$
$$+ \left(\frac{1}{r}\frac{\partial}{\partial r}(rB_\theta) - \frac{1}{r}\frac{\partial B_r}{\partial \theta}\right)\hat{\mathbf{z}}.$$

$$\nabla^2 A = \frac{1}{r}\frac{\partial}{\partial r}\left(r\frac{\partial A}{\partial r}\right) + \frac{1}{r^2}\frac{\partial^2 A}{\partial \theta^2} + \frac{\partial^2 A}{\partial z^2}.$$

$$(\mathbf{B} \cdot \nabla)\mathbf{B} = \left(B_r \frac{\partial B_r}{\partial r} + \frac{B_\theta}{r} \frac{\partial B_r}{\partial \theta} - \frac{B_\theta^2}{r} + B_z \frac{\partial B_r}{\partial z} \right) \hat{\mathbf{r}}$$

$$+ \left(\frac{B_r}{r} \frac{\partial}{\partial r}(r B_\theta) + \frac{B_\theta}{r} \frac{\partial B_\theta}{\partial \theta} + B_z \frac{\partial B_\theta}{\partial z} \right) \hat{\boldsymbol{\theta}}$$

$$+ \left(B_z \frac{\partial B_z}{\partial z} + B_r \frac{\partial B_z}{\partial r} + \frac{B_\theta}{r} \frac{\partial B_z}{\partial \theta} \right) \hat{\mathbf{z}}.$$

Spherical Polar Coordinates (r, θ, ϕ)

$$\nabla A = \frac{\partial A}{\partial r} \hat{\mathbf{r}} + \frac{1}{r} \frac{\partial A}{\partial \theta} \hat{\boldsymbol{\theta}} + \frac{1}{r \sin \theta} \frac{\partial A}{\partial \phi} \hat{\boldsymbol{\phi}}.$$

$$\nabla \cdot \mathbf{B} = \frac{1}{r^2} \frac{\partial}{\partial r}(r^2 B_r) + \frac{1}{r \sin \theta} \frac{\partial}{\partial \theta}(\sin \theta \, B_\theta) + \frac{1}{r \sin \theta} \frac{\partial B_\phi}{\partial \phi}.$$

$$\nabla \times \mathbf{B} = \frac{1}{r \sin \theta} \left(\frac{\partial}{\partial \theta}(\sin \theta B_\phi) - \frac{\partial B_\theta}{\partial \phi} \right) \hat{\mathbf{r}} + \left(\frac{1}{r \sin \theta} \frac{\partial B_r}{\partial \phi} - \frac{1}{r} \frac{\partial}{\partial r}(r B_\phi) \right) \hat{\boldsymbol{\theta}}$$

$$+ \left(\frac{1}{r} \frac{\partial}{\partial r}(r B_\theta) - \frac{1}{r} \frac{\partial B_r}{\partial \theta} \right) \hat{\boldsymbol{\phi}}.$$

$$\nabla^2 A = \frac{1}{r^2} \frac{\partial}{\partial r} \left(r^2 \frac{\partial A}{\partial r} \right) + \frac{1}{r^2 \sin \theta} \frac{\partial}{\partial \theta} \left(\sin \theta \frac{\partial A}{\partial \theta} \right) + \frac{1}{r^2 \sin^2 \theta} \frac{\partial^2 A}{\partial \phi^2}.$$

$$(\mathbf{B} \cdot \nabla)\mathbf{B} = \left(B_r \frac{\partial B_r}{\partial r} + \frac{B_\theta}{r} \frac{\partial B_r}{\partial \theta} - \frac{B_\theta^2 + B_\phi^2}{r} + \frac{B_\phi}{r \sin \theta} \frac{\partial B_r}{\partial \phi} \right) \hat{\mathbf{r}}$$

$$+ \left(B_r \frac{\partial B_\theta}{\partial r} + \frac{B_\theta}{r} \frac{\partial B_\theta}{\partial \theta} + \frac{B_r B_\theta}{r} + \frac{B_\phi}{r \sin \theta} \left(\frac{\partial B_\theta}{\partial \phi} - \cos \theta B_\phi \right) \right) \hat{\boldsymbol{\theta}}$$

$$+ \left(B_r \frac{\partial B_\phi}{\partial r} + \frac{B_r B_\phi}{r} + \frac{B_\phi}{r \sin \theta} \frac{\partial B_\phi}{\partial \phi} + \frac{B_\theta}{r \sin \theta} \frac{\partial}{\partial \theta}(B_\phi \sin \theta) \right) \hat{\boldsymbol{\phi}}.$$

Index